D0706670

NONPARAMETRICS

HOLDEN-DAY SERIES IN PROBABILITY AND STATISTICS

E. L. Lehmann, Editor

Bickel & Doksum: Mathematical Statistics
Carlson: Statistics
Freedman: Markov Processes
Freedman: Approximating Countable Markov Chains
Freedman: Brownian Motion and Diffusion
Hájek: Nonparametric Statistics
Hodges & Lehmann: Elements of Finite Probability
Hodges & Lehmann: Basic Concepts of Probability and Statistics
Lehmann: Nonparametrics: Statistical Methods Based on Ranks
Nemenyi: Statistics From Scratch
Neveu: Mathematical Foundations of the Calculus of Probability
Parzen: Stochastic Processes
Rényi: Foundations of Probability
Tanur, Mosteller, Kruskal, Link, Pieters, Rising: Statistics: A Guide to the Unknown

NONPARAMETRICS

Statistical Methods Based on Ranks

E. L. LEHMANN
*University of California,
Berkeley*

With the special assistance of
H. J. M. D'ABRERA
*University of California,
Berkeley*

HOLDEN-DAY, INC.
San Francisco

McGRAW-HILL INTERNATIONAL BOOK COMPANY
New York
Düsseldorf Johannesburg London Mexico Panama
São Paulo Singapore Sydney Toronto

NONPARAMETRICS

500 Sansome Street, San Francisco, California 94111

Library of Congress Catalog Card Number: 73-94384
ISBN: 0-8162-4996-6 (Holden-Day)
ISBN: 0-07-037073-7 (McGraw-Hill)

Printed in the United States of America

1234567890 HA 80798765

To my mother

PREFACE

1. History

Methods based on ranks form a substantial body of statistical techniques that provide alternatives to the classical parametric methods. Individual rank tests were proposed much earlier [the earliest use may be that of the sign test by Arbuthnot in 1710; for some additional history see, for example, Kruskal (1957 and 1958)*]; the modern development of the subject may be said to have begun with the papers by Hotelling and Pabst (1936), Friedman (1937), Kendall (1938), Smirnov (1939), and those of Wald and Wolfowitz in the early 1940s. An interesting survey of this work was given by Scheffé (1943).

A full-scale development of rank-based methods seems to have been sparked by the publication in 1945 of a paper by Wilcoxon in which he discussed the two tests, now bearing his name, for comparing two treatments, and by the book of Kendall (1948). Since then there has been a flood of publications that has not yet abated. A bibliography of nonparametric statistics (of which rank-based methods constitute the methodologically most important part) by Savage (1962) lists about 3,000 items. If brought up to date, it probably would contain twice that many entries.

* *See References following Preface.*

The feature of nonparametric methods mainly responsible for their great popularity (and to which they owe their name) is the weak set of assumptions required for their validity. Although it was believed at first that a heavy price in loss of efficiency would have to be paid for this robustness, it turned out, rather surprisingly, that the efficiency of the Wilcoxon tests and other nonparametric procedures holds up quite well under the classical assumption of normality and that these procedures may have considerable advantages in efficiency (as well as validity) when the assumption of normality is not satisfied. These facts were first brought out clearly by Pitman (1948) and were strengthened by results of Hodges and Lehmann (1956) and Chernoff and Savage (1958).

In the early stages, rank-based methods were essentially restricted to testing procedures. They thus did not provide a flexible array of methods, which would include not only tests but point and interval estimates as well as various simultaneous inference procedures. This difficulty is gradually being overcome, although rank-based methods do not yet have the flexibility and the wide applicability to complex linear models that make least squares and normal theory so attractive.

2. The present book

The purpose of this book is to provide an introduction to nonparametric methods for the analysis and planning of comparative studies. Only a relatively small number of basic techniques are presented in detail: These are mainly tests of the Wilcoxon type (which can be obtained from the corresponding classical tests by replacing the observations by their ranks) and the estimation and simultaneous inference procedures based upon these tests. These methods are simple, have good efficiency properties, and most are well tabled. They are treated rather fully here, with emphasis on the assumptions under which they are appropriate, the accuracy of the various approximations that are required, and the modifications needed for tied observations. For the simplest cases there is also a discussion of power or accuracy and the determination of the sample sizes required to achieve a given accuracy. The use of the methods is exemplified in the text, and numerous problems furnish opportunities for the student to try them for himself. In many cases the data for these illustrations are the results of actual studies reported in the literature.

An indication of some alternatives to and extensions of the above procedures, and of some additional properties, is provided by sections of further developments at the end of each chapter, which give an introduction to the literature on these subjects. Among the topics treated this way are the Normal Scores procedures, permutation tests, sequential methods, and optimum theory. Two topics that are not covered are multivariate techniques (because of lack of space) and goodness-of-fit tests (because both the problem and the data are quite different from those considered here). On the other hand, a discussion of some tests for two-way contingency tables is included in the text because they can be viewed as special cases of rank tests with tied observations.

As mentioned above, an important advantage of nonparametric tests is the simplicity of the assumptions required for their validity. It is not necessary to postulate a population from which the subjects in a study have been obtained by random sampling, but only that the treatments being compared have been assigned to the subjects at random. All techniques are first discussed in terms of such a *randomization model.* This material (which constitutes Chapters 1 and 3 and some parts of Chapters 5 to 7) requires only the simplest mathematical background for its understanding. All that is needed is an elementary introduction to probability, such as that provided by mathematics courses in many high schools or the first lectures in an introductory course on probability or statistics.

It is possible within this framework to describe the tests, illustrate their use, and discuss the computation of significance or critical values. However, randomization models do not permit an evaluation of power that could be used to plan the size of a study. This is best discussed in terms of a population from which the subjects have been sampled. Unfortunately, an adequate treatment of *population models* (such as those underlying the normal distribution) requires some knowledge of the calculus.

The determination of sample size and the evaluation of the power of a test (which plays an analogous role as the variance of an estimate) seemed too important to omit. So as to include these somewhat more advanced topics, I have allowed the level of the book to vary with the requirements of the material. Despite the obvious disadvantages of such inconsistency, it is my hope that even the reader with little mathematical background will be able to follow the main ideas of the more advanced parts (Chapters 2 and 4 and portions of Chapters 5 to 7) and that the reader whose background is stronger will not be put off by the slow pace of the more elementary sections. I am encouraged in this hope by the fact that courses I have taught along these lines to students with very disparate backgrounds seem to have been reasonably successful.

The main text is followed by an appendix that provides the large-sample theory underlying the many approximations required where tables are not available and exact computations are too laborious. This material is at an intermediate level. It requires substantially more mathematical sophistication than the rest of the book, but it is much less advanced than the books by Hájek and Sidak (1967) and Puri and Sen (1971). A number of standard limit theorems from probability theory are stated and discussed but not proved, and on this basis the needed results are derived with relatively little effort.

3. Acknowledgments

There remains the pleasant task of acknowledging the many debts that I incurred during the writing of the book. I should like to thank Peter Bickel, Kjell Doksum, Gus Haggstrom, Joe Hodges, Bill Kruskal, Vida Lehmann, and Juliet Popper Shaffer who at various stages read portions of the manuscript and gave me their moral support and the benefit of their advice. Of the reviewers who examined the manuscript for the publishers, I am especially grateful to Ralph D'Agostino and Alfred Forsyth for their many valuable criticisms and ideas for improvements, and to Gottfried Noether for his crucial support and suggestions.

To two friends I owe a special debt. Ellen Sherman read critically, checked the computations, and prepared the tables for an early version of the first chapters. Above all, Howard D'Abrera performed the same task for later versions of the whole manuscript and prepared the answers for the problems. He corrected innumerable errors of style, thought, and arithmetic. Without his help I could not have brought this project to its completion.

The aspect of the book that caused me the most difficulty was to find suitable live examples and problems. Authors typically do not publish their data, and when a set of published data is potentially suitable, it usually turns out that the sample size is too large or small, there are too many or too few ties, the results are too obviously significant or too obviously not, or that the design or sampling procedure is not what is required to illustrate the particular point in question. I am grateful to the colleagues who put their unpublished data at my disposal as well as those who published data that I was able to use. To all I would like to extend an apology. When some minor modification of the data made them more suitable for my purpose, I have carried out such modifications (always with an acknowledgment). More seriously, I have used the data to illustrate the point I was trying to make, though this may have borne little relation to the purpose for which they were collected, and I have asked questions of the data that may seem foolish to someone more familiar with the actual situation. I hope that I will be forgiven for these violations of the authors' intentions, and I should like to ask readers who have available more suitable data to illustrate the techniques discussed here to let me know about them for a possible later revision.

Finally, it is a pleasure to thank the Office of Naval Research for their generous support of the research that has gone into this book.

E. L. Lehmann

Berkeley
1974

PREFACE REFERENCES

Arbuthnot, J. (1710): "An Argument for Divine Providence, Taken from the Constant Regularity Observed in the Births of Both Sexes," *Philos. Trans.* **27**:186–190.

Chernoff, Herman and Savage, I. Richard (1958): "Asymptotic Normality and Efficiency of Certain Nonparametric Test Statistics," *Ann. Math. Statist.* **29**:972–994.

Friedman, Milton (1937): "The Use of Ranks to Avoid the Assumption of Normality Implicit in the Analysis of Variance," *J. Amer. Statist. Assoc.* **32**:675–701.

Hájek, J. and Sidak, Z. (1967): *Theory of Rank Tests*, Academic Press.

Hodges, J. L., Jr. and Lehmann, E. L. (1956): "The Efficiency of Some Nonparametric Competitors of the *t*-Test," *Ann. Math. Statist.* **27**:324–335.

Hotelling, Harold and Pabst, Margaret Richards (1936): "Rank Correlation and Tests of Significance Involving No Assumptions of Normality," *Ann. Math. Statist.* **7**:29–43.

Kendall, Maurice G. (1938): "A New Measure of Rank Correlation," *Biometrika* **30**:81–93.

——— (1948): *Rank Correlation Methods*, 4th edition (1970), Griffin, London.

Kruskal, William H. (1957): "Historical Notes on the Wilcoxon Unpaired Two-sample Test," *J. Amer. Statist. Assoc.* **52**:356–360.

——— (1958): "Ordinal Measures of Association," *J. Amer. Statist. Assoc.* **53**:814–861.

Pitman, E. J. G. (1948): *Lecture Notes on Nonparametric Statistics*, Columbia Univ., New York.

Puri, M. L. and Sen, P. K. (1971): *Nonparametric Methods in Multivariate Analysis*, John Wiley.

Savage, I. Richard (1962): *Bibliography of Nonparametric Statistics*, Harvard Univ. Press.

Scheffé, Henry (1943): "Statistical Inference in the Nonparametric Case," *Ann. Math. Statist.* **14**:305–332.

Smirnov, N. V. (1939): "On the Estimation of the Discrepancy Between Empirical Curves of Distribution for Two Independent Samples," *Bull. Math. Univ. Moscow 2*, No. 2, 3–14.

Wilcoxon, Frank (1945): "Individual Comparisons by Ranking Methods," *Biometrics* **1**:80–83.

CONTENTS

1 RANK TESTS FOR COMPARING TWO TREATMENTS **1**

1. Ranks in the comparison of two treatments, 1
2. The Wilcoxon rank-sum test, 5
3. Asymptotic null distribution of the Wilcoxon statistic, 13
4. The treatment of ties, 18
5. Two-sided alternatives, 23
6. The Siegel-Tukey and Smirnov tests, 32
7. Further developments, 40
   Other approximations to the distribution of W_s; Censored observations; Early termination; Power; Permutation tests.
8. Problems, 43
9. References, 52

2 COMPARING TWO TREATMENTS OR ATTRIBUTES IN A POPULATION MODEL **55**

1. Population models, 55
2. Power of the Wilcoxon rank-sum test, 65
3. Asymptotic power, 69
4. Comparison with Student's t-test, 76
5. Estimating the treatment effect, 81
6. Confidence procedures, 91
7. Further developments, 95
   The Behrens-Fisher problem; The Normal Scores test; Increasing the number of levels to improve sensitivity; Small-sample power; Large-sample power and efficiency; Efficiency in the presence of ties; Optimality properties; Additional properties of $\hat{\Delta}$; Efficiency of the Siegel-Tukey test; The scale tests of Capon and Klotz; The Savage (or exponential scores) test; Scale tests with unknown location; Power and efficiency of the Smirnov test; Sequential rank tests; The permutation t-test.
8. Problems, 106
9. References, 114

3 BLOCKED COMPARISONS FOR TWO TREATMENTS **120**
1. The sign test for paired comparisons, 120
2: The Wilcoxon signed-rank test, 123
3. Combining data from several experiments or blocks, 132
4. A balanced design for paired comparisons, 141
5. Further developments, 143
Power of the sign and Wilcoxon tests; Alternative treatment of zeros; Tests against omnibus alternatives; Efficiency and generalizations of the blocked comparisons test W_s.
6. Problems, 146
7. References, 153

4 PAIRED COMPARISONS IN A POPULATION MODEL AND THE ONE-SAMPLE PROBLEM **156**
1. Power and uses of the sign test, 156
2. Power of the signed-rank Wilcoxon test, 164
3. Comparison of sign, Wilcoxon, and t-tests, 171
4. Estimation of a location parameter or treatment effect, 175
5. Confidence procedures, 181
6. Further developments, 185
Power and efficiency of the sign test; The absolute Normal Scores test; Power and efficiency of the Wilcoxon and absolute Normal Scores test; Tests of symmetry; A generalized set of confidence points; Bounded-length sequential confidence intervals for θ; Robust estimation; Some optimum properties of tests and estimators; Departures from assumption.
7. Problems, 191
8. References, 199

5 THE COMPARISON OF MORE THAN TWO TREATMENTS **202**
1. Ranks in the comparison of several treatments, 202
2. The Kruskal-Wallis test, 204
3. $2 \times t$ Contingency tables, 210
4. Population models, 219
5. One-sided procedures, 226
Comparing several treatments with a control; Testing equality against ordered alternatives.
6. Selection and ranking procedures, 238
Ranking several treatments; Selecting the best of several treatments.
7. Further developments, 247
Power and efficiency; Estimation of several differences in location; The estimation of contrasts; Normal Scores and Smirnov tests for the s-sample problem.

8. Problems, 250
9. References, 257

6 RANDOMIZED COMPLETE BLOCKS 260
1. Ranks in randomized complete blocks, 260
2. The tests of Friedman, Cochran, and McNemar, 262
3. Aligned ranks, 270
4. Population models and efficiency, 273
5. Further developments, 279
More general blocks; One-sided tests and ranking procedures; Estimation of treatment differences and other contrasts; Combination of independent tests.
6. Problems, 281
7. References, 285

7 TESTS OF RANDOMNESS AND INDEPENDENCE 287
1. The hypothesis of randomness, 287
2. Testing against trend, 290
3. Testing for independence, 297
4. $s \times t$ Contingency tables, 303
5. Further developments, 311
Pitman efficiency of D; Estimating the regression coefficient β; Tests of randomness based on runs; Other tests of independence; Power and efficiency of tests of independence; Contingency tables.
6. Problems, 317
7. References, 322

APPENDIX 327
1. Expectation and variance formulas, 327
2. Some standard distributions, 339
The binomial distribution; The hypergeometric distribution; The normal distribution; The Cauchy, logistic, and double-exponential distributions; The rectangular (uniform) and exponential distributions; The χ^2-distribution; Order statistics.
3. Convergence in probability and in law, 345
4. Sampling from a finite population, 352
5. U-statistics, 362
6. Pitman efficiency, 371
7. Some multivariate distributions, 380
The multinomial distribution; The multiple hypergeometric distribution; The multivariate normal distribution.

8. Convergence of random vectors, 386
9. Problems, 396
10. References, 405

TABLES **407**

A Number of combinations $\binom{N}{n}$, 407
B Wilcoxon rank-sum distribution, 408
C Area under the normal curve, 411
D Square roots, 412
E Smirnov exact upper-tail probabilities, 413
F Smirnov limiting distribution, 415
G Distribution of sign-test statistic, 416
H Wilcoxon signed-rank distribution, 418
I Kruskal-Wallis upper-tail probabilities, 422
J(a) χ^2 upper-tail probabilities for $\nu = 2, 3, 4, 5$ degrees of freedom, 427
J(b) Critical values c of χ^2 with $\nu = 6(1)40(5)100$ degrees of freedom, 428
K Upper-tail probabilities of Jonckheere's statistic, 429
L Amalgamation probabilities for Chacko's test, 430
M Upper-tail probabilities of Friedman's statistic, 431
N Distribution of Spearman's statistic, 433

ACKNOWLEDGMENTS FOR TABLES, 434
ANSWERS TO SELECTED PROBLEMS, 435
DATA GUIDE (TITLES FOR DATA PRESENTED IN THE TEXT), 445
AUTHOR INDEX, 447
SUBJECT INDEX, 451

CHAPTER 1
RANK TESTS FOR COMPARING TWO TREATMENTS

1. RANKS IN THE COMPARISON OF TWO TREATMENTS

The problem of deciding whether a proposed innovation constitutes an improvement over some standard procedure arises in many different contexts. Does a new "cure" prolong the life of cancer patients? Is the harmful effect of cigarettes reduced by filtering? Does a new expensive gasoline additive increase the mileage? Does cloud-seeding lead to increased precipitation? Or conversely, is televised instruction less effective than live classroom teaching? The following example illustrates the kind of evidence that may be used to obtain at least tentative answers to such questions.

EXAMPLE 1. *A new drug.* A mental hospital wishes to test the effectiveness of a new drug that is claimed to have a beneficial effect on some mental or emotional disorder. There are five patients in the hospital suffering from this disorder to about the same degree. (Actually, this number typically would be too small to provide meaningful results.) Of these five, three are selected at random to receive the new drug, and the other two serve as controls: they are given a placebo, a harmless pill not containing any active ingredients. In this way the patients (and preferably also the staff) do not know which patients are receiving the new treatment. This eliminates the possibility of psychological effects that might result from such knowledge.

After some time, a visiting physician interviews the patients and ranks them according to the severity of their condition. The patient whose condition is judged to be most serious is assigned rank 1, the next most serious rank 2, and so on, up to rank 5. The claim made for the new treatment will be considered warranted if the three treated patients rank sufficiently high in this combined ranking of all five patients. A basis for evaluating the significance of the observed ranking is provided by the following consideration.

Suppose that the treatment has no effect, i.e., that a patient's health is in no way affected by whether or not he receives the new drug. We shall refer to this assumption as the *hypothesis H of no treatment effect*. Since under the assumption of this hypothesis (for short, *under H*) the rank of each patient is determined solely by his state of health, it is clear that the ranking of the patients does not depend on which of them receive the treatment and which serve as controls. We may thus think of each patient's rank as attached to him even before the assignments to treatment and control are made. The selection of three patients to receive the treatment then also selects three ranks: those attached to the selected patients. Each possible such selection divides the ranks into two groups: the ranks of the treated patients and of the controls. These divisions are displayed in (1.1) for all possible cases. Thus, for example, the first box in (1.1) corresponds to the possibility that the three patients who eventually are awarded the highest ranks (3, 4, 5) are those receiving the treatment.

(1.1)

Treated	(3,4,5)	(2,4,5)	(1,4,5)	(2,3,5)	(1,3,5)
Controls	(1,2)	(1,3)	(2,3)	(1,4)	(2,4)
Treated	(2,3,4)	(1,3,4)	(1,2,4)	(1,2,3)	(1,2,5)
Controls	(1,5)	(2,5)	(3,5)	(4,5)	(3,4)

As is seen from (1.1), the patients and hence their ranks can be divided into two groups in 10 different ways. The assumption that the three patients receiving the treatment are selected at random means that these 10 possible divisions are equally likely, i.e., that each of the 10 possibilities displayed in (1.1) has probability $^1/_{10}$. As will be discussed in the next section, this fact provides a basis for assessing the significance of the observed ranking.

At the beginning of the example it was assumed that the five patients are suffering from their disorder to about the same degree. Actually, no use was made of this assumption in the above derivation. The random assignment of the patients to treatment and control implies that the ranks are also assigned to these two groups at random (under the hypothesis of no treatment effect), regardless of the initial states of health of the patients. However, as we shall discuss later (Chap. 2, Sec. 2, and Chap. 3, Sec. 1), increased homogeneity of the patients with respect to initial state of health and to other relevant factors increases the power of the experiment, that is, the likelihood of detecting that the treatment has an effect when this is the case. This is intuitively quite plausible since with nearly identical responses for patients under the same regime any difference must be due to treatment and will stand out clearly, but it will be masked for patients with widely different inherent responses.

The considerations introduced in the context of the above example easily generalize. Suppose that N experimental subjects are available for a comparative study and that n of these are selected at random to receive a new treatment with the remaining $m = N-n$ serving as controls.[1] Let us denote the number of possible choices of n out of N subjects by $\binom{N}{n}$. For $N = 5$ and $n = 3$ we have seen that $\binom{5}{3} = 10$. The number $\binom{N}{n}$ is variously known as *the number of combinations of N things taken n at a time, the number of samples of size n from a population of size N*, or as a *binomial coefficient*, and can be computed from the formula[2]

$$\binom{N}{n} = \frac{N(N-1)\cdot \cdots \cdot(N-n+1)}{1\cdot 2\cdot \cdots \cdot n} \tag{1.2}$$

A table of these coefficients for $N \leqslant 25$ and $n \leqslant 12$ is given as Table A at the end of the book.[3] The values for $N \leqslant 25$ and $n > 12$ can be obtained from the tabulated values by means of the formula (Prob. 70)

$$\binom{N}{n} = \binom{N}{N-n} \tag{1.3}$$

To find, for example, the value of $\binom{20}{16}$ one notes that by (1.3) it is equal to $\binom{20}{4}$; entering Table A in the row $N = 20$ and the column $n = 4$, it is seen that

$$\binom{20}{16} = \binom{20}{4} = 4{,}845$$

By assumption, the n subjects receiving the treatment are selected from the N available subjects *at random*; that is, all $\binom{N}{n}$ possible choices of these subjects are equally likely so that each has probability $1\Big/\binom{N}{n}$. At the termination of the study, the subjects are ranked (preferably by an impartial observer) with respect to the condition at which the treatment is aimed. As before, under the hypothesis H of no treatment effect, the ranking is not affected by which subjects received the treatment. The rank of each subject may be considered as determined (although unknown) before the assignment of subjects to treatment and control is performed, and hence to be assigned to treatment or control together with the subject. Thus,

[1] A convenient aid for making such a random selection is a table of random numbers such as the RAND Corporation's *A Million Random Digits with 100,000 Normal Deviates*. The Free Press, New York, 1955, or the book by Moses and Oakford (1963).

[2] For a proof of Eq. (1.2), see, for example, Goldberg (1960), Hodges and Lehmann (1970), or Mosteller, Rourke, and Thomas (1970). (Consult the References, Sec. 9, at the end of this chapter for detailed bibliographic data.)

[3] A more extensive table giving all values for $N \leqslant 200$ is the *Table of Binomial Coefficients*, Cambridge University Press, London, 1954.

under H, the $\binom{N}{n}$ possible assignments of n of the integers $1, \ldots, N$ as treatment ranks each have probability $1 \Big/ \binom{N}{n}$.

The above result is so fundamental that we shall now restate it more formally. Let the ranks of the treated subjects be denoted by $S_1, \ldots, S_n$, where we shall assume that they are numbered in increasing order so that $S_1 < S_2 < \cdots < S_n$, and let the ranks of the controls be $R_1 < R_2 < \cdots < R_m$. Between them, these $m+n$ ranks are just the integers 1, 2, ..., N. Since the R's are determined once the S's are known, the division of the ranks into the two groups can be specified by the n-tuple $(S_1, \ldots, S_n)$. The $\binom{N}{n}$ possible such n-tuples constitute the possible outcomes of the study.

The basic result derived above states that the probability, under H, of observing any particular n-tuple $(s_1, \ldots, s_n)$ is

$$P_H(S_1 = s_1, \ldots, S_n = s_n) = \frac{1}{\binom{N}{n}} \tag{1.4}$$

for each of the possible n-tuples $(s_1, \ldots, s_n)$.

Let us now illustrate the use of ranks in comparative studies with another example.

EXAMPLE 2. *Effect of discouragement.* To test whether discouragement adversely affects performance in an intelligence test, 10 subjects were divided at random into a control and treatment group of 5 each. Both were given Form L of the revised Stanford-Binet test under the conditions prescribed for this test. Two weeks later they were given Form F, the controls under the prescribed conditions, the treated subjects under conditions of discouragement (you are doing terribly, etc.). The following were the differences in their scores: later value − original value.[1]

Controls: 5 0 16 2 9 Treated: 6 −5 −6 1 4

If the subjects are ranked, with rank 1 going to the subject with the smallest difference, rank 2 to the next smallest, etc., the ranks of the treatment subjects are 1, 2, 4, 6, and 8, and those of the controls 3, 5, 7, 9, and 10. Both the differences and their ranks suggest that, on the whole, the five subjects receiving the discouragement did less well than the other five. An assessment of the significance of this result will be taken up in the next section and will be based on the following consideration. Let H denote the hypothesis that the treatment has no effect, i.e., that the discouragement will have no influence on the score obtained by a subject. If H is

[1] Part of data of Gordon and Durea, "The Effect of Discouragement on the Revised Stanford-Binet Scale," *J. Genetic Psychol.* **73**:201–207 (1948). The original experiment involved 20 subjects in each group. Of these, five were selected at random for the present example. For the original data, see Prob. 48.

true, the difference in the scores of each of the 10 subjects, and hence his rank, is unaffected by the method to which he is assigned. The $\binom{10}{5} = 252$ possible sets of values of the five treatment ranks are therefore equally likely, each having probability $1/252$.

The structure of the two examples is basically the same, but they differ in one important respect. Although in the first example only a ranking of the subjects (patients) was available, the data in Example 2 consisted of a measurement for each subject (the difference in the scores) and the subjects were ranked according to the values of these measurements. The reader may feel that in the second case a test of the hypothesis of no treatment effect should be based on the original measurements rather than on the ranks derived from them. It turns out, however, that tests based on ranks have certain advantages, which will be discussed in Chap. 2.

Throughout this section, we have assumed that the subjects which are available for observation are not chosen but are given and that they are assigned at random, n to treatment and $N-n$ to control. We shall call this model, in which chance enters only through the assignment of the subjects to treatment and control, the *randomization model.* This is to distinguish it from another model, to be considered in Chap. 2, according to which the N subjects are not fixed but are drawn in some specified manner from a population of such subjects. In this case, chance is involved also (in a way that can be taken into account) in the selection of the subjects.

2. THE WILCOXON RANK-SUM TEST

For comparing a new treatment or procedure with the standard method, N subjects (patients, students, etc.) are divided at random into a group of n who will receive a new treatment and a control group of m who will be treated by the standard method. At the termination of the study, the subjects are ranked either directly or according to some response that measures the success of the treatment such as a test score in an educational or psychological investigation, the amount of rainfall in a weather experiment, or the time needed for recuperation in a medical problem. The hypothesis H of no treatment effect is rejected, and the superiority of the new treatment acknowledged, if in this ranking the n treated subjects rank sufficiently high. (Here it is assumed that the success of the treatment is indicated by an increased response; if instead the aim is to decrease the response, H is rejected when the n treated subjects rank sufficiently low.)

To complete the specification of the procedure, it is necessary to decide just when the treatment ranks $(S_1, \ldots, S_n)$ are sufficiently large. Such an assessment is typically made in terms of some test statistic, large values of which correspond to the treatment ranks being large. A simple and effective such statistic is the sum of the treatment ranks

$$W_s = S_1 + \cdots + S_n \tag{1.5}$$

The hypothesis H is then rejected and the treatment judged to be effective when W_s is sufficiently large, say, when

$$W_s \geqslant c \tag{1.6}$$

The test defined by (1.5) and (1.6) is known as the *Wilcoxon rank-sum test.* (The term is used to distinguish this test from another Wilcoxon test that will be discussed in Chap. 3. However, we shall omit the qualification *rank-sum* when there is no possibility of confusion.)

The constant c in (1.6), the *critical value*, is conventionally determined so that under H the probability of getting a value of W_s greater than or equal to c is equal to some specified small number α, the *level of significance.* Common choices of α are .01, .025, .05. The constant c is thus determined by the equation

$$P_H(W_s \geqslant c) = \alpha \tag{1.7}$$

where the subscript H indicates that the probability is computed under H, that is, under the assumption that the treatment has no effect.

Formulas (1.5) to (1.7) make precise just when the treatment ranks will be considered too large for the hypothesis of no treatment effect to remain tenable. On the one hand, values of W_s greater than or equal to c are very unlikely when H is true; in fact, the probability of observing such values just by chance is then α. On the other hand, such values are expected when the treatment has the desired effect. The occurrence of such values therefore leads to the abandonment of H in favor of the alternative that the treatment is effective.

To determine c from Eq. (1.7) it is necessary to learn how to find the probability (under H) that W_s has any specified value. For the case $N = 5$, $n = 3$ discussed in Example 1, these probabilities are easily obtained from (1.1). To each possible set of treatment ranks displayed in (1.1) there corresponds a value w of W_s as shown in (1.8).

(1.8)

Treatment Ranks	3, 4, 5	2, 4, 5	1, 4, 5	2, 3, 5	1, 3, 5
w	12	11	10	10	9

Treatment Ranks	2, 3, 4	1, 3, 4	1, 2, 4	1, 2, 3	1, 2, 5
w	9	8	7	6	8

Since the probability of each set of treatment ranks is $1/10$, it follows for example that

$$P_H(W_s = 9) = P_H(S_1 = 1, S_2 = 3, S_3 = 5) + P_H(S_1 = 2, S_2 = 3, S_3 = 4) = 2/10$$

In this way, one finds the probabilities (under H) of W_s taking on its various possible values, which are displayed in (1.9).

(1.9)

w	6	7	8	9	10	11	12
$P_H(W_s = w)$	.1	.1	.2	.2	.2	.1	.1

These probabilities constitute the distribution of W_s under H. Since H is sometimes called the *null hypothesis* (it states that the treatment effect is zero or null), the distribution of W_s under H is called the *null distribution* of W_s. From (1.9) it follows, in particular, that

$$P_H(W_s \geq 12) = P_H(W_s = 12) = .1$$

Thus if $\alpha = .1$, the null hypothesis is rejected only when $W_s = 12$, that is, when the ranks of the controls are 1 and 2, and those of the treated subjects 3, 4, and 5.

It is seen from (1.9) that the probability $P_H(W_s \geq w)$ takes on only a few values, namely, 0, .1, .2, .4, .6, .8, .9, and 1.0. It is therefore not possible to find a critical value c satisfying (1.7) for every value of α but only for the values just listed. This difficulty is particularly pronounced when N is as small as in the present case, but it persists, although to a lesser degree, for larger N. For $N = 13$ and $n = 8$, for example, the possible values of $P_H(W_s \geq w)$ are 0, .0008, .0016, .0031, .0054, .0093, .0148, .0225, .0326, .0466, .0637, and so on. It is then customary to replace the intended value of α, which in any case tends to be somewhat arbitrary, by the closest value that can be achieved. Thus for $N = 13$, $n = 8$ one could replace an intended .01 by .0093; .03 by .0326; .05 by .0466; and so on. If it seems important not to exceed the intended level, one can instead choose the closest smaller attainable value; for example, .0148 instead of an intended .02, or .0326 instead of .4.

The null distribution of W_s can be obtained quite generally by the method used to calculate the distribution (1.9) for the case $N = 5$ and $n = 3$. Let $\#(w;n,m)$ denote the number of all those divisions of the ranks $1, \ldots, N$ into n treatment and m control ranks for which the sum of the treatment ranks is equal to w. [For example, when $n = 3$ and $m = 2$, it is seen from (1.8) that $\#(8;3,2) = 2$.] Since by (1.4) each such division has probability $1\Big/\binom{N}{n}$ under H, it follows that

$$P_H(W_s = w) = \frac{\#(w;n,m)}{\binom{N}{n}} \tag{1.10}$$

To calculate the distribution of W_s it is therefore only necessary to obtain, by counting, the numbers $\#(w;n,m)$. Actually, to find the critical value c for a given level α one does not need the whole distribution of W_s. To illustrate, let us once more consider Example 2, in which the hypothesis of no effect of discouragement was being tested against the alternative, that discouragement has an adverse effect. Under the alternative, the differences in the scores of the treatment subjects (those being discouraged) will tend to be *lower* than those of the controls. The hypothesis will therefore be rejected when W_s is too small, say $W_s \leqslant c$, where c is determined so as to satisfy as closely as possible the equation

$$P_H(W_s \leqslant c) = \alpha$$

Since $m = n = 5$, each possible set of treatment ranks has probability $1\Big/\binom{10}{5} = 1/252$. Suppose that $\alpha = .05$, and let us enumerate systematically the sets of treatment ranks giving the smallest values to W_s. To see how far this process needs to be carried, note that the number x of such sets required is given approximately by the equation $x/252 = \alpha = .05$, the solution of which is $x = 252/20 = 12.6$. The resulting count is displayed in (1.11).

$$\begin{array}{ll} 1+2+3+4+5 = 15 & 1+2+4+5+6 = 18 \\ 1+2+3+4+6 = 16 & 1+2+3+4+9 = 19 \\ 1+2+3+4+7 = 17 & 1+2+3+5+8 = 19 \\ 1+2+3+5+6 = 17 & 1+2+3+6+7 = 19 \\ 1+2+3+4+8 = 18 & 1+2+4+5+7 = 19 \\ 1+2+3+5+7 = 18 & 1+3+4+5+6 = 19 \end{array} \tag{1.11}$$

This exhausts the cases with $W_s \leqslant 19$ and shows that $P_H(W_s \leqslant 19) = 12/252 = .0476$. This is the closest to $\alpha = .05$ that we can get since $P_H(W_s \leqslant 20) = .0754$ (Prob. 15). Now the sum of the treatment ranks in Example 2 was $1+2+4+6+8 = 21$, which is larger than the critical value $c = 19$. Despite the appearance of a treatment effect noted in Example 2, the results are therefore not significant at level $\alpha = .0476$, nor even at $\alpha = .0754$.

Unfortunately, as is seen from Table A, the number $\binom{N}{n}$ of possible cases increases very rapidly with N and n, and enumerations such as that above then become very cumbersome. It is therefore convenient to have available a table of the distribution of W_s, and such a table covering the values $m = 3,4$; $m \leqslant n \leqslant 12$ and $5 \leqslant m \leqslant 10$; $m \leqslant n \leqslant 10$ is given at the end of the book as Table B.[1] To see how to use this table, note that the distribution of W_s is not the same when the two group sizes are, say, $m = 2$, $n = 4$ as when they are $m = 4$, $n = 2$. It turns out, however, that the distribution of the statistic $W_s - \frac{1}{2}n(n+1)$ is the same in both cases and that more generally for any integers $k_1 \leqslant k_2$ the distribution of $W_s - \frac{1}{2}n(n+1)$ is the same when the group sizes are $m = k_1$, $n = k_2$ as when they are $m = k_2$, $n = k_1$. This

[1] For more extensive tables, see, for example: Buckle, Kraft, and van Eeden (1969b); Milton (1964); and Wilcoxon, Katti, and Wilcox (1970).

result, which suggests tabling $W_s - \frac{1}{2}n(n+1)$ rather than W_s, is proved as (C) at the end of this section.

There is another reason for preferring to table the distribution of $W_s - \frac{1}{2}n(n+1)$. The minimum value of W_s is obtained when the n treatment ranks are 1, 2, ..., n and hence (Prob. 83) is

$$1+2+\cdots+n = \tfrac{1}{2}n(n+1) \tag{1.12}$$

The table becomes more compact by subtracting this minimum value, which depends on n, from W_s. For reasons that will be discussed at the end of the section, the resulting statistic will be denoted by W_{XY} so that

$$W_{XY} = W_s - \tfrac{1}{2}n(n+1) \tag{1.13}$$

Table B gives the probabilities

$$P_H(W_{XY} \leqslant a) \tag{1.14}$$

up to at least the first entry that is $\geqslant .5$, for all combinations of $k_1 \leqslant k_2 \leqslant 10$, and for $k_1 = 3, 4$; $k_2 = 11, 12$, where k_1 denotes the smaller and k_2 denotes the larger of the two group sizes m and n. To illustrate the use of the table, consider the probability $P_H(W_s \leqslant 21)$ of observing a value as small as or smaller than the observed value 21 of Example 2. (For simplicity of notation, we shall suppress the subscript H during the remainder of the section.) Here $m = n = 5$, so that $n(n+1)/2 = 15$ and

$$P(W_s \leqslant 21) = P(W_s - 15 \leqslant 6) = P(W_{XY} \leqslant 6)$$

On entering the table with $k_1 = k_2 = 5$ and $a = 6$, we find this probability to be equal to .1111. Suppose instead that we wish to use the table (instead of the earlier enumeration method) to find the critical value corresponding to an approximate significance level of .05. Looking through the entries for $k_1 = k_2 = 5$, we see that the two probabilities bracketing .05 are .0476 and .0754. The first of these is the closer and corresponds to the value $a = 4$, so that

$$.0476 = P(W_{XY} \leqslant 4) = P(W_s - 15 \leqslant 4) = P(W_s \leqslant 19)$$

Thus we find the critical value 19 as before.

From the table it is easy to obtain a probability ($\leqslant \frac{1}{2}$) that W_s is less than or equal to a given value. However, the hypothesis is frequently rejected for large values of W_s. One then needs to solve Eq. (1.7) and hence requires the probability that W_s is greater than or equal to some constant. These probabilities can be obtained from the table by either of the following two methods.

Method 1. This method is based on the fact, to be proved at the end of this section, that the distribution of W_s is symmetric about the point $n(N+1)/2$. This means that for each k the two values

$$\tfrac{1}{2}n(N+1) - k \qquad \text{and} \qquad \tfrac{1}{2}n(N+1) + k$$

which are equally distant from $n(N+1)/2$ on opposite sides, have the same probability. That is, for all k

$$P[W_s = \tfrac{1}{2}n(N+1)-k] = P[W_s = \tfrac{1}{2}n(N+1)+k] \tag{1.15}$$

and hence also

$$P[W_s \leqslant \tfrac{1}{2}n(N+1)-k] = P[W_s \geqslant \tfrac{1}{2}n(N+1)+k] \tag{1.16}$$

Suppose now, for example, that $m = 4$, $n = 6$ and we wish to find the probability $P(W_s \geqslant 35)$. Here $n(N+1)/2 = 33$, and

$$P(W_s \geqslant 35) = P(W_s \geqslant 33+2) = P(W_s \leqslant 33-2) = P(W_s \leqslant 31)$$

Entering the table with $k_1 = 4$, $k_2 = 6$, we find

$$P(W_s \leqslant 31) = P(W_{XY} \leqslant 10) = .3810$$

which is therefore the desired probability.

Method 2. Let us denote the sum of the control ranks by

$$W_r = R_1 + \cdots + R_m \tag{1.17}$$

Since the totality of the ranks is just the set of integers $1, \ldots, N$ whose sum by (1.12) is $N(N+1)/2$, it follows that

$$W_r + W_s = \tfrac{1}{2}N(N+1) \tag{1.18}$$

The sum of W_r and W_s being constant, large values of W_s correspond to small values of W_r and vice versa. Instead of rejecting H according to (1.6) when W_s is large, it is therefore equivalent to reject when W_r is small, say, when

$$W_r \leqslant c' \tag{1.19}$$

Here c' is determined by the condition that is equivalent to (1.7), namely,

$$P(W_r \leqslant c') = \alpha \tag{1.20}$$

If we wished to table the probabilities on the left-hand side of (1.20), it would again be convenient to subtract from W_r its minimum value, which is $1 + \cdots + m = m(m+1)/2$, and consider the statistic

$$W_{YX} = W_r - \tfrac{1}{2}m(m+1) \tag{1.21}$$

Now at the end of this section it will be proved that the statistics W_{XY} and W_{YX} have the same null distribution, i.e., that

$$P(W_{XY} \leqslant a) = P(W_{YX} \leqslant a) \qquad \text{for all } a \tag{1.22}$$

Thus, Table B can be used to solve Eq. (1.20) just as it was used to solve Eq. (1.7). As an illustration of this method, let us consider again the probability that $W_s \geqslant 35$

when $m = 4, n = 6$, obtained previously by Method 1. Since $N = 10$, it follows from (1.18) that $W_s = 55 - W_r$, and hence that $W_s \geqslant 35$ if and only if $W_r \leqslant 20$. By (1.21), $W_{YX} = W_r - 10$ and

$$P(W_r \leqslant 20) = P(W_{YX} \leqslant 10)$$

Entering Table B with $k_1 = 4$, $k_2 = 6$, and $a = 10$, we find the probability .3810, which agrees with the value for $P(W_s \geqslant 35)$ obtained by the first method.

Consider now once more the situation of Example 2. The sum of the treatment ranks turned out to be $W_s = 21$, and with small values of W_s favoring the alternative it followed that the hypothesis H (discouragement has no effect on test performance) is not rejected at the significance level closest to $\alpha = .05$. Instead of simply reporting the rejection of H at the given significance level, it is more informative to report the probability under H of obtaining a value as extreme as, or more extreme than, the observed value. This probability is called the *significance probability* of the observed result. In the case of Example 2, the significance probability is the probability that W_s is as small as, or smaller than, 21, which was seen above to be equal to $P(W_s \leqslant 21) = .1111$.

The significance probability has the important property of showing, in a single number, whether or not to reject the hypothesis at any attainable level α. For suppose that large values of a test statistic, say W_s, are significant. If w denotes the observed value of W_s, the significance probability is then defined as

$$\hat{\alpha} = P_H(W_s \geqslant w) \tag{1.23}$$

The hypothesis is rejected when $w \geqslant c$, and hence by (1.7) when $\alpha \geqslant \hat{\alpha}$. Conversely, H is accepted for any critical value exceeding w, and hence for any attainable significance level α which is less than $\hat{\alpha}$. Because of this property, when reporting the outcome of a statistical test, one should state not only whether the hypothesis was accepted or rejected at a given significance level; at the least one should also publish the significance probability, thus enabling others to perform the test at a level of their choice.

Beyond this, the significance probability provides an intuitive indication of the strength of the evidence against the hypothesis since it is the probability (under H) of getting a value of the test statistic as extreme as or more than the observed value. The smaller $\hat{\alpha}$ is, the more unlikely it is to observe this extreme a value under H, and the stronger, therefore, is the evidence against H.

Let us now return to the Wilcoxon test and discuss an alternative interpretation of the statistics W_{XY} and W_{YX} which will prove useful in Chap. 2. Denote the control and treatment observations by $X_1, \ldots, X_m$ and $Y_1, \ldots, Y_n$, respectively, and consider all possible pairs of observations (X_i, Y_j) consisting of one X and one Y. Among the mn such pairs there will typically be some for which $X_i < Y_j$ and some with $Y_j < X_i$, and it turns out that

$$W_{XY} = \text{number of pairs } (X_i, Y_j) \text{ with } X_i < Y_j \tag{1.24}$$

and

$$W_{YX} = \text{number of pairs } (X_i, Y_j) \text{ with } Y_j < X_i \tag{1.25}$$

The statistics W_{XY} and W_{YX}, which by (1.13) and (1.21) are equivalent to the Wilcoxon rank-sum statistics W_s and W_r, are known as the *Mann-Whitney* statistics. The notation W_{XY} and W_{YX} reflects the relationships (1.24) and (1.25).

The interpretation (1.24) provides an alternative method for computing W_{XY}, which sometimes is more convenient. To illustrate the method, consider once more the data of Example 2. Instead of ranking the 10 differences, we now count for each of the treatment differences $-6, -5, 1, 4, 6$ the number of control differences that it exceeds, namely, $-6{:}\ 0$, $-5{:}\ 0$, $1{:}\ 1$, $4{:}\ 2$, $6{:}\ 3$. It follows that

$$W_{XY} = 0+0+1+2+3 = 6$$

To conclude this section, we shall now prove relation (1.24), the symmetry of W_s, and relation (1.22).

(A) Proof of (1.24). Denote the ordered Y's by

$$Y_{(1)} < Y_{(2)} < \cdots < Y_{(n)}$$

and for each j, compute the number of X's less than $Y_{(j)}$. Since the rank of $Y_{(1)}$ is S_1, there are $S_1 - 1$ smaller observations. These must all be X's, so that the number of X's less than $Y_{(1)}$ is $S_1 - 1$. Similarly, there are $S_2 - 1$ observations less than $Y_{(2)}$. One of these (namely, $Y_{(1)}$) is a Y, so that the number of X's less than $Y_{(2)}$ is $(S_2 - 1) - 1 = S_2 - 2$. In the same way, it is seen quite generally that since the number of Y's less than $Y_{(j)}$ is $j-1$, the number of X's less than $Y_{(j)}$ is $(S_j - 1) - (j-1) = S_j - j$. The total number of pairs (X_i, Y_j) with $X_i < Y_j$ is therefore

$$(S_1 - 1) + (S_2 - 2) + \cdots + (S_n - n) = W_s - (1+2+\cdots+n) = W_s - \tfrac{1}{2}n(n+1) = W_{XY}$$

as was to be proved.

(B) Proof of the Symmetry of the Null Distribution of W_s about $\tfrac{1}{2}n(N+1)$. Let the N subjects (of which n receive the treatment and m serve as controls) be ranked in inverse order. The subject that previously held rank 1 now has rank N, rank 2 is replaced by rank $N-1$, and in general rank S is replaced by rank $N-S+1$. Let us denote by S_i' this new inverse rank attached to the treatment subject that had previously received rank S_i. Then the argument leading to (1.4) shows that

$$P_H(S_1' = s_1, \ldots, S_n' = s_n) = \frac{1}{\binom{N}{n}}$$

and hence that $W_{s'} = S_1' + \cdots + S_n'$ has the same null distribution as W_s. Now

$$W_{s'} = [(N+1)-S_1]+\cdots+[(N+1)-S_n] = n(N+1)-W_s$$

so that

$$W_{s'}-\tfrac{1}{2}n(N+1) = \tfrac{1}{2}n(N+1)-W_s$$

It follows that $W_s-\frac{1}{2}n(N+1)$ and $\frac{1}{2}n(N+1)-W_s$ have the same distribution, i.e., that

$$P[W_s-\tfrac{1}{2}n(N+1) = k] = P[\tfrac{1}{2}n(N+1)-W_s = k]$$

This is equivalent to (1.15) and hence proves (B).

(C) Proof of (1.22). By (1.18), $W_r+W_s = \frac{1}{2}(m+n)(N+1)$, so that $W_r-\frac{1}{2}m(N+1) = \frac{1}{2}n(N+1)-W_s$. It thus follows from (B) that $W_r-\frac{1}{2}m(N+1)$ has the same distribution as $W_s-\frac{1}{2}n(N+1)$ and, on adding $\frac{1}{2}mn$ to both statistics, that $W_r-\frac{1}{2}m(m+1)$ has the same distribution as $W_s-\frac{1}{2}n(n+1)$, as was to be proved.

3. ASYMPTOTIC NULL DISTRIBUTION OF THE WILCOXON STATISTIC

Critical values and significance probabilities of the Wilcoxon rank-sum test can be obtained from Table B when both m and n are less than or equal to 10. For larger values of m and n, a useful approximation is available for these probabilities. It is a fundamental fact of probability theory, known as the *central limit theorem*, that the sum T of a large number of independent random variables (satisfying some mild restrictions) is approximately normally distributed. (For a discussion of this theorem see Appendix, Sec. 3.) If $E(T)$ and $\text{Var}(T)$ denote the expectation and variance of T,

$$P\left[\frac{T-E(T)}{\sqrt{\text{Var}(T)}} \leqslant a\right] \approx \Phi(a) \tag{1.26}$$

where $\Phi(a)$ denotes the area to the left of a under a standard normal curve as shown in Fig. 1.1, and where $\approx$ indicates that the right-hand side is an approximation to the probability on the left-hand side.

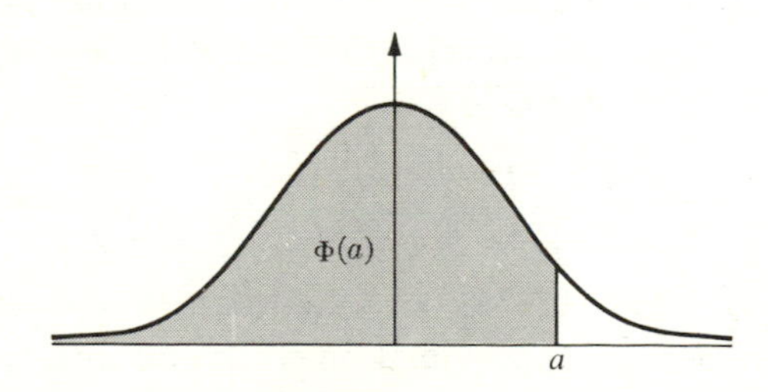

FIGURE 1.1
Standard normal curve.

Now if n is large, the sum of the treatment ranks is the sum of a large number of random variables but these are not independent. However, if m is also large, the dependence turns out to be sufficiently weak for the central limit theorem to continue to hold, so that the normal approximation (1.26) does apply to the Wilcoxon distribution.

Use of the approximation (1.26) requires knowledge of the expectation and variance of W_s. These are determined by the distribution of W_s, which in turn rests on the distribution of the treatment ranks given by (1.4). It is shown in formulas (A.45) and (A.46) of the Appendix that this expectation and variance are

$$E(W_s) = \tfrac{1}{2}n(N+1) \tag{1.27}$$

and

$$\text{Var}(W_s) = \tfrac{1}{12}mn(N+1) \tag{1.28}$$

The corresponding formulas for W_r are obtained by interchanging m and n, and those for W_{XY} from (1.13), so that

$$E(W_r) = \tfrac{1}{2}m(N+1) \qquad \text{and} \qquad E(W_{XY}) = \tfrac{1}{2}mn \tag{1.29}$$

while

$$\text{Var}(W_r) = \text{Var}(W_{XY}) = \tfrac{1}{12}mn(N+1) \tag{1.30}$$

The application of (1.26) requires, in addition to the expectation and variance of W_s, a table of the area $\Phi(a)$ to the left of a under a standard normal curve. Such a table is given as Table C at the end of the book for positive values of a. It can be obtained from this table for negative values of a by using the facts that (i) the standard normal curve is symmetric about the origin and (ii) the total area under the normal curve is equal to 1. By (i), the area $\Phi(-a)$ to the left of $-a$ is equal to the area to the right of a, and by (ii) this latter area is equal to $1-\Phi(a)$. This provides the needed relationship

$$\Phi(-a) = 1-\Phi(a) \tag{1.31}$$

To illustrate both the computation and the accuracy of the normal approximation, let us take an example at the edge of Table B of the Wilcoxon distribution. Suppose that $m = n = 10$ and that we wish to find the probability $P(W_s \leqslant 79)$. Then $E(W_s) = 105$, $\text{Var}(W_s) = 175$, and hence

$$P(W_s \leqslant 79) = P\left(\frac{W_s - 105}{\sqrt{175}} \leqslant \frac{-26}{\sqrt{175}}\right) \approx \Phi\left(\frac{-26}{13.23}\right) = \Phi(-1.965)$$

Table C shows that $\Phi(1.965) = .9753$, so that (to three digits) $\Phi(-1.965) = .025$, which is therefore the normal approximation to the desired probability. In comparison, Table B gives .026 for the same probability.

The accuracy of the approximation is frequently improved by the following refinement, which has its basis in the graphical representation of a distribution by means of a *histogram*. This consists of a series of rectangles or bars, each corresponding to one of the possible values of the random variable in question. The base

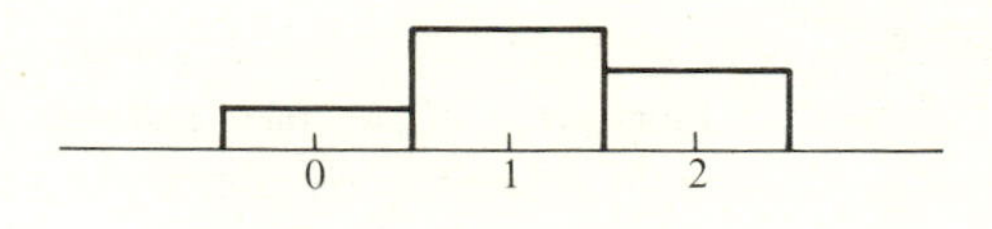

FIGURE 1.2
Histogram for X.

of the rectangle is centered on the value, and its area is equal to the probability with which this value is taken on. Figure 1.2 shows the histogram for a random variable X taking on the values 0, 1, 2 with probabilities $1/6$, $1/2$, and $1/3$, respectively. Since the base of each rectangle in this case is equal to 1, the height is equal to the area of the rectangle, so that the heights of the three rectangles are $1/6$, $1/2$, and $1/3$.

Suppose now that probabilities are being approximated by means of the appropriate area under a smooth curve (Fig. 1.3), so that the approximation to $P(X \leqslant a)$ is the area under the curve to the left of a. Then $P(X \leqslant 1)$, for example, is approximated by the area under the curve to the left of 1. However, since $P(X \leqslant 1)$ is equal to the area under the histogram corresponding to the rectangles centered at 0 and 1, that is, the area under the histogram to the left of 1.5, a more natural and typically more accurate approximation is provided by the area to the left of 1.5.

This modification is called the *continuity correction* because it compensates for the fact that a noncontinuous histogram is being approximated by a continuous curve. As an illustration of its effect, consider the probability $P(W_s \leqslant 12)$ when $m = 11$ and $n = 3$ (although such wide discrepancy between the two group sizes typically is undesirable). The exact probability, which is easily obtained by enumeration, is equal to .063. Since $E(W_s) = 22.5$ and $\sqrt{\text{Var}(W_s)} = \sqrt{41.25} = 6.423$, the normal approximation gives

$$P(W_s \leqslant 12) = P\left(\frac{W_s - 22.5}{6.423} \leqslant \frac{12 - 22.5}{6.423}\right) \approx \Phi(-1.635) = .051$$

Let us now apply the continuity correction. Since the smallest value that W_s can take on is $n(n+1)/2 = 6$, its possible values are 6, 7, ..., 12, 13, ..., and the possible

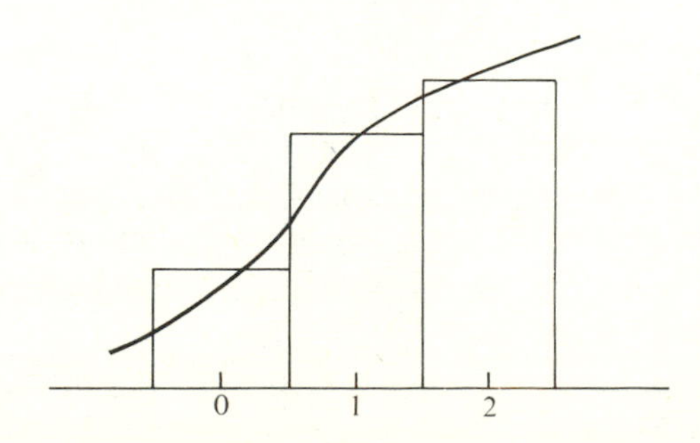

FIGURE 1.3
Approximation to the distribution of X.

values of $(W_s - 22.5)/6.423$ are therefore $(6-22.5)/6.423, \ldots, (12-22.5)/6.423, (13-22.5)/6.423, \ldots$. These are the centers of the bars making up the histogram of $(W_s - 22.5)/6.423$. The normal approximation took in only the area under the curve to the left of $(12-22.5)/6.423$, which is the center of the bar corresponding to $W_s = 12$. However, since the whole area of this bar is counted in the probability $P(W_s \leqslant 12)$ it is better to include the area under the normal curve up to the right-hand endpoint of this bar, which lies midway between $(12-22.5)/6.423$ and $(13-22.5)/6.423$, and is therefore equal to

$$\frac{1}{2}\left(\frac{12-22.5}{6.423} + \frac{13-22.5}{6.423}\right) = \frac{12.5-22.5}{6.423}$$

Thus the continuity correction leads to the approximation

$$\Phi\left(\frac{12.5-22.5}{6.423}\right) = \Phi(-1.557) = .060$$

which is closer to the correct value .063 than the value .051 obtained by the uncorrected normal approximation. In general, the same argument leads to approximating the probability $P(W_s \leqslant c)$ by[1]

$$P(W_s \leqslant c) \approx \Phi\left[\frac{c - \frac{1}{2}n(N+1) + \frac{1}{2}}{\sqrt{mn(N+1)/12}}\right] \tag{1.32}$$

It is difficult to make reliable general statements concerning the accuracy of this and similar approximations, but the following two qualitative statements typically apply not only in the Wilcoxon case but also (with the obvious modifications) to such approximations throughout the book.

(i) The accuracy of the approximation tends to increase as m and n increase.

(ii) If the accuracy is measured not by the absolute error (the difference between the exact and approximate values) but by the relative error (the absolute error divided by the exact value), the approximation tends to become less reliable when the probabilities being approximated are very close to zero, that is, in the tails of the distribution.

To (i) can be added that in the Wilcoxon case an increase in the smaller of the two group sizes typically results in a greater improvement than a corresponding increase in the larger group size.

These statements leave open the crucial question of how large m and n have to be for the approximation to be satisfactory. Some light is thrown on this problem by Table 1.1, which provides a comparison of the Wilcoxon probability $P(W_r \leqslant c)$

[1] Kruskal and Wallis (1952, p. 591) state that the continuity correction tends to improve the approximation when the probability is above .02 and to make it worse when the probability is .02 or less. This is in agreement with the values shown in Table 1.1 on page 17.

with its normal approximation, both with and without continuity correction, for three combinations of the group sizes m and n. The cases $m = 4$, $n = 12$, and $m = n = 8$ suggest that at least for the range $\alpha = .01$ to $\alpha = .15$, the approximation with continuity correction will be adequate for most purposes for all group sizes which are not covered by Table B and in which the smaller group size k_1 is at least 4. Incidentally, it is seen from Table 1.1 that the continuity correction does not always improve the approximation, although on the whole it seems to give very much better results.

TABLE 1.1 Exact and approximate values of $P(W_r \leqslant c)$ with and without continuity correction

c	6	7	8	9	10
Exact	.012	.024	.048	.083	.131
Without	.010	.019	.035	.061	.098
With	.014	.026	.047	.078	.123

$m = 3$, $n = 6$

c	13	15	20	23	25
Exact	.004	.010	.052	.106	.158
Without	.005	.011	.045	.091	.138
With	.006	.012	.051	.102	.151

$m = 4$, $n = 12$

c	44	46	48	52	56	58
Exact	.005	.010	.019	.052	.117	.164
Without	.006	.010	.018	.047	.104	.147
With	.007	.012	.020	.052	.114	.159

$m = 8$, $n = 8$

When the smaller group size k_1 is 1, 2, or 3, the normal approximation is not satisfactory. The case $k_1 = 1$ is covered by Prob. 74, the case $k_1 = 3$ by Table B if the larger group size does not exceed 12. An exact formula for $k_1 = 2$ is given by Buckle, Kraft, and van Eeden (1969a). These authors also propose an alternative, quite different approximation that typically does even better than the normal approximation, and which in particular takes care of the remaining cases with $k_1 = 3$. This approximation is sketched in Sec. 7A. An alternative approach to small values of k_1 is given by Fix and Hodges (1955).

Statements, such as those made above, concerning the validity of an approximation are typically based on two kinds of evidence. There is the numerical evidence obtained by comparing the exact and approximate values for a number of different cases such as those in Table 1.1. This is often supplemented by a theoretical result, a *limit theorem*, which states that the left-hand side of (1.26) tends to the right-hand side as the group sizes increase and tend to infinity. Such a theorem claims only that the left-hand side will get arbitrarily close to the right-hand side *eventually*, that is, for sufficiently large group sizes. However, it also suggests, though it does not prove, that the accuracy of the approximation tends to improve as the group sizes increase. Thus, if computation shows the approximation to be satisfactory for given group sizes and a certain range of significance levels, one may expect it to be even better for larger sample sizes.

The limit theorem relevant to the present situation states that

$$P_H\left[\frac{W_s-\frac{1}{2}n(N+1)}{\sqrt{mn(N+1)/12}}\leqslant a\right]\to\Phi(a) \tag{1.33}$$

as m and n tend to infinity. A proof is given in the Appendix, Sec. 4, Example 16.

4. THE TREATMENT OF TIES

The application of rank tests often meets with a difficulty, which so far we have avoided. We have assumed that either subjective evaluation or numerical observation leads to a complete ranking of the treatment and control subjects. However, this is frequently not the case. An instructor asked to rank two students, or a patient trying to determine on which of two occasions a headache remedy gave greater relief, may feel unable to make the necessary distinction. Alternatively, when measurements or other numerical observations are taken, it often happens that two or more observations are equal, and this leads to the same difficulty. Suppose, for example, that $m=n=2$ and the observations are 1.3, 1.7, 1.7, 2.5. Then the smallest observation has rank 1 and the largest rank 4, but we do not know which of the other two has rank 2 and which rank 3. If no extraneous distinction is introduced between the two tied observations, they should be assigned the same rank, and it then seems most natural to assign to each the *average* of the two tied ranks, in the present case $(2+3)/2=2.5$. The four subjects are then assigned the *midranks* 1, 2.5, 2.5, 4, and we shall denote the midranks assigned to treatment and control by S_1^*, S_2^* and R_1^*, R_2^*, respectively.

Clearly the null distribution (1.4) no longer holds because even the possible values of (S_1^*, S_2^*) are not the same as those of (S_1, S_2). However, the argument leading to (1.4) still applies. Under the hypothesis of no treatment effect, the four subjects will receive the same midranks regardless of which of them are assigned to treatment and which to control. The random selection of two of the subjects to receive the treatment then also selects their midranks, so that each of the six possible choices of two of the four midranks as S_1^* and S_2^* are equally likely. This leads to the following null distribution, which now takes the place of (1.4):

$$P(S_1^* = 1, S_2^* = 2.5) = \tfrac{2}{6}, \qquad P(S_1^* = 1, S_2^* = 4) = \tfrac{1}{6}$$
$$P(S_1^* = S_2^* = 2.5) = \tfrac{1}{6}, \qquad P(S_1^* = 2.5, S_2^* = 4) = \tfrac{2}{6}$$

This distribution determines that of the Wilcoxon midrank-sum statistic

$$W_s^* = S_1^* + S_2^*$$

as before, and we see that W_s^* takes on the values 3.5, 5. 6.5 each with probability $1/3$.

The above approach applies quite generally to situations with tied observations, but instead of giving a general discussion we shall illustrate it with another example. Suppose that $m = n = 3$ and that the control observations are 2, 2, 9 and the treatment observations 4, 9, 9. The six midranks are then 1.5, 1.5, 3, 5, 5, 5, and the sum of the treatment midranks is

$$W_s^* = 3+5+5 = 13$$

What is the significance probability $P(W_s^* \geqslant 13)$ of these observations? There are $\binom{6}{3} = 20$ equally likely choices for the three treatment midranks, and the probabilities of the possible sets of values of (S_1^*, S_2^*, S_3^*) are

(s_1^*, s_2^*, s_3^*)	1.5, 1.5, 3	1.5, 1.5, 5	1.5, 3, 5	1.5, 5, 5	3, 5, 5	5, 5, 5
$P(s_1^*, s_2^*, s_3^*)$	1/20	3/20	6/20	6/20	3/20	1/20

The significance probability is therefore $P(W_s^* \geqslant 13) = 4/20 = .2$.

In certain situtations involving tied observations there arises an ambiguity that is not always resolved correctly. Suppose that the treatment observations are 6, 6, 6, 9 and the control observations 1, 3, 4, 10, with large values of W_s being significant. Since ties occur only among the treatment observations, no question arises concerning the value of W_s. The treatment ranks are 4, 5, 6, 7 and the rank sum W_s thus has the value 22. It is now tempting to compute the significance probability as in Sec. 2 or obtain it from Table B to be $12/70 = .1714$. On the other hand, the approach of the present section gives $W_s^* = 22$ and the significance probability $11/70 = .1571$ [Prob. 45(i)]. Which of these two values is correct?

Recalling the basis for such probability calculations, we see that the actual observations must be compared with all possible ways of assigning the eight numbers 1, 3, 4, 6, 6, 6, 9, 10 to treatment and control. Many of these will split the three tied observations, assigning some to one group and some to the other, and in these cases the value of W_s is not defined. Hence, there is no validity in the first approach, and the correct value of the significance probability is .1571.

Since the distribution of W_s^* depends not only on m and n but also on the number of observations tied at each value (Prob. 44), tables of the distribution are not practicable and the need for an approximation is, if anything, even more pressing than when no ties are present. A normal approximation is again available when m and n are not too small and the maximum proportion of observations tied at any value is not too close to 1.

Use of the normal approximation requires the expectation and variance of W_s^*. The expectation of W_s^* is shown in the Appendix, Example 3, to be the same as that of W_s and hence to be given by

$$E(W_s^*) = \tfrac{1}{2}n(N+1) \tag{1.34}$$

The variance of W_s^* involves the numbers of observations tied at the different values. Suppose that the N observations take on e distinct values, and that d_1 of the N observations are equal to the smallest value, d_2 to the next smallest, ..., d_e to the largest. If the observations are, for example, 2, 2, 4, 9, 9, 9, as in one of the earlier examples, then $e = 3$ and $d_1 = 2$, $d_2 = 1$, and $d_3 = 3$. With this notation, it is shown in the Appendix, Example 3, that

$$\text{Var}(W_s^*) = \frac{mn(N+1)}{12} - \frac{mn \sum_{i=1}^{e} (d_i^3 - d_i)}{12N(N-1)} \tag{1.35}$$

The first term is just the variance of W_s; the second gives the correction for ties. The effect of the correction tends to be quite small, although exceptions to this rule are possible (Prob. 93). In particular, when no ties are present, all the d_i are equal to 1, and the correction term is zero as it should be.

Typically, the normal approximation is not as close in the presence of ties as it is for the same group sizes without ties. This is not surprising since ties tend to decrease the number of values taken on by W_s and thereby to increase the lumpiness of its distribution. A number of comparisons of the exact and approximate distributions with and without continuity correction are given by Lehman (1961). Although these comparisons are on the whole favorable to the continuity correction, in this book—and hence in the present case—we shall use the approximation without the correction whenever the values of the random variable in question are not equally spaced.

The normal approximation is supported by a limit theorem, which states that the null distribution of $[W_s^* - E(W_s^*)]/\sqrt{\text{Var}(W_s^*)}$ tends to the standard normal distribution Φ provided both m and n tend to infinity and

$$\max_{i=1,\dots,e} \left(\frac{d_i}{N}\right) \text{ is bounded away from 1 as } N \to \infty \tag{1.36}$$

Condition (1.36) suggests that the normal approximation should be used only when the largest of the proportions $d_1/N, \ldots, d_e/N$ is not too close to 1. The proof of this limit theorem is given in the Appendix, Sec. 4, Example 17.

The W_s^*-test can also be applied in situations in which no numerical responses are observed but the subjects are divided into a number of categories such as poor, indifferent, or good, and it is observed how many of the treatment and control subjects fall into each of these categories. Let us now illustrate the normal approximation on an example of this kind.

EXAMPLE 3. *Psychological counseling.* In a test of the effect of psychological counseling,[1] 80 boys are divided at random into a control group of 40 to whom only the normal counseling facilities are available, and a treatment group of 40 who receive special counseling. At the end of the study, a careful assessment is made of each boy who is then classified as having made a good, fairly good, fairly poor, or poor adjustment, with the following results:

	Poor	Fairly Poor	Fairly Good	Good	Total
Treatment	5	7	16	12	40
Control	7	9	15	9	40

The data can be treated as if the observations were capable of taking on only four values, with $5+7 = 12$ observations tied at the smallest value, $7+9 = 16$ at the next smallest, $16+15 = 31$ at the third, and $12+9 = 21$ at the largest value. The midrank of the 12 subjects whose adjustment is classified as poor is then $(1+2+\cdots+12)/12 = 6.5$; the midrank of the fairly poor is $(13+\cdots+28)/16 = 20.5$; and the midranks of the subjects in the third and fourth categories are 44 and 70, respectively. [For a general formula giving the value of the midrank in the ith group, see (1.41) below.]

The value of W_s^* is thus

$$W_s^* = 5 \times 6.5 + 7 \times 20.5 + 16 \times 44 + 12 \times 70 = 1{,}720$$

and the significance probability of the observed results is $P_H(W_s^* \geqslant 1{,}720)$. Since $m = n = 40$ and $d_1 = 12, d_2 = 16, d_3 = 31, d_4 = 21$, it follows from (1.34) and (1.35) that $E(W_s^*) = 1{,}620$ and $\sqrt{\text{Var}(W_s^*)} = 99.27$. The normal approximation to the significance probability therefore gives

$$P_H(W_s^* \geqslant 1{,}720) = 1 - \Phi(1.01) = .16$$

The statistic W_s^* is the natural generalization of the rank sum W_s when the observations are not all distinct. In the same situation, it is also possible to generalize the statistic W_{XY}. Recall that W_{XY} counts the number of pairs (X_i, Y_j) for which $X_i < Y_j$; that is, each pair (X_i, Y_j) is assigned the score 1 or 0 as X_i is less or greater than Y_j, and W_{XY} is the sum of these scores. When ties are present, it may happen that Y_j is equal to X_i and it is then natural to assign the score $\frac{1}{2}$ to the pair (X_i, Y_j).

[1] An extensive report of a large-scale study of this question is given by Powers and Witmer, *An Experiment in the Prevention of Juvenile Delinquency*, Columbia University Press, New York, 1951. Since this study used the matched samples design (to be discussed in Chap. 3), the data are not suitable as an example here.

Let us denote the sum of the resulting scores by W_{XY}^*, so that

$$W_{XY}^* = [\text{number of pairs } (X_i, Y_j) \text{ with } X_i < Y_j] + \tfrac{1}{2}[\text{number of pairs } (X_i, Y_j) \text{ with } X_i = Y_j] \tag{1.37}$$

in generalization of (1.24). Then the tests based on W_s^* and W_{XY}^* are again equivalent and in fact the two statistics satisfy the relationship corresponding to (1.13) for the unstarred statistics, namely,

$$W_{XY}^* = W_s^* - \tfrac{1}{2}n(n+1) \tag{1.38}$$

It follows that

$$E(W_{XY}^*) = \tfrac{1}{2}mn \tag{1.39}$$

and

$$\text{Var}(W_{XY}^*) = \text{Var}(W_s^*) \tag{1.40}$$

where the latter variance is given by (1.35). The relation (1.38) provides an alternative method for computing W_s^*.

We conclude this section by proving (1.38). It is convenient to begin by obtaining a general expression for the values of the midranks. If d_1 observations are tied at the smallest value, the midrank of these observations is the average of the ranks $1, \ldots, d_1$, that is,

$$\frac{1 + \cdots + d_1}{d_1} = \tfrac{1}{2}(d_1 + 1)$$

The observations in the second group occupy the positions $d_1+1, d_1+2, \ldots, d_1+d_2$, and their midrank is therefore

$$\frac{(d_1+1) + \cdots + (d_1+d_2)}{d_2} = d_1 + \tfrac{1}{2}(d_2+1)$$

Similarly, the midrank of an observation in the ith group is

$$\frac{(d_1 + \cdots + d_{i-1} + 1) + \cdots + (d_1 + \cdots + d_{i-1} + d_i)}{d_i} = d_1 + \cdots + d_{i-1} + \tfrac{1}{2}(d_i + 1) \tag{1.41}$$

Suppose now that of the d_1 observations tied at the smallest value, A_1 are X's and B_1 are Y's; that of the d_2 observations tied at the next smallest value, A_2 are X's and B_2 are Y's; and so on. The sum of the Y-midranks is then

$$W_s^* = B_1[\tfrac{1}{2}(d_1+1)] + B_2[d_1 + \tfrac{1}{2}(d_2+1)] + B_3[d_1 + d_2 + \tfrac{1}{2}(d_3+1)] + \cdots \tag{1.42}$$

To obtain a similar expression for W_{XY}^*, consider the contribution to W_{XY}^* from an observation Y in the first group. There are no X's smaller than this Y, and the number of X's equal to Y is A_1. Hence the contribution of this Y is $\frac{1}{2}A_1$. For an observation Y in the second group the number of X's smaller than Y is A_1 and the number of X's equal to Y is A_2, so that the contribution from this Y is $A_1 + \frac{1}{2}A_2$.

Continuing in this way and remembering that the number of Y's in the first, second, and subsequent groups is B_1, B_2, and so on, we find that

$$W_{XY}^* = B_1(\tfrac{1}{2}A_1) + B_2(A_1 + \tfrac{1}{2}A_2) + B_3(A_1 + A_2 + \tfrac{1}{2}A_3) + \cdots \tag{1.43}$$

Since $d_i = A_i + B_i$, it follows that

$$W_s^* - W_{XY}^* = B_1[\tfrac{1}{2}(B_1+1)] + B_2[B_1 + \tfrac{1}{2}(B_2+1)] + B_3[B_1 + B_2 + \tfrac{1}{2}(B_3+1)] + \cdots$$

The sum of the terms in this difference not involving the factor $1/2$ is

$$B_2 B_1 + B_3(B_1 + B_2) + \cdots$$

which is equal to

$$\sum_{i>j} B_i B_j = \tfrac{1}{2} \sum_{i \neq j} B_i B_j$$

and hence

$$W_s^* - W_{XY}^* = \tfrac{1}{2}[\sum B_i + \sum B_i^2 + \sum_{i \neq j} B_i B_j] = \tfrac{1}{2}[\sum B_i + (\sum B_i)^2]$$

Since the sum of the B_i is the total number of Y's and hence $\Sigma B_i = n$, this completes the proof of (1.38).

5. TWO-SIDED ALTERNATIVES

The Wilcoxon rank-sum test provides a simple and effective method for comparing a new treatment with a standard one. This comparison is strongly biased in favor of the latter. In fact, the critical value is set so that the probability is $1-\alpha$ (typically .9 or more) of deciding in favor of the standard treatment when there is actually no difference between the two. The decision goes to the new competitor only if it has proved itself beyond a reasonable doubt.

Such an asymmetric procedure is not always appropriate. In particular, a more symmetric approach is indicated for the following two kinds of problems.

(i) When trying to decide which of two treatments is better, one may be faced with a situation in which both treatments are "new" and there is therefore no reason to bias the procedure in favor of one rather than the other. Two equally reputable firms may, for example, simultaneously and at about the same price, introduce two drugs for the treatment of a condition for which previously no treatment was known. Or one may wish to decide with which of two brands of gas a car obtains the better mileage.

(ii) The problem may not be to decide which of two treatments is better but only to determine whether they differ at all, and in such situations the two treatments typically play a symmetric role. In the laboratory of a statistics course, for example, the instructor may have available two different types of computing machines and he may want to know whether students would be handicapped in an examination by being assigned to one type rather than the other. Similarly, in a study comparing a surgical approach to a medical problem with a more conservative treatment, it may be desirable to examine first whether it is necessary to distinguish between two surgical techniques used by different surgeons.

In the present section we shall show how the Wilcoxon test can be adapted to problems of these two types. We shall begin with the second problem as the simpler of the two. To test whether there exists a significant difference between two treatments, say, A and B, the N subjects are assigned as before, m to A and n to B. The responses of the N subjects are ranked, with W_A and W_B denoting the sum of the A-ranks and the B-ranks respectively. The null hypothesis H to be tested is that there is no difference between A and B. While this is the same as the hypothesis H (no treatment effect) considered in Sec. 1, the alternatives no longer specify that B is superior to A, but that it is either superior or inferior to A. The hypothesis is rejected therefore not only when the sum W_B of the B-ranks is too large but also when it is too small, say, when

$$W_B \leqslant c_1 \quad \text{or} \quad W_B \geqslant c_2 \qquad (c_1 < c_2) \tag{1.44}$$

The constants c_1 and c_2 are determined so that under H the probability of rejection is equal to the specified significance level α, that is, so that

$$P_H(W_B \leqslant c_1) + P_H(W_B \geqslant c_2) = \alpha \tag{1.45}$$

Equation (1.45) is not enough to specify c_1 and c_2 since it leaves open the question of how to distribute the total probability α between the first and the second term. If the two treatments play a completely symmetric role, it seems natural to choose c_1 and c_2 so that H is as likely to be rejected when W_B is too large as when it is too small, that is, to put

$$P_H(W_B \leqslant c_1) = P_H(W_B \geqslant c_2) = \tfrac{1}{2}\alpha \tag{1.46}$$

From the fact that the distribution of W_B is symmetric about $\frac{1}{2}n(N+1)$ (Sec. 2), it follows that c_1 and c_2 must be equally distant from $\frac{1}{2}n(N+1)$ and hence must be of the form

$$c_1 = \tfrac{1}{2}n(N+1) - c, \qquad c_2 = \tfrac{1}{2}n(N+1) + c$$

The test defined by (1.44) and (1.46), the *two-sided Wilcoxon rank-sum test*, thus rejects when $W_B - \frac{1}{2}n(N+1)$ is either $\leqslant -c$ or $\geqslant c$, that is, when

$$|W_B - \tfrac{1}{2}n(N+1)| \geqslant c \tag{1.47}$$

Because of the discrete nature of W_B, typically there will not exist a value c for which the equation

$$P_H[|W_B - \tfrac{1}{2}n(N+1)| \geqslant c] = \alpha \tag{1.48}$$

94402

holds exactly. As in Sec. 2, one may then choose the value of c for which the left-hand side of (1.48) is as close to α as possible, or the smallest value for which it does not exceed α. As in the one-sided case, one will typically not only be interested in rejection or acceptance at a fixed level α, but also in the significance probability of the data. For the two-sided test (1.47) this is given by

$$P_H[|W_B - \tfrac{1}{2}n(N+1)| \geqslant |w - \tfrac{1}{2}n(N+1)|] \tag{1.49}$$

where w denotes the observed value of W_B.

The two-sided test (1.47) can of course also be expressed in terms of W_{XY}. From (1.13) it is in fact seen that (1.47) is equivalent to

$$|W_{XY} - \tfrac{1}{2}mn| \geqslant c \tag{1.50}$$

with the constant c being the same in both cases.

For large m and n, the test is carried out with the help of the normal approximation. Since under H

$$\frac{W_B - \frac{1}{2}n(N+1)}{\sqrt{mn(N+1)/12}}$$

has approximately the standard normal distribution, the significance probability (1.49) is approximated by the area under the normal curve to the left of

$$\frac{-|w - \frac{1}{2}n(N+1)| + \frac{1}{2}}{\sqrt{mn(N+1)/12}}$$

plus the area to the right of

$$\frac{|w - \frac{1}{2}n(N+1)| - \frac{1}{2}}{\sqrt{mn(N+1)/12}}$$

These two areas are equal, and the significance probability of the two-sided test (1.47) is therefore approximated by

$$2\left\{1 - \Phi\left[\frac{|w - \frac{1}{2}n(N+1)| - \frac{1}{2}}{\sqrt{mn(N+1)/12}}\right]\right\} \tag{1.51}$$

In the presence of ties, the test (1.44) and conditions (1.46) are modified by replacing W_B by W_B^*. Since the distribution of W_B^* is in general not symmetric, (1.44) can no longer be replaced by (1.47). Instead, c_1 and c_2 are determined so that $P_H(W_B^* \leqslant c_1)$ and $P_H(W_B^* \geqslant c_2)$ each are as close to $\alpha/2$ as possible. For large m and n, a normal approximation can again be applied. The symmetry of the normal curve implies that the distribution of W_B^* is then approximately symmetric, and the argument leading to (1.51) now shows that the significance probability of the two-sided Wilcoxon test can be approximated by

$$2\left\{1 - \Phi\left[\frac{|w - \frac{1}{2}n(N+1)|}{\sqrt{\operatorname{Var}_H(W_B^*)}}\right]\right\} \tag{1.52}$$

Here w is the observed value of W_B^*, the variance $\text{Var}_H(W_B^*)$ is given by (1.35), and the continuity correction of $\frac{1}{2}$ in (1.51) has been omitted since the values of W_B^* are not equally spaced.

It is important to be clear about the difference between the two-sided test (1.47) and the one-sided test (1.6). The earlier test is appropriate when it is desired to reject the hypothesis H of equality only when treatment B is better than treatment A. (If there is a possibility that B is worse than A, this can be included in the hypothesis, which can be replaced by H': B is not better than A. This possibility is discussed more formally in Chap. 2, Sec. 2.) The one-sided test thus provides an answer to the question: "Can we confidently say that B is better than A?" On the other hand, the two-sided test is appropriate if it is desired to reject H when B is either better or worse than A. It provides an answer to the question: "Is either treatment significantly better than the other?"

A confusion as to which of the two tests is appropriate sometimes arises when at the beginning of an investigation it is not clear which of the treatments would be expected to be better should they turn out to differ at all. However, after the observations have been taken, a visual inspection of the data may suggest that, say, B is better and also an explanation of why this might have been expected. It is then tempting to see whether the superiority of B is significant by testing H against the one-sided alternative that B is better than A. Under the given circumstances this is not appropriate. If A had come out ahead, an explanation might also have suggested itself for this finding, and the same line of argument would have led to testing H against the alternative that A is superior. This shows that H would have been rejected if W_B had been either sufficiently large or sufficiently small and that the appropriate procedure is a three-decision procedure of the kind considered below.

If $c^{(1)}$ and $c^{(2)}$ denote the critical values corresponding to the same significance level α in the one- and two-sided cases with rejection rules (1.6) and (1.47), respectively, they are determined by

$$P_H[W_B - \tfrac{1}{2}n(N+1) \geqslant c^{(1)}] = \alpha$$

and

$$P_H[|W_B - \tfrac{1}{2}n(N+1)| \geqslant c^{(2)}] = \alpha$$

Since the second condition is equivalent to

$$P_H[W_B - \tfrac{1}{2}n(N+1) \geqslant c^{(2)}] = \tfrac{1}{2}\alpha$$

it is seen that $c^{(1)} < c^{(2)}$. To be judged significant, the difference must therefore be more striking when H is being tested against alternatives in both directions (B better or worse than A) than when it is tested against a specified direction, say, B better than A, decided upon before the observations are taken. This corresponds to the fact that a striking difference in either direction is more likely to be observed "by pure chance," that is, under H, than the same difference in just one direction.

EXAMPLE 4. *Anticipation of hypnosis.* In a study of the effect of hypnosis, 16 subjects were divided at random into a group of 8 to be hypnotized and 8 controls. When the results of the experiment were analyzed, it was noticed that a measure of ventilation taken on each subject at the beginning of the experiment unexpectedly appeared higher for the experimental subjects than for the controls. A plausible explanation was that such an increase could be caused in the experimental subjects by the anticipation of being hypnotized, and it seemed of interest to test the significance of the effect. The observations were as follows:[1]

Controls:	3.99	4.19	4.21	4.54	4.64	4.69	4.84	5.48
Treated:	4.36	4.67	4.78	5.08	5.16	5.20	5.52	5.74

Although it is tempting here to use a one-sided test, this is not appropriate since a corresponding difference in the other direction would also have been noticed and tested. The hypothesis of no effect would therefore have been rejected for a sufficiently large difference in either direction, and the rejection rule is that of a two-sided test.

To test the hypothesis of no treatment effect at significance level $\alpha = .05$, the critical values c_1 and c_2 are computed from (1.46) or the corresponding equations for W_{XY}, with $\alpha/2 = .025$. For group sizes $m = n = 8$, Table B shows that $P(W_{XY} \leqslant 13) = .025$. Since

$$P(W_{XY} \leqslant 13) = P(W_r \leqslant 49) = P(W_r \geqslant 87)$$

it is seen that at level $\alpha = .05$ the data will be significant provided

$$|W_r - 68| \geqslant 19$$

The observed values are $W_r = 49$ and hence $|W_r - 68| = 19$, and the hypothesis of no difference is therefore rejected at level $\alpha = .05$.

The significance probability of the test (1.47) is the probability $P_H(|W_r - 68| \geqslant |49-68|)$, which happens to be just .05. As an illustration of the normal approximation, let us compute the significance probability using formula (1.51). Since $n = 8$, $N = 16$, this gives the value $2[1-\Phi(2.001)] = .0454$. (See Prob. 53.)

The two-sided Wilcoxon test accepts the hypothesis of no difference between A and B when

$$|W_B - \tfrac{1}{2}n(N+1)| < c \tag{1.53}$$

and rejects H in the contrary case. Typically, when rejecting the hypothesis one will not be satisfied with asserting the existence of a significant difference but will wish to know which of the two treatments is better. One is thus concerned with choosing between the three decisions: D_0: accepting H; D_1: deciding that B is better than A; or D_2: deciding that A is better than B. The obvious procedure, suggested by (1.53), is to take decision D_1 if $W_B \geqslant \frac{1}{2}n(N+1)+c$, decision D_2 if $W_B \leqslant \frac{1}{2}n(N+1)-c$, and decision D_0 when (1.53) holds. For this procedure the significance level α retains its interpretation as the probability of falsely rejecting H (and therefore declaring one of the treatments superior to the other) when in fact there is no difference.

[1] Data from Agosti and Camerota (1965). "Some Effects of Hypnotic Suggestion on Respiratory Function." *Intern. J. Clin. Expt. Hypnosis* **13**:149–156.

There is a more natural way of looking at the present problem. Rather than emphasizing the somewhat artificial hypothesis of no difference between A and B, one may stress the problem of determining which of the two treatments is better. If, in addition, one permits the possibility of remaining in doubt when the data are not sufficiently conclusive, the procedure may be reformulated as

$$\text{(1.54)} \qquad \begin{array}{ll} \text{Choose } B \text{ if} & W_B \geqslant \tfrac{1}{2}n(N+1)+c \\ \text{Choose } A \text{ if} & W_B \leqslant \tfrac{1}{2}n(N+1)-c \\ \text{Suspend judgment if} & |W_B - \tfrac{1}{2}n(N+1)| < c \end{array}$$

This is seen to be a solution to problem (i) stated at the beginning of this section.

How in such a procedure would one specify the critical value c? Let α' denote the common probability

$$\text{(1.55)} \qquad \alpha' = P[W_B \leqslant \tfrac{1}{2}n(N+1)-c] = P[W_B \geqslant \tfrac{1}{2}n(N+1)+c]$$

computed under the assumption that there is no difference between the two treatments. Then $2\alpha'$ is just the level α defined by (1.48); that is, it is the probability of concluding that one of the treatments is better than the other (either B better than A, or A better than B) when, in fact, they are equal. An interpretation which is closer to the present point of view is obtained by considering the two errors to which the procedure (1.54) can lead: choosing A when B is the better treatment, or conversely, choosing B when A is better. Suppose, for example, that B is better but that $W_B \leqslant \frac{1}{2}n(N+1)-c$, so that erroneously A is selected. It is intuitively plausible that the probability of this error increases as the quality of B decreases and is largest when B is essentially equal to A, being better by just a hair's breadth; that is, the probability of erroneously selecting A is always less than or equal to α' and comes arbitrarily close to α' when the two treatments get sufficiently close. [A more precise statement and proof of this result requires a model for the case that one treatment is better than the other, together with the distribution of the ranks in such a model, in generalization of the distribution (1.4) for the case that there is no treatment difference. We shall postpone the development of such a model and its consequences to Chap. 2, Secs. 1 and 2, and for the moment be content with the above intuitive argument.] The same argument applies when A is better than B (but note that only one of these errors is possible in any given situation). Together, the two results show that α' has the interesting property of being the maximum probability of coming to the wrong conclusion, that is, of selecting the worse of the two treatments. The critical value c thus can be determined from (1.55) by specifying the maximum error probability α' that one is willing to tolerate.

EXAMPLE 5. *Two diets.* In a comparison of a number of diets, 12 rats were assigned at random, 7 to diet A and 5 to diet B. After 7 weeks they showed the following weight gains.[1]

$$\begin{array}{lccccccc} A: & 156 & 183 & 120 & 113 & 138 & 145 & 142 \\ B: & 130 & 148 & 117 & 133 & 140 & & \end{array}$$

Here the B ranks are 2, 4, 5, 7, and 10, and $W_B = 28$. Since this is less than $\frac{1}{2}n(N+1) = 32.5$, the data favor diet A. To see how strongly they support A, let us determine the smallest level α' at which the procedure defined by (1.54) and (1.55) would declare the superiority of A over B significant. This is $P(W_B \leqslant w)$, where $w = 28$ is the observed value of W_B and where the probability is computed under the assumption that $A = B$. From Table B it is found that

$$\alpha' = P(W_B \leqslant 28) = P[W_B - \tfrac{1}{2}n(n+1) \leqslant 13] = .265$$

Even when A is very slightly worse than B, there is thus a probability of more than $^1/_4$ that A would come out ahead. At any value of $\alpha' < .265$ that is a possible value of the Wilcoxon distribution, procedure (1.54) would suspend judgment as to which of the two diets is superior.

The reader may feel that procedure (1.54), as illustrated in Example 5, contradicts the admonition made following (1.52) and again in Example 4. It was stressed there that a one-sided error probability is appropriate only when the direction of the difference or effect is clear before any observations have been made. Yet, in Example 5, where it is not known a priori whether A or B is better, the error probability, $\alpha' = P(W_B \leqslant 28)$ is one-sided. The explanation lies of course in the different interpretations as to what constitutes an *error*. In the two-sided Wilcoxon test (1.44), the error to be controlled is falsely declaring one treatment better than the other if in fact they do not differ at all. On the other hand, in the three-decision procedure (1.54) the error to be controlled is that of declaring one treatment better than the other if in fact it is worse—if there is no difference, we now do not care which of the two treatments is preferred.

Procedure (1.54) provides a symmetric choice between two treatments in situations where the choice is to be made only when the data are reasonably conclusive, with judgment being suspended otherwise. An analogous procedure is available when the choice lies between a new and a standard treatment or in other asymmetric cases, for example, when one treatment is less expensive than the other or has fewer undesirable side effects. The resulting three-decision procedure will then take the form

$$\begin{array}{lll} & \text{Choose } B \text{ if} & W_B \geqslant c_2 \\ (1.56) & \text{Choose } A \text{ if} & W_B \leqslant c_1 \\ & \text{Suspend judgment if} & c_1 < W_B < c_2 \end{array}$$

[1] Data from tables 3 and 4 of Nieman, Groot, and Jansen, "The Nutritive Value of Butter Fat Compared with That of Vegetable Fats," *Koninkl. Ned. Akad. Wetenschap. Proc.* **55**:582–604 (1952). (The diets A and B are diets B and C of the original paper.)

Here c_1 and c_2 can be determined so as to give preassigned values to the maximum error probabilities

$$\text{(1.57)} \qquad \alpha_1 = P(W_B \leqslant c_1) \quad \text{and} \quad \alpha_2 = P(W_B \geqslant c_2)$$

where the probabilities are computed under the assumption of no difference (with respect to the response being observed) between the treatments, and where, because of the discrete nature of the distribution involved, one will typically be satisfied with values c_1 and c_2 for which the equations hold as closely as possible. If A is the standard treatment or that which is a priori more desirable, one will prefer A to B when the two treatments are equally effective, and hence will specify a smaller value for α_2 than for α_1.

The procedure (1.56) generalizes both the symmetric three-decision procedure (1.54) and the test (1.6). It reduces to the former when $\alpha_1 = \alpha_2 = \alpha'$ and to the latter when $c_1 = c-1$ and $c_2 = c$ and when A and B are identified with the standard and new treatment, respectively, so that $\alpha_2 = \alpha$ and $\alpha_1 = 1-\alpha$. There is no difficulty in extending (1.56) to the case in which some observations are tied. It is only necessary for this purpose to replace W_B by the sum W_B^* of the B midranks. The presence of ties does, however, create a difficulty with respect to the symmetric procedure (1.54) since the distribution of W_B^* need no longer be symmetric. One will in this case determine the critical values c_1 and c_2 so as to make α_1 and α_2 as close to α' (and hence to each other) as possible. If the demands of fairness make the equality of α_1 and α_2 imperative, this can be achieved through the otherwise undesirable device of randomization.

The three-decision procedure (1.54) assumes the possibility of not coming to a decision as to which of the two treatments is better. This will be the case if a decision can be postponed until further information (for example, additional observations) becomes available. However, the decision sometimes cannot be delayed and an immediate choice between A and B is required. A procedure suggested by (1.54) is then to choose treatment B when $W_B > \frac{1}{2}n(N+1)$ and treatment A when $W_B < \frac{1}{2}n(N+1)$. This still leaves the question what to do when $W_B = \frac{1}{2}n(N+1)$ or equivalently when $W_{XY} = \frac{1}{2}mn$. In this case (which cannot arise when both m and n are odd), exactly half of the pairs $Y_j - X_i$ are positive and half are negative, and there seems no reason to prefer either one of the treatments to the other. If a decision must be taken, it may be possible to base it on some auxiliary consideration such as the price of the treatment. If no other recourse is available, it may become necessary to toss a coin (as is sometimes done in tournaments which cannot be completed because of time or weather).

The above procedure for making a definite choice between A and B can be given a slightly more intuitive form by noting that by (1.18)

$$W_A + W_B = \tfrac{1}{2}(m+n)(N+1)$$

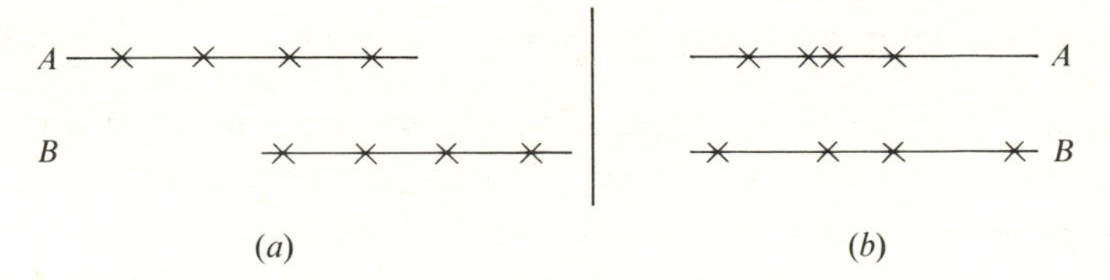

FIGURE 1.4
Observational patterns under alternatives to H.

It is then easy to see (Prob. 97) that

$$W_B \begin{matrix} < \\ = \\ > \end{matrix} \tfrac{1}{2}n(N+1) \qquad \text{if and only if} \qquad \frac{W_B}{n} \begin{matrix} < \\ = \\ > \end{matrix} \frac{W_A}{m}$$

The procedure therefore chooses A or B as W_A/m is greater than or less than W_B/n.

In discussing the different situations in which each of the three procedures (1.6), (1.47), and (1.54) is appropriate, it is important to be aware of a common assumption underlying all three. This is the assumption that if there is a difference between the two treatments being compared, the difference takes the form of one treatment being better than the other. More specifically, this means that the responses under one treatment tend to be larger than under the other [see Fig. 1.4(a)]. We may say that this assumption describes the *alternatives* against which the hypothesis H of equality of the two treatments is being tested. These are, of course, by no means the only possible alternatives to H. Instead, the responses under one treatment might for example be more variable or spread out than under the other as illustrated in Fig. 1.4(b). A test for a comparison in such a case is the Siegel-Tukey test, which will be discussed in the next section. Or the responses under the two treatments might differ in some other, more complicated ways such as those illustrated in Fig. 1.5. A test of equality of the two treatments against the alternatives that there is a difference without any specification of the form that this difference takes is the Smirnov test, which will also be discussed in the next section.

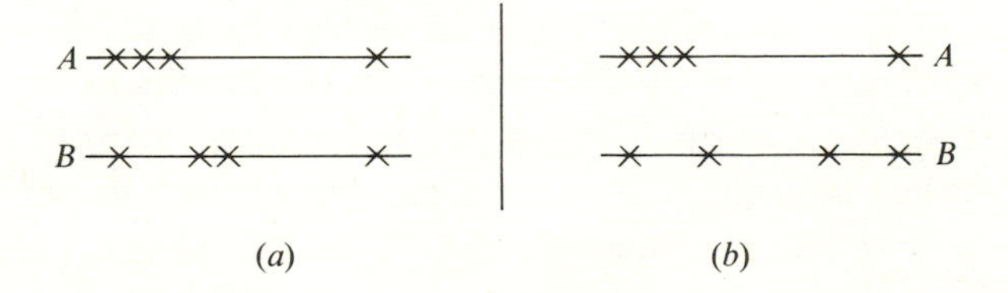

FIGURE 1.5
Observational patterns under other alternatives to H.

6. THE SIEGEL-TUKEY AND SMIRNOV TESTS

In the comparison problems discussed so far, we have assumed that the effect of the treatment, if any, is to increase (or decrease) the responses over what they would have been without the treatment. As was pointed out at the end of the last section, these are not the only possible alternatives to the hypothesis of no treatment effect. Suppose, for example, that the "treatment" in question is a new method of measuring or weighing some physical or biological quantity. This method is believed to measure the same value as the standard method in the sense that neither gives systematically higher or lower values than the other, so that measurements taken with either will be centered roughly on the same value. It is, however, hoped that the new method is less variable, and hence that the resulting measurements will be less dispersed about this value. The effect of the treatment will then be to bring the response closer to the value being measured, hence to lower the responses that exceed this value and raise those that fall short of it. Is the Wilcoxon rank-sum test appropriate in such a situation?

As before, the hypothesis being tested states that there is no difference between the two treatments. If the critical value c is determined by (1.7), the significance level of the test will therefore continue to be α, and the same is true if the test is based on (1.19) and (1.20), or if the two-sided test (1.47) or (1.50) is used. Suppose now that the treatment has the desired effect. If the lowering of the dispersion is pronounced, the smallest as well as the largest observations may be expected to be control measurements. When the control ranks are summed to form W_r, the occurrence of extremes at both ends will tend to prevent the sum from achieving a significantly large (or small) value and hence lead to acceptance of the hypothesis. The Wilcoxon test thus has little power to detect a treatment effect of this kind. (As a technical term, *power* denotes the probability of rejecting the hypothesis when it is false. A test is good if its power is large. The power of the Wilcoxon test will be studied in Chap. 2.)

To obtain a suitable test, note that in the present situation the least desirable responses are those at the two extremes, those which tend to be furthest away from the value being measured. This suggests assigning low ranks to these extreme observations and having the ranks increase toward the center. One possible such scheme is the following: Assign rank 1 to the smallest observation, rank 2 to the largest, rank 3 to the second largest, rank 4 to the second smallest, and so on, as indicated in Fig. 1.6. If the treatment has the desired effect, the treatment ranks will tend to be large and the control ranks small. Let $S_1, \ldots, S_n$ denote the treatment ranks and W_s their sum. Then large values of W_s will be significant, and the hypothesis of no treatment effect will be rejected when

$$W_s \geqslant c \tag{1.58}$$

The resulting test is known as the *Siegel-Tukey test*.

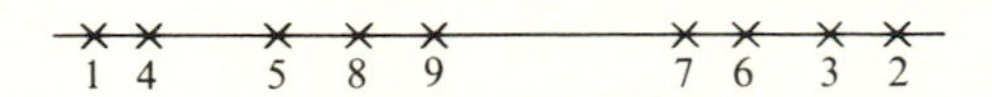

FIGURE 1.6
Ranking for dispersion.

What is the null distribution of W_s? In view of the assumed random assignment, the argument that led to (1.4) shows that the null distribution of $S_1, \ldots, S_n$ is again given by (1.4), even with this new definition of the ranks. Thus, the null distribution of W_s is that discussed in Sec. 2, or, for large group sizes, in Sec. 3. If ties are present, the distribution is that given in Sec. 4. As an example, suppose that $m = 5, n = 6$ and that the control and treatment observations are, respectively,

$$.8 \quad .1 \quad .14 \quad .6 \quad .34 \qquad \text{and} \qquad .4 \quad .38 \quad .64 \quad .26 \quad .31 \quad .55$$

The dispersion ranks, defined as in Fig. 1.6, are then

Observation	.1	.14	.26	.31	.34	.38	.4	.55	.6	.64	.8
Rank	1	4	5	8	9	11	10	7	6	3	2

and the values of W_r and W_s are

$$W_r = 1+2+4+6+9 = 22 \qquad \text{and} \qquad W_s = 3+5+7+8+10+11 = 44$$

Small values of W_r are significant, and Table B shows that $P(W_r \leqslant 22) = .089$.

The test described above requires an important assumption: Both methods must measure the same quantity, so that the observations will cluster around the same value. However, this assumption often is not justified. If the difference Δ between the quantities being measured by the new and standard method is known, the problem can be reduced to that for the case $\Delta = 0$ by subtracting Δ from each of the treatment observations. If Δ is unknown, it is tempting instead to subtract from each treatment observation an estimate of Δ (for example, the estimate to be discussed in Sec. 5 of Chap. 2) and to apply the Siegel-Tukey test to the observations modified in this way. When m and n are sufficiently large the significance probability of the resulting test will often be close to that computed for the Siegel-Tukey test with $\Delta = 0$ (see Chap. 2, Sec. 7L for references). Alternative procedures that may be preferable are discussed by Nemenyi (1969) and Shorack (1969).

A slight blemish of the Siegel-Tukey test is its lack of symmetry. Another test with exactly the same properties can be obtained by reversing the pattern of Fig. 1.6, assigning rank 1 to the largest observation, rank 2 to the smallest, and so on. Typically, both of these rankings will lead to the same decision (although not to exactly the same significance probability). It can, however, happen (Prob. 99) that they will lead to opposite decisions (one to accept, the other to reject the hypothesis),

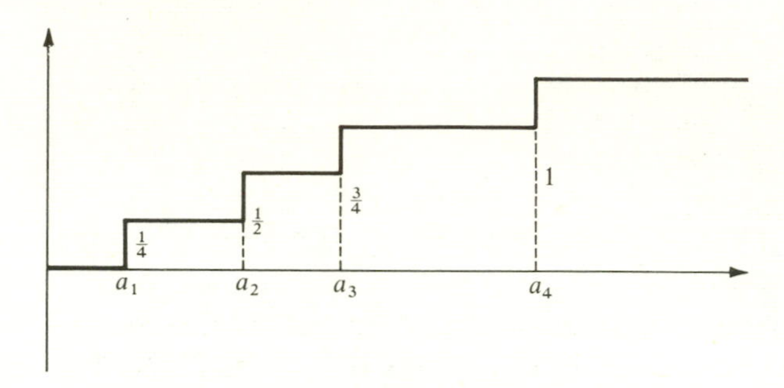

FIGURE 1.7
Sample cumulative distribution function.

which might leave the user in somewhat of a quandary. The asymmetry can be removed by averaging the ranks assigned to each observation by these two procedures and rejecting H when the sum of the average ranks for the treatment observations is sufficiently large. This sum, however, no longer has the Wilcoxon distribution. Tables of its null distribution are provided by Ansari and Bradley (1960).

We have now considered tests of the hypothesis H of no treatment effect against two kinds of alternatives: The Wilcoxon rank-sum test, if the treatment effect tends to increase (or decrease) the response; the Siegel-Tukey test, if it tends to make the response less (or more) variable. As pointed out in the preceding section (see Fig. 1.5), these two rather simple types of alternatives by no means exhaust the possibilities, and for any other given type of alternative one could again design a suitable rank test. We shall, however, now consider the somewhat different situation that one wishes to test H without having a clear idea of the likely effect of the treatment in case H is false.

A test of H against the general alternative of an unspecified treatment effect, the *omnibus* alternative, can be based on the following representation of a set of observations, say, $a_1 < \cdots < a_k$. For any number x, the *sample cumulative distribution function* (sample c.d.f.) $F_k(x)$ of the set is defined as

$$F_k(x) = \frac{\text{number of } a\text{'s} \leqslant x}{k} \tag{1.59}$$

Note that $F_k(x)$ is zero for all $x < a_1$; it jumps to $1/k$ at $x = a_1$ and stays at that level until x reaches a_2; at a_2 it jumps to $2/k$, and so on, until it finally reaches its maximum value 1 at $x = a_k$, where it remains for all $x > a_k$. Figure 1.7 shows the sample c.d.f. of $a_1, \ldots, a_k$ for the case $k = 4$.

Consider now the case of m control and n treatment observations, and denote their sample c.d.f.'s by F_m and G_n, respectively. What do these tell us about the relative position of the two groups of observations? Suppose, for example, that F_m lies above G_n. Then it follows that the treatment observations tend to be larger

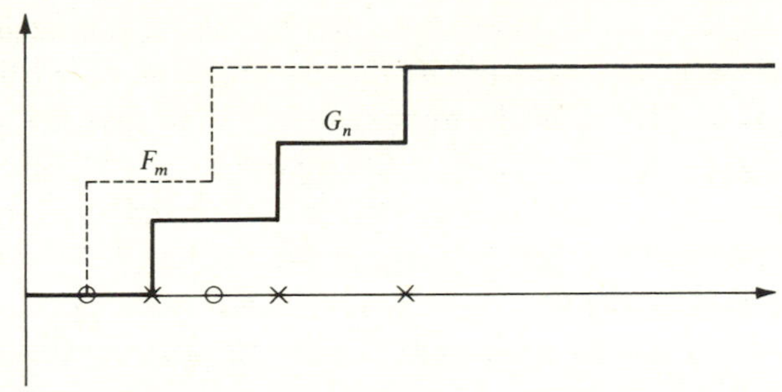

FIGURE 1.8
Two sets of observations and their sample c.d.f.'s (dots indicate control, crosses indicate treatment).

than the controls, as is seen in Fig. 1.8, in which the two sets of observations are shown below the sample c.d.f.'s to which they correspond. Similarly, if G_n is steeper than F_m, the treatment observations are less dispersed than the controls. More generally, to any difference in the two sets of observations will correspond some difference between F_m and G_n. A natural measure of the extent to which the two sets of observations differ is the largest difference between F_m and G_n

$$D_{m,n} = \max_x |G_n(x) - F_m(x)| \tag{1.60}$$

If $D_{m,n}$ is small, the graphs of F_m and G_n are close together and there is little difference between the two sets of observations. If it is large, the two graphs differ substantially, at least at some point, and this reflects the fact that in some way the two sets of observations must differ considerably. It is thus natural to reject the hypothesis when the difference $D_{m,n}$ is sufficiently large, say, when

$$D_{m,n} \geqslant c \tag{1.61}$$

This test was proposed by Smirnov (1939).

Is this *Smirnov test*, like those considered so far, a rank test? To see whether the statistic $D = D_{m,n}$ depends only on the ranks of the observations, that is, on their order but not on their actual values, consider a simple example. Let $m = 2$, $n = 1$ and suppose that the ranks of the three observations, in the usual notation, are

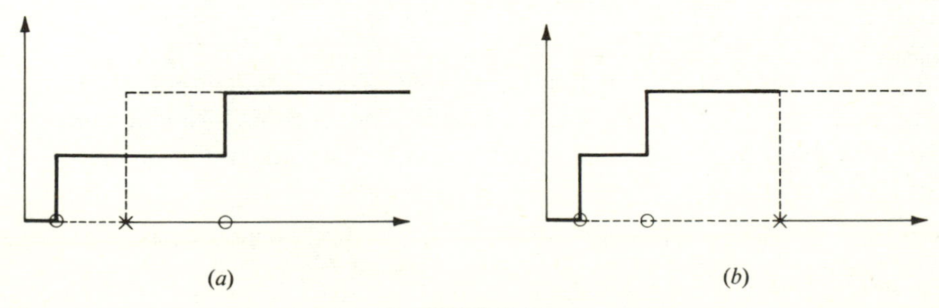

FIGURE 1.9
Sample c.d.f.'s for two different rankings. (a) $R_1 = 1$, $R_2 = 3$, $S_1 = 2$; (b) $R_1 = 1$, $R_2 = 2$, $S_1 = 3$.

$R_1 = 1$, $R_2 = 3$, and $S_1 = 2$. This means that the single treatment observation lies between the two control observations; the associated sample c.d.f.'s are shown in Fig. 1.9(a). It is seen from this figure that $D = \frac{1}{2}$, and that this will always be the case when $R_1 = 1$, $R_2 = 3$, and $S_1 = 2$, regardless of the exact values of the observations that lead to these ranks. On the other hand, if the ranks are changed, this may (and often will) affect the value of D. In Fig. 1.9(b), for example, which corresponds to the case $R_1 = 1$, $R_2 = 2$, $S_1 = 3$, the value of D is 1. This value is again seen to be determined by the ranks and to be independent of the values of the observations that lead to these ranks.

Quite generally, the ranks of the two groups of observations determine the *relative* positions of the steps of the two graphs with respect to each other. A change in the values of the observations that does not affect the ranks changes the position at which the vertical distance $G_n(x) - F_m(x)$ occurs but not the value of this distance, and hence leaves unchanged the maximum difference D. Explicit expressions for $D_{m,n}$ in terms of the ranks are given in, for example, Hajek (1969).

An important consequence of the fact that the value of D depends only on the ranks of the observations is that its null distribution is determined by (1.4). To see how it can be computed, consider once more the case $m = 2$, $n = 1$. The ranks can then take on three possible sets of values: $R_1 = 1$, $R_2 = 2$, $S_1 = 3$; $R_1 = 1$, $R_2 = 3$, $S_1 = 2$; $R_1 = 2$, $R_2 = 3$, $S_1 = 1$. Under H, these are equally likely, so that each has probability $\frac{1}{3}$. The value of D in the three cases is respectively 1, $\frac{1}{2}$, and 1, and the null distribution of D is therefore $P(D = \frac{1}{2}) = \frac{1}{3}$, $P(D = 1) = \frac{2}{3}$.

The distribution is particularly simple in the case $m = n$. Then each of the sample cumulative distribution functions F_m and G_n takes on the values 0, $1/n$, $2/n, \ldots, (n-1)/n$, $n/n = 1$, and the possible values of D are therefore also 0, $1/n, \ldots, n/n = 1$. (A comparison with Table B shows that the number of available significance levels in this case is much smaller than for the Wilcoxon test.) If $d = a/n$, it can be shown that when $m = n$

$$P(D \geqslant d) = \frac{2\left[\binom{2n}{n-a} - \binom{2n}{n-2a} + \binom{2n}{n-3a} - \cdots\right]}{\binom{2n}{n}} \tag{1.62}$$

where the series is continued as long as $n-a$, $n-2a$, $n-3a$, and so on are greater than or equal to zero. A proof will not be given here of this formula, which is due to Gnedenko and Korolyuk (1951),[1] but in Table E we give the probabilities (1.62) for $m = n = 1(1)30$.†

[1] For a proof see, for example, Drion (1952). "Some Distribution-free Tests for the Difference between Two Empirical Distribution Functions," *Ann. Math. Statist.* **23**:563–574, or Feller, *An Introduction to Probability Theory and its Applications*. John Wiley & Sons. New York. 1966.

† For an extensive table not restricted to the case $m = n$ see Kim and Jennrich (1970).

EXAMPLE 6. *Two drugs for pain relief.*[1] In a comparison of twc drugs used for the relief of postoperative pain, the following numbers of hours of relief were observed for 16 patients, of which 8 had been assigned to a standard drug A and the other 8 to an experimental drug B.

A: 6.8 3.1 5.8 4.5 3.3 4.7 4.2 4.9

B: 4.4 2.5 2.8 2.1 6.6 0.0 4.8 2.3

Suppose it is thought possible that B might give greater or lesser relief than A, or that its effect might be more or less variable than that of A, or that the two drugs might differ in other ways, for example, by A being more effective for patients who are highly susceptible to pain but less effective for patients with a greater tolerance for pain. As a first step in the investigation, it is therefore decided to test whether there are any differences at all, and for this purpose to apply the Smirnov test.

The ranks of A and B are, respectively,

A: 6 7 8 10 11 13 14 16

B: 1 2 3 4 5 9 12 15

To find the maximum difference between the sample cumulative distribution functions F_m of A and G_n of B it is clearly enough to determine the difference between F_m and G_n at the 16 points at which one or the other of the two graphs has a jump. The first jump occurs in G_n (since the observation with rank 1 belongs to B). At this point, G_n takes on the value $^1/_8$ while F_m is still zero. The next four jumps also all occur in G_n (since the observations with ranks 2 to 5 all belong to B); at the last of these, $G_n = {}^5/_8$, $F_m = 0$. The values of F_m and G_n at all 16 points are listed below, multiplied by 8.

	1	2	3	4	5	6	7	8	9	10	11	12	13	14	15	16
G_n	1	2	3	4	5	5	5	5	6	6	6	7	7	7	8	8
F_m	0	0	0	0	0	1	2	3	3	4	5	5	6	7	7	8

It is seen at a glance that the maximum absolute difference between F_m and G_n is $^5/_8$, which occurs at point 5. The significance probability is therefore $P(D \geqslant {}^5/_8)$, and from Table E this is seen to be equal to .0870. This probability can, of course, also be obtained directly from (1.62). Since $n = 8$ and $a = 5$, it is seen that (1.62) gives

$$P(D \geqslant 5/8) = 2\left[\binom{16}{3}\right] \Big/ \binom{16}{8}$$

Reference to Table A shows this to be equal to .0870.

For large m and n, the standard approximation to the null distribution of D is not a normal distribution. The probability that $\sqrt{mn/(m+n)}\,D$ is greater than or equal to any given value z is approximated by a value $K(z)$

$$P\left(\sqrt{\frac{mn}{m+n}}D \geqslant z\right) \approx K(z) \tag{1.63}$$

[1] Part of the data of Table 1a of Meier, Free, and Jackson (1958), "Reconsideration of Methodology in Studies of Pain Relief," *Biometrics* **14**:330–342. (Two numbers were changed in the second digit to avoid ties; the drugs A and B are labeled D and T_3 in the original paper.) The problems treated in this paper are considered further in Meier and Free (1961), "Further Consideration of Methodology in Studies of Pain Relief," *Biometrics* **17**:576–583.

which is shown in Table F, and which was first obtained for a related but slightly different problem by Kolmogorov. The approximation is supported by a limit theorem, which for the present problem is due to Smirnov, and which states that the left-hand side of (1.63) tends to the right-hand side as m and n tend to infinity. [For a more detailed discussion of the accuracy of the approximation, see Hodges (1958). An alternative approximation is discussed by Kim (1969); however, his results do not apply to the present problem since he studies the probabilities $P(D > d)$ instead of the probabilities $P(D \geqslant d)$ and the approximations work quite differently in the two cases.] For a recent survey of the Smirnov test, see Durbin (1973).

As an illustration of the approximation, let us use it to evaluate the probability $P(D \geqslant 5/8)$ when $m = n = 8$. Since

$$P(D \geqslant 5/8) = P\left(\sqrt{\frac{mn}{m+n}}D \geqslant \sqrt{\frac{mn}{m+n}}\,\frac{5}{8}\right)$$

we enter the table with $z = (5/8)\sqrt{mn/(m+n)} = 1.25$ to find $K(z) = .0879$, which is satisfactorily close to the value .0870 given in the table.

Table 1.2 gives the exact and approximate values of $P(D \geqslant d)$ for $m = n = 10$, 20, 30, and a number of values of d. The approximation is seen to work reasonably well when $m = n = 10$ in the range .01 to .15 but is less satisfactory for values less

TABLE 1.2 Exact and approximate values of $P(D \geqslant d)$

a	5	6	7	8
d	.5	.6	.7	.8
Exact	.1678	.0524	.0123	.0021
Approximate	.1641	.0546	.0149	.0033

$m = n = 10$

a	7	8	9	10	11	12
d	.35	.40	.45	.50	.55	.60
Exact	.1745	.0811	.0335	.0123	.0040	.0011
Approximate	.1725	.0815	.0348	.0135	.0047	.0015

$m = n = 20$

a	9	10	11	12	13	14
d	.3000	.3333	.3667	.4000	.4333	.4667
Exact	.1350	.0709	.0346	.0156	.0065	.0025
Approximate	.1344	.0713	.0354	.0165	.0072	.0030

$m = n = 30$

than .01. It slowly improves as the common group size increases and at $m = n = 30$ is adequate for most purposes.

The situation is quite different when the group sizes are unequal. Although the Smirnov limit theorem continues to hold in this case, the left-hand side of (1.63) tends to its limit $K(z)$ very slowly, even if the difference between the group sizes is only 1. For this reason, a very extensive table is required to supplement the large-sample approximation. Such a table is provided by Kim and Jennrich (1970).

Since the Smirnov test (1.61) was designed as a test of H against all possible alternatives, it is of course also applicable against the particular alternatives that the treatment has the effect of increasing the response. However, it tends to be less powerful against such alternatives than the Wilcoxon rank-sum test. This is not surprising. The Wilcoxon rank-sum test specializes in these alternatives and thus needs to reject H only for those of the $\binom{m+n}{n}$ possible arrangements which suggest that the treatment increases the response. On the other hand, the Smirnov test must divide the allotment of arrangements it is allowed to reject in such a way as to guard against many different types of alternatives. In general, if one has an idea of the way in which a treatment effect would manifest itself, it is preferable to use a test designed against the specific alternatives.

Ties present no new problems for the Siegel-Tukey test since the null distribution of the midrank statistic continues to be the same as that of the Wilcoxon statistic $W_s{}^*$. However, the presence of ties does raise difficulties for the Smirnov test. Not only are the tables of the null distribution of $D_{m,n}$ no longer applicable to the midrank statistic

$$D^*_{m,n} = \max |G^*_n(x) - F^*_m(x)|$$

but even the asymptotic distribution (1.63) no longer applies. It is therefore of interest to note that the value of $P_H(D_{m,n} \geqslant d)$ given in Table E and for large m and n approximated by (1.63) provides an upper bound to the corresponding probability for $D^*_{m,n}$, that is,

$$P_H(D^*_{m,n} \geqslant d) \leqslant P_H(D_{m,n} \geqslant d) \tag{1.64}$$

It follows from (1.64) that if a critical value d is obtained from the Smirnov null distribution to provide a nominal significance level α, the actual level of the test in the presence of ties does not exceed α. Of course, if it is much less than α, the test is also considerably less powerful (i.e., is less likely to detect a treatment effect when it exists) than if it had been carried out at level α.

To prove the inequality (1.64), suppose that all ties are broken at random; that is, if for example the second, third, and fourth smallest observations are tied, the ranks 2, 3, and 4 are assigned to these three observations at random. Then the argument

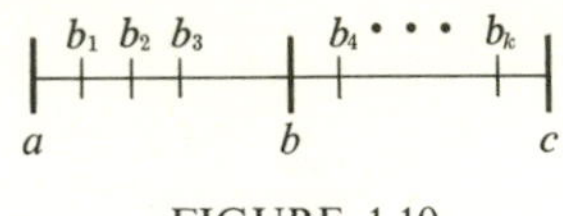

FIGURE 1.10
Breaking of ties at b.

leading to (1.4) shows that the null distribution of the resulting ranks $S_1, \ldots, S_n$ is still given by (1.4). If $D_{m,n}$ is the value of the Smirnov statistic obtained by breaking all ties at random, it follows that $D_{m,n}$ has the Smirnov distribution. If, furthermore, $D^*_{m,n}$ denotes the statistic computed without breaking ties, the inequality (1.64) will follow from the fact that $D^*_{m,n}$ is always less than or equal to $D_{m,n}$.

Suppose that there are ties at a point b (among, possibly, other places), with the nearest observations to the left and right occurring at, say, a and c ($a < b < c$). If k observations are tied at b, the breaking of their tied ranks is equivalent to replacing the observations with one observation each at k points $b_1 < b_2 < \cdots < b_k$, all between a and c as indicated in Fig. 1.10. If F^*_m, G^*_n and F'_m, G'_n denote the sample cumulative distribution functions before and after the breaking of the ties, we have two facts:

(i) For all $x < b_1$ and $x > b_k$

$$F'_m(x) = F^*_m(x) \qquad \text{and} \qquad G'_n(x) = G^*_n(x)$$

(ii) The functions F^*_m and G^*_n are constant in the intervals $a \leqslant x < b$ and $b \leqslant x < c$.

It follows that every value taken on by $G^*_n - F^*_m$ is also taken on by $G'_n - F'_m$ (but the latter difference may take on some additional values) and hence that

$$D^*_{m,n} = \max|G^*_n(x) - F^*_m(x)| \leqslant \max|G'_n(x) - F'_m(x)| = D'_{m,n}$$

Application of the above argument consecutively to all points at which ties occur provides the desired proof.

7. FURTHER DEVELOPMENTS

7A. Other Approximations to the Distribution of W_s

A better approximation to the null distribution of W_{XY} than the normal approximation of Sec. 3 can be obtained by means of an Edgeworth series. The necessary formulas are provided by Fix and Hodges (1955), who also present a comparison of the accuracy of the two approximations. The approximations are studied further by Verdooren (1963) and Bickel (1974). The corresponding problem in the presence of ties is investigated by Klotz (1966).

A quite different approximation, proposed by Buckle, Kraft, and van Eeden (1969a), gives good results not only when both m and n are large but also if only one of the group sizes is large. To fix ideas, suppose that $m = 2$ and that n is large. Then W_r is the sum of two integers selected at random from the integers $1, 2, \ldots, N$ where $N = n+2$. Since N is large, the dependence of these two is small, and it is natural to approximate W_r/N by the sum of two independent random variables, each uniformly distributed over the points $1/N, 2/N, \ldots, 1$. The desired approximation is obtained by replacing the discrete uniform distribution over these N points by the continuous uniform distribution over the interval (0, 1).

7B. Censored Observations

In certain situations, there is a point T beyond which the values of the observations are not recorded. For example, the measuring device may only go up to T, or an observation may be the time until a piece of equipment fails, a treated patient has a relapse, or a rat has learned some task, and the study is terminated after time T. For the values that are not recorded, it is then only known that they exceed T; they are said to be *censored* at T. If a rank test is used, and if there are no ties among the observations less than T, the situation is one in which all the ties occur at one end. The Wilcoxon test for this case is discussed by Sugiura (1963) and by Hodges and Lehmann (1970), and a slight modification of the test by Halperin (1960).

Halperin's test is extended by Gehan (1965a) to the situation in which the study terminates at a fixed time T but the subjects enter the study at different times, and in Gehan (1965b) to the case that observations are censored both on the left and the right. Other papers dealing with related problems are Rao, Savage, and Sobel (1960), Saw (1966), Gehan and Thomas (1969), and Johnson and Mehrotra (1972).

7C. Early Termination

If the observations are times to failure or to success, the Wilcoxon statistic may at a certain moment be so large or so small that rejection or acceptance of H is assured regardless of the values of the remaining observations. The procedure that in this case discontinues observation is made explicit by Alling (1963), who also discusses the expected saving in time of observation. It is further shown how to adapt the procedure through "backdating" to the case in which the subjects enter the study at varying times.

7D. Power

It is important to consider the performance of a test not only when the hypothesis is true but also when it is false. By controlling the significance level, one makes sure of a small probability of rejecting the hypothesis H of no treatment effect when it is true but says nothing about the probability of recognizing that the treatment has an effect when this is in fact the case. This latter probability, that is, the probability of rejecting H when the treatment is effective, is called the *power* of the test. It depends of course not only on which test is being used but also on the nature and extent of the effect.

Clearly, one would like the power of the test to be high. If the treatment is effective, one wants to know that this is so. To make sure that a test has enough power for the purpose at hand, one must be able to compute the power. Unfortunately, the framework of the randomization model is not well suited for the assessment of power because of the difficulty of formulating appropriate alternatives to the hypothesis. A more suitable model will be described in Chap. 2, where there is a full discussion of how to evaluate power and how to determine the group sizes m and n needed to achieve a given level of significance and power.

Here we shall only briefly describe two alternatives to H, which are related to the randomization model of the present chapter. The first of these, the *constant effect model*, assumes that the treatment has the effect of adding a fixed amount, say, Δ, to the response the subject would have given as a control. This model is discussed by Hodges and Lehmann (1970), where some examples of the associated power computations are given. The model has the disadvantage that the value of the power depends not only on Δ, which is a natural measure of the effectiveness of the treatment, but also on the control responses of the $m+n$ subjects.

A different model, which is conceptually less simple but has the advantage of depending on only one parameter, is proposed by Barton, David, and Mallows (1958). Consider the situation of Example 1, in which only the ranks of the $m+n$ subjects are observed and are obtained through a subjective judgment. As a basis for a model, the authors propose that the mental process leading to this judgment is equivalent to a pairwise comparison of the subjects. For each pair of patients, the physician asks himself which of the two is in better condition. Such judgments, of course, involve an element of randomness. The assumption is made that in the comparison of two controls, the probability of either being judged to be in better condition than the other is $1/2$, and that the same is true in the comparison of two treated patients. On the other hand, in the comparison of a treated patient with a control, the probability of the former appearing to be in better condition is assumed to have a fixed value $p \geqslant 1/2$, which is equal to $1/2$ when H is true and otherwise greater than $1/2$. For these assumptions to be even approximately true, the group of patients would clearly have to be very homogeneous. Under these assumptions, the authors derive the probability of any ranking as a function of p, and give some power values of the Wilcoxon test. It is also shown that in this model the Wilcoxon test enjoys the property of having greater power than any other test carried out at the same significance level.

Finally, a complication in the model must be mentioned: The pairwise comparisons may lead to judgments that contradict each other. In this case, it is assumed that the whole process is repeated until an unambiguous ranking is obtained.

7E. Permutation Tests

Suppose that there are available not only the ranks for the $m+n$ subjects but also scores (or some other response measurements) as in Example 2. Let these be denoted by $X_1, \ldots, X_m$ (for the controls) and $Y_1, \ldots, Y_n$ (for the treated subjects) and suppose that the treated and control subjects are to be compared in terms not of their ranks but of a statistic such as $\bar{Y}-\bar{X}$, where $\bar{Y}$ and $\bar{X}$ denote the average of the Y's and X's, respectively. Such a comparison is possible on the basis of the random assignment of the subjects to treatment and control.

To see this, recall that the hypothesis of no treatment effect implies that the response of each subject can be assumed to be determined before the assignment to treatment and control was made. The selection of n subjects to be treated means the selection of n of these N fixed responses to be called Y's, the remaining m responses to be labeled X's. Each of these $\binom{N}{n}$ equally likely selections results in a value of $\bar{Y}-\bar{X}$. If these $\binom{N}{n}$ values are distinct, each of them has probability $1\Big/\binom{N}{n}$. Suppose that large values of the Y's are significant and that k of the $\binom{N}{n}$ possible divisions of the N responses into X's and Y's lead to values of $\bar{Y}-\bar{X}$ larger than or as large as the one observed. Then the significance probability according to this test will be $k\Big/\binom{N}{n}$.

A similar test could be obtained if $\bar{Y}-\bar{X}$ were replaced by any other appropriate test statistic. Unfortunately, the computations required make it impracticable to carry out such *permutation tests* exactly (except for very small N). For discussion of an approximation and further properties of the permutation version of the $\bar{Y}-\bar{X}$ test, see Chap. 2, Sec. 7O. The idea of such permutation tests was introduced by Fisher (1935) and was developed more fully by Pitman (1937, 1938).

8. PROBLEMS

(A)

Section 1

1. Display all possible divisions of the ranks $1, \ldots, N$ into m controls and n treatment ranks as in (1.1) for the cases (i) $m=2$, $n=4$; (ii) $m=2$, $n=5$; (iii) $m=n=3$.

2. Use (1.2) to find (i) $\binom{50}{2}$; (ii) $\binom{35}{3}$.

3. Use (1.2) and (1.3) to find (i) $\binom{40}{37}$; (ii) $\binom{23}{19}$.

4. Use Table A to find (i) $\binom{10}{4}$; (ii) $\binom{10}{5}$; (iii) $\binom{12}{6}$. Check the results by means of (1.2).

5. Use Table A and (1.3) to find (i) $\binom{25}{18}$; (ii) $\binom{20}{15}$; (iii) $\binom{16}{10}$.

6. If $m = 2, n = 3$, use (1.4) to find the probabilities (i) $P_H(S_1 = 1, S_2 = 2, S_3 = 3)$; (ii) $P_H(S_1 = 1, S_2 = 3, S_3 = 5)$; (iii) $P_H(S_1 = 1)$; (iv) $P_H(R_1 = 1)$; (v) $P_H(S_1 = 2)$. [*Hint:* List all triples which give rise to the event in question.]

7. If $m = 2, n = 3$, use (1.4) to find the probabilities (i) $P_H(S_1+S_2+S_3 = 7)$; (ii) $P_H(S_1+S_2+S_3 \leqslant 7)$; (iii) $P_H(S_1+2S_2+3S_3 \leqslant 20)$; (iv) $P_H(S_1{}^2+S_2{}^2+S_3{}^2 \leqslant 26)$.

8. In a study of a new approach to elementary mathematics, 10 children are selected at random from a group of 21 to be taught by the new method with the other 11 serving as controls. Find the probability that by chance the treatment group (i) consists of the 10 most intelligent children; (ii) contains the most intelligent child; (iii) contains the three most intelligent children.

9. A population consists of six subjects A, B, C, D, E, F of which three are drawn at random for observation. Of these three, two are selected at random and assigned to treatment while the other one serves as control. Show that each of the $\binom{6}{2} = 15$ possible pairs that can be formed from the six subjects is equally likely to be chosen as the pair of treatment subjects. [*Hint:* To find the probability that AB is the chosen pair, note that this will be the case only if the three subjects selected for observation are ABC, ABD, ABE, or ABF, and if the pair chosen from these three is AB.]

Section 2

10. Obtain the null distribution of W_s by the method leading to (1.9) for the cases (i) $m = 2$, $n = 4$; (ii) $m = 2, n = 5$; (iii) $m = n = 3$.

11. Use (1.9) to obtain the *cumulative distribution function* of W_s defined by

$$F_{W_s}(w) = P(W_s \leqslant w)$$

as a function of w for the case $m = 2, n = 3$, and graph this function.

12. Use the results of Probs. 10(i) and (iii) to find all possible significance levels of the Wilcoxon test when (i) $m = 2, n = 4$; (ii) $m = n = 3$.

13. When $m = 2, n = 5$, use the results of Prob. 10(ii) to find the attainable significance level closest to (i) .01; (ii) .05; (iii) .1.

14. In the preceding problem, find the closest attainable significance level which is less than or equal to (i) .01; (ii) .05; (iii) .1.

15. Continue the enumeration (1.11) to the values $W_s = 20$ and $W_s = 21$, and verify the probabilities $P_H(W_s \leqslant 20)$ and $P_H(W_s \leqslant 21)$ given in the text.

16. Use Table B to find the probabilities $P_H(W_s = w)$ when (i) $m = 5$, $n = 6$, $w = 21$; (ii) $m = 6$, $n = 4$, $w = 23$; (iii) $m = n = 7$, $w = 34$.

17. Use Table B to find $P_H(W_s \leqslant w)$ for the three cases of the preceding problem.

18. Use Table B to find $P_H(W_s > w)$ for the three cases of Prob. 16.

19. Use Table B to find $P_H(W_s \geqslant w)$ for the three cases of Prob. 16.

20. Obtain the null distribution of W_r for the three cases of Prob. 10.

21. Use Table B to find $P_H(W_r = w)$ when (i) $m = 5$, $n = 6$, $w = 23$; (ii) $m = 6$, $n = 5$, $w = 23$; (iii) $m = 6$, $n = 4$, $w = 22$; (iv) $m = 4$, $n = 6$, $w = 22$.

22. From a group of nine rats available for a study of the transfer of learning, five were selected at random and were trained to imitate leader rats in a maze. They were then placed together with four untrained control rats in a situation where imitation of the leaders enabled them to avoid receiving an electric shock. The results (the number of trials required to obtain ten correct responses in ten consecutive trials) were as follows:[1]

Trained rats:	78	64	75	45	82
Controls:	110	70	53	51	

Find the significance probability of these results when the Wilcoxon test is used.

23. The effectiveness of vitamin C in orange juice and in synthetic ascorbic acid was compared in 20 guinea pigs (divided at random into two groups of 10) in terms of the length of the odontoblasts after 6 weeks, with the following results:[2]

Orange juice:	8.2	9.4	9.6	9.7	10.0	14.5	15.2	16.1	17.6	21.5
Ascorbic acid:	4.2	5.2	5.8	6.4	7.0	7.3	10.1	11.2	11.3	11.5

Find the significance probability when the hypothesis of no difference is being tested against the alternative that the orange juice tends to give rise to larger values.

24. Suppose that a new postsurgical treatment is being compared with a standard treatment by observing the recovery times of n treatment subjects and of m controls. State for what values of W_s you would reject the hypothesis of no treatment difference at the closest attainable significance level to .05 if (i) $m = n = 8$; (ii) $m = 8$, $n = 9$; (iii) $m = 9$, $n = 8$.

25. Under the conditions of the preceding problem, suppose that $m = n = 9$ and the observed recovery times are (in days)

Controls:	20	21	24	30	32	36	40	48	54
Treatment:	19	22	25	26	28	29	34	37	38

Find the significance probability of these results when the Wilcoxon rank-sum test is used.

26. To determine whether a certain additive prolongs the life of batteries, $2n$ batteries are divided into a control and treatment group of n each. State for what values of W_s you would reject the

[1] From Siegel, *Nonparametric Statistics*, McGraw-Hill Book Company, New York, 1956, p. 119. Original data from Solomon and Coles (1954). "A Case of Failure of Generalization of Imitation across Drives and across Situations." *J. Abnorm. Soc. Psychol.* **49**:7–13.

[2] Data from Crampton, "The Growth of the Odontoblasts of the Incisor Tooth as a Criterion of the Vitamin C Intake of the Guinea Pig." *J. Nutr.* **33**:491–504 (1947). quoted in Bliss. *Statistics in Biology*. McGraw-Hill Book Company. New York, 1967, vol. 1, p. 228. Three of the observations have been increased by .1 to avoid ties.

hypothesis of no treatment effect at the closest attainable significance level which does not exceed .02 when (i) $n = 6$; (ii) $n = 8$; (iii) $n = 10$.

Section 3

27. Find the area under the normal curve to the left of (i) 1.23; (ii) $-.36$; (iii) .893; (iv) -2.051.

28. Find the area under the normal curve to the right of (i) .95; (ii) -1.16; (iii) 2.087; (iv) $-.038$.

29. Find the area under the normal curve between (i) .13 and 1.72; (ii) -3.04 and 1.11; (iii) -1.60 and 1.97; (iv) $-.92$ and .92; (v) -2.034 and 1.002.

30. Find the value x such that the area under the normal curve (i) to the left of x is .90; (ii) to the right of x is .85; (iii) between $-x$ and x is .80.

31. Find the normal approximation to the probability $P_H(W_s \geqslant w)$ and compare it with the value given in Table B when $m = n = 8$ and (i) $w = 41$; (ii) $w = 44$; (iii) $w = 47$; (iv) $w = 50$; (v) $w = 53$; (vi) $w = 56$.

32. Find the normal approximation to the probability $P_H(|W_s - 52.5| \geqslant w)$ and compare it with the value obtained from Table B when $m = n = 7$ and (i) $w = 19.5$; (ii) $w = 20.5$; (iii) $w = 21.5$; (iv) $w = 22.5$; (v) $w = 23.5$; (vi) $w = 24.5$.

33. For each combination of $m = 4, 6, 8, 10$ and $n = 4, 6, 8, 10$ with $m \leqslant n$, use Table B to find the value of w for which $P_H(W_s \geqslant w)$ is closest to .05, and compare this probability with its normal approximation.

34. Draw a histogram of the distribution of W_s for the cases (i) $m = n = 3$; (ii) $m = 3, n = 5$; (iii) $m = 3, n = 7$.

35. Draw a histogram of the distribution of W_s for the cases (i) $m = n = 4$; (ii) $m = n = 5$; (iii) $m = n = 6$.

36. For each combination of $m = 3, 4, 5$ and $n = 3, 4, 5$ with $m \leqslant n$, use Table B to find the value of w for which $P_H(W_s \geqslant w)$ is closest to .08, and compare the resulting probabilities with their normal approximations obtained (i) with and (ii) without continuity correction.

37. Find the normal approximation to the probability $P_H(W_s \geqslant w)$ both with and without continuity correction, and compare it with the value given in Table B, when $m = 5, n = 3$, and (i) $w = 9$; (ii) $w = 11$; (iii) $w = 13$.

38. In the situation of Example 2, use the normal approximation to determine for what values of W_s to reject the hypothesis H of no treatment effect at significance level $\alpha = .01$ when (i) $m = n = 20$; (ii) $m = 30, n = 20$; (iii) $m = n = 50$.

39. In the situation of Prob. 23, use the normal approximation to determine for what values of W_s to reject the hypothesis of no difference at significance level $\alpha = .05$ when (i) $m = 20, n = 15$; (ii) $m = 15, n = 20$; (iii) $m = n = 25$; (iv) $m = 20, n = 30$. (Here n is the number of guinea pigs receiving the orange juice.)

40. Use the normal approximation to obtain the critical values c corresponding to $\alpha = .02$ for the three cases of Prob. 26 and compare (i) the values c with the critical values, say c', obtained in Prob. 26; (ii) the probabilities $P_H(W_s \geqslant c)$ as read from Table B with the nominal significance level .02.

Section 4

41. Let the control and treatment observations be

$-5 \quad -5 \quad -2 \quad 0 \quad 1 \quad 1 \quad 1 \qquad \text{and} \qquad -4 \quad -2 \quad 1 \quad 2 \quad 2 \quad 3$

respectively. Find the midranks of the treatment observations.

42. Suppose that 20 treatment patients are being compared with 20 controls, and that the progress of each patient is classified as very poor, poor, indifferent, good, or very good. If the data are given in the following table, find the midranks of the treatment subjects.

	Very Poor	Poor	Indifferent	Good	Very Good
Control	2	2	11	4	1
Treatment	0	1	9	7	3

43. Let the treatment and control observations be -3, -3, 0, 1 and -3, 1, 2, 2, respectively. Find the significance probability if small values of W_s^* are significant.

44. Find the null distribution of W_s^* and graph its histogram when $m = n = 3$ and (i) $d_1 = 2$, $d_2 = 2$, $d_3 = 1$, $d_4 = 1$; (ii) $d_1 = 1$, $d_2 = 2$, $d_3 = 3$; (iii) $d_1 = d_2 = d_3 = 2$; (iv) $d_1 = d_2 = d_3 = 1$, $d_4 = 3$; (v) $d_1 = d_2 = 3$.

45. By enumeration find (i) $P_H(W_s^* \geqslant 22)$ when $m = n = 4$ and $d_1 = d_2 = d_3 = 1$, $d_4 = 3$, $d_5 = d_6 = 1$; (ii) $P_H(W_r^* \leqslant 28)$ when $m = n = 6$ and $d_1 = 2$, $d_2 = 1$, $d_3 = 3$, $d_4 = 2$, $d_5 = 1$, $d_6 = 2$, $d_7 = 1$; (iii) $P_H(W_r^* \leqslant 86.5)$ when $m = n = 10$ and $d_1 = 5$, $d_2 = 9$, $d_3 = 6$.

46. To test the effectiveness of vitamin B_1 in stimulating growth in mushrooms, vitamin B_1 was applied to 13 mushrooms selected at random from a group of 24, while the remaining 11 did not receive this treatment. The weights of the mushrooms at the end of the period of observation were[1]

Controls: 18 14.5 13.5 12.5 23 24 21 17 18.5 9.5 14

Treated: 27 34 20.5 29.5 20 28 20 26.5 22 24.5 34 35.5 19

Use the normal approximation to find the significance probability of these results.

47. In an investigation of a new drug for postoperative pain relief, it was desired to determine (among other things) whether the relief brought by 3 mg of the drug is significantly higher than that resulting from a dose of 1 mg. In one phase of the study, 15 freshly operated-upon patients were assigned at random, 7 to the lower dose (T_1) and the remaining 8 to the higher dose (T_3). The responses (number of hours of pain relief) were[2] T_1: 2, 0, 3, 3, 0, 0, 8 and T_3: 6, 4, 4, 0, 1, 8, 2, 8. How significant are these results?

48. The following were the differences of all 40 subjects in the experiment described in Example 2:

Controls: -1 8 3 13 0 1 6 2 16 3 14 1 4 1 -3 9 3 3 5 2

Treated: 7 -5 4 -4 -5 -7 -2 0 -6 6 -3 -3 -3 2 -4 -3 1 -9 -4 0

Use the normal approximation to find the approximate significance probability of these results.

[1] From Linder, *Statistische Methoden*, 2d ed., Birkhäuser, Basel, 1951, p. 91. Original data from Schopfer and Blumer, "Zur Wirkstoffphysiologie von Trichophyton album Sab.," *Ber. Schweiz. Botan Ges.* **53**:409–456 (1943).

[2] From Meier, Free, and Jackson, "Reconsideration of Methodology in Studies of Pain Relief," *Biometrics* **14**:330–342 (1958).

49. In the context of Prob. 42, suppose that $m = n = 10$ and the data are given in the following table:

	Very Poor	Poor	Indifferent	Good	Very Good
Control	2	2	5	1	0
Treatment	0	2	4	3	1

Find the critical value c giving the significance level closest to .01, and compare $P(W_s^* \geqslant c)$ with its normal approximation.

50. Letting $m = 7, n = 5$, suppose that large values of W_s^* are significant, and that

(i) $d_1 = d_2 = d_3 = d_4 = d_5 = d_6 = 2$

(ii) $d_1 = d_2 = 1, d_3 = d_4 = d_5 = 3, d_6 = 1$

(iii) $d_1 = d_2 = d_3 = d_4 = 3$

In each case, find the critical value which gives the significance level closest to .05, and compare the actual significance level with its normal approximation.

Section 5

51. Find the critical value c for which the significance level of the two-sided Wilcoxon test given by (1.47) and (1.48) is as close as possible to .05 if (i) $m = 7$, $n = 10$; (ii) $m = n = 8$; (iii) $m = 9$, $n = 8$.

52. Find the critical value c for which the significance level of the two-sided Wilcoxon test given by (1.47) and (1.48) is the largest attainable level not exceeding .03 if (i) $m = 9$, $n = 10$; (ii) $m = 10$, $n = 6$; (iii) $m = 5$, $n = 8$.

53. Verify the exact and approximate values given in the text as the significance probability of the data of Example 4.

54. Use the normal approximation to obtain the critical values c corresponding to $\alpha = .05$ for the three cases of Prob. 51 and compare (i) the values c with the critical values, say c', obtained in Prob. 51; (ii) the probabilities given by the left-hand side of (1.48) as read from Table B, with the nominal significance level .05.

55. In the context of Example 5, suppose that the group sizes are $m = 3$, $n = 4$. Find the critical values c_1 and c_2 so that $P_H(W_A^* \leqslant c_1)$ and $P_H(W_A^* \geqslant c_2)$ are as close to $\frac{1}{2}\alpha = .05$ as possible if the growth figures are (i) diet A: 240, 230, 240; diet B: 230, 250, 245, 240; (ii) diet A: 225, 220, 240; diet B: 240, 230, 230, 240.

56. In Example 5, suppose that $m = n = 5$ and that the B-ranks are 3, 4, 5, 8, 10. What decision will be taken by procedure (1.54) if the maximum allowable error probability is $\alpha' = .1$?

57. In the context of Example 5, find the critical value c which in procedure (1.54) gives the value of α' closest to .05 if (i) $m = n = 8$; (ii) $m = 6, n = 10$; (iii) $m = 8, n = 10$.

58. With procedure (1.56), determine the critical values c_1 and c_2 which give error probabilities as close as possible to $\alpha_1 = .1, \alpha_2 = .01$ when (i) $m = n = 9$; (ii) $m = n = 10$; (iii) $m = 7, n = 10$.

59. Suppose that in the comparison of two headache remedies A and B, the degree of relief is reported as $0, 1, \ldots, 10$. To what decision does procedure (1.56) lead if $m = n = 4$, the maximum allowable error probability is $\alpha' = .05$, and the observations are A: 3, 5, 4, 3 and B: 6, 5, 3, 5.

Section 6

60. Let $m = 5, n = 6$ and suppose that the treatment and control observations are, respectively, 1.5, 1.8, 2.8, 2.5, 1.9, 2.3 and 2.1, 1.4, 3.1, .8, 2.7. Find the significance probability of the Siegel-Tukey test if the ranking is (i) that given in Fig. 1.6; (ii) the mirror image of that ranking, which assigns rank 1 to the largest observation, ranks 2 and 3 to the smallest and next smallest, and so on.

61. For the two parts of the preceding problem, obtain an approximate significance probability using the normal approximation, and compare the results with those of Prob. 60.

62. A symmetric alternative to the Siegel-Tukey statistic is obtained by assigning score 1 to both the smallest and largest observation, score 2 to both the second smallest and second largest, and so on. Let T be the sum of the scores of the treatment observations. Find the null distribution of this *Ansari-Bradley statistic* if $m = 3, n = 4$.

63. Graph the sample cumulative distribution function for each of the following sets of observations:

(i) 3.1 1.6 2.1 2.4 3.3 2.8
(ii) 1.5 1.1 1.5 1.5 1.3 2.1 1.4 1.4
(iii) −.7 −.3 −.4 −.7 0 .1 −.3 −.7 .1 −.7

64. Compute the value of $D_{m,n}$ for the control and treatment observations of (i) Prob. 22; (ii) Example 2; (iii) Prob. 25; (iv) Prob. 60.

65. Use Table E to determine the significance probability of the value $D_{m,n}$ obtained in Prob. 64 (iii), and compare it with the significance probability obtained in Prob. 25.

66. Use formula (1.4) to determine the null distribution of $D_{m,n}$ if (i) $m = 3, n = 2$; (ii) $m = 3, n = 4$.

67. Use formula (1.62) to compute $P(D_{m,n} \geqslant d)$ when (i) $m = n = 13, d = 4/13$; (ii) $m = n = 18, d = 1/3$. Check your answers against the values in Table E.

68. Use the approximation (1.63) and Table F to obtain an approximate value for the probability $P(D_{m,n} \geqslant d)$ when (i) $m = n = 15, d = 3/15$; (ii) $m = n = 22, d = 8/22$; (iii) $m = n = 27, d = 8/27$ In each case, compare your answer with the value given in Table E.

(B)

Section 1

69. If $k! = 1 \cdot 2 \cdot \cdots \cdot k$ (read $k!$ as "k factorial") denotes the product of the first k integers, show that (1.2) is equivalent to

$$\binom{N}{n} = \frac{N!}{n!(N-n)!} \tag{1.65}$$

70. Prove (1.3) by (i) using (1.65); (ii) using the definition of $\binom{N}{n}$ as the number of samples of size n from a population of size N. [*Hint* for (ii): Each sample of size n leaves a remainder of size $N-n$.]

71. Use (1.65) to show that

$$\binom{N}{n} = \frac{N}{n}\binom{N-1}{n-1} \tag{1.66}$$

72. From both (1.2) and by using the definition of $\binom{N}{n}$ as in Prob. 70(ii), find (i) $\binom{N}{N}$; (ii) $\binom{N}{1}$; (iii) $\binom{N}{N-1}$.

73. Prove the recursion formula

$$\binom{N}{n} = \binom{N-1}{n} + \binom{N-1}{n-1} \tag{1.67}$$

(i) from (1.65); (ii) by using the definition of $\binom{N}{n}$ as in Prob. 70(ii). [*Hint* for (ii): Single out one of the N subjects from which the sample of n is being drawn, and consider the number of samples of n which do and which do not contain this particular subject.]

74. Suppose that $n = 1$ and $m = N-1$, and let S denote the rank of the single treatment subject. Find the possible values that S can take on together with the probabilities of these values.

75. For arbitrary m and n find (i) $P_H(S_1 = 1)$; (ii) $P_H(S_1 = 2)$; (iii) $P_H(S_1 = k)$. Show that this last probability is nonincreasing in k. [*Hint:* Once the value s_1 of S_1 is fixed, it is only necessary to count the number of choices for $(S_2, \ldots, S_n)$ from the integers exceeding s_1.]

76. Generalize the result of the preceding problem to find $P_H(S_r = k)$ for any $r \leqslant n$. Check your formula by comparing it with the special case $r = 1$ of the preceding problem.

77. Show that the inequalities $W_s \geqslant c$ and $W_r \leqslant c'$ are equivalent when $c' = \frac{1}{2}N(N+1) - c$.

78. If $\bar{S} = (S_1 + \cdots + S_n)/n$ and $\bar{R} = (R_1 + \cdots + R_m)/m$, find a constant k such that the inequality $W_s \geqslant c$ is equivalent to $\bar{S} - \bar{R} \geqslant k$.

79. Determine whether W_s and W_r have the same null distribution when (i) $m = n$; (ii) $m \neq n$.

80. In analogy to (1.23) let $\hat{\alpha}_- = P_H(W_s > w)$. Show that the hypothesis H is accepted by the Wilcoxon rank-sum test at all significance levels $\alpha > \hat{\alpha}_-$.

81. The minimum value of W_s is $n(n+1)/2$. Find the minimum value of W_r and the maximum value of both W_s and W_r.

82. For arbitrary m and n find (i) $P_H[W_s = \frac{1}{2}n(n+1)]$; (ii) $P_H[W_s = \frac{1}{2}n(n+1)+1]$; (iii) find $P_H[W_s = \frac{1}{2}n(n+1)+2]$ under the restriction that $m \geqslant 2$.

83. Show that $\sum_{k=1}^{n} k$ is given by (1.12) by noting that on the one hand,

$$\sum_{k=2}^{n+1} k^2 - \sum_{k=1}^{n} k^2 = n^2 + 2n$$

while the left-hand side is on the other hand equal to

$$\sum_{k=1}^{n} [(k+1)^2 - k^2] = 2\sum_{k=1}^{n} k + n$$

84. Show that

(1.68) $$1^2 + \cdots + n^2 = \tfrac{1}{6}n(n+1)(2n+1)$$

[*Hint:* In the proof of the preceding problem, replace k^2 by k^3.]

85. For $n = 2$ and arbitrary m, find the null distribution of W_s by enumeration. [*Hint:* Try the cases $m = 1, 2, 3, 4$ and establish a pattern for (*a*) m = odd, (*b*) m = even.]

86. For the number $\#(w;n,m)$ occurring on the right-hand side of (1.10), prove the recursion formula

(1.69) $$\#(w;n,m) = \#(w-N;n-1,m) + \#(w;n,m-1)$$

[*Hint:* Consider separately the number of ways of selecting the n treatment ranks when these do and when they do not include the maximum rank N.]

87. Use (1.69) to prove the recursion formula

(1.70) $$P_H(W_s = w|n,m) = \frac{n}{N}P_H(W_s = w-N|n-1,m) + \frac{m}{N}P_H(W_s = w|n,m-1)$$

where the two numbers behind the vertical bar indicate the group sizes for which the probability is being computed.

88. Use (1.70) and Prob. 74 to add to Table B the distribution of W_s when $n = 11$ and $m = 5$.

Section 3

89. In a study involving $2m$ subjects, m are assigned to treatment and m to control. Let the control observations be denoted by X and the treatment observations by Y and suppose that the ordered set of observations turns out to have the pattern $XYXY \ldots XY$, so that the treatment ranks are $2, 4, \ldots, 2m$. If large values of W_s are significant, use the normal approximation to find the approximate significance probability when m is large.

90. Solve the preceding problem for the case of $3m$ observations, of which $2m$ are assigned to treatment and m to control, when the observations form the pattern $XYYXYY \ldots XYY$.

91. Generalize the result of Prob. 90 to the case of $(k+1)m$ observations, of which km are assigned to treatment and m to control.

Section 4

92. (i) Prove that the distribution of W_s^* is symmetric whenever $m = n$. (ii) With $m \neq n$, give examples of where the distribution of W_s^* is and is not symmetric. (iii) Make a conjecture concerning a class of situations in which W_s^* is symmetrically distributed for arbitrary m and n, and prove your conjecture.

93. Show that for large N, the second term in (1.35) (which is the correction term) can be approximately as large as the first term (the variance without ties).

Section 5

94. Let $\hat{\hat{\alpha}}$ be the significance probability (1.49) of the two-sided Wilcoxon test and let

$\hat{\alpha} = P_H(W_B \geqslant w)$ be the significance probability of the one-sided test. Show that $\hat{\alpha}$ and $\hat{\hat{\alpha}}$ are related by the equations

$$\hat{\hat{\alpha}} = \begin{cases} 2\hat{\alpha} & \text{if} \quad \hat{\alpha} \leqslant \frac{1}{2} \\ 2(1-\hat{\alpha}) & \text{if} \quad \hat{\alpha} \geqslant \frac{1}{2} \end{cases}$$

95. Suppose that a one-sided test is used incorrectly as described on page 26 when a two-sided test is appropriate. Denote by $\hat{\alpha}$ the stated significance probability of the incorrect one-sided test and by $\hat{\hat{\alpha}}$ the significance probability of the correct two-sided test. Show that $\hat{\alpha} = \frac{1}{2}\hat{\hat{\alpha}}$. [*Hint:* Use the result of Prob. 94.]

96. Show that the second line in procedure (1.54) is equivalent to choosing A if $W_A \geqslant \frac{1}{2}m(N+1)+c$. [*Hint:* $W_A + W_B = \frac{1}{2}(m+n)(N+1)$.]

97. Show that W_B is greater than, equal to, or less than $\frac{1}{2}n(N+1)$ as W_B/n is greater than, equal to, or less than W_A/m.

98. If w denotes the observed value of W_B, and if $\hat{\alpha} = P_H(W_B \geqslant w)$ and $\hat{\alpha}_- = P_H(W_B > w)$, show that procedure (1.56) leads to the selection of treatment B when $\alpha_2 \geqslant \hat{\alpha}$, to the selection of treatment A when $\alpha_1 \geqslant 1-\hat{\alpha}_-$, and to the suspension of judgment when $\alpha_2 < \hat{\alpha}$ and $\alpha_1 < 1-\hat{\alpha}_-$. Show that these three decisions are mutually exclusive provided $\alpha_1 + \alpha_2 \leqslant 1$.

Section 6

99. Give an example of values $m, n, \alpha < .1$ and a pattern of X's and Y's such that the Siegel-Tukey test rejects the hypothesis when the ranking of Fig. 1.6 is used, and accepts the hypothesis when the mirror image of this ranking is used.

9. REFERENCES

Alling, David W. (1963): "Early Decision in the Wilcoxon Two-sample Test," *J. Am. Statist. Assoc.* **58**:713–720.

Ansari, A. R. and Bradley, R. A. (1960): "Rank Sum Tests for Dispersion," *Ann. Math. Statist.* **31**:1174–1189. [Provides tables of critical values for the symmetrized version of the Siegel-Tukey test, and discusses the normal approximation to the null distribution.]

Barton, D. E., David, F. N., and Mallows, L. L. (1958): "Non-randomness in a Sequence of Two Alternatives. I. Wilcoxon's and Allied Test Statistics," *Biometrika* **45**:166–180.

Bickel, P. J. (1974): "Edgeworth Expansions in Nonparametric Statistics," *Ann. Statist.* **2**:1–20.

Buckle, N., Kraft, C. H., and van Eeden, C. (1969a): "An Approximation to the Wilcoxon-Mann-Whitney Distribution," *J. Am. Statist. Assoc.* **64**:591–599.

———, ———, and ——— (1969b): *Tables Prolongées de la Distribution de Wilcoxon-Mann-Whitney*, Presse de l'Université de Montreal, Montreal. [Gives the cumulative distribution function to 8 decimals for $1 \leqslant m \leqslant n \leqslant 25$.]

Durbin, J. (1973): "Distribution Theory for Tests Based on the Sample Distribution Function," Regional Conference Series in Applied Math. No. 9.

Fisher, Ronald A. (1935): *The Design of Experiments*, Sec. 21, Oliver & Boyd, Ltd., Edinburgh.

Fix, Evelyn and Hodges, Jr., J. L. (1955): "Significance Probabilities of the Wilcoxon Test," *Ann. Math. Statist.* **26**:301–312.

Gehan, Edmund A. (1965a): "A Generalized Wilcoxon Test for Comparing Arbitrarily Singly Censored Samples," *Biometrika* **52**:203–223.

——— (1965b): "A Generalized Two-sample Wilcoxon Test for Doubly Censored Data," *Biometrika* **52**:650–653.

——— and Thomas, Donald G. (1969): "The Performance of Some Two-sample Tests in Small Samples with and without Censoring," *Biometrika* **56**:127–132.

Gnedenko, B. V. and Korolyuk, V. C. (1951): "On the Maximal Deviation between Two Empirical Distributions," *Dokl. Akad. Nauk SSSR* **80**:525–528.

Goldberg, S. (1960): *Probability—An Introduction*, Prentice-Hall, Englewood Cliffs, N.J.

Hajek, Jaroslav (1969): *Nonparametric Statistics*, Holden-Day, Inc., San Francisco.

Halperin, Max (1960): "Extension of the Wilcoxon-Mann-Whitney Test to Samples Censored at the Same fixed Point," *J. Am. Statist. Assoc.* **55**:125–138.

Hodges, J. L., Jr. (1958): "The Significance Probability of the Smirnov Two-sample Test," *Arkiv. for Mathematik* **3**:469–486.

——— and Lehmann, E. L. (1970): *Basic Concepts of Probability and Statistics*, 2d ed., Holden-Day, Inc., San Francisco.

Jacobson, James A. (1963): "The Wilcoxon Two-sample Statistic: Tables and Bibliography," *J. Am. Statist. Assoc.* **58**:1086–1103. [Surveys the tables of the Wilcoxon distributions available at the time.]

Johnson, Richard A. and Mehrotra, K. G. (1972): "Locally Most Powerful Rank Tests for the Two-sample Problem with Censored Data," *Ann. Math. Statist.* **43**:823–831.

Kim, P. J. (1969): "On the Exact and Approximate Sampling Distribution of the Two-sample Kolmogorov-Smirnov Criterion $D_{m,n}, m \leqslant n$," *J. Am. Statist. Assoc.* **64**:1625–1638.

——— and Jennrich, R. I. (1970): "Tables of the Exact Sampling Distribution of the Two-sample Kolmogorov-Smirnov Criterion, $D_{m,n}, m \leqslant n$," *Selected Tables in Mathematical Statistics* (Harter and Owen, eds.), Markham Publishing Co., Chicago, vol. 1, pp. 79–170.

Klotz, Jerome (1966): "The Wilcoxon, Ties, and the Computer," *J. Am. Statist. Assoc.* **61**:772–787.

Kolmogorov, A. (1941): "Confidence Limits for an Unknown Distribution Function," *Ann. Math. Statist.* **12**:461–463. [Points out the conservative nature of the Smirnov test when F and G are not necessarily continuous.]

Kraft, Charles H. and van Eeden, Constance (1968): *A Nonparametric Introduction to Statistics*, The Macmillan Co., New York.

Kruskal, William H. (1952): "A Nonparametric Test for the Several Sample Problem," *Ann. Math. Statist.* **23**:525–540. [Proves the asymptotic normality of the midrank statistic W_s^*.]

——— (1957): "Historical Notes on the Wilcoxon Unpaired Two-sample Test," *J. Am. Statist. Assoc.* **52**:356–360. [Traces the history of the test.]

——— and Wallis, W. Allen (1952): "Use of Ranks in One-criterion Variance Analysis," *J. Am. Statist. Assoc.* **47**:583–621; —"Errata," *Ibid.*, **48**:910.

Lehman, Shirley Young (1961): "Exact and Approximate Distributions for the Wilcoxon Statistic with Ties," *J. Am. Statist. Assoc.* **56**:293–298. [Compares the exact distribution with its normal approximation for a number of cases.]

Mann, Henry B. and Whitney, D. R. (1947): "On a Test of Whether One of Two Random Variables is Stochastically Larger Than the Other," *Ann. Math. Statist.* **18**:50–60. [Proposes the statistic W_{XY} and provides the first proof of asymptotic normality of the null distribution.]

Milton, Ray C. (1964): "An Extended Table of Critical Values for the Mann-Whitney (Wilcoxon) Two-sample Statistic," *J. Am. Statist. Assoc.* **59**:925–934. [Gives critical values for significance levels .0005, .0025, .005, .001, .01, .025, .1 and $n = 1(1)20$, $m = 1(1)40$.]

Moses, L. E. and Oakford, R. V. (1963): *Tables of Random Permutations*, Stanford University Press, Stanford.

Mosteller, F., Rourke, R. E., and Thomas, G. B. (1970): *Probability with Statistical Applications*, 2d ed., Addison-Wesley, New York.

Nemenyi, P. (1969): "Variances: An Elementary Proof and a Nearly Distribution-free Test," *The Am. Statistician* **23**:5, 35–37.

Pitman, E. J. G. (1937, 1938): "Significance Tests Which May Be Applied to Samples from Any Populations," *J. Roy. Statist. Soc.* **B4**:119–130, 225–237; *Biometrika* **29**:322–335.

Rao, U. V. R., Savage, I. R. and Sobel, M. (1960): "Contributions to the Theory of Rank Order Statistics: The Two-sample Censored Case," *Ann. Math. Statist.* **31**:415–426.

Saw, J. G. (1966): "A Nonparametric Comparison of Two Samples One of Which is Censored," *Biometrika* **53**:599–602.

Shorack, G. (1969): "Testing and Estimating Ratios of Scale Parameters," *J. Am. Statist. Assoc.* **64**:999–1013.

Siegel, Sidney and Tukey, John W. (1960): "A Nonparametric Sum of Ranks Procedure for Relative Spread in Unpaired Samples," *J. Am. Statist. Assoc.* **55**:429–444. [Proposes the test (1.58).]

Smirnov, N. V. (1939): "On the Estimation of the Discrepancy between Empirical Curves of Distribution for Two Independent Samples," *Bull. Univ. Moscow* **2**(2):3–14. [Proposes the test statistic (1.60) and derives its asymptotic null distribution in generalization and simplification of Kolmogorov's work for the goodness-of-fit problem.]

Sugiura, Nariaki (1963): "On a Generalization of the Wilcoxon Test for Censored Data," *Osaka Math. J.* **15**:257–268.

Verdooren, L. R. (1963): "Extended Tables of Critical Values for Wilcoxon's Test Statistic," *Biometrika* **50**:177–186. [Gives critical values for significance levels .001, .005, .01, .025, .05, and sample sizes $m < n = 1(1)25$.]

Wilcoxon, Frank (1945): "Individual Comparisons by Ranking Methods," *Biometrics* **1**, 80–83. [One of the many original sources of this test and the one from which it derives its name.]

———, Katti, S. K., and Wilcox, Roberta A. (1970): "Critical Values and Probability Levels for the Wilcoxon Rank Sum Test and the Wilcoxon Signed Rank Test," *Selected Tables in Mathematical Statistics* (Harter and Owen, eds.), Markham Publishing Co., Chicago, vol. 1, pp. 171–259. [Gives critical values for significance levels .005, .01, .025, .05 and sample sizes $m < n = 3(1)50$.]

CHAPTER 2
COMPARING TWO TREATMENTS OR ATTRIBUTES IN A POPULATION MODEL

1. POPULATION MODELS

The tests of Chap. 1 were derived under the sole assumption of random assignment of subjects to treatment and control. This derivation has the advantage of simplicity, and the required randomization is easy to carry out, thereby ensuring that the assumptions are satisfied. On the other hand, its narrow basis limits the scope of the resulting inference. Since no assumptions are made concerning the nature or provenance of the subjects, any inference regarding the effectiveness of the treatment will refer only to the particular subjects in the study. In some situations this may be all that is required. For example, a farmer who owns several fields may wish to know the value of a fertilizer for the particular soil conditions on these fields and may not be interested in the effectiveness of the fertilizer for other fields. However, most investigations of any significance are concerned with the effectiveness of the treatment under study for a large population of potential users. An inference of this latter kind clearly requires a relationship between the particular subjects in the study and this population. This relationship may take the tenuous form of a judgment by the investigator that his group of subjects is representative of some larger population. However, such judgments require great care and are notoriously unreliable. Children in a particular school, for example, may not be representative even of the city in which the school is located, much less of the population of school children at large. The most reliable way of establishing the desired relationship consists in drawing the subjects from the population according to some well-defined random-sampling scheme. The theoretically simplest such scheme is the method of *simple* (or *unrestricted*) *random sampling* according to which all sets of N subjects from the population have the same chance of being drawn. This may be relatively easy to carry out when the subjects are items coming from a conveyor belt or pieces of fruit on a truck, but it may be difficult for large human populations.

Besides the great practical difficulties of drawing random samples from large populations, there is another problem that tends to weaken the assumption of random sampling. Most populations of interest are continuously undergoing change. The new treatment being investigated will be applied to a *future* population that it is impossible to sample today. Thus, even under the most favorable circumstances, judgment will be required whether the findings of a study made at a given time and place are applicable to populations in other parts of the world or at a later time.

Suppose now that $N = m+n$ subjects are drawn from a given population by means of simple random sampling and that n of these receive the new treatment while the other m serve as controls. (Because of the difficulties just mentioned, it is desirable to make the assignment of the N subjects to treatment and control at random.) Let the response of a subject drawn from the population at random be denoted by Y if it receives the treatment and by X if it serves as control. Then X and Y are random variables with cumulative distribution functions, say, F and G, defined for all real numbers x and y by the equations

$$P(X \leqslant x) = F(x) \qquad \text{and} \qquad P(Y \leqslant y) = G(y) \tag{2.1}$$

The hypothesis of no treatment effect states that X and Y have the same distribution, i.e., that $G = F$. If the responses of the m controls are $X_1, \ldots, X_m$ and those of the n treatment subjects $Y_1, \ldots, Y_n$, it follows from the assumption of simple random sampling that each of the X's is distributed according to F and each of the Y's according to G.

To the assumption of simple random sampling we shall now add a second assumption. The population from which the sample is drawn will be supposed to be sufficiently large that the dependence of the X's and Y's (which results from the finiteness of the population) can be neglected and they can be treated as independent. The two assumptions together lead to the *population model* according to which $X_1, \ldots, X_m$ and $Y_1, \ldots, Y_n$ are independent random variables, the X's with distribution F and the Y's with distribution G. The hypothesis to be tested is the hypothesis of no treatment effect

$$H: \quad G = F \tag{2.2}$$

The problem of testing H in the above population model is frequently called the *two-sample problem* since the X's and Y's constitute samples from the populations characterized by the distributions F and G, respectively. It is this problem with which we shall be mainly concerned in the present chapter.

The reason for introducing the population model was the dependence of the inference on the particular subjects in the study. In the randomization model, the conclusion that the new treatment does or does not appear to be an improvement cannot legitimately be extended to subjects outside of the study. There is, however, a second related reason. An important aspect of a test is its *power*, that is, the probability of rejecting the hypothesis when it is false—in the present context, when the new treatment is better than the standard one. Both the determination of the size of the study and the choice of the particular test to be used involve considerations of power. Unfortunately, in the randomization model of Chap. 1, the power (unlike the significance probability) depends, not only on the sample sizes m and n and the degree by which the new treatment is better than the standard, but also on the responses that would result if all N subjects were used as controls and hence on the particular subjects at hand (Probs. 1 and 2). However, these N responses cannot all be known, since some of the subjects must be assigned to treatment, and none of them are known at the beginning of the study when the decisions regarding sample size and choice of test must be made. It is then natural to base these decisions on the kind of subjects or responses that one would *expect*. This is exactly the information which is furnished by the population model, since by (2.1) the distributions F and G provide the probabilities of obtaining various responses under the two treatments. The use of this information to compute power, select an appropriate test, and determine the sample sizes m and n for rank tests such as the Wilcoxon rank-sum test constitutes the principal topic of the present chapter. However, it is first necessary to consider the significance level of these tests in the population model.

We must now distinguish two cases. The ranks were seen earlier to be defined unambiguously only if there are no ties among the observations; tied observations require special treatment. It is therefore convenient at this point to introduce the additional assumption that there will be no ties, and to postpone the consideration of tied observations until later in this section. Since we shall only be concerned with probability calculations, it is enough to assume that the absence of ties is an event which has probability 1. A sufficient condition for this is the continuity of the distribution functions F and G, and apart from some obvious and trivial exceptions this condition is also necessary. (See Appendix, Theorem 1 and Probs. 9 to 11.) In particular, under the hypothesis $H: F = G$, continuity of the common distribution F is both necessary and sufficient, and the joint distribution of the ranks under H is then given by the following theorem.

Theorem 1. Let $X_1, \ldots, X_m$; $Y_1, \ldots, Y_n$ be independently distributed according to a common continuous distribution F, and let $S_1 < \cdots < S_n$ denote the ranks of the Y's in the combined ranking of all $N = m+n$ observations. Then for each of the $\binom{N}{n}$ possible n-tuples $(s_1, \ldots, s_n)$ the probability of observing $(s_1, \ldots, s_n)$ is equal to

$$P_H(S_1 = s_1, \ldots, S_n = s_n) = \frac{1}{\binom{N}{n}} \tag{2.3}$$

Proof. Denote the variables $X_1, \ldots, X_m$; $Y_1, \ldots, Y_n$ by $Z_1, \ldots, Z_N$ in this order. Then (2.3) is the probability that the ranks of $Z_{m+1}, \ldots, Z_{m+n}$ among all Z's are $s_1, \ldots, s_n$ (not necessarily in this order). Since $s_1, \ldots, s_n$ must be the ranks of some n of the Z's and since the joint distribution of the N Z's is symmetric in its N variables, the members of each subset of n Z's are equally likely to have ranks $(s_1, \ldots, s_n)$. There are $\binom{N}{n}$ such subsets, the ranks of each of which therefore have probability $1\Big/\binom{N}{n}$ of being equal to $(s_1, \ldots, s_n)$. Thus, in particular, this is the probability that $(s_1, \ldots, s_n)$ are the ranks of $Z_{m+1}, \ldots, Z_{m+n}$, that is, of $Y_1, \ldots, Y_n$, as was to be proved.

The theorem shows that Eq. (1.4) remains valid in the population model, and so therefore do all results derived from (1.4). In particular, the null distribution of the Wilcoxon rank-sum statistic (also known as the *Wilcoxon two-sample* statistic) continues to be determined by (1.4); that is, it is the distribution discussed in Chap. 1, Secs. 2 and 3, and tabulated in Table B. More generally, the null distribution of any rank statistic continues to be determined by (1.4), and is therefore independent of F. It is for this reason that rank tests are frequently called *distribution-free* or *nonparametric*, that is, free of the assumption that F belongs to some specified parametric family of distributions.

Between the randomization model of Chap. 1 and the population model of the present section there is an intermediate model, which assumes that the N subjects for the study have been drawn at random from a population of, say, A members without assuming A to be so large that the dependence of the different members of the sample can be neglected. This is the finite population model, of which the earlier population model is the limiting case as A tends to infinity, and which reduces to the randomization model when $A = N$. The proof of Theorem 1, which uses the symmetry of the variables $Z_1, \ldots, Z_N$ but not their independence, shows that (2.3) holds also for this more general finite model. However, we shall not consider this model any further.

In Theorem 1, the distribution F was assumed to be continuous. Let us now drop this assumption, so that ties do occur with positive probability, and let us suppose that tied ranks are replaced by their midranks. As a simple example, suppose that F is a discrete distribution which assigns to the two values a and b $(a < b)$ the probabilities p and q $(p+q = 1)$, and that $m = 2$ and $n = 1$. Then it is easily seen (Prob. 3) that the midrank S_1^* of the single treatment observation Y has the following distribution:

s	1	1.5	2	2.5	3
$P(S_1^* = s)$	pq^2	$2p^2q$	p^3+q^3	$2pq^2$	p^2q

The distribution of S_1^* thus depends on p and hence on F.

The dependence on F of the distribution of the midranks in the population model is quite general. It is related to the fact that in the randomization model this distribution depends on the numbers $d_1, d_2, \ldots, d_e$ of observations tied at the smallest value, the next smallest, etc. (These numbers are sometimes called the *configuration* of the ties.) In the earlier model, these numbers were assumed given; now they are random variables (Prob. 5), and their distribution depends on F. However, it follows from the argument of Theorem 1 that, given e and $d_1, \ldots, d_e$, the *conditional distribution* of the midranks is independent of F and is in fact the distribution derived for the midranks in Chap. 1, Sec. 4. For the configuration $(e; d_1, \ldots, d_e)$ determines the N midranks through Eq. (1.41), and with these N numbers given, by the argument of Theorem 1 the probability that the midranks of $(Y_1, \ldots, Y_n)$ coincide with any subset of n of them is $1\Big/\binom{N}{n}$.

Consider now some particular midrank statistic, for example, the sum W_s^* of the treatment midranks. Since the conditional distribution of the midranks given e and $d_1, \ldots, d_e$ is independent of F, a critical value $C = C(e; d_1, \ldots, d_e)$ can be determined so that the conditional significance level

$$P_H[W_s^* \geqslant C(e; d_1, \ldots, d_e) | e; d_1, \ldots, d_e] \tag{2.4}$$

of the rejection rule

$$W_s^* \geqslant C(e; d_1, \ldots, d_e)$$

is as close to the desired significance level α as possible. Clearly C is the critical value of the W_s^*-test discussed in terms of the randomization model in Chap. 1, Sec. 4. The W_s^*-test for the population model is thus performed by determining the values of e and $d_1, \ldots, d_e$ and carrying out the W_s^*-test for the randomization model corresponding to these values.

For this test, the probability (2.4) is the conditional significance level, given the configuration of ties; the unconditional overall significance level of the test is

$$\Sigma P_H(e; d_1, \ldots, d_e) \, P_H[W_s^* \geqslant C(e; d_1, \ldots, d_e) | e; d_1, \ldots, d_e]$$

This unconditional level cannot be expected to be exactly equal to α but can only be guaranteed to lie between the smallest and largest value of (2.4).

In Chap. 1, Sec. 4, it was pointed out (in the context of the randomization model) that the computation of significance probabilities, or critical values, becomes difficult in the presence of ties. It is then impracticable to tabulate the exact distribution of W_s^* and the accuracy of the normal approximation is hard to assess. Ties are now seen to cause additional complications in the population model. The null distribution of the Wilcoxon statistic then depends on unknown parameters, so that the test is no longer distribution-free. Although this difficulty can be overcome by conditioning on the configuration of ties, the resulting significance level is conditional; that is, it refers only to situations with the same configuration of ties. This largely vitiates the advantage of the population model since the inference is no longer valid for the whole population from which the sample was drawn. A way out of these difficulties is the one already recommended in Chap. 1: to avoid ties through more detailed observation or through supplementary criteria.

The above difficulties diminish as the sample sizes increase since the conditional distribution then tends to become more refined so that it is possible to obtain conditional levels that are close to the desired level α. Use of the normal approximation in fact shows the test to become asymptotically unconditional. To see this, write the rejection rule as

$$\frac{W_s^* - E(W_s^*)}{\sqrt{\text{Var}(W_s^*)}} \geqslant C'(e; d_1, \ldots, d_e)$$

With the normal approximation, the critical value C' is determined so that the area to the right of C' under the standard normal curve is equal to α, and the test rejects when

$$\frac{W_s^* - E(W_s^*)}{\sqrt{\text{Var}(W_s^*)}} \geqslant C' \tag{2.5}$$

Up to this point, we have considered the *conditional* distribution of the left-hand side given e and the d's, and determined C' so that the conditional level is approximately α. However, the unconditional distribution of the left-hand side of (2.5) is also approximately standard normal (see the Appendix, Prob. 40), and hence the unconditional level of (2.5) is also approximately α. Thus, asymptotically, the rejection rule (2.5) defines an unconditional level α test, as was to be shown.

The two-sample problem described at the beginning of this section arises not only when two treatments are compared by observing their effect on subjects drawn at random from a large population but also in other comparison problems. An important class of such problems does not fit the assumptions made so far because the conditions being compared are not treatments that can be assigned to the subjects at will but attributes inseparably attached to the subjects. For example, we may wish to know whether women tend to be more musical than men, whether voters favoring candidate A tend to be wealthier than those favoring candidate B, whether first borns are more independent than second borns, etc. In each of these cases, the condition—man or woman, favoring candidate A or B, and so on—is an integral attribute of the subject and therefore cannot be assigned at random.

Suppose that we are interested in comparing the two conditions in question for some population of subjects: the houses of a city or part of a city; the cars of a particular make manufactured during a given year; the voters of a district qualified to vote in a particular election. The two conditions then divide the members of this population into two categories or *subpopulations*: the houses in question which are on the east or west side of the street; the cars with two or four doors; the students at a university who do or do not smoke; and so on.

In principle, it is now easy to answer the question: Do smokers tend to be better students than nonsmokers? We ask each member of the student population under consideration whether or not he or she smokes and compare the grades of the smokers with those of the nonsmokers. Similarly, if we wish to determine whether the houses on the east side of the streets in some city tend to be larger than those on the west side, it is only necessary to obtain the number of rooms (or some other measure of size) for each house, and to compare the two sets of results. However, unless the populations are quite small, this procedure is usually not practicable. Fortunately, an alternative is available: it is enough to take a random sample from each of the subpopulations. If the sample sizes are m and n, and if the responses of the subjects are $X_1, \ldots, X_m$ and $Y_1, \ldots, Y_n$, respectively, the problem reduces to the two-sample problem described earlier. The distributions F and G of the two-sample model here represent the distributions of the responses over the two subpopulations being compared, and the hypothesis of no difference between the subpopulations with respect to the responses under consideration is given by (2.2).

EXAMPLE 1. *Psychological factors and cancer.* To see whether the wide variations found in the behavior of the same type of cancer in different individuals are linked to psychological factors, a comparison was made between a number of patients with a rapidly progressing, uncontrollable form of the disease (Group I) and another group whose disease is progressing slowly and is easily controllable (Group II). Each subject was given a psychological test, with the test scores displayed in the following table. Highly negative values of the scores were considered indicative of high defensiveness or a strong tendency to present the appearance of serenity while under great stress.[1]

[1] From Blumberg, West, and Ellis, "A Possible Relationship between Psychological Factors and Human Cancers," *Psychosomat. Med.* **16**:277–286 (1954), Tables 2 and 3.

Scores of Group I: −14 3 1 −16 −21 7 −7 −13 −22 −17 −14 −8 7
−18 −13 −18 −9 −22 −25 −24 −18 −13 −13 −18 −5

Scores of Group II: −18 −16 −9 −14 3 −9 −16 10 −11 −3 −13 −21
−2 −11 −16 −12 −13 −6 −9 −7 −11 −9

The intention here is clearly the comparison of two subpopulations: patients whose disease is progressing rapidly and those in whom progress of the disease is slow. However, the populations are not clearly delineated. Patients from what area or time period, of what ages and backgrounds? Furthermore, it is doubtful whether the subjects were obtained as a random sample from some population. Rather, they were patients accessible to the investigators so that the required observations could be made.

This difficulty is not peculiar to the present example, but is typical for studies of this kind. Nevertheless, although the basis of the inference is then somewhat shaky, it is often not too farfetched to suppose that the chance mechanisms which have brought the subjects to the investigator's attention, and hence into the study, have essentially provided him with random samples from the populations of responses with which he is concerned. Of course, this remark should not be construed as encouragement to dispense with efforts to obtain subjects by random sampling, or at least to guard carefully against biases that might be introduced through the use of a particular group of subjects.

On the assumption that the two groups of responses in the study constitute random samples from the two subpopulations of interest, it is now possible to test the hypothesis of no difference in the scores of these populations by means of the Wilcoxon test. Ranking all $N = 47$ observations, the midranks of the $m = 25$ and $n = 22$ members of the two groups are found to be

Midranks of Group I: 1 2 3.5 3.5 5.5 9 9 9 9 12 14.5 18 18 22.5 22.5
22.5 22.5 32 35 36.5 39 42 43.5 45.5 45.5

Midranks of Group II: 5.5 9 14.5 14.5 14.5 18 22.5 22.5 26 28 28 28 32
32 32 32 36.5 38 40 41 43.5 47

The sum of the midranks of Group II is $W_{\text{II}}^* = 605$. Computation of the significance probability requires the expectation and variance of W_{II}^* given by formulas (1.34) and (1.35). The values are

$$E_H(W_{\text{II}}^*) = 528 \qquad \text{and} \qquad \text{Var}_H(W_{\text{II}}^*) = 2{,}188$$

Suppose it was assumed at the outset of the investigation that if the two types of patients differed at all in their distribution of scores, the difference would be in the direction of lower (more highly negative) scores for Group I. A one-sided test would then be appropriate, with large values of W_{II}^* being significant. The significance probability is thus

$$P_H(W_{II}^* \geqslant 605) = P_H\left(\frac{W_{II}^* - 528}{\sqrt{2{,}188}} \geqslant \frac{605-528}{\sqrt{2{,}188}}\right)$$

This can be approximated by the area under the normal curve to the right of $77/\sqrt{2{,}188} = 1.646$, or about .05. If the direction of the difference could not have been assumed a priori, the significance probability would of course be twice as large, i.e., equal to .1.

Although the computations for the comparison of two attributes are the same as those for the comparison of two treatments, there is an important aspect in which the comparison of two attributes is less satisfactory than the corresponding comparison of two treatments that can be assigned at random. Consider, for example, the classical question whether smoking increases the risk of cancer. A natural procedure is to take a sample of smokers and nonsmokers and to observe them for many years to determine the incidence of cancer (or the age of onset of cancer, etc.) for the two groups. However, such a study cannot establish a *causal* relationship between cancer and smoking but only an *association* between the two. Such an association could be caused, at least theoretically, by certain traits or environmental factors which would predispose a person toward both smoking and cancer. Nor can such a study rule out the logical possibility of cancer causing smoking rather than conversely, for example, by producing an irritation which is relieved by smoking.

The situation would be quite different if smoking and nonsmoking could be handled as two treatments; that is, if it were possible to assign a number of people at random to smoking or nonsmoking regardless of their desires in the matter. (Such an approach is possible in animal experiments and sometimes in human situations through the use of volunteers.) Then even if there were, for example, certain traits which would cause both a desire to smoke and a tendency toward cancer, these would be balanced through random assignment, and a strong preponderance of cancer cases in the treatment group (those assigned to smoking) would clearly indicate smoking as the cause.

In the comparison of two subpopulations such as nonsmokers and smokers, we have assumed that samples of size m and n were drawn from these subpopulations at random. It may, however, be much easier to draw a sample from the total population than from the two separate subpopulations. It is, for example, relatively easy to draw a random sample of students at a university, but it would be quite difficult to draw a sample from those students who smoke. Suppose, therefore, that for the sake of convenience a sample of size N is drawn from the combined population. The numbers m and n of subjects in this sample belonging to the two subpopulations will then be random variables. However, conditionally, given the values of m and n, the distribution of the responses $X_1, \ldots, X_m$ and $Y_1, \ldots, Y_n$ will be exactly as if samples of these two fixed sizes had been drawn from the two subpopulations. The method discussed above for the comparison of two populations is therefore applicable without any changes.

A difficulty that could result from this approach is that undesirable values of m and n might be obtained. (Typically, one would prefer the sample sizes to be equal or at least nearly so, and a random sample from the combined population can be expected to yield values of m and n that are approximately proportional to the sizes of the two subpopulations.) This difficulty can be avoided when it is possible to draw the subjects for the sample one at a time and when it is easy to contact the sampled subjects. After the sample has reached its quota of, say, $N/2$ subjects from one of the subpopulations, any further subjects that are drawn from this subpopulation are then discarded and sampling is continued until the required quota of $N/2$ has been obtained from the other subpopulation.

The null distribution (1.4) (and hence all the procedures of Chap. 1) have now been shown to be applicable to four different comparison problems, which may be summarized in the following four models.

Model 1. *(Randomization model for the comparison of treatments.)* The N subjects are assumed given and are assigned at random, m to one treatment and n to the other.

Model 2. *(Population model for the comparison of two treatments.)* The N subjects are drawn as a simple random sample from the population of potential users of the two treatments and are assigned, m to one treatment and n to the other.

Model 3. *(Comparison of two attributes or subpopulations through a sample from each.)* Random samples of m and n are drawn, respectively, from the first and second subpopulation.

Model 4. *(Comparison of two attributes or subpopulations through a sample from the total population.)* A sample of N is drawn from the combined population, leading to samples from the two subpopulations with random sample sizes m and n $(m+n = N)$.

If there are no ties, the distribution of the n ranks corresponding to the second treatment or subpopulation, under the hypothesis H of no difference between the treatments or subpopulations, is given by (1.4) for Models 1 to 3. The same distribution applies in Model 4 as the conditional distribution of the ranks, given the values of m and n. In the presence of ties, the null distribution of the midranks under Model 1 is given in Chap. 1, Sec. 4. The same distribution obtains in Models 2 and 3 as the conditional distribution of the midranks given the configuration of ties, and in Model 4 as the conditional distribution of the midranks given m, n and the configuration of ties.

Besides the many similarities among the four models, there is the important distinction discussed earlier in the section: in Models 1 and 2 it is possible to point to the treatment difference as the cause of the observed difference in the two sets of responses. In Models 3 and 4, it is only possible to establish an association between the two attributes in question.

To conclude this section, we shall consider a class of problems in which neither randomization nor deliberate sampling appears possible. Suppose, for example, that two observatories are each making a number of independent determinations of a distance, speed, or temperature. Here no possibility of random assignment seems to exist. Also, a population model is not applicable because the measurements cannot be obtained as purposeful random samples from a population of such measurements. Nevertheless, the values in a sequence of measurements will vary and will give the appearance of randomness. The sources of this variation are fluctuations in the conditions (such as temperature and humidity) under which the measurements are performed, fluctuations in the quantity being measured, and physiological and psychological fluctuations in the observer. Although randomness in these fluctuations cannot be instituted purposefully, their combined effect—like that of the fluctuations responsible for producing varying outcomes in the throwing of a die—produces measurements that portray the features of random variation and which may reasonably be represented by random variables. The $m+n$ measurements, say, $X_1, \ldots, X_m$ and $Y_1, \ldots, Y_n$, can often be assumed to be independent, the X's with a common distribution F and the Y's with a common distribution G. The hypothesis being tested is that of no difference between the distributions F and G of the two kinds of measurements. In analogy with Models 1 to 4 the present *measurement model* can be summarized as:

Model 5. *(Model for the comparison of two sets of measurements.)* Independent sets of m and n measurements are obtained from two sources or by two methods.

The mathematical representation of this model is once more the two-sample problem that was described at the beginning of the section and which was seen above to be appropriate for the population Models 2 to 4.

2. POWER OF THE WILCOXON RANK-SUM TEST

The *power* of a test is the probability of rejecting the hypothesis when it is false. In a comparative study, this is the probability of recognizing the new treatment to be superior when this is in fact the case. To evaluate the power, it is necessary to specify the *alternatives* to the hypothesis, that is, the models representing the situation in which one of two treatments is more effective than the other. A general two-treatment model is that of the two-sample problem of the last section, in which F and G characterize the response distributions of subjects receiving the two treatments. In this model the hypothesis $G = F$ corresponds to the hypothesis of no treatment difference, while the alternatives $G \neq F$ correspond to situations in which the two treatments do differ in their effect.

However, not all pairs of distributions $G \neq F$ are of equal relevance to the present comparison problem since the hypothesis $H: G = F$ is to be tested only against the alternatives that the new treatment is *more effective* than the standard one. This implies a relationship between F and G that we shall now try to make precise. Suppose for the sake of definiteness (as in Chap. 1, Sec. 2) that the success of the treatment is indicated by an increase (rather than decrease) in the response. Then under the alternative that the new treatment is more effective, G will lead to high response values more frequently than F and to low values less frequently. This relationship may be expressed by the inequality

$$(2.6) \qquad G(x) \leqslant F(x) \qquad \text{for all } x \text{, with strict inequality for at least some } x$$

At first glance it appears that the inequality goes in the wrong direction. However, $G(x)$ and $F(x)$ are the probabilities of responses $\leqslant x$, that is, of small responses, and (2.6) expresses the fact that small values are less probable under G than under F. If X and Y are random variables with distributions F and G satisfying (2.6), Y is said to be *stochastically larger* than X or also G is said to be stochastically larger than F.

A simple but important special case of (2.6) is the *shift model*, which assumes that for some $\Delta > 0$

$$(2.7) \qquad G(x) = F(x-\Delta) \qquad \text{for all } x$$

so that the distribution G is obtained by shifting F by an amount Δ (Prob. 29). This model is the mathematical expression of the assumption of *additivity*, which supposes that the treatment adds a constant amount Δ to the control response of the subject independent of the value of that response. This assumption is frequently satisfied at least approximately (sometimes after a suitable transformation of the responses), and Δ then measures the effect of the treatment.[1]

Consider now some fixed continuous distribution F and the power $\Pi_F(\Delta)$ of the Wilcoxon test against the shift alternative (2.7). For the Mann-Whitney form of the test, it is given by

$$(2.8) \qquad \Pi_F(\Delta) = P_\Delta(W_{XY} \geqslant c)$$

where P_Δ indicates that the probability is calculated for the model (2.7). It is useful to extend the definition (2.8) to $\Delta = 0$ and to negative values of Δ. Since $\Delta = 0$ corresponds to the case of no treatment difference, the value $\Pi_F(0)$ is just the significance level of the test. Negative values of Δ correspond to the possibility that the treatment, by decreasing the response, has a deleterious rather than a beneficial effect. The function $\Pi_F(\Delta)$, considered as a function of Δ over the range $-\infty < \Delta < \infty$, is called the *power function* of the test.

It is intuitively plausible that the probability of rejecting the hypothesis and thus recognizing the new treatment as superior will increase as the treatment effect Δ increases. This finds its formal expression in the following theorem.

[1] For a detailed discussion of additivity, which underlies much of classical statistical analysis, see, for example, Cox (1958, Chap. 2).

Theorem 2. The power function $\Pi_F(\Delta)$ defined by (2.8) is a nondecreasing function of Δ.

Proof. Let $\Delta_0 < \Delta_1$, and let $X_1, \ldots, X_m$ be independently distributed with distribution $F(x)$ and $Y_1, \ldots, Y_n$ with distribution $G_0(y) = F(y-\Delta_0)$. If $V_j = Y_j + (\Delta_1 - \Delta_0)$, the distribution of the random variables $V_1, \ldots, V_n$ is $G_1(y) = F(y-\Delta_1)$ since

$$P(V_j \leqslant y) = P[Y_j \leqslant y-(\Delta_1-\Delta_0)] = F(y-\Delta_1)$$

It follows that

$$\Pi_F(\Delta_0) = P(W_{XY} \geqslant c) \qquad \text{and} \qquad \Pi_F(\Delta_1) = P(W_{XV} \geqslant c)$$

where W_{XY} denotes the number of pairs (X_i, Y_j) with $X_i < Y_j$ and W_{XV} the number of pairs (X_i, V_j) with $X_i < V_j$. From the fact that $Y_j < V_j$ for all j, it is seen that $W_{XY} \leqslant W_{XV}$ and hence that $\Pi_F(\Delta_0) \leqslant \Pi_F(\Delta_1)$ as was to be proved.

This theorem has a number of interesting consequences.

(i) If $\Pi_F(0) = \alpha$ is the significance level of the Wilcoxon test, it follows from the theorem that for every continuous F

$$\Pi_F(\Delta) \geqslant \alpha \qquad \text{for all } \Delta > 0 \tag{2.9}$$

A test whose power against a class of alternatives never falls below the significance level is said to be *unbiased* against these alternatives. The inequality (2.9) thus shows that the Wilcoxon test is unbiased against the shift alternatives (2.7).

(ii) In formulating the problem of testing the hypothesis of no treatment effect against the alternatives that the new treatment is better than the standard one, we tacitly excluded the case in which instead it is worse. This is, however, an ever present possibility. Patients suffering from pneumonia used to be treated with great confidence by means of leeches and emetics, treatments that more than doubled the death rate rather than increasing the number of cures. In the shift model (2.7), this possibility is represented by negative values of Δ, and it is reassuring to note that as a consequence of Theorem 2

$$\Pi_F(\Delta) \leqslant \alpha \qquad \text{for all } \Delta < 0 \tag{2.10}$$

This inequality guarantees that the Wilcoxon test will lead to adoption of the new treatment when it is actually worse than the standard one, with a probability which never exceeds the level of significance. More formally, the hypothesis of no treatment effect $H{:}\Delta = 0$ may be replaced by the hypothesis

$$H'{:}\quad \Delta \leqslant 0 \tag{2.11}$$

that the new treatment is no better than the standard one. It follows from (2.10) that the level α Wilcoxon test of H is also a level α test of H'.

(iii) The shift model and Theorem 2 provide a rigorous answer to a problem raised in Chap. 1, Sec. 5, concerning the symmetric comparison between two treatments A and B. The procedure (1.54) of that section chooses treatment B when $W_B \geqslant \frac{1}{2}n(N+1)+c$, treatment A when $W_B \leqslant \frac{1}{2}n(N+1)-c$, and otherwise suspends judgment as to which of the two treatments is better. The critical value c is determined so that

$$P(\text{choosing } A) = P(\text{choosing } B) = \alpha'$$

when there is actually no difference between the treatments. The intuitive interpretation of α' as the maximum error probability can now be made rigorous. Let $F(x)$ and $F(y-\Delta)$ denote the (continuous) response distributions under A and B, respectively. When $\Delta = 0$, there is no difference between the treatments and none of the three choices is considered an error. If $\Delta \neq 0$, suppose without loss of generality that A is better than B, so that $\Delta < 0$. Then an error is committed if and only if $W_B \geqslant \frac{1}{2}n(N+1)+c$. It follows from Theorem 2 (and from the equivalence of the Wilcoxon and Mann-Whitney forms of the test) that the probability of this inequality is a nondecreasing function of Δ, and hence for any negative Δ is less than or equal to α', its value for $\Delta = 0$. This shows that the error probability can never exceed α'. On the other hand, it tends to α' as Δ tends to zero (Prob. 38), and this establishes α' as the maximum error probability.

The shift model (2.7) is based on the assumption of a constant treatment effect. Consider now the more general case in which the size of the treatment effect depends on the response that the subject would give without the treatment. (A supplementary diet, for example, will clearly do more for an undernourished person than for one who is well fed.) Suppose the treatment adds the amount $\Delta(x)$ when the response of the untreated subject would be x. Then the distribution G of the treatment responses is that of the random variable $X+\Delta(X)$, where X is distributed according to F. [Some properties of $\Delta(x)$ and the problem of estimating it are discussed by Doksum (1974).]

The treatment will have a beneficial effect if

$$\Delta(x) \geqslant 0 \qquad \text{for all } x \tag{2.12}$$

with strict inequality holding for at least some x. Suppose, for example, that X is a positive random variable such as the amount of rainfall produced by a storm, that $\Delta(x)$ is the effect of cloud seeding, and that $\Delta(x)$ is proportional to x, say, $\Delta(x) = ax$, $a > 0$. Then (2.12) clearly holds. In this example

$$X+\Delta(X) = (1+a)X$$

and the effect of the treatment is to multiply the response by $1+a$, so that F and G differ only in scale. (A better test for this case than the Wilcoxon test is discussed in Sec. 7K.)

When (2.12) holds, so does (2.6), and G is stochastically larger than F. Conversely, (2.6) implies the existence of a function $\Delta(x)$ satisfying (2.12) and such that G is the distribution of $X+\Delta(X)$ when F is the distribution of X, provided F and G

are not only continuous but also strictly increasing (Prob. 39). The following generalizations of properties (i) to (iii) are easily proved (Prob. 40).

(i′) The Wilcoxon test is unbiased against the alternatives (2.12).

(ii′) The level α Wilcoxon test is not only a level α test of H and of the hypothesis H' given by (2.11) but also of the still more general hypothesis

$$H'': \quad \Delta(x) \leqslant 0 \qquad \text{for all } x \tag{2.13}$$

when the treatment effect is not assumed to be constant.

(iii′) In the symmetric comparison of two treatments, the error probability can never exceed the value α' defined by (1.55), provided the treatment effect $\Delta(x)$ has the same sign for all x.

Properties (i′) to (iii′) hold even without the restriction that F and G are strictly increasing, but we shall not prove this. [It follows by the argument given here from Lehmann (1959, Chap. 3, Lemma 1).]

3. ASYMPTOTIC POWER

The results on the power of the Wilcoxon rank-sum test obtained in the preceding section were only qualitative. The computation of numerical values of the power requires the distribution of the ranks when F and G are continuous distributions with $F \neq G$. This is considerably more complicated than the distribution (2.3) for the case $F = G$. and we shall not discuss it here (but see Sec. 7D). Useful results can, however, be obtained from the normal approximation to the power. This approximation is most conveniently expressed in terms of the Mann-Whitney form of the test and corresponds to the fact that the distribution of

$$\frac{W_{XY} - E(W_{XY})}{\sqrt{\text{Var}(W_{XY})}} \tag{2.14}$$

tends to the standard normal distribution as m and n tend to infinity for any fixed distributions F and G of X and Y for which

$$0 < P(X < Y) < 1 \tag{2.15}$$

(proved in the Appendix, Example 20). When $P(X < Y)$ is 0 or 1, the distribution of Y lies entirely to the left or entirely to the right of the distribution of X. In either case, W_{XY} reduces to a constant and the limit problem no longer arises.

Application of this approximation requires the expectation and variance of W_{XY}. The first of these depends on the probability

$$p_1 = P(X < Y) \tag{2.16}$$

where X and Y indicate two independent random variables with distributions F

and G and is shown in the Appendix, Example 5, to be given by

$$E(W_{XY}) = mnp_1 \tag{2.17}$$

As a check, consider the case $F = G$. The variables X and Y then have the same distribution, so that $P(X < Y) = P(Y < X)$. If this common distribution is continuous, $P(X = Y) = 0$, and hence $p_1 = 1/2$. The resulting value $E(W_{XY}) = mn/2$ agrees with that given by (1.29).

The probability p_1 defined by (2.16) provides a measure of the degree of improvement of the new treatment over the standard one and it may be of interest to estimate it. (Under the shift model, Δ provides another such measure.) Equation (2.17) shows that

$$\hat{p}_1 = \frac{W_{XY}}{mn} \tag{2.18}$$

is an estimator of p_1, which is *unbiased* in the sense that its expectation is identically equal to the quantity being estimated. If F and G are completely unknown it is in fact true, although we shall not prove it here, that among all unbiased estimators of p_1, the estimator $\hat{p}_1$ has the smallest variance.

The variance of W_{XY} is somewhat more complicated than the expectation. It depends, besides p_1, on the two quantities p_2 and p_3, where

$$p_2 = P(X < Y \text{ and } X < Y') \tag{2.19}$$

X, Y, and Y' being independently distributed, X with distribution F, and Y and Y' each with distribution G, and

$$p_3 = P(X < Y \text{ and } X' < Y) \tag{2.20}$$

X, X', and Y being independently distributed, X and X' with distribution F and Y with distribution G. With this notation, the variance formula is

$$\text{Var}(W_{XY}) = mnp_1(1-p_1)+mn(n-1)(p_2-p_1{}^2)+nm(m-1)(p_3-p_1{}^2) \tag{2.21}$$

as is proved in the Appendix, Example 5. It is useful for later applications to note that $p_2 = p_3$ in the shift model (2.7) provided the distribution F is symmetric (Prob. 41).

Again, it is interesting to check the formula for the case that $F = G$ and this common distribution is continuous. Then p_2 becomes the probability that of three independent variables X, Y, and Y' with the same continuous distribution, X is the smallest. Since each of the three is equally likely to be the smallest, it follows that $p_2 = 1/3$ and by the same argument that $p_3 = 1/3$. Using the fact that $p_1 = 1/2$ the resulting value $mn(N+1)/12$ is seen to agree with the null variance given by (1.30) (Prob. 7).

If the hypothesis $G = F$ is rejected when

$$W_{XY} \geqslant c \tag{2.22}$$

the power $\Pi(F, G)$ of the test against any fixed alternative $F \neq G$ is

$$P(W_{XY} \geqslant c) = P\left[\frac{W_{XY} - mnp_1}{\sqrt{\text{Var}(W_{XY})}} \geqslant \frac{c - mnp_1}{\sqrt{\text{Var}(W_{XY})}}\right]$$

and hence the normal approximation with continuity correction gives

$$\Pi(F, G) \approx 1 - \Phi\left[\frac{c - \frac{1}{2} - mnp_1}{\sqrt{\text{Var}(W_{XY})}}\right] \tag{2.23}$$

EXAMPLE 2. *Effect of background music.* To test whether playing background music would speed up work, an office employing several typists decides to use 20 consecutive days for an experiment. On 10 of the days selected at random, soft background music will be provided, and the average numbers of pages typed will be compared with the numbers for the other 10 days, which serve as controls. The Wilcoxon test is to be used at a significance level near .05, and Table B shows that the critical value $c = 72$ gives significance level $\alpha = .0526$. What is the power of this test against the alternative that the playing of music adds 5 pages to the average daily output?

Suppose that a shift model seems appropriate, and that it is reasonable to assume the average output to be approximately normally distributed, with the same variance σ^2 for control and treatment response. Let us therefore compute the power against the *normal shift alternative*

$$F(x) = \Phi\left(\frac{x - \xi}{\sigma}\right) \qquad G(y) = \Phi\left(\frac{y - \xi - \Delta}{\sigma}\right) = F(y - \Delta) \tag{2.24}$$

with $\Delta = 5$. If $N(a, b^2)$ denotes the normal distribution with mean a and variance b^2, this alternative can be written more compactly as

$$F = N(\xi, \sigma^2) \qquad G = N(\xi + \Delta, \sigma^2) \tag{2.25}$$

It would appear that the power will also depend on the values of the parameters ξ and σ^2. Actually, it is independent of ξ but does depend on σ^2 because it is easier to detect a given shift Δ when σ^2 is small so that the X's are concentrated about their mean ξ and the Y's about $\xi + \Delta$ than when σ^2 is large and the X's and Y's are widely spread out. Suppose then that past efficiency studies suggest a value of $\sigma^2 = 32$.

Application of formula (2.23) requires the values of p_1, p_2, and p_3. To determine p_1, note that

$$p_1 = P(X - Y < 0) = P\left[\frac{X - (Y - \Delta)}{\sigma\sqrt{2}} < \frac{\Delta}{\sigma\sqrt{2}}\right]$$

Since X and Y are independent with means ξ and $\xi + \Delta$ and common variance σ^2, the random variable $[X - (Y - \Delta)]/\sigma\sqrt{2}$ has zero mean and unit variance, and it is normally distributed since X and Y are normal [see Appendix, formula (A.94).] It follows that

$$p_1 = \Phi\left(\frac{\Delta}{\sigma\sqrt{2}}\right) \tag{2.26}$$

In our example $\Delta = 5$ and $\sigma^2 = 32$, so that $\Delta/\sigma\sqrt{2} = .625$ and p_1 is read from Table C to be $p_1 = .734$.

Analogously it is seen that

$$p_2 = P\left[\frac{X-(Y-\Delta)}{\sigma\sqrt{2}} < \frac{\Delta}{\sigma\sqrt{2}} \text{ and } \frac{X-(Y'-\Delta)}{\sigma\sqrt{2}} < \frac{\Delta}{\sigma\sqrt{2}}\right] \tag{2.27}$$

where X, Y, and Y' are independently distributed, X as $N(\xi,\sigma^2)$ and Y and Y' as $N(\xi+\Delta,\sigma^2)$. The random variables

$$Z = \frac{X-(Y-\Delta)}{\sigma\sqrt{2}} \quad \text{and} \quad Z' = \frac{X-(Y'-\Delta)}{\sigma\sqrt{2}} \tag{2.28}$$

are both normal with mean zero and unit variance, and their correlation coefficient is $\frac{1}{2}$ (Prob. 43).

The probability

$$P(Z \leqslant z,\ Z' \leqslant z)$$

can be obtained from the National Bureau of Standards "Tables of the bivariate normal distribution function and related functions."[1] For $z = .625$ these tables show the value to be $p_2 = .6$. Because of the symmetry of the normal distribution p_2 and p_3 are equal, so that $p_3 = .6$ also. Substitution of these values in (2.21) gives $\text{Var}(W_{XY}) = 129.76$, and for (2.23) we find

$$\Pi(F,G) \approx 1-\Phi(-.167) = .57$$

Under the assumptions made, the probability of detecting an increase of 5 pages as a result of the "treatment" is thus less than .6. For an increase of $\Delta = 10$ pages, the analogous computation gives an approximate power of .99.

By the same method, the approximate power can in principle be computed against any alternative (F,G), but the calculation of p_2 and p_3 may present difficulties. These can be avoided by an alternative approximation, which is applicable in the shift model (2.7) when Δ is small. Let the power $\Pi(F,G)$ in the shift model (2.7) be denoted by $\Pi_F(\Delta)$ and let F^* denote the distribution of the difference of two independent random variables, each with distribution F. Then if $f^*(0)$ is the density of F^* evaluated at zero, the approximation in question is

$$\Pi_F(\Delta) \approx \Phi\left[\sqrt{\frac{12mn}{N+1}}\, f^*(0)\,\Delta - u_\alpha\right] \tag{2.29}$$

where u_α is the upper α point of the standard normal distribution determined by

$$\Phi(u_\alpha) = 1-\alpha \tag{2.30}$$

A heuristic derivation of (2.29) is given at the end of this section. A slightly different form of this approximation is derived from a different point of view as formula (A.229) (see Appendix, Prob. 47).

To apply (2.29), it is necessary to find $f^*(0)$. Consider in particular the case

[1] National Bureau of Standards, *Appl. Math. Ser.* 50 (1959).

that F is the normal distribution $N(\xi,\sigma^2)$. Then F^* is the distribution of the difference of two independent random variables each distributed according to $N(\xi,\sigma^2)$ and hence is the normal distribution $N(0,2\sigma^2)$. It follows from the expression of the normal density [given as formula (A.92)] that $f^*(0) = 1/(2\sigma\sqrt{\pi})$ so that (2.29) becomes

$$\Pi(\Delta) \approx \Phi\left[\sqrt{\frac{3mn}{(N+1)\pi}}\frac{\Delta}{\sigma} - u_\alpha\right] \tag{2.31}$$

As an illustration, consider once more Example 2, where $\Delta = 5$, $\sigma^2 = 32$, $m = n = 10$, and $\alpha = .05$. Then $u_\alpha = 1.645$ and (2.31) gives $\Pi(\Delta) \approx .595$, compared with the value .57 obtained from the earlier approximation.

In the case of a shift model, what can be said about the accuracy of the approximations (2.23) and (2.29)? It was seen in Chap. 1, Sec. 3, that (2.23) works quite well, even for relatively small values of m and n, when $G = F$. As G moves away from F, the distribution of W_{XY} tends to become less symmetric, and the normal approximation therefore less accurate. In particular, for the shift model (2.7), the approximation (2.23) may be expected to give the best results when Δ is small. This is likely to hold even more strongly for (2.29), where the true expectation and variance have been replaced by their first-order terms in the neighborhood of $\Delta = 0$, and which therefore would typically be expected to be less accurate than (2.23). Table 2.1 below shows the two approximations together with the true values for a normal shift model and sample sizes $m = n = 7$. Unfortunately, little is known about the closeness of the approximation for nonnormal distributions. Some numerical results given in Chap. 4, Sec. 2, for a related problem suggest that both approximations may then be quite inaccurate unless the sample sizes are fairly large.

TABLE 2.1 Power of the Wilcoxon rank-sum test for normal shift alternatives; $m = n = 7$, $\alpha = .049$

Δ/σ	0	.2	.4	.6	.8	1.0	1.5	2.0
Exact	.049	.094	.165	.264	.386	.520	.815	.958
(2.23)	.048	.094	.160	.249	.359	.484	.806	.981
(2.29)	.049	.098	.177	.287	.423	.568	.861	.977

A power of less than .6 for an increase as large as five pages in Example 2 may be too low to be satisfactory. Higher power against the same alternative can be achieved by increasing the size of the sample (Prob. 12). Before embarking on a study, it is desirable to determine the sample size needed to achieve a given power. An approximation to this sample size is easily obtained from (2.29). Since it is usually desirable to have the two sample sizes about equal, put $m = n$ in (2.29), and replace $N+1$ by $N = 2n$. Then (2.29) becomes

$$\Pi_F(\Delta) \approx \Phi[\sqrt{6n}\,\Delta f^*(0) - u_\alpha] \tag{2.32}$$

Suppose we wish to find the value of n for which $\Pi_F(\Delta)$ has a specified value Π against a given alternative Δ. If u_Π is defined by $\Phi(u_\Pi) = 1-\Pi$, so that $\Phi(-u_\Pi) = \Pi$, it follows that

$$u_\Pi \approx u_\alpha - \sqrt{6n}\,\Delta f^*(0)$$

Solving this equation for n, we find the desired sample size to be approximately

$$n \approx \frac{(u_\alpha - u_\Pi)^2}{6\,\Delta^2 f^{*2}(0)} \tag{2.33}$$

When F is a normal distribution with variance σ^2, it was seen earlier that $f^*(0) = 1/(2\sigma\sqrt{\pi})$ so that (2.33) reduces to

$$n \approx \frac{2\pi\sigma^2(u_\alpha - u_\Pi)^2}{3\,\Delta^2} \tag{2.34}$$

EXAMPLE 3. *Cultural influences on IQ.* As part of a study of cultural influences on IQ scores, a group of culturally underprivileged children are to meet on an individual basis with college students for 2 hours each week of a school year, and at the end of the year will be compared with a control group without this experience. It has been decided that the two groups should be of equal size n. How large should n be if at level of significance $\alpha = .01$ the power is to be $\Pi = .95$ against the alternative that the weekly meetings add about two points to the IQ scores?

In analogy with Example 2, the study could be set up by selecting a sample of $2n$ children of whom n are assigned at random to treatment and the other n to control, and by comparing the IQ scores of the two groups at the end of the year. However, in the present situation, it is possible to achieve the same power with fewer observations (or higher power with the same number of observations) by giving each child an IQ test at the beginning as well as at the end of the year. The response of interest will then be the difference between the scores obtained on these two tests.

The reason for the greater efficiency of this approach is easy to see. Since the control scores may be expected to change only slightly during the year, the difference between the two control scores will typically be close to zero. Similarly, if the treatment adds Δ points to the score, the differences between the two scores for the treatment subjects may be expected to be close to Δ. Since the differences within both the control group and the treatment group thus show little variation, it is easy to detect even a relatively small treatment effect. On the other hand, the scores at the end of the year (rather than the differences) will be much more variable since they reflect the range of intelligence of the groups, and the same effect Δ will therefore show up much less clearly.

It might seem that with an observation on each child both at the beginning and at the end of the year one could dispense with the controls since through the initial observation each child provides his own control. This would certainly be the case if it were possible to decide at random which of the two occasions (beginning or end) to assign to treatment and which to control. Examples in which this is possible will be given in Chap. 3. However, in the present situation, without controls, one could not be sure that an observed effect is the result of the treatment rather than of some other cause affecting the group during the year, such as an exceptionally stimulating teacher or the attaining of a particular age level.

Let us now return to the problem stated at the beginning of the example: The determination of the sample size $m = n$ which at significance level $\alpha = .01$ will give power .95 for detecting an effect of $\Delta = 2$. To complete the specification of the alternative, it is necessary to specify the distribution F of the control responses. Suppose that other IQ studies on similar subjects suggest that the differences of the IQ scores at the beginning and end of the year are approximately normally distributed with variance $\sigma^2 = 2$. Then formula (2.34) with $u_\alpha = 2.33$ and $u_\Pi = -1.645$ gives $n \approx 16.5$ or an approximate sample size of $n = 17$.

To conclude the section, we shall now give a heuristic derivation of the power approximation (2.29). Suppose that the sample size is large enough so that the critical value is determined from the normal approximation rather than the exact null distribution, and write the rejection rule as

$$\frac{W_{XY} - \frac{1}{2}mn}{\sqrt{\frac{1}{12}mn(N+1)}} \geqslant u_\alpha \tag{2.35}$$

where u_α is determined by (2.30). Comparison of (2.35) with (2.22) shows the critical values c and u_α to be related through the approximate equation

$$c = \tfrac{1}{2}mn + u_\alpha \sqrt{\tfrac{1}{12}mn(N+1)} \tag{2.36}$$

Substituting this value into (2.23), neglecting the continuity correction, and using the symmetry of the normal distribution, we can rewrite (2.23) as

$$\Pi(F,G) \approx \Phi\left[\frac{mn(p_1 - \frac{1}{2}) - u_\alpha \sqrt{mn(N+1)/12}}{\sqrt{\operatorname{Var}(W_{XY})}}\right] \tag{2.37}$$

As in the argument leading to (2.26), it is seen in the case of the shift alternatives $G(y) = F(y - \Delta)$ that

$$p_1 = P[X - (Y - \Delta) < \Delta] = F^*(\Delta)$$

If F^* has a density f^*, an expansion of F^* about zero gives the approximation

$$p_1 \approx F^*(0) + \Delta f^*(0)$$

Since F^* is the distribution of the difference of two identically distributed independent random variables, it is symmetric about zero and hence

$$p_1 - \tfrac{1}{2} \approx \Delta f^*(0)$$

The variance of W_{XY} also depends on Δ, but for small Δ it is close to its value for $\Delta = 0$, which is $mn(N+1)/12$. Application of these two approximations to (2.37) leads to the desired approximation (2.29).

4. COMPARISON WITH STUDENT'S t-TEST

The common use of rank tests is of fairly recent origin. Although the Wilcoxon rank-sum test was proposed independently by a number of writers throughout the first half of the century [see Kruskal (1957) for the history of the test], it was not accepted as a standard statistical technique until after the appearance of the papers by Wilcoxon (1945) and Mann-Whitney (1947).

The classical test for comparing two treatments is not a rank test but Student's t-test, which rejects the hypothesis of no treatment difference when

$$\frac{\bar{Y}-\bar{X}}{S\sqrt{1/m+1/n}} \geqslant c \tag{2.38}$$

where

$$S^2 = \frac{\Sigma(X_i-\bar{X})^2+\Sigma(Y_j-\bar{Y})^2}{m+n-2} \tag{2.39}$$

and where the critical value c is determined from the t-distribution with $m+n-2$ degrees of freedom. This test is derived for testing the hypothesis $\Delta = 0$ in the normal shift model (2.24), that is, when $X_1, \ldots, X_m$ and $Y_1, \ldots, Y_n$ are independently, normally distributed with common unknown variance σ^2 and with unknown means $E(X_i) = \xi$ and $E(Y_j) = \xi+\Delta$. Under these assumptions, the null distribution of the statistic (2.38) is Student's t-distribution with $m+n-2$ degrees of freedom. The t-test, besides being the likelihood ratio test, has various optimum properties within this model. It is, for example, unbiased, and among all unbiased tests it maximizes the power against each alternative ξ, σ^2 and $\Delta > 0$. (A test with this property is called *uniformly most powerful unbiased.*)

Since the two tests were developed for different situations—the t-test under the assumption of normality, the Wilcoxon test without this assumption—there appears to be little point in comparing them. However, situations commonly arise in which it is not clear which of the two tests is more appropriate. The ambiguity is caused by the difficulty of deciding whether, or to what degree, an assumption such as normality is satisfied in a given situation. There may, for example, be theoretical reasons for believing that certain observations are approximately normally distributed but one may not be willing to trust this belief fully. In particular, it is quite common for observations obtained under standard conditions to follow an approximately normal distribution but for some of the observations—without knowledge of the person performing the study—to have been obtained under nonstandard conditions. Thus, measurements can be misread, a decimal point can be entered incorrectly, through careless handling a specimen may pick up an impurity, or a patient being treated for some disease may be suffering from another rare and unsuspected condition which distorts the results of the treatment. Or, to give a more extreme example, it occasionally happens in a true-false examination with a long list of questions and a separate answer sheet that the two sheets are not correctly lined up, which produces results completely at variance with the intentions of the candidate and with his usual performance. Even a relatively small proportion of such "wild" observations, or *gross errors*, may seriously affect the shape of the distribution and invalidate the assumption of normality for the complete sample. To see which of the two tests is more appropriate in such situations, it is necessary to compare their behavior both under the hypothesis (i.e., with respect to the significance level) and under the alternatives (i.e., with respect to power).

As was seen in Sec. 1, the Wilcoxon test is distribution-free for testing the hypothesis $H: G = F$ in the two-sample problem; that is, the null distribution of the test statistic and hence the significance level are independent of F. This is not true of the t-test. There the significance level is computed under the assumption of normality and when F is not normal cannot be expected to be equal to the stated value α. A clue to the strength of the dependence of the level on F is obtained by considering the case that m and n are large. If σ^2 denotes the variance of F, it then follows from the central limit theorem that under H

$$\frac{\bar{Y}-\bar{X}}{\sigma\sqrt{1/m+1/n}} \tag{2.40}$$

is approximately normally distributed with zero mean and unit variance. Furthermore, the quantity S/σ tends in probability to 1; that is, the probability that S/σ differs from 1 by less than any preassigned quantity tends to 1 as m and n tend to infinity. Combining these two statements, it is intuitively plausible, and it follows rigorously from the Appendix, Theorem 4, that the left-hand side of (2.38) is asymptotically normally distributed with zero mean and unit variance. If we denote the critical value c on the right-hand side of (2.38) more precisely by $c_{m,n}(\alpha)$ and if, as in (2.30), u_α denotes the critical value of the standard normal curve, it follows that

$$c_{m,n}(\alpha) \to u_\alpha \qquad \text{as } m \text{ and } n \to \infty \tag{2.41}$$

and hence that the probability of the inequality (2.38) tends to $1-\Phi(u_\alpha) = \alpha$.

The only implicit assumption in the above argument is that the variance σ^2 of F is finite. With this restriction, it has now been shown that the asymptotic null distribution of the t-statistic on the left-hand side of (2.38) is the standard normal distribution and hence is independent of F. In this sense, the t-test is *asymptotically distribution-free*. For sufficiently large m and n, the significance level of the t-test for any F with finite variance is thus close to its significance level α for normal F. While this does not prove that the level is close to α for any particular m, n, and F, it does suggest that the dependence of the level on F will not be too serious for moderate m and n and for distributions which are not too far from normal.

A comparison of the null distributions of the Wilcoxon and t-statistics should be concerned with the sensitivity of the distributions not only to the assumption of normality but also to other assumptions such as the independence of the observations $X_1, \ldots, X_m, Y_1, \ldots, Y_n$ and the equality of the variances of the X's and Y's. Some discussion of these aspects will be given in Sec. 7A, and we now turn to a comparison of the power of the two tests. For the power Π of the Wilcoxon test, we have found an approximation in (2.29). Let us now obtain an approximation for the power of the t-test (2.38) in the shift model (2.7). This can be written as

$$\Pi' = P\left(\frac{\bar{Y}-\bar{X}-\Delta}{\sigma\sqrt{1/m+1/n}} \geqslant \frac{cS}{\sigma} - \frac{\Delta}{\sigma\sqrt{1/m+1/n}}\right)$$

where the X's and $(Y-\Delta)$'s are independently distributed, each with distribution F. The facts, that the distribution of the left-hand side of the inequality tends to the standard normal distribution, that $c = c_{m,n}(\alpha)$ tends to u_α by (2.41), and that S/σ tends to 1 in probability, suggest the approximation

$$\Pi' \approx 1 - \Phi\left(u_\alpha - \frac{\Delta}{\sigma\sqrt{1/m+1/n}}\right) \tag{2.42}$$

By using (2.29) and (2.42), the values of Π and Π' can now be compared for any given m, n, and F. Consider, for instance, Example 2 where at level $\alpha = .05$ and for $m = n = 10$, F normal with variance $\sigma^2 = 32$, and $\Delta = 5$, the approximation (2.29) or (2.31) to the power of the Wilcoxon test was .595. In the same circumstances, (2.42) gives an approximate value of .63 for the power of the t-test; the actual power turns out to be .60.[1] Whether such an increase in power, in case the assumption of normality is justified, is worth the uncertainty in the significance level in case it is not, depends on the confidence placed in this assumption. The answer would also have to take into account the power of the two tests in case F is not normal. We shall return to this aspect of the comparison below.

As an alternative to the comparison of Π and Π', consider the situation that presents itself when the study is being planned. Suppose that, as in Example 3, the sample size $m = n$ has been determined for which the Wilcoxon test will achieve a specified power for a given F and Δ. One would then wish to know what sample size $m' = n'$ is required by the t-test to achieve the same power against the same alternative. The ratio n'/n is called the *efficiency* of the Wilcoxon test relative to the t-test for the given situation, i.e., for the given values of α, Π, F, and Δ. An efficiency of $1/2$, for example, means that the Wilcoxon test requires twice as many observations as the t-test to achieve the same power Π under the same circumstances. As an illustration consider Example 3, where the Wilcoxon test was seen to require

[1] Convenient tables of this power are given by Owen (1965), "The Power of Student's t-test," *J. Amer. Stat. Assoc.* **60**:320–333, and by Hodges and Lehmann (1968), "A Compact Table for the Power of the t-test," *Ann. Math. Statist.* **39**:1629–1637.

$m = n = 17$ observations to achieve power $\Pi = .95$ at level $\alpha = .01$ when $\Delta = 2$ and F is normal with variance $\sigma^2 = 2$. The approximate size $m' = n'$ required by the t-test for the same purpose can be determined by substituting n' for m and n and putting $\Pi' = .95$ in (2.42), and solving the resulting equation for n'. The result is seen to be $n' = 15.8$, so that $e = 15.8/16.5 = .95$.

To see how the efficiency of the Wilcoxon to the t-test changes in the normal case with varying Δ or Π, we show the approximation to this efficiency in Table 2.2, as it is obtained from (2.31) and (2.42) for normal F with unit variance, $\alpha = .01$ and $n = 25$. In each case, $m' = n'$ is the sample size required by the t-test to attain the same power as that achieved by the Wilcoxon test with $m = n = 25$ observations. Here the fractional value obtained by solving (2.42) has been retained. In practice one would instead take either the closest or the next higher sample size, but the (approximate) power of the two tests would then no longer be the same. (Alternatively, it is possible to realize a fractional sample size n' by randomizing between the two sample sizes bracketing n'.)

TABLE 2.2 Approximate efficiency of Wilcoxon to t-test; $\alpha = .01$, $m = n = 25$, F = normal

$\Delta/\sqrt{2}$	.10	.20	.30	.40	.50	.60	.70	.80	.90
Π	.03	.08	.18	.32	.50	.68	.83	.93	.98
n'	22.6	22.1	21.7	21.5	21.6	21.8	22.2	22.9	23.9
e	.91	.88	.87	.85	.86	.87	.89	.92	.96

In view of the optimum properties of the t-test in the normal case, and the fact that the Wilcoxon test is a rank test and hence utilizes only the order relationships between the observations and not their actual values, the reader may be surprised at the high efficiencies of Table 2.2. Actually, the situation is even more favorable to the Wilcoxon test than indicated. The approximation (2.42) overestimates the power of the t-test in the normal case, and the efficiency loss of the Wilcoxon test in this case is only about 5 percent rather than the 10 percent shown in the major part of the table. An exact efficiency has been computed for the case $m = n = 5$ and is shown in Table 2.3.

It is remarkable how little the efficiency varies as Π ranges from below .1 to above .99. This stability is not peculiar to the normal shift model. When the X's and Y's are distributed according to (2.1) and (2.7) with any F, the efficiency of the Wilcoxon test to the t-test considered as a function of Π over any fixed range of Π-values

TABLE 2.3* Efficiency of Wilcoxon to t-test; $\alpha = 4/126$, $m = n = 5$, F normal

Δ	.5	1.0	1.5	2.0	2.5	3.0	3.5	4.0
Π	.072	.210	.431	.674	.858	.953	.988	.998
e	.968	.978	.961	.956	.960	.960	.964	.976

* Taken from Dixon (1954) and Hodges and Lehmann (1956).

bounded away from 1 always stabilizes for sufficiently large m and n. More precisely, for any fixed Π satisfying $\alpha < \Pi < 1$, the efficiency tends to a limit independent of Π as m and n tend to infinity. The limiting efficiency, which turns out to be independent not only of Π but also of α, is called the *Pitman efficiency* (or *asymptotic relative efficiency*) of the Wilcoxon test to the t-test. We denote it by $e_{W,t}(F)$ to indicate its dependence on the underlying distribution F. An expression for $e_{W,t}(F)$ is given as formula (A.240) of the Appendix. It is also shown there that for normal F this efficiency has the value $3/\pi = .955$. Table 2.3 shows that the actual efficiency is quite close to this limit value even for samples as small as $m = n = 5$.

Since consideration of the Wilcoxon test as an alternative to the t-test is motivated by uncertainty concerning the assumption of normality, it is important to compare the two tests also for nonnormal distributions, and the Pitman efficiency $e_{W,t}(F)$ provides a simple basis for such comparisons. As examples, consider the four distributions (logistic, double exponential, rectangular, and exponential), the densities of which are given in the Appendix, Secs. 2D and 2E. The values of $e_{W,t}(F)$ in these four cases are shown in Table 2.4.

TABLE 2.4 Pitman efficiency of Wilcoxon to t-test

F	Logistic	Double Exponential	Rectangular	Exponential
$e_{W,t}(F)$	$\pi^2/9 = 1.097$	1.5	1	3

The values for the double-exponential and exponential distributions show that for distributions that are sufficiently far from normal and for sufficiently large sample sizes the Wilcoxon test can be considerably more efficient than the t-test. It might be asked whether such a comparison is relevant in situations such as that of exponential F in which neither test is at all appropriate. The comparison is being made, however, under the assumption that the true distribution F is unknown, and it therefore seems interesting to explore the relative efficiency of the two tests for a variety of distributions. There is in addition another less theoretical reason for such a comparison. Simple statistical methods are applied in many different situations by persons with little statistical training who apply whatever method is readily accessible to them without much regard as to how appropriate it is for the particular distribution at hand. Although this may be regrettable, the t-test has undoubtedly been applied on more than one occasion to distributions such as the exponential. It therefore seems desirable not to exclude such distributions from the comparison.

The efficiency values in the normal case and those displayed in Table 2.4 raise the question whether there exist distributions F for which the t-test is much more efficient than the Wilcoxon test. It turns out that in fact

$$e_{W,t}(F) \geqslant .864 \tag{2.43}$$

for all F with finite variance. (Without the assumption of finite variance the comparison becomes meaningless since the asymptotic significance level of the t-test is then in doubt.) For a proof see formula (A.241) of the Appendix.

The above and similar results (Appendix, Probs. 49 to 51), while suggesting that the Wilcoxon test tends to be superior to the t-test in sufficiently nonnormal situations, must be taken with a certain amount of caution since the sample sizes required for the Pitman efficiency to be close to the actual efficiency may be quite large (this appears to be the case for example when F is the exponential distribution), and the asymptotic result may then be rather misleading for the values of m and n with which one is dealing.

An overall comparison of the Wilcoxon and t-tests may be summarized as follows.

(i) *Significance level.* The Wilcoxon test has the advantage that its significance level is independent of the true F and hence is known exactly. On the other hand, if F is close to normal or the sample sizes are sufficiently large, the level of the t-test will tend to be close to the nominal level, so that this advantage is not as marked as might at first be expected.

(ii) *Power.* The t-test is more powerful when F is normal, but the efficiency loss of the Wilcoxon test in this case is slight (about 5 percent). For distributions close to normal such as a normal distribution with a small proportion of moderate gross errors, there will typically be little difference in the power of the two tests. When nothing is known about the shape of the distribution, and particularly for distributions whose tails are very much heavier than those of the normal distribution [i.e., whose density $f(x)$ tends to zero, as x tends to plus or minus infinity, much more slowly than the normal density], the Wilcoxon test may be considerably more efficient than the t-test.

5. ESTIMATING THE TREATMENT EFFECT

Testing the hypothesis of no treatment effect is only one aspect of the comparison of a new treatment with a standard one. Frequently, it is important to estimate the degree of the improvement rather than to establish its significance. A natural measure of the effect of a treatment is available when this effect is additive. The shift model (2.7) is then appropriate, and the treatment effect can be measured by the amount Δ by which the treatment shifts the response distribution.

A possible approach to the problem of estimating Δ is suggested by the fact that under the assumptions made, the variables $X_1, \ldots, X_m$ and $Y_1-\Delta, \ldots, Y_n-\Delta$ have the same distribution. If the observations are plotted as in Fig. 2.1, one may then estimate Δ by the amount $\hat{\Delta}$ by which the Y-values must be shifted to give the best possible agreement with the X-values. To complete the definition of the estimator $\hat{\Delta}$, it is necessary to specify what is meant by two sets of observations being

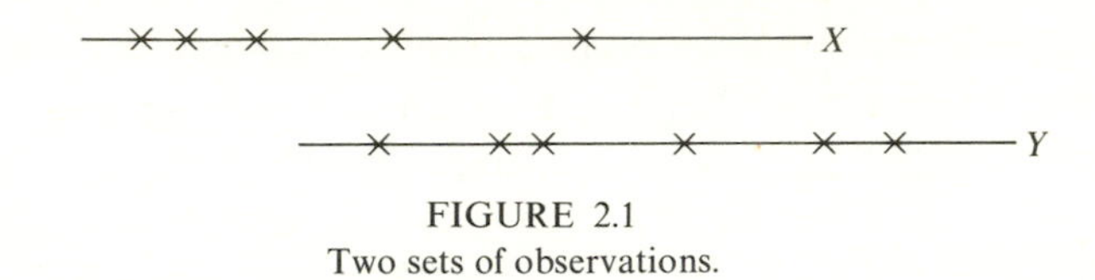

FIGURE 2.1
Two sets of observations.

in the best possible agreement. Such a specification can be made in terms of a test of the hypothesis $H: \Delta = 0$ and then provides a desirable relationship between this test and the resulting estimator. Suppose that under H the test statistic, say the Wilcoxon statistic W_{XY}, is symmetrically distributed about some value μ. Let the two-sided test reject H when $W_{XY} \geqslant \mu + c$ and when $W_{XY} \leqslant \mu - c$. Then the value of W_{XY} most supportive of H is the value closest to the center of the distribution, that is, the value equal to or as close as possible to μ. The two sets $X_1, \ldots, X_m$ and $Y_1 - \hat{\Delta}, \ldots, Y_n - \hat{\Delta}$ will therefore be considered as being in best agreement if they assign to the Wilcoxon statistic $W_{X,Y-\hat{\Delta}}$ this closest value.

It is easy to get an approximate idea of the resulting estimator $\hat{\Delta}$. For any number a, the statistic $W_{X,Y-a}$ is by definition the number of pairs (i,j) for which $Y_j - a > X_i$, or equivalently $Y_j - X_i > a$. Thus, $W_{X,Y-a}$ will take on its central value if half of the differences $Y_j - X_i$ are greater than a and half are less than a. The value $\hat{\Delta}$ of a for which this is achieved is the median of the mn values $Y_j - X_i$. Although this is in fact the desired estimator, a somewhat more careful argument is required to establish it rigorously.

Denote the ordered set of mn differences $Y_j - X_i$ by $D_{(1)} < D_{(2)} < \cdots < D_{(mn)}$, and consider two cases depending on the parity of mn.

(i) Suppose first that the product mn of the sample sizes is even (this will be the case if at least one of the two sample sizes is even), say, $mn = 2k$. The center of symmetry of the distribution of $W_{X,Y}$ is $mn/2$ and hence in the present case is equal to k. We are therefore looking for the value of a for which $W_{X,Y-a} = k$, that is, for which exactly k of the differences $Y_j - X_i$ exceed a. This is clearly the case if and only if $D_{(k)} < a < D_{(k+1)}$. Thus, $W_{X,Y-a}$ takes on its central value k, not for a single value of a, but for an interval of such values. The natural estimator is then the midpoint of this interval

$$\hat{\Delta} = \tfrac{1}{2}(D_{(k)} + D_{(k+1)}) \tag{2.44}$$

which is the median of the D's.

(ii) Suppose on the other hand that mn is odd, say, $mn = 2k+1$. Then the center of symmetry of the Wilcoxon distribution is $k + \frac{1}{2}$. Since this is not a possible value of the integer-valued Wilcoxon statistic, let us consider for which values of a the statistic $W_{X,Y-a}$ is greater than or less than $k + \frac{1}{2}$. It is seen that

$$W_{X,Y-a} > k + \tfrac{1}{2} \qquad \text{if and only if} \qquad a < D_{(k+1)}$$

since the inequality on the left-hand side states that at least $k+1$ of the differences $Y_j - X_i$ exceed a, and similarly that

$$W_{X,Y-a} < k + \tfrac{1}{2} \qquad \text{if and only if} \qquad D_{(k+1)} < a$$

since the inequality now states that at most k of the differences $Y_j - X_i$ exceed a. The value of $W_{X,Y-a}$ thus jumps down from $k+1$ to k as a moves through the point $D_{(k+1)}$ and, if (X_i, Y_j) is the pair whose difference is equal to $D_{(k+1)}$, a tie occurs between X_i and $Y_j - a$. Taking the contribution of a tie to $W_{X,Y-a}$ to be $\frac{1}{2}$ as suggested in Chap. 1, Sec. 4, it follows that the associated value of $W_{X,Y-a}$ is exactly $k+\frac{1}{2}$. We therefore take as estimator in this case

$$\hat{\Delta} = D_{(k+1)} \tag{2.45}$$

which is again the median of the D's.

The computation of $\hat{\Delta}$ requires finding the median of the mn differences $Y_j - X_i$, which we shall denote by $\text{med}\,(Y_j - X_i)$. Since the ordering of a large set of numbers is rather tedious, in the following example we shall illustrate a short-cut method of obtaining $\hat{\Delta}$.

EXAMPLE 4. *Two routes.* A man who had just bought a new house found that he had a choice of two routes for driving to work. His driving times $X_1, \ldots, X_{11}$ for 11 trips on route A and $Y_1, \ldots, Y_5$ for 5 trips on route B are given below.[1]

Route A:	6.0	5.8	6.5	5.8	6.3	6.0	6.3	6.4	5.9	6.5	6.0
Route B:	7.3	7.1	6.5	10.2	6.8						

Since $mn = 55$, the estimate $\hat{\Delta}$ is the twenty-eighth smallest of the 55 differences $Y_j - X_i$. To determine the value of this difference, order the 11 observations for route A (the X's) and the 5 observations for route B (the Y's), and arrange the ordered sets as borders of a rectangle as shown in Fig. 2.2.

	5.8	5.8	5.9	6.0	6.0	6.0	6.3	6.3	6.4	6.5	6.5
6.5		.7									
6.8				.8							
7.1						1.1					
7.3								1.0			
10.2										3.7	

FIGURE 2.2
Computation of $\hat{\Delta}$.

The entries in the body of the rectangle are the differences $Y - X$, but it is not necessary to compute all of them. A reduction is possible because of certain monotonicities: the entries

[1] Provided by G. W. Haggstrom.

decrease as one moves to the right of Fig. 2.2 and also as one moves upward. Consider now a diagonal connecting the upper-left-hand and lower-right-hand corners of the rectangle. Roughly half of the entries lie to the right and above such a diagonal and roughly half to the left and below it. The monotonicity of the entries then suggests that their median should lie on or at least close to the diagonal.

Let us approximate the diagonal by the five cells indicated in the figure. One can obtain an idea of the value of $\hat{\Delta}$ by computing the ranks of the values in the five cells. The number of entries $\leqslant .7$ is seen to be equal to 11 in the first row, to 5 in the second row, to 3 in the third row, and to zero in the last two rows. The largest rank occupied by .7 is therefore $11+5+3 = 19$, so $\hat{\Delta}$ must be larger than .7. Similarly, the number of observations $\leqslant .8$ is 26 and the number $\leqslant 1.0$ is 32. Thus, .8 is still too small, but 1.0 is a possibility. Inspection shows the number of entries equal to 1.0 to be $2+2 = 4$, so that 1.0 occupies the ranks 29 to 32 and hence is too large. The median thus must lie strictly between .8 and 1.0. In the present case, the only value between .8 and 1.0 is .9, which therefore occupies the ranks 27 and 28 and is the desired median.

For other values of m and n it may not be possible to find as symmetric a choice for a diagonal as in the present case. Fortunately, it is not crucial which particular set of roughly diagonal elements is chosen, since these only represent trial values from which, hopefully, it is easy to arrive at the correct value. [A graphical procedure for obtaining $\hat{\Delta}$ is given by Høyland (1964).]

It is interesting to compare the value of $\hat{\Delta}$ with that of the classical estimator $\bar{\Delta} = \bar{Y} - \bar{X}$, which in the present case is equal to $7.58 - 6.14 = 1.44$. The unusually large difference ($\bar{\Delta}$ exceeds $\hat{\Delta}$ by more than 50 percent) is mainly due to a single observation, the value 10.2 among the Y's. If this observation is deleted, the values of $\hat{\Delta}$ and $\bar{\Delta}$ become .8 and .79, respectively, and hence show good agreement. Some insight can be gained into the relative advantages and disadvantages of the two estimators by considering a number of different situations that might have led to such a large observation. (This discussion should not, of course, be understood as a suggestion to look at the data and then decide which estimator to use; the choice of estimator should always be made before the observations are taken.)

(i) The figure 10.2 may be the result of an error. Perhaps the driver left $1\frac{1}{2}$ min after the hour but when recording the driving time remembered (incorrectly) leaving $1\frac{1}{2}$ min before the hour, so that the figure should have been 7.2 rather than 10.2. In such cases, it is clearly preferable to use an estimator which is not too sensitive to the outlying observations.

(ii) A second possible explanation for the observation 10.2 is a delay, which was unrelated to the route taken. For example, if it was raining and the windshield wiper was not working properly, it might have been necessary to stop the car and adjust the wiper. Such incidents, as well as incorrectly recorded observations, affect the normality of the distribution (which might otherwise be a reasonable assumption) by giving the distribution a heavy tail. For distributions of this kind, $\hat{\Delta}$ tends to be more efficient than $\bar{\Delta}$ in the sense of being more likely to be close to the true value Δ. The reason is the greater sensitivity of $\bar{\Delta}$ to the few outliers which may not balance (in our example, one outlying observation occurred on route B but none on route A), and which may then unduly affect the value of $\bar{\Delta}$.

To illustrate a further aspect of the problem of outliers, imagine that the delay on this particular day was not 3 min but 5, 10, or 15 min. Let us even be extravagant and suppose that it was half an hour. (Perhaps it became necessary to change a tire.) At this point, the person comparing the two routes on the basis of $\bar{\Delta}$ would presumably declare the day's drive "out of the ordinary," discard this particular observation, and compute $\bar{\Delta}$ for the remaining days. A procedure of this kind provides a reasonable alternative to estimators such as $\hat{\Delta}$, but it is legitimate only if a precise rule for discarding observations is formulated *before* any observations are taken. Discarding observations according to vague subjective criteria as to whether an observed value does or does not constitute a "normal" observation opens the way to the suppression of data that do not fit sufficiently well into preconceived ideas or do not produce desired results. It also deprives the statistical procedure of any basis for assessing its performance (the significance or power of a test, the accuracy of an estimator, etc.). It is an advantage of the estimator $\hat{\Delta}$ that there is less temptation to discard such outliers, because of the relatively small effect they have on the estimated value. In the present example, for instance, the value of $\hat{\Delta}$ is unchanged if the observation 10.2 is replaced by 40.2 or 70.2.

(iii) As a last possibility, consider a delay such as that caused by a traffic jam or an accident, which may be more likely on one of the routes than on the other. In this case a shift model is no longer appropriate since the distribution with the higher proportion of these delays will have a heavier tail, and it is then not clear what quantity is being estimated by $\hat{\Delta}$. [In this connection, see Prob. 62 (ii).] On the other hand, $\bar{\Delta} = \bar{Y} - \bar{X}$ continues to estimate $E(Y) - E(X)$, which represents the long-run average difference in the time required by the two routes.

The above considerations were motivated by the effect of an outlying observation on the estimators $\hat{\Delta}$ and $\bar{\Delta}$. Another question is raised by the many ties occurring among the differences $Y_j - X_i$, since the derivation of $\hat{\Delta}$ assumed these differences to be untied. The presence of ties, in fact, puts in doubt the appropriateness of a shift model. If the possible values of the observations are multiples of some fixed unit, the only possible shifts would be multiples of this unit, and such a restriction on the amount of shift appears unrealistic. Suppose, however, that the ties are caused by the rounding of the observations to the nearest unit, in the example the nearest tenth of a minute, and that the responses before rounding can be considered to be continuous variables for which a shift model is appropriate. Let the possible values of the (rounded) observations be $0, \pm\varepsilon, \pm 2\varepsilon, \ldots$. If the original responses giving rise to the (rounded) observations X_i and Y_j are X_i' and Y_j', then $|X_i' - X_i| \leqslant \varepsilon/2$ and $|Y_j' - Y_j| \leqslant \varepsilon/2$, and hence

$$|(Y_j' - X_i') - (Y_j - X_i)| \leqslant \varepsilon \tag{2.46}$$

Now it is easy to see (Prob. 54) that if each of a set of numbers is modified by an amount less than ε, then the median of this set is also modified by an amount less than ε, so that med $(Y_j - X_i)$ differs from med $(Y_j' - X_i')$ by less than ε. Suppose that we should like to estimate the shift Δ by $\hat{\Delta}' = \text{med}(Y_j' - X_i')$. Unfortunately the variables X_i' and Y_j' are not observable, but the estimator $\hat{\Delta} = \text{med}(Y_j - X_i)$ differs from $\hat{\Delta}'$ by at most ε. If ε is sufficiently small (relative to Δ), $\hat{\Delta}$ will therefore provide an estimated value that will always be close to what would have been obtained had the responses been observed exactly, i.e., without rounding.

Let us next consider more formally the performance of the estimator $\hat{\Delta}$ in the shift model (2.7) when F is continuous. It is convenient first to note that in this case the distribution of $\hat{\Delta}$ is continuous (Prob. 58), so that

$$P(\hat{\Delta} = d) = 0 \qquad \text{for any given number } d$$

This may seem surprising since $\hat{\Delta}$ is derived from the statistic $W_{X,Y}$ whose distribution is not continuous, but it is important to keep this fact in mind in some of the considerations below.

Another useful property is described in the following lemma.

Lemma 1. The distribution of the difference $\hat{\Delta} - \Delta$, the error of the estimator, is independent of Δ.

Proof. The difference $\hat{\Delta} - \Delta = \text{med}(Y_j - X_i) - \Delta$ is equal to $\text{med}(Y_j - \Delta - X_i)$. Since the distribution of the variables X_i and $Y_j - \Delta$ is independent of Δ, the same is true of the distribution of $\hat{\Delta} - \Delta$, as was to be proved.

Two aspects of the performance of an estimator are of particular interest: how well it is centered on the true value Δ and how widely it is dispersed about this value. An estimator of Δ is well centered if the center of its distribution (suitably defined) is equal to Δ, and hence in particular whenever this distribution is symmetric about Δ. The following theorem gives two conditions under which this is the case (the proof will be given at the end of this section).

Theorem 3. The estimator $\hat{\Delta} = \text{med}(Y_j - X_i)$ of the shift parameter Δ in (2.7) is distributed symmetrically about Δ if either of the following two conditions hold:

(i) The distribution F is symmetric about some point μ.

(ii) The two sample sizes are equal, that is, $m = n$.

This theorem shows in particular that under the stated conditions, the estimator $\hat{\Delta}$ is unbiased, i.e., it satisfies

$$E(\hat{\Delta}) = \Delta \tag{2.47}$$

It also follows that Δ is the median of the distribution of $\hat{\Delta}$ (an estimator with this property is called *median unbiased*) and that in fact

$$P(\hat{\Delta} < \Delta) = P(\hat{\Delta} > \Delta) \tag{2.48}$$

so that $\hat{\Delta}$ is as likely to overestimate Δ as to underestimate it. Since $P(\hat{\Delta} = \Delta) = 0$, the common probability (2.48) is equal to $1/2$.

When the conditions of the theorem are not satisfied, $\hat{\Delta}$ will typically no longer be symmetrically distributed about Δ, nor even be unbiased. However, (2.48) continues to hold when mn is odd, and to hold approximately when mn is even. This and other properties of $\hat{\Delta}$ are consequences of the following theorem, which provides a basis for relating the distribution of $\hat{\Delta}$ to the power of the Wilcoxon test.

Theorem 4. Suppose that the differences $Y_j - X_i$ are distinct. If $D_{(1)} < \cdots < D_{(mn)}$ denote the ordered differences $Y_j - X_i$, then for any integer l between 1 and mn and any real number a

(2.49) $$D_{(l)} \leqslant a \quad \text{if and only if} \quad W_{X,Y-a} \leqslant mn - l$$

and hence

(2.50) $$D_{(l)} > a \quad \text{if and only if} \quad W_{X,Y-a} \geqslant mn - l + 1$$

Proof. The first inequality in (2.49) holds if and only if at least l of the differences $(Y_j - a) - X_i$ are less than or equal to zero and hence if at most $mn - l$ of these differences are greater than zero; that is, if $W_{X,Y-a}$ does not exceed $mn - l$.

Consider now Eq. (2.48) and suppose first that mn is odd, $mn = 2k+1$ say, so that $\hat{\Delta} = D_{(k+1)}$. Then

$$P_\Delta(\hat{\Delta} < \Delta) = P_\Delta(\hat{\Delta} \leqslant \Delta) = P_0(\hat{\Delta} \leqslant 0) = P_0(D_{(k+1)} \leqslant 0) = P_0(W_{X,Y} \leqslant k) = \tfrac{1}{2}$$

Here the second equality follows from Lemma 1, the fourth from (2.49), and the last from the fact that the null distribution of $W_{X,Y}$ is symmetric about $k + \frac{1}{2}$. Similarly,

$$P_\Delta(\hat{\Delta} > \Delta) = P_0(W_{X,Y} \geqslant k+1) = \tfrac{1}{2}$$

which proves (2.48) when mn is odd.

Suppose next that mn is even, $mn = 2k$ say, so that $\hat{\Delta} = \frac{1}{2}[D_{(k)} + D_{(k+1)}]$. Then

$$P_0[D_{(k)} \leqslant 0] = P_0(W_{X,Y} \leqslant k) \quad \text{and} \quad P_0[D_{(k+1)} \leqslant 0] = P_0(W_{X,Y} < k)$$

and hence

(2.51) $$P_0(W_{X,Y} < k) \leqslant P_\Delta(\hat{\Delta} \leqslant \Delta) \leqslant P_0(W_{X,Y} \leqslant k)$$

Since the null distribution of $W_{X,Y}$ is symmetric about k, the first probability in (2.51) is less than or equal to $\frac{1}{2}$ and the third one is greater than or equal to $\frac{1}{2}$. Typically, their difference is small and tends to zero as mn tends to infinity, so that $P_\Delta(\hat{\Delta} < \Delta) = P_\Delta(\hat{\Delta} \leqslant \Delta)$ is close to $\frac{1}{2}$ and hence close to $P_\Delta(\hat{\Delta} > \Delta)$. When $m = 8$ and $n = 9$, for example, it is seen from (2.51) and Table B that

$$.48 \leqslant P_\Delta(\hat{\Delta} < \Delta) \leqslant .52$$

Having discussed the centering of $\hat{\Delta}$, let us now turn to its dispersion about Δ. The dispersion of a random variable is traditionally measured in terms of its variance. This is particularly easy to compute for many classical estimators but less convenient for the type of estimator with which we are concerned here. Instead we shall use as measures of dispersion the probabilities $P(|\hat{\Delta}-\Delta| \leqslant a)$ that $\hat{\Delta}$ differs from the true value Δ by less than a given amount a, which is more convenient and provides a more direct interpretation. As is seen from Theorem 4, these probabilities are closely related to the power of the Wilcoxon test, and are therefore also difficult to compute exactly. However, (2.23) furnishes the basis for an approximation when m and n are not too small, which will be derived at the end of this section:

$$(2.52) \qquad P(|\hat{\Delta}-\Delta| \leqslant a) \approx \Phi\left[\frac{mn(\frac{1}{2}-p_1)}{\sqrt{\operatorname{Var}(W_{X,Y-a})}}\right] + \Phi\left[\frac{mn(\frac{1}{2}-p_1)}{\sqrt{\operatorname{Var}(W_{X,Y+a})}}\right] - 1$$

where $p_1 = P(X < Y-a)$ and where $\operatorname{Var}(W_{X,Y-a})$ is given by the right-hand side of (2.21) with

$$p_2 = P(X < Y-a \text{ and } X < Y'-a)$$

and

$$p_3 = P(X < Y-a \text{ and } X' < Y-a)$$

where X, X', Y, and Y' are independent random variables distributed according to F.

When either $m = n$ or F is symmetric, $\operatorname{Var}(W_{X,Y-a}) = \operatorname{Var}(W_{X,Y+a})$ (Prob. 61) so that (2.52) can be simplified by combining the first two terms. The approximation (2.52) can be further simplified for small a by applying the argument that led to (2.29). This suggests the approximation

$$p_1 - \tfrac{1}{2} \approx -af^*(0)$$

and the replacement of $\operatorname{Var}(W_{X,Y-a})$ and $\operatorname{Var}(W_{X,Y+a})$ by the null variance $mn(N+1)/12$. With these additional approximations (2.52) becomes

$$(2.53) \qquad P(|\hat{\Delta}-\Delta| \leqslant a) \approx 2\Phi\left[\sqrt{\frac{12mn}{N+1}}f^*(0)a\right] - 1$$

When F is a normal distribution with variance σ^2, it was seen earlier that $f^*(0) = 1/(2\sigma\sqrt{\pi})$, and (2.53) then reduces to

$$(2.54) \qquad P(|\hat{\Delta}-\Delta| \leqslant a) \approx 2\Phi\left[\sqrt{\frac{3mn}{(N+1)\pi}}\left(\frac{a}{\sigma}\right)\right] - 1$$

To illustrate the approximations (2.52) and (2.54), suppose that F is normal so that p_1 and $p_2 = p_3$ are given by (2.26) and (2.27) with $\Delta = -a$, and that $\sigma^2 = 2$. Then Table 2.5 gives p_1, p_2, and the approximations (2.52) and (2.54) for $m = n = 15$ and for a number of values of a.

The approximations (2.52) and (2.54) can also be used to determine the sample size required for $\hat{\Delta}$ to have a preassigned probability of falling within a stated distance a of the true value Δ. In particular, putting $m = n$, it is easy to solve (2.54) for an approximation to the unknown sample size n.

TABLE 2.5 $P(|\hat{\Delta}-\Delta| \leqslant a)$ according to (2.52) and (2.54); F = Normal. $\sigma^2 = 2$, $m = n = 15$

a	.2	.4	.6	.8	1.0	1.2
p_1	.4602	.4207	.3821	.3446	.3085	.2743
p_2	.2944	.2577	.2235	.1920	.1633	.1376
(2.52)	.29	.55	.75	.88	.95	.98
(2.54)	.29	.54	.74	.86	.94	.97

Even though (2.52) was derived only as an approximation, it is also true that the difference between the right- and left-hand sides of (2.52) tends to zero as m and n tend to infinity. This is not difficult to show using the present Theorem 3, and Theorem 3 and Example 20 of the Appendix. An evaluation of the right-hand side of (2.52) shows further that it tends to 1 and hence that

$$P(|\hat{\Delta}-\Delta| \leqslant a) \rightarrow 1 \qquad \text{as } m \text{ and } n \rightarrow \infty \tag{2.55}$$

for any positive a. A sequence of estimators of a parameter Δ, which differs from the true value Δ by less than any given amount a with probability tending to 1, is said to be *consistent* for estimating Δ. (In probabilistic terms, consistency states that the sequence of estimators tends to the true value Δ in probability.) The limit relation (2.55) thus shows that $\hat{\Delta}$, which should really be denoted by $\hat{\Delta}_{m,n}$ to indicate the dependence on the sample sizes, is consistent for estimating Δ as m and n tend to infinity.

Earlier, we compared the effect of gross errors on $\hat{\Delta}$ with the effect they have on the classical estimator $\bar{\Delta} = \bar{Y}-\bar{X}$. Let us now make a more formal comparison in terms of the centering and dispersion properties of the two estimators. It is easily seen (Probs. 56 and 57) that the distribution of $\bar{\Delta}$ is also symmetric about Δ under the conditions of Theorem 3. When these conditions do not hold, the distribution of $\bar{\Delta}$ is typically no longer symmetric, but, instead of continuing to be median unbiased, as was essentially the case for $\hat{\Delta}$, $\bar{\Delta}$ continues to be unbiased in the sense of satisfying $E(\bar{\Delta}) = \Delta$. Of these two unbiasedness properties, neither appears more compelling than the other, and it seems more important to ask which of the two estimators tends to be closer to the true value.

It follows from the relationship between the probabilities $P(|\hat{\Delta}-\Delta| \leqslant a)$ and the power of the Wilcoxon test [for example, from (2.37) with $u_\alpha = 0$ and (2.52)] and the corresponding relationship between $P(|\bar{\Delta}-\Delta| \leqslant a)$ and the asymptotic power of the t-test, that the efficiency results for the Wilcoxon to the t-test essentially carry over to the efficiency of $\hat{\Delta}$ relative to $\bar{\Delta}$. Here the latter efficiency is defined as the ratio of sample sizes required for the estimators to have the same probability of falling within a stated distance of Δ. The conclusions concerning the power of the two tests summarized in the preceding section therefore also apply to the accuracy of the two estimators and supplement the comparison made in connection with Example 4. It should be noted, however, that there is no corresponding carry-over of the remarks concerning the significance levels of the two tests. In particular, although the estimator $\hat{\Delta}$ was derived from a nonparametric test, it cannot be called nonparametric because this term applies to the significance level of a test, a concept that has no parallel in the problem of point estimation.

To conclude the section, we shall now (A) prove Theorem 3 and (B) derive the approximation (2.52).

(A) Proof of Theorem 3. Consider the probabilities $P_\Delta(\hat{\Delta}-\Delta < a)$, where the subscript Δ indicates that the probability is being computed under the assumption that the shift is Δ. By Lemma 1 we can assume, without loss of generality, that Δ is zero. To simplify the writing, let us introduce the notation X for $(X_1, \ldots, X_m)$ and $aX+b$ for $(aX_1+b, \ldots, aX_m+b)$, and similarly for Y. Then it is easily checked (Prob. 57) that $\hat{\Delta} = \hat{\Delta}(X, Y)$ satisfies the equation

$$\hat{\Delta}(aX+b, aY+b) = a\hat{\Delta}(X, Y) \tag{2.56}$$

(i) By applying (2.56) with $a = 1$, $b = \mu$, it is seen that in proving the first part of Theorem 3 we can assume without loss of generality that $\mu = 0$. Since F is then symmetric about zero, it follows that $(-X, -Y)$ has the same joint distribution as (X, Y), and hence that $\hat{\Delta}(X, Y)$ has the same distribution as $\hat{\Delta}(-X, -Y) = -\hat{\Delta}(X, Y)$. Thus, $\hat{\Delta}$ and $-\hat{\Delta}$ have the same distribution, and $\hat{\Delta}$ is distributed symmetrically about zero, as was to be proved.

(ii) If $m = n$ and $\Delta = 0$, then (X, Y) and (Y, X) have the same joint distribution. Hence $\hat{\Delta}(X, Y)$ has the same distribution as $\hat{\Delta}(Y, X) = -\hat{\Delta}(X, Y)$, and this completes the proof as before.

(B) Derivation of the Approximation (2.52). Consider first the case $mn = 2k+1$. Then (2.49) with $l = k+1$ and (2.23) lead to the approximation

$$P(\hat{\Delta}-\Delta \leqslant a) \approx \Phi\left[\frac{k+\frac{1}{2}-mnp_1}{\sqrt{\operatorname{Var}(W_{X,Y-a})}}\right] = \Phi\left[\frac{mn(\frac{1}{2}-p_1)}{\sqrt{\operatorname{Var}(W_{X,Y-a})}}\right]$$

Replacing a by $-a$ in this approximation, we get further

$$P(\hat{\Delta}-\Delta < -a) = P(\hat{\Delta}-\Delta \leqslant -a) \approx \Phi\left[\frac{mn(\frac{1}{2}-p_1')}{\sqrt{\operatorname{Var}(W_{X,Y+a})}}\right]$$

with $p_1' = P(X < Y+a)$. It is not difficult to see (Prob. 61) that $p_1' = 1-p_1$ and hence to obtain (2.52).

In the case $mn = 2k$, the approximation is obtained by noting that in generalization of (2.51), $P_\Delta(\hat{\Delta}-\Delta < a)$ is bounded below and above by

(2.57) $P_0(D_{(k+1)} < a) = P_0(W_{X,Y-a} < k)$ and $P_0(D_{(k)} \leqslant a) = P_0(W_{X,Y-a} \leqslant k)$

Application of (2.23) leads to

$$P_0(W_{X,Y-a} < k) \approx \Phi\left[\frac{k-\frac{1}{2}-mnp_1}{\sqrt{\operatorname{Var}(W_{X,Y-a})}}\right] \quad \text{and} \quad P_0(W_{X,Y-a} \leqslant k) \approx \Phi\left[\frac{k+\frac{1}{2}-mnp_1}{\sqrt{\operatorname{Var}(W_{X,Y-a})}}\right]$$

which suggests the approximation

$$P(\hat{\Delta}-\Delta < a) \approx \Phi\left[\frac{k-mnp_1}{\sqrt{\operatorname{Var}(W_{X,Y-a})}}\right] = \Phi\left[\frac{mn(\frac{1}{2}-p_1)}{\sqrt{\operatorname{Var}(W_{X,Y-a})}}\right]$$

and hence for $P(|\hat{\Delta}-\Delta| \leqslant a)$ again the approximation (2.52).

6. CONFIDENCE PROCEDURES

A point estimate, such as the estimate $\hat{\Delta}$ discussed in the preceding section, provides one type of solution to the problem of estimating the treatment effect Δ. It gives a single value, which hopefully will be close to Δ. A different formulation of the estimation problem does not ask for a definite value for Δ; it is willing to settle for a less specific statement regarding the position of Δ but one which can be made with a prescribed degree of certainty or confidence. At first glance, it might be thought that (2.52) provides such a statement since it gives (at least approximately) the probability that Δ lies between $\hat{\Delta}-a$ and $\hat{\Delta}+a$. However, this probability while independent of Δ does depend [through p_1 and $\operatorname{Var}(W_{X,Y-a})$] on the unknown distribution F of the variables X_i and $Y_j-\Delta$. What is required instead are limits $\underline{\Delta}$ and $\bar{\Delta}$ such that Δ lies between $\underline{\Delta}$ and $\bar{\Delta}$ with specified probability independent of both Δ and F.

Such *distribution-free confidence intervals* $(\underline{\Delta}, \bar{\Delta})$ for Δ are easily obtained from Theorem 4. It follows from this theorem that for $i = 0, 1, \ldots, mn$

(2.58) $$D_{(i)} \leqslant \Delta < D_{(i+1)} \quad \text{if and only if} \quad W_{X,Y-\Delta} = mn-i$$

where

$$D_{(0)} = -\infty \quad \text{and} \quad D_{(mn+1)} = \infty$$

Since the distribution of $W_{X,Y-\Delta}$, computed under the assumption that Δ is the true value of the shift parameter, is independent of Δ, we find that for any value of Δ

(2.59) $$P_\Delta(D_{(i)} \leqslant \Delta < D_{(i+1)}) = P_0(W_{X,Y} = mn-i) = P_0(W_{X,Y} = i)$$

for all $i = 0, 1, \ldots, mn$, where the subscript on P indicates the value of Δ for which the probability is being computed.

Thus the order statistics $D_{(1)}, \ldots, D_{(mn)}$ divide the real line into $mn+1$ random intervals, which contain Δ with known probabilities, independent of Δ and F and given exactly by the Wilcoxon null distribution. The mn values $D_{(1)}, \ldots, D_{(mn)}$, together with the probabilities on the right-hand side of (2.59), will typically provide more information about Δ than can easily be assimilated. It may then be preferable to summarize some of the principal features of this information by giving the D's corresponding to a number of fixed probability points such as, for example, 1, 5, 10, 25, 50, 75, 90, 95, and 99 percent; these may be said to constitute a set of *standard confidence points.*[1]

EXAMPLE 5. *Augmenters and reducers.* Petrie has developed an interesting classification of persons into augmenters (who perceive an exaggerated impression of sensory stimuli), reducers (whose perception tends to diminish such stimuli), and an intermediate category of moderates. As a check on his classification, Petrie tested the reactions of a number of subjects, once with audio analgesia (an acoustic treatment which tends to increase the tolerance to pain) and once without. According to his theory, the reduction in response due to analgesia should be more marked for the augmenters, so that the differences (without-with) should tend to be larger for augmenters than for reducers. Let us assume that the distribution of these differences for the population of augmenters is shifted to the right by an amount Δ from the corresponding distribution for the reducers, and compute a set of confidence points for Δ from the following sample of $n = 10$ augmenters (Y's) and $m = 7$ reducers (X's).[2]

Augmenters:	17.9	13.3	10.6	7.6	5.7	5.6	5.4	3.3	3.1	.9
Reducers:	7.7	5.0	1.7	0.0	−3.0	−3.1	−10.5			

It is seen from (2.50) or (2.59) that

$$P_\Delta(\Delta \leqslant D_{(i)}) = P_0(W_{XY} \leqslant i-1) \tag{2.60}$$

The values of i for which

$$\gamma = P_\Delta(\Delta \leqslant D_{(i)}) \tag{2.61}$$

is closest to the values 1,...,99 percent when $n = 10$, $m = 7$ are seen from Table B to be as shown in Table 2.6.

To compute the order statistics $D_{(12)}, D_{(19)}, \ldots, D_{(59)}$, let us exhibit the differences $D_{ij} = Y_j - X_i$ for all $mn = 70$ combinations of i and j and ignore the fact that a few of these differences are tied.

	17.9	13.3	10.6	7.6	5.7	5.6	5.4	3.3	3.1	.9
7.7	10.2	5.6	2.9	−.1	−2.0	−2.1	−2.3	−4.4	−4.6	−6.8
5.0	12.9	8.3	5.6	2.6	.7	.6	.4	−1.7	−1.9	−4.1
1.7	16.2	11.6	8.9	5.9	4.0	3.9	3.7	1.6	1.4	−.8
0.0	17.9	13.3	10.6	7.6	5.7	5.6	5.4	3.3	3.1	.9
−3.0	20.9	16.3	13.6	10.6	8.7	8.6	8.4	6.3	6.1	3.9
−3.1	21.0	16.4	13.7	10.7	8.8	8.7	8.5	6.4	6.2	4.0
−10.5	28.4	23.8	21.1	18.1	16.2	16.1	15.9	13.8	13.6	11.4

[1] The use of such standard points was proposed by Tukey (1949) and is also discussed by Cox (1958) and Birnbaum (1961).

[2] Data from Petrie, *Individuality in Pain and Suffering*, University of Chicago Press, Chicago, 1967. The original data are given to two digits and have $n = 11$. One of the 11 treatment differences has been deleted (at random) to bring the data within the compass of Table B.

There are 11 differences less than 0, 4 differences between 0 and 1, 4 between 1 and 3, and so on, and the desired confidence points are now easily obtained as

$$\begin{array}{lllll} D_{(12)} = .4 & D_{(19)} = 2.9 & D_{(22)} = 3.7 & D_{(28)} = 5.6 & D_{(35)} = 6.3 \\ D_{(36)} = 6.4 & D_{(43)} = 8.7 & D_{(49)} = 10.7 & D_{(52)} = 12.9 & D_{(59)} = 16.1 \end{array}$$

For graphical procedures to obtain the order statistics $D_{(i)}$, see Moses (1965) or Noether (1971, page 140).

In the preceding discussion, the confidence points were obtained through (2.60) from Table B. If m and n are too large for Table B, the normal approximation of Chap. 1, Sec. 3, can be used instead. Applying this approximation with continuity correction to the right-hand side of (2.61), we find from (2.60) the approximation

$$\gamma \approx \Phi\left[\frac{i-\frac{1}{2}(mn+1)}{\sqrt{mn(N+1)/12}}\right] \tag{2.62}$$

Applying (2.62) to the case $n = 10$, $m = 7$, one finds for $\gamma = .01$, .05, and .1, respectively, the values $i = 12$, 18, and 22 in good agreement with Table 2.6.

TABLE 2.6 Values of i satisfying (2.61) for $n = 10$, $m = 7$

γ	.009	.054	.097	.237	.481	.519	.763	.903	.946	.991
i	12	19	22	28	35	36	43	49	52	59

Let us now return to the problem of obtaining confidence intervals $(\underline{\Delta}, \bar{\Delta})$ that have a prescribed probability of containing Δ. Suppose we would like this probability, the so-called *confidence coefficient*, to be equal to .9. By (2.59), it is then only necessary to combine sufficiently many of the intervals $(D_{(i)}, D_{(i+1)})$, so that their total probability is equal to .9. If these intervals are

$$(D_{(i)}, D_{(i+1)}), (D_{(i+1)}, D_{(i+2)}), \ldots, (D_{(j-1)}, D_{(j)})$$

it is perhaps most natural to choose them symmetrically, i.e., in such a way that the probabilities of the interval $(D_{(i)}, D_{(j)})$ falling entirely to the left or to the right of Δ are equal, or

$$P(\Delta < D_{(i)}) = P(D_{(j)} < \Delta) \tag{2.63}$$

Each of these probabilities should then be equal to half of $1-.9$ and hence equal to .05. It is seen from Table B that the closest we can come to this for $n = 10$, $m = 7$ is to take $i = 19$ and $j = 52$. Then

$$P(\Delta < D_{(19)}) = P(D_{(52)} < \Delta) = .054$$

and hence

$$D_{(19)} \leqslant \Delta \leqslant D_{(52)}$$

with probability $1-.108 = .892$. In Example 5, this interval extends from 2.9 to

12.9. Analogously, the interval $(D_{(12)}, D_{(59)})$ is a confidence interval for Δ with confidence coefficient $1-.108 = .892$. Quite generally, it follows from (2.60) and the symmetry of the Wilcoxon distribution that (2.63) will hold when $j = mn+1-i$.

Instead of an interval for Δ, one may wish to find an upper bound $\bar{\Delta}$, which will be reasonably sure (with specified probability independent of F) not to fall below Δ. Such a bound is provided by (2.60) and (2.61). If we again take as illustration $n = 10$ and $m = 7$, Table 2.6 shows that the probability is .903 of $D_{(49)}$ not falling below Δ, so that in this case $D_{(49)}$ constitutes an *upper confidence bound* for Δ with confidence coefficient .903. Analogous lower confidence bounds $\underline{\Delta}$ can be found by noting that (2.60) implies

$$P_\Delta(D_{(i)} \leqslant \Delta) = P_0(W_{XY} \geqslant i) = P_0(W_{XY} \leqslant mn-i)$$

Thus, if $n = 10$, $m = 7$, a lower confidence bound for Δ with confidence coefficient .991 is provided by $D_{(12)}$.

If one is interested in computing not a whole set of confidence points but only a confidence bound or confidence interval, it will typically be more convenient to use the shortcut method described in Example 4 rather than to obtain all the differences D_{ij}. For instance, to find the lower confidence bound $D_{(12)}$ in Example 5, one might start by determining a diagonal that has about 12 entries to the right and above it. For example, the diagonal connecting the cell in the first row and sixth column with that in the fifth row and last column has five entries directly on it and ten entries to the right and above it, so that one would expect $D_{(12)}$ to be close to this diagonal and can proceed as in Example 4. In fact, it is seen from the completed rectangle that the value $D_{(12)} = .4$ happens to lie on this diagonal. This can, of course, not be expected in general.

So far in this section we have assumed that there are no tied observations. When there are ties, the order statistics $D_{(1)}, D_{(2)}, \ldots$ can be computed as before; however, the associated confidence statements such as (2.60) will no longer be valid. To obtain the necessary modifications, suppose as in the preceding section that the ties are the result of rounding to the nearest multiple of ε, and that the unobserved control and treatment responses X_i' and Y_j' before rounding are continuous random variables for which a shift model is appropriate. If the ordered differences $Y_j' - X_i'$ are denoted by $D'_{(1)} < \cdots < D'_{(mn)}$, relations (2.60) and (2.61) hold with $D_{(i)}$ replaced by $D'_{(i)}$. On the other hand, by (2.46)

$$D'_{(i)} - \varepsilon \leqslant D_{(i)} \leqslant D'_{(i)} + \varepsilon$$

Suppose now that i is determined from (2.60) or approximately from (2.62) so that $D'_{(i)}$ is an upper confidence bound for Δ with confidence coefficient γ. It then follows from the last displayed equation that $D_{(i)} + \varepsilon$ is an upper confidence bound for Δ with confidence coefficient $\geqslant \gamma$. Similarly, if $D'_{(i)}$ is a lower confidence bound for Δ with confidence coefficient γ, then $D_{(i)} - \varepsilon$ is a lower confidence bound with confidence coefficient $\geqslant \gamma$. Finally, if $(D'_{(j)}, D'_{(k)})$ is a confidence interval for Δ with confidence coefficient γ, the same argument shows that

$$(D_{(j)} - \varepsilon, D_{(k)} + \varepsilon)$$

is a confidence interval with confidence coefficient $\geqslant \gamma$.

7. FURTHER DEVELOPMENTS

7A. The Behrens-Fisher Problem

The shift model (2.7) may be an oversimplification because the response distributions F and G under the two treatments will often differ not only in location but also in scale and possibly shape. Two hypotheses may then be of interest in the comparison of two treatments. One may still be concerned with the hypothesis

(2.64) $$H_1\colon \quad G = F$$

that there is no difference between the effects of the two treatments. Of greater interest, typically, is the hypothesis that the location of the two distributions is the same but that they need not necessarily agree otherwise. To make this more precise, suppose that F and G do not differ in shape, so that they can be represented as

(2.65) $$F(x) = E\left(\frac{x-\xi}{\sigma}\right) \qquad G(y) = E\left(\frac{y-\eta}{\tau}\right)$$

for some distribution E. The second problem then reduces to that of testing the hypothesis

(2.66) $$H_2\colon \quad \eta = \xi$$

Both when E is arbitrary or when it is given to be normal, this problem is known as the *Behrens-Fisher problem*. The hypotheses H_1 and H_2 differ principally in the decision that is desirable when $\eta = \xi$ but $\tau \neq \sigma$. In the first problem, one would then wish to reject H_1; in the second problem, one would still wish to accept H_2.

For testing H_1, no question arises regarding the significance level of a rank test such as the Wilcoxon test. However, it is no longer clear that the Wilcoxon test will have good power against the more general alternative that F and G differ not only in location but also in scale. A test of H_1 against this more general alternative is given by Lepage (1971, 1973). On the other hand, with respect to H_2 it is important to realize that in spite of the description of the Wilcoxon test as nonparametric or distribution-free, its significance level under H_2 will depend on σ, τ, and E. The important question as to how sensitive the Wilcoxon and Student's t-test are to differences in scale (and possibly shape) is investigated by Wetherill (1960), Van der Vaart (1961), and Pratt (1964).

Since the Wilcoxon test is neither exactly nor even asymptotically distribution-free for testing H_2, one may try to modify it so as to achieve at least this latter property. If E is symmetric, the expectation of the normal limit distribution of W_{XY}/mn under H_2 is $\frac{1}{2}$, and the variance depends on σ, τ, and E. A test with the desired property can then be obtained by estimating this variance. Such a test was proposed by Zaremba (1962) as a test of the more general hypothesis $H: P(X < Y) = \frac{1}{2}$.

Another assumption of the model (2.7) is the independence of the X's and Y's. Some aspects of the performance of the Wilcoxon test when this assumption is violated are investigated by Serfling (1968) and Hollander, Pledger, and Lin (1974).

7B. The Normal Scores Test

An alternative to the Wilcoxon test is provided by the Normal Scores test, which is the rank analogue of Student's t-test. Let the combined sample $X_1, \ldots, X_m$; $Y_1, \ldots, Y_n$ be denoted by $Z_1, \ldots, Z_N$ and suppose for a moment that the Z's are independently normally distributed, say, with zero mean and unit variance. Imagine now that only the ranks of the Z's are preserved, the original observations having been lost, and let us try and reconstruct as well as possible the original observations from their ranks. Denote the ordered Z's by $Z_{(1)} < \cdots < Z_{(N)}$ and consider, for example, the random variable $Z_{(s)}$. It has rank s and hence is known to be the sth smallest of a sample of size N from a standard normal distribution. What value do we expect such a random variable to take on? A natural estimate is

$$a_N(s) = E_\Phi(Z_{(s)}) \tag{2.67}$$

the expectation of $Z_{(s)}$ when the Z's are a sample from the standard normal distribution Φ. The expectations (2.67) are known as *Normal Scores*. Their use was first proposed by Fisher and Yates (1938) to replace the original observations in the standard normal theory tests. Extensive tables of the Normal Scores are provided by David et al. (1968) and Harter (1961, 1969). A table for $N \leqslant 50$ is given, for example, by Owen (1962).

Let the reconstructed observations be substituted into the t-statistic to obtain a test resembling the t-statistic but depending only on the ranks $R_1, \ldots, R_m$ of the X's and $S_1, \ldots, S_n$ of the Y's. The resulting test, after some simplification, rejects when

$$T_s \geqslant c \tag{2.68}$$

where

$$T_s = a_N(S_1) + \cdots + a_N(S_n) \tag{2.69}$$

The null distribution of T_s is determined by (2.3) and some critical values are given by Klotz (1964). For large N, the distribution of T_s is approximately normal. The accuracy of the normal approximation and an alternative t-approximation are studied by Klotz (1964). A discussion of the asymptotic normality of T_s for general scores

$a_N(s)$, and a review of the literature on this problem, are given by Dupac and Hajek (1969).

The asymptotic efficiency $e_{N,t}(F)$ of the Normal Scores test relative to Student's t-test in the shift model (2.7) has the remarkable property of being greater than or equal to 1 for all F, and of being equal to 1 when F is normal. Asymptotically, and with the comparison restricted to shift models, the Normal Scores test is thus always at least as powerful as the t-test. This result was first proved by Chernoff and Savage (1958); a simpler proof was later found by Gastwirth and Wolff (1968).

The asymptotic efficiency of the Normal Scores to the Wilcoxon test is discussed by Hodges and Lehmann (1960). This comparison suggests that there is little difference in the power of these two tests when F is close to normal, that the Normal Scores test is preferable when the distribution breaks off abruptly (as does the exponential distribution at zero and the uniform distribution at its two endpoints), but that the Wilcoxon test will tend to be more powerful for distributions with heavy tails such as a normal distribution contaminated by a substantial proportion of gross errors.

A test that is asymptotically equivalent to the Normal Scores test is obtained by replacing the scores (2.67) in T_s by

$$a_N(s) = \Phi^{-1}\left(\frac{s}{N+1}\right) \tag{2.70}$$

where $\Phi^{-1}(z)$ denotes that number t for which $\Phi(t) = z$. The test based on these scores, introduced by Van der Waerden (1952, 1953), typically differs little from the Normal Scores test. Its use is discussed in detail in the books by Van der Waerden and Nievergelt (1956) and Hajek (1969), both of which include the necessary tables.

7C. Increasing the Number of Levels to Improve Sensitivity

An advantage of the Normal Scores test over the Wilcoxon test is the greater number of values taken on by the Normal Scores statistic. This not only makes it easier to get close to any desired level, but it also tends to improve the significance probability. To see how an increase in levels improves the significance of a sample, suppose that S and T are two statistics such that S breaks up the sample more finely than T by taking on a larger number of distinct values. To be specific, suppose that S takes on the values $1, \ldots, 10$ with probability $1/10$ each (under the hypothesis), while T takes on only three values, namely, $T = 0$ when $S = 1, 2, 3$; $T = 1$ when $S = 4, 5, 6$; and $T = 2$ when $S = 7, 8, 9, 10$. Let large values of S and T be significant and suppose that $S = 9$ and hence $T = 2$. Then with S the significance probability is $P(S \geq 9) = .2$, but with T it is $P(T \geq 2) = .4$. Similarly, it is seen that the significance probability with T is always at least as large as it is with S and that it is actually larger in most cases (Prob. 66). Thus, by assigning an unnecessarily high value to the significance probability, the statistic T will tend to underestimate the significance of the data (recall that small significance probabilities are an indication of high significance, and conversely). With a rank statistic T, this phenomenon will occur whenever T has relatively few values since it must then take on the same value for many different sets of ranks.

Instead of turning to the Normal Scores test one can also increase the number of levels of the Wilcoxon statistic $W_s = \Sigma S_i$ through a supplementary criterion such as ΣS_i^2 or ΣiS_i. The idea is simple. Determining the significance of a set of observations involves an ordering of all possible sets of observations (with equality permitted). The Wilcoxon test orders the sample points according to the value of W_s. If ΣS_i^2 is used as supplementary criterion, two points with the same value of W_s are ordered according to the value of ΣS_i^2. Suppose, for example, that $m = 3$, $n = 2$, and $W_s = 7$. If large values are significant, the significance probability of the Wilcoxon statistic is $P(W_s \geqslant 7) = P(W_s = 7, 8, \text{or } 9) = 4/10$. With $W_s = 7$, the treatment ranks can be $(2, 5)$ or $(3, 4)$. If $S_1{}^2 + S_2{}^2$ is used as a supplementary criterion, the significance probability of $(2, 5)$ is

$$P(W_s > 7) + P(W_s = 7 \text{ and } S_1{}^2 + S_2{}^2 \geqslant 29) = \tfrac{3}{10}$$

while the significance probability of $(3, 4)$ is

$$P(W_s > 7) + P(W_s = 7 \text{ and } S_1{}^2 + S_2{}^2 \geqslant 25) = \tfrac{4}{10}$$

This procedure is proposed in an unpublished report by Tukey (1962); a different approach to the problem is discussed by Lancaster (1961).

7D. Small-sample Power

The computation of the exact distribution of rank statistics under alternatives to the hypothesis $F = G$ tends to be difficult, and only few such computations have been carried out. Milton (1970) provides the probabilities of all $\binom{m+n}{n}$ possible rankings of the X's and Y's for $1 \leqslant n \leqslant m \leqslant 7$ for normal shift alternatives and uses these results to obtain exact values of the power of the Wilcoxon and Smirnov tests for these sample sizes. Bell, Moser, and Thompson (1966) give the power of the Wilcoxon test for rectangular shift alternatives and Haynam and Govindarajulu (1966) do so for exponential shift alternatives, for certain combinations of m and n. Some power values for the Normal Scores test against normal alternatives are given by Klotz (1964). For double exponential shift alternatives, sample sizes $m = n = 5$, and $\alpha = .05$, Ramsey (1971) obtains the power of a number of tests including the Wilcoxon, Normal Scores, and Smirnov tests.

The difficulty of these computations for shift alternatives is related to the fact that the probability of different rankings under an alternative (F, G) depends only on

$G(F^{-1})$, which is typically rather complicated for shift alternatives. This difficulty disappears, however, if one considers alternatives for which $G(F^{-1})$ is a simple function. In particular, for the case that $G[F^{-1}(u)] = u^k$ so that

$$G(x) = [F(x)]^k \tag{2.71}$$

an explicit expression of the probability of any ranking is obtained in Lehmann (1953).

In view of the difficulty of evaluating the probabilities of different rankings for most alternatives of interest, one may wish to know what order relationships exist among such probabilities. This problem has been investigated for a number of classes of alternatives by Savage (1956) and by Savage, Sobel, and Woodworth (1966).

7E. Large-sample Power and Efficiency

The normal approximation to the power of the Wilcoxon rank-sum test can be improved through an Edgeworth expansion. The necessary third and fourth moments of W_{XY} are provided by Sundrum (1954), and the corresponding terms of the expansion are given by Witting (1960). For the mathematical justification of the expansion, see Bickel (1974).

At a fixed alternative of interest, say, a shift model with shift parameter Δ, the power tends to 1 as the sample sizes m and n tend to infinity. A different approximation to the power is obtained by considering a fixed value Π of the power instead of fixing the alternative. By the preceding remark, the alternative Δ at which this power is attained will tend to zero as m and n tend to infinity. The Edgeworth approximation to the power of such a sequence of alternatives is discussed by Witting (1960), as is the associated Pitman efficiency of the Wilcoxon test relative to the Normal test (2.38) with σ in place of S.

A different approach to efficiency was suggested by Cochran (1952). To understand the difference, recall that the Pitman efficiency is defined relative to a fixed value α of the significance level and a fixed value Π of the power. Consider a shift model with shift parameter Δ. As was pointed out above, the value of Δ at which a fixed power Π is achieved tends to zero as the sample sizes tend to infinity. Instead of fixing α and Π and thereby forcing Δ to tend to zero, one may fix Δ and Π, which forces α to tend to zero. The resulting efficiencies in many cases turn out to coincide with a type of efficiency developed from a different point of view by Bahadur (1967). They have been worked out for the Wilcoxon, Normal Scores, and t-tests by Hoadley (1967), Stone (1967) and (1968), and Woodworth (1970). A systematic exposition of this approach is given by Bahadur (1971).

The reader may feel that of two such different approaches, (i) fixing α and Π so that $\Delta \to 0$ and (ii) fixing Δ and Π so that $\alpha \to 0$, only one can be correct. In fact, however, neither corresponds to the actual situation in which not only α, Π, and Δ

but also m and n are fixed. A sequence with m and n tending to infinity is a mathematical device for obtaining an approximation. Any particular situation, with particular values of m and n and α, Π, and Δ, can be considered a member of many different sequences. Which sequence is most useful depends on which provides the best approximation. No general results on the relative accuracy of the Pitman and Bahadur efficiencies are available at present.

So far, we have discussed efficiency only within the framework of the shift model (2.7). Interesting efficiency results in the wider model (2.6) are given by Behnen (1972).

7F. Efficiency in the Presence of Ties

An alternative to the method of midranks in the presence of ties is to break the ties at random, as described in Chap. 1, Sec. 6. The relative efficiency of these two methods as applied to the Wilcoxon test is found by Putter (1955) and Bühler (1967), and it is shown that the method of midranks is more efficient, as one would expect. The efficiency of the Wilcoxon midrank statistic W_{XY}^* is computed for a number of discrete situations by Chanda (1963).

7G. Optimality Properties

In the normal shift model [i.e., the model (2.7) with F normal], the Normal Scores test is asymptotically as good as the optimum test for this model, which is Student's t-test (see Sec. 7B). In the same model, the Normal Scores test also has an optimum property for any fixed sample sizes m and n. Consider the power function (2.8) for $F = \Phi$. For any given m and n the Normal Scores test maximizes, among all rank tests of the same size, the power $\Pi_\Phi(\Delta)$ for all Δ that are sufficiently close to zero, the value under the hypothesis. As a limiting case, it also maximizes the slope of the power curve $\Pi_\Phi(\Delta)$ at $\Delta = 0$.

The Normal Scores test is thus the *locally most powerful* rank test in the normal shift model [Hoeffding (1951) and Terry (1952)]. {Similarly, the Wilcoxon test is the locally most powerful rank test in a logistic shift model [Lehmann (1953)].} It is also asymptotically efficient in this model in the sense that its asymptotic efficiency relative to the best test for this model is 1 [see Capon (1961)]. Most powerful and locally most powerful tied-rank tests for the case of discrete distributions are considered by Vorličková (1970) and by Krauth (1971).

The optimum properties mentioned so far are for parametric models such as the normal shift model. It is, of course, of interest to see what the best rank test is in such a specific situation, and how well it performs under the assumptions of this model. On the other hand, rank tests are employed just when no reliable parametric model is available, and it may then be more appropriate to look for a test with optimum properties relative to a broad nonparametric class of distributions.

The following approach differs from those discussed above not only in that the alternatives are nonparametric but also that they are at a certain distance from the hypothesis instead of in its vicinity as in the earlier local definitions. Let $\delta(F,G)$ be a measure of the distance between F and G; for example, in the case of the one-sided alternatives satisfying $G(x) \leqslant F(x)$ for all x, the measure

$$\delta(F,G) = \max\left[F(x) - G(x)\right] \tag{2.72}$$

Consider the class of all alternatives to the hypothesis for which this measure has at least a certain value Δ_0, that is, satisfies

$$\delta(F,G) \geqslant \Delta_0 \tag{2.73}$$

Unless Δ_0 is very small, (2.73) means that the hypothesis is far from being correct, and we should therefore like the power of the test to be high against all these alternatives. It may then be of interest to find the test that guarantees as high a power as possible. This would be the test that maximizes the minimum power it achieves for the alternatives (2.73). It is shown by Doksum (1966) that for the distance function (2.72) and in a certain asymptotic sense, the Wilcoxon test maximizes this minimum power among a large class of rank tests.

7H. Additional Properties of $\hat{\Delta}$

In analogy with the questions discussed in Sec. 7A, one may be interested in the performance of the estimate $\hat{\Delta}$ when the distributional assumptions of Sec. 5 are not satisfied. A first problem then is to determine what quantity is being estimated in this case. If the distributions F and G are symmetric about points ξ and η, respectively, it turns out that the value being estimated is the difference $\eta - \xi$; without these assumptions, the answer is less simple. A second problem is to find the accuracy of the estimate and to study, for example, its efficiency relative to $\bar{Y} - \bar{X}$. These and related questions are considered by Høyland (1965), Ramachandramurty (1966), and Knüsel (1969). [See also Kraft and Van Eeden (1972).]

Under the assumption $G(x) = F(x - \Delta)$ for all x, the estimate $\hat{\Delta}$ was defined in Sec. 5 as the value of Δ which, when subtracted from the Y's, brings these shifted observations into good agreement with the X's. The following is a slightly different approach. Since $G(x)$ is approximated by $G_n(x)$ and $F(x)$ by $F_m(x)$, with F_m and G_n as defined in Chap. 1, Sec. 6, one would expect the differences between $F_m(x - \Delta)$ and $G_n(x)$ to be small. Let δ be a measure of the distance between two distribution functions, and denote by $F_{m,\Delta}$ the cumulative distribution function defined by

$$F_{m,\Delta}(x) = F_m(x - \Delta)$$

Then the quantity Δ^* minimizing $\delta(F_{m,\Delta}, G_n)$ with respect to Δ should be a reasonable estimate of Δ. A distance function δ for which the resulting estimate Δ^* turns out to agree with $\hat{\Delta}$ is obtained by Fine (1966) and Knüsel (1969).

A very different derivation of $\hat{\Delta}$ and of the intervals $(D_{(i)}, D_{(i+1)})$ for Δ discussed in Sec. 6 is given by Beran (1971). He shows that in a certain sense, these intervals contain all the information about Δ provided by the observations, and obtains $\hat{\Delta}$ as the median of the distribution of Δ defined by these intervals. However, his probabilities refer to a quite different concept of replication of the experiment from the usual one, and they do not agree with those given by (2.62).

Restriction to the confidence points $D_{(i)}$ is justified from a different point of view by Bauer (1972), who shows that any rank test, when converted to a confidence statement, leads to confidence points belonging to the set of points $D_{(i)}$.

7I. Efficiency of the Siegel-Tukey Test

Let $X_1, \ldots, X_m$ and $Y_1, \ldots, Y_n$ be measurements of a quantity μ, made, for example, by two different methods or in two different laboratories. Suppose for simplicity that μ is known. Then the usual normal model (the *normal scale model*) assumes the $m+n$ measurements to be independently distributed, say, with normal distributions

$$N(\mu, \sigma^2) \quad \text{and} \quad N(\mu, \tau^2) \tag{2.74}$$

respectively. For testing $H: \tau^2 \leqslant \sigma^2$, the best test in this model rejects when

$$\frac{\Sigma (Y_j - \mu)^2/n}{\Sigma (X_i - \mu)^2/m} \geqslant c \tag{2.75}$$

The null distribution of the left-hand side of (2.75) in the model (2.74) is the so-called F-distribution with n and m degrees of freedom. Unfortunately, the asymptotic (Pitman) efficiency of the Siegel-Tukey test relative to the F-test (2.75) has the disconcertingly low value of $6/\pi^2 = .608$ [Klotz (1962)].

In the more general scale model in which the X's and Y's are respectively distributed according to

$$G\left(\frac{x-\mu}{\sigma}\right) \quad \text{and} \quad G\left(\frac{y-\mu}{\tau}\right) \tag{2.76}$$

but where G need no longer be normal, the F-test (in contrast to the corresponding behavior of Student's t-test) is not asymptotically distribution-free. The probability of (2.75) depends very heavily on the underlying common distribution G, and the test should not be used if there is any doubt regarding the assumption of normality, particularly if there is a possibility of gross errors.[1]

[1] For a review of various alternative procedures not suffering from this defect, and a discussion of their properties, see Miller (1968), "Jackknifing Variances," *Ann. Math. Statist.* **39**:576–582; Nemenyi (1969), "Variances: An Elementary Proof and a Nearly Distribution-free Test," *Amer. Statist.*, **23**:**5**:35–37; and Shorack (1969), "Testing and Estimating Ratios of Scale Parameters," *J. Amer. Statist. Assoc.*, **64**:999–1013.

7J. The Scale Tests of Capon and Klotz

In view of the low asymptotic efficiency of the Siegel-Tukey test relative to the F-test, it is of interest to construct a rank analogue of the F-test using the methods of Sec. 7B. The resulting test statistic, given by (2.69) with $a_N(s) = E_\Phi[(Z_{(s)})^2]$, the expected values of the square of a normal order statistic, was proposed by Capon (1961). The values of the scores for $N \leqslant 20$ can be found in Teichroew (1956). An asymptotically equivalent test, with the $a_N(s)$ given by the squares of the scores (2.67), is discussed by Klotz (1962), who also provides tables of the scores and critical values. The asymptotic efficiency of these tests relative to the F-test (2.75) is 1. However, in the more general model (2.76) it is not always $\geqslant 1$ (as was the case for the Normal Scores test relative to the t-test) but can get arbitrarily close to zero [Raghavachari (1965b); see also Scholz (1970)].

7K. The Savage (or Exponential Scores) Test

The tests of Siegel-Tukey, Capon, and Klotz permit the comparison of two sets of measurements of a quantity μ that are thought to differ in their degree of dispersion about μ. Such measurements are typically distributed fairly symmetrically about μ, for example, with an approximately normal distribution, perhaps disturbed by some gross errors. However, the testing of two sets of observations for a possible difference in dispersion arises also in a very different context. Consider the length of life (or time to failure) of some material or a complex piece of equipment. A particularly simple type of distribution for this kind of observation, suggested by theoretical considerations and frequently providing a reasonable fit, is the exponential distribution with density

$$\frac{1}{a} e^{-x/a} \qquad \text{for } x > 0 \tag{2.77}$$

In reliability theory,[1] which is concerned with such times to failure, the exponential distribution plays a role somewhat analogous to that of the normal distribution in the theory of measurements.

Consider now the investigation of a treatment that, it is hoped, will prolong the time to failure. Let $X_1, \ldots, X_m$ and $Y_1, \ldots, Y_n$ denote the lengths of life of samples of size m and n observed, respectively, under standard conditions and under the new treatment. These $m+n$ random variables are assumed to be independent, the X's distributed according to (2.77) and the Y's according to the same distribution but with a replaced by b. As in the measurement problem discussed in Sec. 7I, the distributions of X and Y differ only by a scale factor. There is, however, this difference between the two situations: Although before $\tau < \sigma$ meant that the Y's

[1] For a survey of the use of the exponential distribution in this area. see the book by Barlow and Proschan, *Mathematical Theory of Reliability*, John Wiley & Sons, New York, 1965.

were more concentrated than the X's about μ, the common center of the two distributions, in the present case $b < a$ means that the Y's tend to be closer than the X's to zero, which is now the common left-hand endpoint of the two distributions. As a result, in the earlier situation, values $\tau < \sigma$ implied that one would expect to find the X's at both extremes with the Y's tending to be concentrated toward the center of the pattern of observations. On the other hand, in the present case values $b < a$ will tend to place the X's to the right and the Y's to the left of the pattern, while values $b > a$ (corresponding to an increase in length of life) will conversely tend to place the X's to the left and the Y's to the right.

The classical test of the hypothesis $H: b = a$ against the alternatives $b > a$ in the exponential model just described rejects H when

(2.78) $$\frac{\Sigma Y_j}{\Sigma X_i} \geqslant c$$

This test possesses strong optimum properties in this model,[1] but the assumption of an exponential distribution is an idealization on which one should not rely too heavily. To make the test less sensitive to this assumption, one may again try to reconstruct the observations from their ranks. To this end, the sth smallest of the N observations is replaced by

(2.79) $$a_N(s) = E_\varepsilon[Z_{(s)}]$$

its expectation under the exponential distribution (2.77), where without loss of generality a can be taken to be equal to 1. The scores (2.79) can be evaluated explicitly and shown to be equal to

$$a_N(s) = \frac{1}{N} + \frac{1}{N-1} + \cdots + \frac{1}{N-s+1}$$

A table for $N \leqslant 20$ is provided by Basu (1968) and for $N \leqslant 120$ by Harter (1969). Critical values for the resulting test for $6 \leqslant m \leqslant n \leqslant 10$ are given by Hajek (1969, Tables X and XI). Approximations to the null distribution are discussed by Cox (1964). For other aspects of the test, which was first discussed by Savage (1956), see James (1966), Basu and Woodworth (1967), Hajek (1969), Davies (1971), and Woinsky (1972).

7L. Scale Tests with Unknown Location

As was pointed out in Chap. 1, Sec. 6, the assumption made there—and in the preceding Secs. 7I and 7J—that the two methods being compared measure the same quantity, often cannot be trusted. If ξ and η denote the quantities measured by the two methods, it is then natural to estimate ξ and η by estimates, say, $\hat{\xi}$ and $\hat{\eta}$ and

[1] See for example, Lehmann, *Testing Statistical Hypotheses*, John Wiley & Sons, New York, 1959.

to apply the scale test to the adjusted observations $X_1-\hat{\xi},\ldots,X_m-\hat{\xi};\ Y_1-\hat{\eta},\ldots,Y_n-\hat{\eta}$. The significance level of the resulting test will be affected by the substitution of $\hat{\xi}$ and $\hat{\eta}$ for ξ and η, but one may hope that the effect will not be serious and that asymptotically the significance level will retain its value. Conditions under which this is the case are given by Sukhatme (1958), Raghavachari (1965a), and Gross (1966).

7M. Power and Efficiency of the Smirnov Test

The Smirnov test, defined by (1.60) and (1.61), was there considered as a test of the hypothesis of no treatment effect against the alternative of the existence of an effect, the nature of which is completely unspecified. Analogously, in the population model of the present chapter, it is a test of the hypothesis $H: F = G$ against the alternatives

$$F \neq G \tag{2.80}$$

which (except for the assumption of continuity) are not delimited further. It is shown by Massey (1950) that the Smirnov test is consistent against the class of alternatives (2.80). This is a desirable feature from one point of view, but it also suggests the possibility that a test which tries to please all alternatives may end up pleasing none. That is, by having to guard against all possible deviations from the hypothesis, the test may spread itself too thin to do a really efficient job against any alternative.

Some exact power values for $m = n \leqslant 5$ are provided by Dixon (1954), and a useful lower bound to the power is given by Massey (1950). He notes that the power of the test, which is equal to

$$P_{F,G}\left[\max_x |F_m(x)-G_n(x)| \geqslant c\right]$$

is certainly no less than

$$P_{F,G}[|F_m(x_0)-G_n(x_0)| \geqslant c]$$

where x_0 is any fixed value of x, for example, the value that maximizes $|F(x)-G(x)|$. This bound is used by Capon (1965) to obtain lower bounds to the Pitman efficiencies of the test. Other bounds are obtained by Yu (1971), and Bahadur efficiencies (discussed in Sec. 7E above) are studied by Abrahamson (1967) and Klotz (1967).

For normal shift alternatives, the various approaches give comparable numerical results, which suggest that for these alternatives the efficiency of the Smirnov test relative to Student's t-test typically varies between .64 and .75 depending on the sample sizes and the values of α, Π, and Δ.

7N. Sequential Rank Tests

Application of Wald's theory of sequential probability ratio tests requires the probability distribution of the ranks both under the hypothesis and under the particular alternative that is under consideration. Since an explicit expression for this latter distribution is available for the alternatives (2.71), much of the work on sequential rank tests has been concerned with these alternatives. Tests have been proposed, and their properties discussed, among others by Wilcoxon, Rhodes, and Bradley (1963), Bradley, Merchant, and Wilcoxon (1964), and Savage and Sethuraman (1966). Some more general considerations can be found in Hall, Wijsman, and Ghosh (1965) and Berk and Savage (1968). For an interesting new development see Darling and Robbins (1968) and Robbins (1970).

7O. The Permutation *t*-Test

The permutation version of $\bar{Y}-\bar{X}$ described in Chap. 1, Sec. 7E, is asymptotically equivalent to the two-sample *t*-test and can in fact be viewed as a distribution-free modification of the *t*-test. Since its power is asymptotically equal to that of the *t*-test [see Hoeffding (1952)], the efficiency results of Sec. 4 also apply when the *t*-test is replaced by this modification. For further discussion of this and related permutation tests see, for example, Lehmann (1959).

8. PROBLEMS

(A)

Section 1

1. In the randomization model of Chap. 1, Sec. 1, with $m = 2$, $n = 2$, let the hypothesis of no treatment effect be rejected when $W_s = 7$, that is, when the two largest ranks are those of the two subjects assigned to treatment. If the treatment adds Δ to the response a subject would give without treatment, find the power $\Pi(\Delta)$ of this test for all values of Δ not leading to ties, under the assumption that the four subjects without treatment would give the responses (i) $-1, 1, 2, 5$; (ii) $-1, 0, 2, 3$; (iii) $-2, 0, 4, 6$. [*Hint* for (i): Note that the four observations will be $(-1+\Delta, 1+\Delta, 2, 5)$, $(-1, 1+\Delta, 2+\Delta, 5)$, and so on, each with probability $1/6$.]

2. Solve the preceding problem when $m = n = 3$, the hypothesis is rejected when $W_s \geq 14$, and the six responses without treatment are (i) $-2, -1, 0, 3, 5, 6$; (ii) $-6, -3, 0, 9, 15, 18$.

3. Let X_1, X_2, Y be independent, each taking on the values a and b $(a < b)$ with probabilities p and q $(p+q = 1)$, respectively. Show that the midrank S_1^* of Y has the distribution given in the text.

4. Let X_1, X_2, Y_1, Y_2 be independently distributed according to the distribution of the preceding problem, and let S_1^* and S_2^* denote the midranks of Y_1 and Y_2, respectively. Find the distribution of $S_1^* + S_2^*$.

5. Let $d_1, \ldots, d_e$ denote the configuration of ties of the variables X_1, X_2, Y of Prob. 3. Find (i) the joint distribution of e and $d_1, \ldots, d_e$; (ii) the distribution of e.

6. (i) In a business administration course, a set of lectures was given televised to one group and live to another. In each case an examination was given prior to the lectures and immediately following them. The differences between the two examination scores for the women in the two groups were as follows:[1]

Live: 20.3 23.5 4.7 21.9 15.6 20.3 26.6 21.9 −9.4 4.7 −1.6 25.0

TV: 6.2 15.6 25.0 4.7 28.1 17.2 14.1 31.2 12.6 9.4 17.2 23.4

Use the two-sided Wilcoxon test to test the hypothesis of no difference at significance level $\alpha = .05$.

(ii) Carry out the corresponding test for the scores of the men, which were as follows:[1]

Live: 12.5 7.8 21.9 −7.8 −3.1 3.1 29.7 18.8 28.1 36.0 4.7 −3.2 45.4 6.3 18.8 9.4 −3.1

TV: 28.1 −7.8 17.1 14.1 18.8 17.2 1.5 20.3 4.7 15.6 34.4 18.8 29.7 39.1 9.4 20.3 14.1 18.7 17.2 1.5 25.0 29.7 25.0 23.4 9.3 1.5 4.7 15.6 26.7 12.5 10.9 32.8 −6.2 3.1 28.1 37.5 20.3

Section 3

7. Show that (2.21) reduces to $mn(N+1)/12$ when $p_1 = \frac{1}{2}$ and $p_2 = p_3 = \frac{1}{3}$.

8. In Chap. 1, Example 2, use (2.31) to find the probability of detecting at significance level $\alpha = .05$ that discouragement has an adverse effect if it lowers the score by three points, if the scores are assumed to be normally distributed with variance $\sigma^2 = 8$, and when the sample sizes are (i) $m = n = 20$; (ii) $m = 15, n = 25$; (iii) $m = 10, n = 30$.

9. Solve the preceding problem when the significance level is .01 instead of .05.

10. In the context of Chap. 1, Prob. 24, when the recovery times are normally distributed with variance $\sigma^2 = 4$ and when the sample sizes are $m = n = 25$, use (2.31) to find the power of the Wilcoxon test at level .02 if the new treatment decreases the recovery time by (i) $\Delta = 1$ day; (ii) $\Delta = 2$ days; (iii) $\Delta = 3$ days.

11. In Prob. 8 find the sample size $m = n$ required at significance level $\alpha = .02$ to achieve power $\Pi = .9$ against the alternative that discouragement lowers the score by (i) $\Delta = 1$ point; (ii) $\Delta = 2$ points; (iii) $\Delta = 4$ points.

12. In Example 2, against the alternative $\Delta = 5$, use (2.34) to find the approximate sample size $m = n$ required to achieve power (i) $\Pi = .7$; (ii) $\Pi = .8$; (iii) $\Pi = .9$.

13. Under the assumptions of Prob. 10, find the sample size $m = n$ required to achieve power $\Pi = .95$ for detecting a decrease of $\Delta = 2$ in the recovery time at significance level (i) $\alpha = .01$; (ii) $\alpha = .02$; (iii) $\alpha = .05$.

Section 4

14. From a table of the t-distribution, find the critical value $c_{m,n}(\alpha)$ occurring on the left-hand side of (2.41) for $m = n = 6, 11, 16, 21, 31$ and compare it with u_α, when (i) $\alpha = .01$; (ii) $\alpha = .05$.

[1] Data from MacLachlan (1965), "Variations in Learning Behavior in Two Social Situations According to Personality Type," unpublished master's thesis at University of California at Berkeley.

[*Hint:* A suitable table can be found in Pearson and Hartley (eds.), *Biometrika Tables for Statisticians*, vol. 1, Cambridge University Press, London, 1966.]

15. From a table of the t-distribution, find the probability (under the hypothesis) that the t-statistic on the left-hand side of (2.38) is greater than the approximate critical value u_α for the values of $m = n$ and α of Prob. 14.

16. (i) For $\alpha = .05$, $m = n = 25$, and F normal, use (2.31) to find the approximate power of the Wilcoxon test against the alternatives $\Delta/\sigma = .1(.1).9$. [*Note:* The notation $a(\delta)b$ means the set of values starting at a and increasing at intervals of δ to b.]

(ii) For the same alternatives obtain the power of the t-test from an appropriate table or graph, and compare the results.

Section 5

17. In Example 4, replace the observation 10.2 by x and compute the estimators $\hat{\Delta}$ and $\bar{\Delta}$ as functions of x, for x varying from $-\infty$ to $+\infty$.

18. Assuming a shift model, use $\hat{\Delta}$ and $\bar{\Delta}$ to estimate the effect of discouragement in Chap. 1, Example 2.

19. At the beginning of a study of the effect of hypnotism, the following measurements of ventilation were taken on eight treatment subjects (to be hypnotized) and eight controls.[1]

Controls:	4.69	4.19	3.99	4.21	4.84	4.54	5.48	4.64
Treatment:	5.52	4.36	5.08	5.20	4.78	5.74	4.67	5.16

Assuming a shift model, use $\hat{\Delta}$ and $\bar{\Delta}$ to estimate the effect which the anticipation of being hypnotized has on the subjects.

20. In a comparison of the effect on growth of two diets B and C, a number of growing rats were placed on these two diets, and the following growth figures were observed after 7 weeks.[2]

B:	156	183	120	113	138	145	142			
C:	109	107	119	162	121	123	76	111	130	115

(i) Assuming a shift model, use $\hat{\Delta}$ and $\bar{\Delta}$ to estimate the amount of shift.

(ii) Use a two-sided Wilcoxon test at significance level $\alpha = .05$ to test the significance of the difference.

Section 6

21. (i) For $m = n = 8$, determine confidence intervals $(\underline{\Delta}, \bar{\Delta})$ for Δ with confidence coefficient as close as possible to .95.

(ii) Find the numerical values of $\underline{\Delta}$ and $\bar{\Delta}$ determined in part (i) for the data of Prob. 19, and compare them with the corresponding values based on Student's t-test.

22. Solve the preceding problem for $m = 11$, $n = 5$, and for the data of Example 4. [*Hint:* To extend Table B to the case $m = 11$, $n = 5$, see Chap. 1, Probs. 87 and 88.]

[1] From Agosti and Camerota, "Some Effects of Hypnotic Suggestion on Respiratory Function," *Intern. J. Clin. Exptl. Hypnosis* **13**:149–156 (1965).

[2] From Nieman, Groot, and Jansen, "The Nutritive Value of Butter Fat Compared with that of Vegetable Fats, I," *Koninkl. Ned. Akad. Van. Wetenschap. Ser.* **C55**:588–604 (1952). Original data from Boer and Jansen, *Arch. Neerl. Physiol.* **26**:1 (1952).

23. Use (2.62) to determine approximate confidence intervals for Δ satisfying (2.63) for (i) $\gamma = .90$, $m = n = 25$; (ii) $\gamma = .90$, $m = 20$, $n = 30$; (iii) $\gamma = .95$, $m = n = 25$; (iv) $\gamma = .95$, $m = 20$, $n = 30$.

24. In Prob. 20 suppose that an additional observation had been taken on diet B and had produced the value 188. Compute the 1 percent confidence point $D_{(15)}$ by the method discussed at the end of Sec. 6.

25. (i) In analogy with Table 2.6, determine the confidence points with confidence probability as close as possible to 1, 5, 10, 25, 50, 75, 90, 95, and 99 percent for $m = n = 8$.
(ii) Find the numerical values of the confidence points of part (i) for the data of Prob. 19, and compare these values with the corresponding ones based on Student's t.

(B)

Section 1

26. Let $X_1, \ldots, X_m$ and $Y_1, \ldots, Y_n$ be independently distributed, each according to the distribution of Prob. 3, and let A_1, B_1 and A_2, B_2 denote the number of X's equal to a, b and the number of Y's equal to a, b, respectively:

	a	b	
X	A_1	B_1	m
Y	A_2	B_2	n
	d_1	d_2	

(i) Find the joint distribution of A_1, B_1, A_2, B_2.
(ii) Find the conditional distribution of B_2 given d_1 and d_2.

27. Let B_1 and B_2 be the number of successes in two independent sequences of m and n binomial trials with success probability p_1 and p_2, respectively, and consider the hypothesis H: $p_2 = p_1$. Show that the computation of the null distribution of B_1 and B_2 can be viewed as an example of the situation described in the preceding problem and evaluate W_s^* in terms of B_2. [*Hint:* Use the formula for W_s^* given in Chap. 1, Sec. 4, and the fact that $B_1 + B_2 = d_2$.]

28. (i) In Prob. 4, find the critical value $C(e; d_1, \ldots, d_e)$ giving the significance level closest to .2 for each possible set of values $(e; d_1, \ldots, d_e)$, assuming large values of $S_1^* + S_2^*$ to be significant.
(ii) Find the overall unconditional level of (2.5) as a function of p, and determine its minimum and maximum value for $0 < p < 1$.
(iii) Let F denote the number of a's among X_1, X_2, Y_1, Y_2, and Z the number of a's among Y_1, Y_2. Find the unconditional level of the test which accepts when $F = 0, 2, 4$; when $F = 1, Z = 0$; and when $F = 3, Z \leqslant 1$ as a function of p and compare it with the result of (ii).

Section 2

29. Show that (2.7) and $\Delta > 0$ imply (2.6).

30. Let Ψ be a cumulative distribution function and let

$$F(x) = \Psi\left(\frac{x-\xi}{\sigma}\right) \quad \text{and} \quad G(x) = \Psi\left(\frac{x-\eta}{\tau}\right)$$

with $\xi < \eta$.

(i) If Ψ is strictly increasing from $-\infty$ to ∞, show that G is stochastically larger than F if and only if $\tau = \sigma$.

(ii) Show that the result of (i) need no longer hold when Ψ is not strictly increasing. [*Hint:* Consider, for example, the case that Ψ is a uniform distribution.]

31. Let F_1 be stochastically larger than F_0. Show that $p'F_1+(1-p')F_0$ is stochastically larger or smaller than $pF_1+(1-p)F_0$ as $p' > p$ or $p' < p$.

32. Let E be a cumulative distribution function with $E(0) = 0$, and let

$$F(x) = E\left(\frac{x}{\sigma}\right) \quad \text{and} \quad G(x) = E\left(\frac{x}{\tau}\right)$$

Show that G is stochastically larger than F if $\tau > \sigma$.

33. If X_3 is stochastically larger than X_2, and X_2 is stochastically larger than X_1, show that X_3 is stochastically larger than X_1.

34. If Y is stochastically larger than X, and Z is independent of both X and Y, show that $Y+Z$ is stochastically larger than $X+Z$.

35. Let E be symmetric about zero, and let F and G be defined in terms of E as in Prob. 32.

(i) Discuss whether either of the tests defined by the rejection rules (*a*) $W_{XY} \geqslant c$, (*b*) $|W_{XY} - \frac{1}{2}mn| \geqslant c$, is reasonable for testing $\tau = \sigma$ against $\tau > \sigma$.

(ii) Describe a rank test which might be preferable to (*a*) and (*b*).

36. Prove that the power function $\Pi_F(\Delta)$ defined by (2.8) tends to 1 as $\Delta \to \infty$ and to zero as $\Delta \to -\infty$, provided

$$\frac{1}{\binom{N}{n}} \leqslant \alpha \leqslant 1 - \frac{1}{\binom{N}{n}}$$

[*Hint:* The probability of the ranking $s_1 = m+1, \ldots, s_n = m+n$ tends to 1 as $\Delta \to \infty$. This can be seen by comparing it with the probability $P(X_1, \ldots, X_m < \frac{1}{2}\Delta < Y_1, \ldots, Y_n)$.]

37. (i) If $\Pi_F(\Delta)$ denotes the power function of the test (*b*) of Prob. 35 in the shift model (2.7), find the limit of $\Pi_F(\Delta)$ as $\Delta \to \infty$.

(ii) Find the limits of the power of the tests (*a*) and (*b*) of Prob. 35 as $\tau/\sigma \to \infty$.

38. If F is continuous, prove that the power function (2.8) is a continuous function of Δ. [*Hint:* The joint distribution

$$P_\Delta(X_1 \leqslant x_1, \ldots, X_m \leqslant x_m;\ Y_1 \leqslant y_1, \ldots, Y_n \leqslant y_n) = F(x_1) \ldots F(x_m)F(y_1-\Delta) \ldots F(y_n-\Delta)$$

is a continuous function of Δ, and the result therefore follows from Appendix, Theorem 16.]

39. Let F and G be continuous, strictly increasing, and satisfy (2.6). Prove the existence of a function $\Delta(x) \geqslant 0$ such that $X+\Delta(X)$ is distributed according to G when the distribution of X is F. {*Hint:* For each x, determine $\Delta(x)$ by the equation $F(x) = G[x+\Delta(x)]$.}

40. Assuming (2.12), prove the properties (i′) to (iii′) stated at the end of Sec. 2. [*Hint:* The proofs are completely analogous to those given in the text for (i) to (iii) under the shift model (2.7).]

Section 3

41. (i) Prove that $p_2 = p_3$ when G and F are related by (2.7) and when in addition F is symmetric.

(ii) Show that p_2 and p_3 need not be equal when F is not symmetric. [*Hint:* (i) In the definition of p_2 and p_3 put $Y = V+\Delta$, and $Y' = V'+\Delta$, so that the variables X, X', V, and V' have the same distribution F. Assume without loss of generality that F is symmetric about zero. Then p_2 is unchanged if X, X', V, V' are replaced by their negatives. (ii) Let $P(-1 < X < 0) = 3/4$, $P(3 < X < 4) = 1/4$, and $\Delta = -2$.]

42. For fixed total sample size $m+n = N$, show that (2.29) is maximized by putting $m = n = k$ when $N = 2k$ and $m = k$, $n = k+1$, or $m = k+1$, $n = k$ when $N = 2k+1$.

43. Show that the correlation coefficient of the variables Z and Z' defined by (2.28) is $1/2$.

44. In the shift model (2.7) with continuous F, prove that: (i) $p_1 > 1/2$ when $\Delta > 0$; (ii) $p_1 < 1/2$ when $\Delta < 0$. {*Hint* for (i): Show that

$$p_1 - \tfrac{1}{2} = \tfrac{1}{2}P(|X_2 - X_1| < \Delta)$$

where X_1 and X_2 are independent variables with distribution F, and use the fact that the probability on the right-hand side is greater than or equal to

$$\sum_{i=-\infty}^{\infty} P[i\Delta < X_1, X_2 < (i+1)\Delta].\}$$

[*Note:* It can be shown that $p_1 > 1/2$ for any continuous distribution satisfying (2.6).]

45. Let $X_1, \ldots, X_m$ and $Y_1, \ldots, Y_n$ be distributed with exponential densities e^{-x}, $x \geqslant 0$ and $e^{-(y-\Delta)}$, $y \geqslant \Delta$, respectively, and let $m = 11$, $n = 10$. The power $\Pi(\Delta)$ of the Wilcoxon test $W_{XY} \geqslant c$ is given in the following table for a number of shifts Δ, where the value for $\Delta = 0$ is the significance level.[1]

Δ	0	.5	1.0	1.5	2.0
$\Pi(\Delta)$	.021	.316	.736	.930	.985

Compare the tabled values with the approximate values obtained from (2.37). [*Hint:* The required formulas for p_1, p_2, and p_3 are given in Appendix, Probs. 35 and 36.] [*Note:* The approximations to p_1 and $\text{Var}(W_{XY})$ leading to (2.29) are very poor in the present case because of the abrupt behavior of the exponential density at $x = 0$, and the approximation (2.29) therefore gives misleading results.]

46. Show that the approximate power $\Pi(F,G)$ given by (2.23), (2.29), or (2.37) is unchanged if (*a*) the same constant is added to both the X's and Y's; (*b*) the distributions F and G are related by (2.7) and the X's, Y's, and Δ are all multiplied by the same positive constant.

47. Show that the exact power of the Wilcoxon test is unchanged under the transformations (*a*) and (*b*) of Prob. 46.

[1] These power values are given by Haynam and Govindarajulu (1966).

48. (i) If the power of the Wilcoxon test $W_{XY} \geqslant c$ is equal to Π, show that approximately

$$c = mnp_1 + u_\Pi \sqrt{\operatorname{Var}(W_{XY})}$$

(ii) Equating this expression for c with (2.36), show that approximately

$$mn(p_1 - \tfrac{1}{2}) = u_\alpha \sqrt{\tfrac{1}{12}mn(N+1)} - u_\Pi \sqrt{\operatorname{Var}(W_{XY})}$$

(iii) Putting $m = n$, neglecting terms of order mn in the expression (2.21) for $\operatorname{Var}(W_{XY})$, and replacing $N+1$ by $2n$ in (2.37), show that an alternative to the approximate formula (2.33) for the sample size required to obtain power Π is given by

$$n \approx \frac{(u_\alpha/\sqrt{6} - u_\Pi\sqrt{p_3 - p_1{}^2 + p_2 - p_1{}^2})^2}{(p_1 - \frac{1}{2})^2} \tag{2.81}$$

Section 4

49. Show that the efficiency of the Wilcoxon to the t-test for any given α and n depends on Δ and the variance σ^2 of F only through the ratio Δ/σ. [*Hint:* See Probs. 46 and 47.]

50. Show that when F is normal, (2.42) is exactly the power of the test which rejects when

$$\frac{\bar{Y} - \bar{X}}{\sigma\sqrt{1/m + 1/n}} \geqslant u_\alpha$$

51. For any fixed numbers $a_1, \ldots, a_m, b_1, \ldots, b_n$ let $x_1 = a_1, \ldots, x_m = a_m$; $y_1 = b_1, \ldots, y_{n-1} = b_{n-1}$, $y_n = b_n + A$. Show that as A tends to infinity, the t-statistic on the left-hand side of (2.38) tends to a finite limit, which is equal to 1 when $m = n$.

52. (i) Show that the Wilcoxon test has the following monotonicity property. If the test rejects for the observations $x_1, \ldots, x_m$; $y_1, \ldots, y_n$ and if $y'_j \geqslant y_j$ for all j, then the test also rejects for $x_1, \ldots, x_m$; $y'_1, \ldots, y'_n$.

(ii) Show that this monotonicity property is not shared by the t-test. [*Hint:* Show that there exist values $x_1, \ldots, x_m; y_1, \ldots, y_n$ for which t is arbitrarily large (it is only necessary to make S small) and apply the result of the preceding problem.]

Section 5

53. Show that for $m = n = 2$ the two estimators $\hat{\Delta}$ and $\bar{\Delta}$ are always equal.

54. Let $a_1, \ldots, a_N$ and $a'_1, \ldots, a'_N$ be any two sets of N numbers and let M and M' be their medians. If $|a'_i - a_i| \leqslant \varepsilon$ for all i, prove that $|M' - M| \leqslant \varepsilon$. [*Hint:* Note that $M' \leqslant \operatorname{med}(a_i + \varepsilon) = M + \varepsilon$.]

55. In the notation used in the proof of Theorem 3, let $\delta(X, Y)$ be an estimator of Δ in the shift model (2.7), and suppose that δ satisfies

$$\delta(X + b, Y + c) = \delta(X, Y) + (c - b) \tag{2.82}$$

and is nondecreasing in each Y coordinate and nonincreasing in each X coordinate. If $|X'_i - X_i| \leqslant \varepsilon/2$ and $|Y'_j - Y_j| \leqslant \varepsilon/2$ for all i and j, prove that

$$|\delta(X', Y') - \delta(X, Y)| \leqslant \varepsilon$$

[*Hint:* Use (2.82) and the inequalities

$$\delta(X' + \tfrac{1}{2}\varepsilon, Y' - \tfrac{1}{2}\varepsilon) \leqslant \delta(X, Y) \leqslant \delta(X' - \tfrac{1}{2}\varepsilon, Y' + \tfrac{1}{2}\varepsilon)$$]

56. In the notation used in the proof of Theorem 3, let $\delta(X, Y)$ be any estimator of Δ satisfying (2.82). Prove that the distribution of $\delta(X, Y)$ is symmetric about Δ in the shift model (2.7) provided either (i) F is symmetric and $\delta(-X, -Y) = -\delta(X, Y)$; or (ii) $m = n$ and $\delta(Y, X) = -\delta(X, Y)$.

57. Prove that the estimators $\hat{\Delta}$, $\bar{\Delta}$, and $\tilde{\Delta} = \text{med}(Y_j) - \text{med}(X_i)$ satisfy

$$\delta(aX+b, aY+c) = a\,\delta(X, Y)+(c-b) \tag{2.83}$$

and hence in particular (2.82).

58. If $X_1, \ldots, X_m$ and $Y_1, \ldots, Y_n$ are independently distributed according to continuous distributions F and G, prove that the distribution of $\hat{\Delta}$ is continuous. [*Hint:* See Appendix, Theorem 1, and use the inequality

$$P(\hat{\Delta} = d) \leqslant \sum_i \sum_j P(Y_j - X_i = d) \text{]}$$

59. Let $X_1, \ldots, X_m$ and $Y_1, \ldots, Y_n$ be independently distributed according to continuous distributions F and G, and suppose that F is symmetric about μ and G about ν. If $\delta(X, Y)$ satisfies (2.83) prove that $\delta(X, Y)$ is distributed symmetrically about $\Delta = \nu - \mu$.

60. Under the assumptions of the preceding problem, discuss the suitability of Δ as a measure of superiority of the Y's over the X's for a number of applicational situations including that of Example 4.

61. Prove (i) the relationship $p_1' = 1 - p_1$ stated at the end of Sec. 5; (ii) that $\text{Var}(W_{X, Y-a}) = \text{Var}(W_{X, Y+a})$ if either $m = n$ or F is symmetric. [*Hint* for (ii): For $m = n$ the result follows directly from (2.21). For symmetric F use also the fact that $p_2 = p_3$ by Prob. 41.]

62. Let $X_1, \ldots, X_m$ and $Y_1, \ldots, Y_n$ be independently distributed according to distributions F and G.

- (i) Prove that $\bar{Y} - \bar{X}$ is consistent for estimating $E(Y_1) - E(X_1)$ as m and n tend to infinity, if these expectations exist.
- (ii) If the distribution of $Y_1 - X_1$ has a unique median M, prove that $\hat{\Delta}$ is consistent for estimating M as m and n tend to infinity.

[*Hint:* (i) See Khinchine's result mentioned in the Appendix, Example 7. (ii) Without loss of generality suppose that $M = 0$. Then $P(Y_1 - X_1 > a) < \frac{1}{2}$ when $a > 0$, and the result follows by the argument used to prove (2.55).]

63. Show that the consistency of a sequence of estimators T_N of a parameter θ neither implies that

(i) $E(T_N) \to \theta$ (asymptotic unbiasedness)

nor that

(ii) $P(T_N < \theta) - P(T_N > \theta) \to 0$ (asymptotic median-unbiasedness)

[*Hint:* (i) Let $X_1, \ldots, X_N$ be independently normally distributed with mean θ and variance 1, and let $T_N = \bar{X}_N$ with probability $1 - \varepsilon_N$ and $T_N = A_N$ with probability ε_N where ε_N and A_N are suitable sequences of constants. (ii) Let $X_1, \ldots, X_N$ be independently distributed according to the uniform distribution over the interval $(\theta, \theta + 1)$ and let $T_N = \min(X_1, \ldots, X_N)$.]

64. Show that the estimator, obtained from the t-statistic by the method that led from the Wilcoxon statistic to $\hat{\Delta}$, is $\bar{Y} - \bar{X}$.

65. Let $X_1', \ldots, X_m'$ and $Y_1', \ldots, Y_n'$ be independently distributed according to (2.7) with F

continuous, and suppose that each observation is rounded to the nearest value among $0, \pm\varepsilon, \pm 2\varepsilon, \ldots$. Denote the resulting rounded observations by $X_1, \ldots, X_m$ and $Y_1, \ldots, Y_n$. Determine $i < j$ from a table of the Wilcoxon distribution so that

$$P_0(W_{X'Y'} < i) + P_0(W_{X'Y'} \geqslant j) = \alpha$$

and let $D_{(i)}$ and $D_{(j)}$ be the ith and jth smallest of the mn differences $Y_l - X_k$.

(i) Show that

$$P(D_{(i)} \leqslant \Delta \leqslant D_{(j)}) \geqslant 1 - \alpha \tag{2.84}$$

for any value of Δ which is a multiple of ε.

(ii) Show that

$$P(D_{(i)} - \varepsilon \leqslant \Delta \leqslant D_{(j)} + \varepsilon) \geqslant 1 - \alpha \tag{2.85}$$

for all Δ.

[*Hint:* Assume without loss of generality that $\Delta = 0$, and note that $X_i' < Y_j'$ implies $X_i \leqslant Y_j$, and that $X_i' > Y_j'$ implies $X_i \geqslant Y_j$. If $D_{(i)}'$ denotes the ith smallest of the differences $Y_l' - X_k'$, it follows that $D_{(i)}' < 0$ implies $D_{(i)} \leqslant 0$ and $D_{(i)}' > 0$ implies $D_{(i)} \geqslant 0$.]

Section 7C

66. For each value $s = 1, \ldots, 10$ and the associated value of t, compute the probabilities $P(S \geqslant s)$ and $P(T \geqslant t)$ defined in Sec. 7C.

9. REFERENCES

Abrahamson, Innis G. (1967): "Exact Bahadur Efficiencies for the Kolmogorov-Smirnov and Kuiper One- and Two-sample Statistics," *Ann. Math. Statist.* **38:**1475–1490.

Ansari, A. R. and Bradley, R. A. (1960): "Rank Sum Tests for Dispersion," *Ann. Math. Statist.* **31:**1174–1189. [Establishes the value $6/\pi^2$ for the efficiency of the symmetrized Siegel-Tukey test relative to the F-test.]

Bahadur, R. R. (1967): "Rates of Convergence of Estimates and Test Statistics," *Ann. Math. Statist.* **38:**303–324.

——— (1971): "Some Limit Theorems in Statistics," *Regional Conference Series in Applied Mathematics No. 4*, Society for Industrial and Applied Mathematics.

Basu, A. P. (1968): "On a Generalized Savage Statistic with Applications to Life Testing," *Ann. Math. Statist.* **39:**1591–1604.

———and Woodworth, George (1967): "A Note on Nonparametric Tests for Scale," *Ann. Math. Statist.* **38:**274–277.

Bauer, D. F. (1972): "Constructing Confidence Sets Using Rank Statistics," *J. Am. Statist. Assoc.* **67:**687–690.

Behnen, Konrad (1969): "Asymptotische Relative Effizienz von Tests und Folgen Benachbarter Verteilungen," unpublished Ph.D thesis, University of Münster.

———(1972): "A Characterization of Certain Rank-order Tests with Bounds for the Asymptotic Relative Efficiency, *Ann. Math. Statist.* **43:**1839–1851.

Bell, C. B., Moser, J. M. and Thompson, Rory (1966): "Goodness Criteria for Two-sample Distribution-free Tests," *Ann. Math. Statist.* **37**:133–142.

Beran, R. J. (1971): "On Distribution-free Statistical Inference with Upper and Lower Probabilities," *Ann. Math. Statist.* **42**:157–168, 1943–1948.

Berk, Robert H. and Savage, I. Richard (1968): "The Information in Rank-order and the Stopping Time of Some Associated SPRT's," *Ann. Math. Statist.* **39**:1661–1674.

Bickel, P. J. (1974): "Edgeworth Expansions in Nonparametric Statistics," *Ann. Statist.* **2**:1–20.

Birnbaum, Allan (1961): "Confidence Curves: An Omnibus Technique for Estimation and Testing Statistical Hypotheses," *J. Am. Statist. Assoc.* **56**:246–249.

Bradley, Ralph A., Merchant, Sarla D., and Wilcoxon, F. (1964): "Sequential Rank Tests. II. Modified Two-sample Procedures," *Technometrics* **8**:615–623.

Bühler, Wolfgang J. (1967): "The Treatment of Ties in the Wilcoxon Test," *Ann. Math. Statist.* **38**:519–523.

Capon, J. (1961): "Asymptotic Efficiency of Certain Locally Most Powerful Rank Tests," *Ann. Math. Statist.* **32**:88–100.

——— (1965): "On the Asymptotic Efficiency of the Kolmogorov-Smirnov Test," *J. Am. Stat. Assoc.* **60**:843–853.

Chanda, K. C. (1963): "On the Efficiency of Two-sample Mann-Whitney Test for Discrete Populations," *Ann. Math. Statist.* **34**:612–617.

Chernoff, Herman and Savage, I. Richard (1958): "Asymptotic Normality and Efficiency of Certain Nonparametric Test Statistics," *Ann. Math. Statist.* **29**:972–994.

Cochran, Wm. G. (1952): "The χ^2 Test of Goodness of Fit," *Ann. Math. Statist.* **23**:315–345.

Cox, D. R. (1958): *Planning of Experiments*, John Wiley & Sons, New York.

——— (1964): "Some Applications of Exponential Ordered Scores," *J. Roy. Statist. Soc.* **B26**:103–110.

Darling, D. A. (1957): "The Kolmogorov-Smirnov, Cramer-Von Mises Tests," *Ann. Math. Statist.* **28**:823–838. [A survey of the Smirnov test and related procedures.]

——— and Robbins, Herbert (1968): "Some Nonparametric Sequential Tests with Power One," *Proc. Nat. Acad. Sci.* **61**:804–809.

David, F. N., Barton, D. E., Ganeshalingam, S., Harter, H. L., Kim, P. J., and Merrington, M. (1968): *Normal Centroids, Medians and Scores for Ordinal Data*, Cambridge University Press, London.

Davies, R. B. (1971): "Rank Tests for 'Lehmann's Alternative'," *J. Am. Statist. Assoc.* **66**:879–883.

Dixon, W. J. (1954): "Power under Normality of Several Nonparametric Tests," *Ann. Math. Statist.* **25**:610–614.

Doksum, Kjell (1966): "Asymptotically Minimax Distribution-free Procedures," *Ann. Math. Statist.* **37**:619–628.

——— (1974): "Empirical Probability Plots and Statistical Inference for Nonlinear Models in the Two-sample Case," *Ann. Statist.* **2**:267–277.

Dupac, Vaclav and Hajek, Jaroslav (1969): "Asymptotic Normality of Simple Linear Rank Statistics under Alternatives, II," *Ann. Math. Statist.* **40**:1992–2017.

Fine, Terrence (1966): "On the Hodges and Lehmann Shift Estimator in the Two-sample Problem," *Ann. Math. Statist.* **37**:1814–1818.

Fisher, R. A. and Yates, F. (1938): *Statistical Tables for Biological, Agricultural and Medical Research*, Oliver and Boyd, Edinburgh. [Proposes the substitution of Normal Scores for the observations in a t-test. The resulting procedure is a forerunner of the Normal Scores test.]

Gastwirth, J. L. and Wolff, S. (1968): "An Elementary Method for Obtaining Lower Bounds on the Asymptotic Power of Rank Tests," *Ann. Math. Statist.* **39**:2128–2130.

Gross, Shulamith (1966): "Nonparametric Tests When Nuisance Parameters are Present," unpublished Ph.D. thesis, University of California at Berkeley.

Hajek, Jaroslav (1969): *A Course in Nonparametric Statistics*, Holden-Day, San Francisco.

——— and Sidak, Zbynek (1967): *Theory of Rank Tests*, Academic Press, New York.

Hall, W. J., Wijsman, R. A. and Ghosh, J. K. (1965): "The Relationship between Sufficiency and Invariance with Applications in Sequential Analysis," *Ann. Math. Statist.* **36**:575–614.

Harter, H. Leon (1961): "Expected Values of Normal Order Statistics," *Biometrika* **48**:151–165. [Provides 5-decimal tables of the expected Normal Order Statistics for samples of sizes $N = 2(1)100(25)250(50)400$ and discusses the history and uses of these quantities.]

——— (1969): *Order Statistics and Their Use in Testing and Estimation*, U.S. Government Printing Office, Washington D.C., vol. 2.

Haynam, George E. and Govindarajulu, Z. (1966): "Exact Power of Mann-Whitney Tests for Exponential and Rectangular Alternatives," *Ann. Math. Statist.* **37**:945–953.

Hoadley, A. B. (1967): "On the Probability of Large Deviations of Functions of Several Empirical c.d.f.'s," *Ann. Math. Statist.* **38**:360–381.

Hodges, J. L., Jr. and Lehmann, E. L. (1956): "The Efficiency of Some Nonparametric Competitors of the t-test," *Ann. Math. Statist.* **27**:324–335. [Proves that the asymptotic efficiency of the Wilcoxon to the t-test in a shift model is always $\geqslant .864$.]

——— and ——— (1960): "Comparison of the Normal Scores and Wilcoxon Tests," *Proceedings of the Fourth Berkeley Symposium on Mathematical Statistics and Probability*, University of California Press, Berkeley, vol. 1, 307–317.

——— and ——— (1963): "Estimates of Location Based on Rank Tests," *Ann. Math. Statist.* **34**:598–611. [Proposes the estimator $\hat{\Delta}$. See also Sen (1963).]

Hoeffding, Wassily (1948): "A Class of Statistics with Asymptotically Normal Distribution," *Ann. Math. Statist.* **19**:293–325.

— (1951): "'Optimum' Nonparametric Tests," *Proceedings of the Second Berkeley Symposium on Mathematical Statistics and Probability*, University of California Press, Berkeley, pp. 83–92.

——— (1952): "The Large-sample Power of Tests Based on Permutations of Observations," *Ann. Math. Statist.* **23**:169–192.

Hollander, Myles; Pledger, Gordon; and Lin, Pi-Erh (1974): "Robustness of the Wilcoxon Test to a Certain Dependency between Samples," *Ann. Statist.* **2**:177–181.

Høyland, Arnljot (1964): "Numerical Evaluation of Hodges-Lehmann Estimates," *Norske Vid. Selsk. Fork.* **37**:42–47.

——— (1965): "Robustness of the Hodges-Lehmann Estimates for Shift," *Ann. Math. Statist.* **36**:174–197.

James, B. (1966): "On Pitman Efficiency of Some Tests for Scale for the Gamma Distribution," *Proceedings of the Fifth Berkeley Symposium on Mathematical Statistics and Probability*, University of California Press, Berkeley, pp. 389–393.

Klotz, Jerome (1962): "Nonparametric Tests for Scale," *Ann. Math. Statist.* **33**:498–512.

——— (1964): "On the Normal Scores Two-sample Rank Test," *J. Am. Statist. Assoc.* **59**:652–664. [Gives critical values of the Normal Scores test for significant levels .001, .005, .010, .025, .050, .075, .100, and sample sizes $6 \leqslant N \leqslant 20$, $1 < m < n$.]

——— (1967): "Asymptotic Efficiency of the Two-sample Kolmogorov-Smirnov Test," *J. Am. Statist. Assoc.* **62**:932–938.

Knüsel, Leo Franz (1969): "Über Minimum-distance-Schätzungen," thesis at the Federal Institute of Technology, Zürich.

Kraft, C. H. and van Eeden, C. (1972): "Asymptotic Efficiencies of Quick Methods of Computing Efficient Estimates Based on Ranks," *J. Am. Statist. Assoc.* **67**:199–202.

Krauth, J. (1971): "A Locally Most Powerful Tied Rank Test in a Wilcoxon Situation," *Ann. Math. Statist.* **42**:1949–1956.

Kruskal, William H. (1957): "Historical Notes on the Wilcoxon Unpaired Two-sample Test," *J. Am. Statist. Assoc.* **52**:356–360.

Lancaster, H. O. (1961): "Significance Tests in Discrete Distributions," *J. Amer. Statist. Assoc.* **56**:223–234.

Lehmann, E. L. (1951): "Consistency and Unbiasedness of Certain Nonparametric Tests," *Ann. Math. Statist.* **22**:165–179. [Proves the asymptotic normality of the Wilcoxon distribution when $F \neq G$, using the theory of U-statistics developed by Hoeffding (1948).]

——— (1953): "The Power of Rank Tests," *Ann. Math. Statist.* **24**:28–43.

——— (1959): *Testing Statistical Hypotheses*, John Wiley & Sons, New York.

——— (1963): "Nonparametric Confidence Intervals for a Shift Parameter," *Ann. Math. Statist.* **34**:1507–1512. [Discusses the intervals (2.59) that were first proposed by Moses (1953).]

Lepage, Yves (1971): "A Combination of Wilcoxon's and Ansari-Bradley's Statistics," *Biometrika* **58**:213–217.

——— (1973): "A Table for a Combined Wilcoxon Ansari-Bradley Statistic," *Biometrika* **60**:113–116.

Mann, Henry and Whitney, D. R. (1947): "On a Test of Whether One of Two Random Variables is Stochastically Larger Than the Other," *Ann. Math. Statist.* **18**:50–60. [Obtains the formulas (2.17) and (2.21) for the expectation and variance of W_{XY}.]

Massey, F. J., Jr. (1950): "A Note on the Power of a Nonparametric Test," *Ann. Math. Statist.* **21**:440–443.

Milton, Roy C. (1970): *Rank Order Probabilities: Two-sample Normal Shift Alternatives*, John Wiley & Sons, New York.

Moses, Lincoln E. (1953): "Nonparametric Methods," *Statistical Inference* (Walker and Lev), Henry Holt, New York, chap. 18, pp. 426–450. [Proposes the intervals (2.59).]

——— (1965): "Confidence Limits from Rank Tests (Query)," *Technometrics* **7**:257–260.

Noether, Gottfried E. (1955): "On a Theory of Pitman," *Ann. Math. Statist.* **26**:64–68.

——— (1971): *Introduction to Statistics*, Houghton Mifflin, Boston.

Owen, D. B. (1962): *Handbook of Statistical Tables*, Addison-Wesley, Reading, Mass.

Pitman, E. J. G. (1948): *Lecture Notes on Nonparametric Statistics*, Columbia University Press, New York. [Develops the efficiency concept that carries his name. A convenient reference to his approach is Noether (1955).]

Pratt, John W. (1964): "Robustness of Some Procedures for the Two-sample Location Problem," *J. Am. Statist. Assoc.* **59**:665–680.

Putter, Joseph (1955): "The Treatment of Ties in Some Nonparametric Tests," *Ann. Math. Statist.* **26**:368–386.

Raghavachari, M. (1965a): "The Two-sample Scale Problem When Locations Are Unknown," *Ann. Math. Statist.* **36**:1236–1242.

——— (1965b): "On the Efficiency of the Normal Scores Test Relative to the *t*-test," *Ann. Math. Statist.* **36**:1306–1307.

Ramachandramurty, P. V. (1966): "On Some Nonparametric Estimates for the Shift in the Behrens-Fisher Situation," *Ann. Math. Statist.* **37**:593–610.

Ramsey, Fred L. (1971): "Small Sample Power Functions for Nonparametric Tests of Location in the Double Exponential Family," *J. Am. Statist. Assoc.* **66**:149–151.

Robbins, Herbert (1970): "Statistical Methods Related to the Law of the Iterated Logarithm," *Ann. Math. Statist.* **41**:1397–1409.

Savage, I. Richard (1956): "Contributions to the Theory of Rank Order Statistics—The Two-sample Case," *Ann. Math. Statist.* **27**:590–615.

——— and Sethuraman, J. (1966): "Stopping Time of a Rank Order Sequential Probability Ratio Test Based on Lehmann Alternatives," *Ann. Math. Statist.* **37**:1154–1160.

———, Sobel, Milton and Woodworth, George (1966): "Fine Structure of the Ordering of Probabilities of Rank Orders in the Two-sample Case," *Ann. Math. Statist.* **37**:98–112.

Scholz, F. (1970): "Note on Efficiencies of Rank Tests Relative to the *F*-test in Testing Differences of Scale," *Ann. Math. Statist.* **41**:1156.

Sen, Pranab Kumar (1963): "On the Estimation of Relative Potency in Dilution (-direct) Essays by Distribution-free Methods," *Biometrics* **19**:532–552. [Proposes the estimate $\hat{\Delta}$ and the confidence intervals (2.59). See also Hodges and Lehmann (1963) and Moses (1953).]

Serfling, R. J. (1968): "The Wilcoxon Two-sample Statistic on Strong Mixing Processes," *Ann. Math. Statist.* **39**:1202–1209.

Stone, M. (1967): "Extreme Tail Probabilities for the Null Distribution of the Two-sample Wilcoxon Statistics," *Biometrika* **54**:629–640.

——— (1968): "Extreme Tail Probabilities for Sampling without Replacement and Exact Bahadur Efficiency of the Two-sample Normal Scores Test," *Biometrika* **55**:371–375.

Sukhatme, B. V. (1957): "On Certain Two-sample Nonparametric Tests for Variances," *Ann. Math. Statist.* **28**:188–194.

——— (1958): "Testing the Hypothesis that Two Populations Differ Only in Location," *Ann. Math. Statist.* **29**:60–78.

Sundrum, R. M. (1954): "A Further Approximation to the Distribution of Wilcoxon's Statistic in the General Case," *J. Roy. Statist. Soc.* **B16**:255–260.

Teichroew, D. (1956): "Tables of Expected Values of Order Statistics and Products of Order

Statistics for Samples of Size Twenty and Less from the Normal Distribution," *Ann. Math. Statist.* **27**:410–426.

Terry, Milton E. (1952): "Some Rank Order Tests Which Are Most Powerful against Specific Parametric Alternatives," *Ann. Math. Statist.* **23**:346–366.

Tukey, J. W. (1949): "Standard Confidence Points," *Unpublished Report 16*, Statistical Research Group, Princeton University.

——— (1962): "Reduction of Wasted Stringency in Ranking and Counting Procedures Through the Use of Supplementary Criteria," Unpublished report, Statistical Research Group, Princeton University.

Van der Vaart, H. R. (1961): "On the Robustness of Wilcoxon's Two-sample Test," in *Quantitative Methods in Pharmacology*, Interscience, New York, pp. 140–158.

Van der Waerden, B. L. (1952, 1953): "Order Tests for the Two-sample Problem and Their Power," *Indagationes Math.* **14**:453–458, **15**:303–316. "Errata," *ibid.* (1953), p. 80.

——— and Nievergelt, E. (1956): *Tables for Comparing Two Samples by X-test and Sign Test*, Springer, Berlin.

Vorličková, D. (1970): "Asymptotic Properties of Rank Test under Discrete Distributions," *Z. Wahrscheinlichkeitsth Verw. Gebiete* **14**:275–289.

Wetherill, G. B. (1960): "The Wilcoxon Test and Non-null Hypotheses," *J. Roy. Statist. Soc.* **B27**:402–418.

Wilcoxon, Frank (1945): "Individual Comparisons by Ranking Methods," *Biometrics* **1**:80–83.

———, Rhodes, L. J., and Bradley, R. A. (1963): "Two Sequential Two-sample Grouped Rank Tests with Applications to Screening Experiments," *Biometrics* **19**:58–84.

Witting, Hermann (1960): "A Generalized Pitman Efficiency for Nonparametric Tests," *Ann. Math. Statist.* **31**:405–416.

Woinsky, Melvin N. (1972): "A Composite Nonparametric Test for a Scale Slippage Alternative," *Ann. Math. Statist.* **43**:65–73.

Woodworth, George G. (1970): "Large Deviations and Bahadur Efficiency of Linear Rank Statistics," *Ann. Math. Statist.* **41**:251–283.

Yu, C. S. (1971): "Pitman Efficiencies of Kolmogorov-Smirnov Tests," *Ann. Math. Statist.* **42**:1595–1605.

Zaremba, S. K. (1962): "A Generalization of Wilcoxon's Test," *Monatsh. Math.* **66**:359–370.

CHAPTER 3
BLOCKED COMPARISONS FOR TWO TREATMENTS

1. THE SIGN TEST FOR PAIRED COMPARISONS

The first two chapters were concerned with the comparison of two treatments when the N subjects available for observation are either given and are divided at random into a treatment and control group (Chap. 1) or when they are obtained through random sampling from a large population (Chap. 2). As has been pointed out, this type of comparison is likely to be ineffective when the subjects are highly variable since the inherent variation may then swamp any difference that may exist between the treatments. In such cases, the effectiveness of the comparison can often be increased by dividing the subjects into more homogeneous subgroups and comparing treatment and control subjects only within each subgroup.

It will typically be easier to obtain small homogeneous subgroups than large ones, and particularly good opportunities are often available for dividing the subjects into homogeneous subgroups of size two. An obvious example is provided by studies of twins, where each subgroup consists of a pair of twins. In other situations, the two eyes or hands of the same person constitute a natural subgroup. Sometimes a subject can serve as his own control; for example, in taste tests in which two preparations or products are being compared or in a comparison of two computing machines through the speeds with which a number of subjects perform the same type of calculation on both. Divisions into homogeneous subgroups of size two are not restricted to situations in which there exists a natural pairing but also can often be achieved through careful matching of subjects, for example, of patients who are alike with respect to age, sex, and severity of disease, of communities that have the same size, urban-rural character, geographic location, and so forth.

Suppose that N pairs of subjects are available for such a *paired comparison* and that within each pair one subject is selected at random, i.e., with probability $\frac{1}{2}$, to receive the treatment while the other serves as control. (If a subject serves as his own control, it is typically the order of application of the two treatments that is randomized.) The selections for different pairs are assumed to be independent of each other. In the present section, it will be assumed that the N available pairs of subjects are not chosen but are given and that chance enters only through the random assignments within each pair to treatment and control. In analogy with the model of Chap. 1, Sec. 1, we shall call this the *randomization model for paired comparisons*. A corresponding population model, in which the subjects are not fixed but are drawn from a population of such subjects, will be discussed in Chap. 4.

A simple test of the hypothesis H of no treatment effect in the above randomization model can be based on the number S_N of pairs for which the treated subject comes out ahead of the control, for example, gives the higher response when it is hoped that the treatment will increase the response. The hypothesis is rejected when S_N is sufficiently large.

The statistic S_N depends on the responses of the N pairs of subjects only through the differences between the treatment and control response in each pair. It is in fact equal to the number of positive such differences and the resulting test is therefore called the *sign test*. Since the test does not require the values of the responses or even of their differences, but only the sign of each difference, it is applicable when only qualitative comparisons of the two treatments are available, such as "Drug B gave greater relief than drug A," or "Candidate B made a better impression than candidate A." Its applicability in such cases is possibly the most valuable feature of the test. (For further discussion of this point, see Chap. 4, Secs. 1 and 3.)

The null distribution of S_N is found by an argument analogous to that of Chap. 1, Sec. 2. Under H, the responses of the two subjects of a given pair depend only on the subjects, not on which of them is assigned to treatment and which to control. The treated subject will therefore come out ahead of the control provided the subject with the better response happens to be assigned to treatment, an event which occurs with probability $\frac{1}{2}$. Each pair thus constitutes a trial which with probability $\frac{1}{2}$ ends in success (treated subject comes out ahead) or failure (control subject comes out ahead). The test statistic S_N is the number of successes in N independent such trials, and the null distribution of S_N is therefore the binomial distribution corresponding to N trials with success probability $\frac{1}{2}$.

This distribution is given by the formula

$$P_H(S_N = a) = \binom{N}{a}\frac{1}{2^N} \quad \text{for } a = 0, 1, \ldots, N \tag{3.1}$$

A table of the distribution giving the probabilities $P_H(S_N \leqslant a)$ for $N \leqslant 40$ up to at least the first entry that is $\geqslant .5$ is found as Table G at the end of this book. For larger values of a, the probabilities can be obtained from the table using the fact that the distribution is symmetric about $N/2$ (Prob. 58).

EXAMPLE 1. *A headache remedy.* To test the effectiveness of a new remedy for tension headaches, each of 15 sufferers from such headaches is given a number of the new pills and an equal number of a standard brand, in bottles labeled at random A and B. The subjects are asked to take one of the pills at the onset of each headache, alternating between A and B, and when the pills are exhausted to report which of the two drugs was more effective. Suppose that 10 of the subjects report in favor of the bottle that contains the new drug (a list is of course kept which for each of the subjects states whether A or B contains the new drug). The significance probability of this result is obtained from Table G to be

$$P_H(S_{15} \geqslant 10) = P_H(S_{15} \leqslant 5) = .1509$$

For large N, the distribution of S_N is approximately normal. The required expectation and variance of S_N are shown in formulas (A.24) and (A.26) of the Appendix with $p = \frac{1}{2}$ to be given by

$$E_H(S_N) = \frac{N}{2} \quad \text{and} \quad \mathrm{Var}_H(S_N) = \frac{N}{4} \tag{3.2}$$

The approximation is supported by a limit theorem which states that the distribution of

$$\frac{S_N - \frac{1}{2}N}{\frac{1}{2}\sqrt{N}} \tag{3.3}$$

tends to the standard normal distribution. (See Appendix, Example 10.) To illustrate the normal approximation, consider once more the significance probability $P_H(S_N \geqslant 10)$ for $N = 15$. Using the continuity correction and the values $E_H(S_N) = 7.5$ and $\mathrm{Var}_H(S_N) = 3.75$, we find

$$P_H(S_N \geqslant 10) = P_H\left(\frac{S_N - \frac{1}{2}N}{\frac{1}{2}\sqrt{N}} \geqslant \frac{9.5 - 7.5}{1.937}\right) \approx \Phi\left(\frac{-2}{1.937}\right) = .151$$

which agrees well with the result obtained earlier from the table.

In the description of the sign test it was tacitly assumed that each difference is either positive or negative. It may however happen that in one or more pairs the treatment and control observations are equal and the resulting differences are then zero. Similarly, in qualitative evaluations like those of Example 1, a zero may result from the inability to decide whether the new or standard treatment is more effective. A natural test statistic in the presence of such ties, analogous to the statistic W^*_{XY} considered in Chap. 1, is obtained by assigning to a pair of subjects the score 1, $\frac{1}{2}$, or 0 as the difference between the treated and control response is positive, zero, or negative. If N_+, N_0, and N_- denote the numbers of positive, zero, and negative differences among the N pairs, the test statistic is then $N_+ + \frac{1}{2}N_0$, and the test rejects when this statistic is sufficiently large.

Recall now that the subjects and their responses are considered as given and fixed, and only the assignment of subjects within each pair to treatment and control as variable. This implies that under the hypothesis of no treatment effect the number N_0 of zero differences is fixed, i.e., is a constant rather than a random variable. A zero difference occurs only when the responses of the two subjects in that pair are equal, and the difference of their responses (treated minus control) will then be zero regardless of which is assigned to treatment and which to control. Thus, under H the statistics $N_+ + \frac{1}{2}N_0$ and N_+ are equivalent, and the test can be based on the simpler N_+, which has the binomial distribution corresponding to $N' = N - N_0$ trials with success probability $\frac{1}{2}$. The resulting test, which rejects the hypothesis when N_+ is sufficiently large, consists in discarding the zeros and applying the sign test to the $N' = N - N_0$ untied pairs.

2. THE WILCOXON SIGNED-RANK TEST

The sign test, which was discussed in the preceding section, utilizes only the signs of the differences of the N pairs. More effective tests are possible when the values of the differences are available, as is indicated by the following example.

EXAMPLE 2. *Testing a new fertilizer.* To test whether a new fertilizer will give a higher yield than the fertilizer that has been used in the past, three strawberry fields on a farm are each divided into two parts which are assigned to the two fertilizers at random. There are then three pairs, the two parts of each field constituting a pair. Suppose that the yields are 76 and 78 lb on one field, 82 and 91 on the second, and 80 and 86 on the third, so that the differences are 2, 9, and 6 lb, respectively. The sign test will look only at the number of fields in which the new fertilizer gave the bigger yield. However, suppose that this is the case for two of the fields; it is then also of interest to know just which two. The results will appear more significant if they were the fields with differences 6 and 9, than if they were those with differences 2 and 6. This suggests that the test should be based not only on the signs of the differences but also on the values or ranks of the differences to which these signs are attached. In (3.4) are displayed all possible sign combinations of the three differences and below each their *signed ranks*, that is, the ranks of the absolute values of the differences together with their signs.

(3.4)

$-2,-6,-9$ $-1,-2,-3$	$-2,-6,+9$ $-1,-2,+3$	$-2,+6,-9$ $-1,+2,-3$	$-2,+6,+9$ $-1,+2,+3$
$+2,-6,-9$ $+1,-2,-3$	$+2,-6,+9$ $+1,-2,+3$	$+2,+6,-9$ $+1,+2,-3$	$+2,+6,+9$ $+1,+2,+3$

The signed ranks $-1,+2,+3$ in the fourth box, for example, correspond to the fact that in this case the difference with the smallest absolute value (2) is negative, but those with the larger absolute values (6 and 9) are positive.

Let us now find the null distribution of the signed ranks under the hypothesis H that there is no difference between the fertilizers. If H is true, the yields of the two halves of the first field will be 76 and 78 lb, regardless of which half is given the new and which the standard fertilizer. The difference new $-$ standard will therefore be equal to either $+2$ or -2, each with probability $\frac{1}{2}$. Similarly, with probability $\frac{1}{2}$ each, the second difference will be $+6$ or -6, and the third difference will be $+9$ or -9. Furthermore, the signs of the three differences will be independent. It follows that the $2 \times 2 \times 2 = 2^3 = 8$ sign combinations displayed in (3.4) each have probability $\frac{1}{8}$, and the same is therefore true of the 8 possible sets of values of the signed ranks. This specifies the desired null distribution.

The above considerations easily generalize and provide the basis for the signed-rank test that will be discussed below. Let there be N pairs of subjects, with random assignment to treatment and control within each pair, and suppose that $N_+ = n$ of the differences between the treated and control response are positive and the remaining $N_- = m = N-n$ are negative. Rank the absolute values of the differences and to the rank of each absolute value attach the sign of the difference as in (3.4). Let us denote the ranks whose signs are negative by $R_1 < \cdots < R_m$ and those with positive signs by $S_1 < \cdots < S_n$, so that the R's and S's between them are the integers $1, \ldots, N$.

There are 2^N possible sign combinations $\pm 1, \pm 2, \ldots, \pm N$ and each is characterized by the subset $(S_1, \ldots, S_n)$ of ranks with positive signs. If, for example, $n = 2$ and $S_1 = 1$, $S_2 = N$, the signed ranks are $[+1, -2, \ldots, -(N-1), +N]$. This corresponds to the case that the differences with the smallest and largest absolute value are positive and all others negative. Similarly, if $n = N-1$ and $S_1 = 1$, $S_2 = 3$, $S_3 = 4, \ldots, S_n = N$, the signed ranks are $(+1, -2, +3, +4, \ldots, +N)$, and the difference with the second smallest absolute value is negative and all others are positive. A special case occurs when all differences are negative. Then $n = 0$ and the set $(S_1, \ldots, S_n)$ is the empty set. A check on the above accounting is obtained by noting that for any fixed value of n, there are $\binom{N}{n}$ possible choices for $(S_1, \ldots, S_n)$. Since

n can take on the values $0, 1, \ldots, N$, the total number of possible choices for the S's is therefore

$$\binom{N}{0}+\binom{N}{1}+\cdots+\binom{N}{N} = (1+1)^N = 2^N$$

just the same as the total number of sign combinations, as it should be. In obtaining the signed ranks, we have assumed that no ties occur among the absolute values of the differences and that none of the differences are zero. The modifications required in the presence of such ties or zeros will be discussed later in this section.

Under the hypothesis H of no treatment effect, it follows as before from the random assignment within each pair that each difference is positive or negative with probability $\frac{1}{2}$, and that the N signs are independent, so that each of the 2^N possible sign combinations has probability $1/2^N$. Since each sign combination corresponds to exactly one value of n and $(S_1, \ldots, S_n)$, the joint null distribution of these variables is given by

$$P_H(N_+ = n; S_1 = s_1, \ldots, S_n = s_n) = 1/2^N \tag{3.5}$$

Here the left-hand side denotes the probability that the number of positive differences is n and that their ranks among the absolute values of the differences are $(s_1, \ldots, s_n)$. This formula plays the same fundamental role for paired comparisons as did formula (1.4) of Chap. 1 for completely randomized comparisons.

Consider now testing the hypothesis H of no treatment effect against the alternatives that the treatment tends to increase the response. Rejection of H is indicated if for many of the pairs the treated subject comes out ahead and if the differences tend to be larger for the pairs in which the treated subject comes out ahead than for those in which it is surpassed by the control; that is, if n is large and if the positive signed ranks $(S_1, \ldots, S_n)$ tend to be larger than the ranks $(R_1, \ldots, R_m)$. A simple statistic that combines these two criteria is the sum of the positive signed ranks

$$V_s = S_1 + \cdots + S_n \tag{3.6}$$

The hypothesis of no treatment difference is rejected when V_s is sufficiently large, say, when

$$V_s \geqslant c \tag{3.7}$$

The resulting test is the *Wilcoxon signed-rank* test. We shall omit the qualification *signed-rank* when there is no possibility of confusion with the Wilcoxon rank-sum test of Chap. 1. The rejection rule $V_s = S_1 + \cdots + S_n \geqslant c$ formally looks exactly like that of the Wilcoxon rank-sum test which rejects H when $W_s = S_1 + \cdots + S_n \geqslant c$. However, in the earlier test n was fixed, while in the present case n is the value of a random variable, which can take on the value $0, 1, \ldots, N$. As a consequence, the statistics V_s and W_s have quite different null distributions and different critical values c corresponding to the same significance level α.

The computation of the null distribution of V_s may be illustrated by the case $N = 3$. To each set of signed ranks corresponds a value v of V_s, as shown in (3.8):

(3.8)

Signed ranks	$-1, -2, -3$	$-1, -2, +3$	$-1, +2, -3$	$-1, +2, +3$
v	0	3	2	5

Signed ranks	$+1, -2, -3$	$+1, -2, +3$	$+1, +2, -3$	$+1, +2, +3$
v	1	4	3	6

Since the probability of each set of signed ranks is 1/8, the null distribution of V_s is given by

(3.9)

v	0	1	2	3	4	5	6
$P_H(V_s = v)$	$\frac{1}{8}$	$\frac{1}{8}$	$\frac{1}{8}$	$\frac{2}{8}$	$\frac{1}{8}$	$\frac{1}{8}$	$\frac{1}{8}$

This method of computing the null distribution by enumeration is quite general. Let $\#(v;N)$ denote the number of sign combination of the ranks $1, \ldots, N$ for which the sum of the positive signed ranks is equal to v. For example, when $N = 3$, it is seen from (3.8) that

$$\#(2;N) = 1 \qquad \text{and} \qquad \#(3;N) = 2$$

Since under H the probability of any particular sign combination is $1/2^N$, it follows that

$$P_H(V_s = v) = \frac{\#(v;N)}{2^N} \tag{3.10}$$

and the null distribution can be obtained by determining, through enumeration, the numbers $\#(v;N)$. Let us illustrate the method on another example.

EXAMPLE 3. *Sutured vs. taped wounds.* In a study of the comparative tensile strength of tape-closed and sutured wounds, the following results were obtained on 10 rats, 40 days after incisions made on their backs had been closed by suture or by surgical tape.[1]

Rat	1	2	3	4	5	6	7	8	9	10
Tape	659	984	397	574	447	479	676	761	647	577
Suture	452	587	460	787	351	277	234	516	577	513
Difference	207	397	−63	−213	96	202	442	245	70	64
Signed rank	6	9	−1	−7	4	5	10	8	3	2

Suppose it is desired to test the hypothesis of no difference against the alternatives that the

[1] Data from Ury and Forrester, "More on Scale Dependence of the Wilcoxon Signed-rank Test," *Amer. Statist.* **24**(3):25–26 (June 1970).

tape-closed wounds are stronger.[1] Since there are fewer negative than positive signed ranks, it will be more convenient to work with the sum

$$V_r = R_1 + \cdots + R_m \tag{3.11}$$

of the ranks corresponding to the negative differences. The sum of V_r and V_s is equal to the sum of the integers from 1 to N, so that

$$V_r + V_s = \tfrac{1}{2}N(N+1) \tag{3.12}$$

The test statistics V_r and V_s are therefore equivalent, with small values of V_r corresponding to large values of V_s. The observed value of V_r is $1+7=8$ and since small values of V_r are significant, the significance probability is $P_H(V_r \leqslant 8)$. To compute this probability, let us list the values of m and $(r_1, \ldots, r_m)$ for which $V_r \leqslant 8$:

$m = 0$ empty set
$m = 1$ $r_1 = 1, 2, \ldots, 8$
$m = 2$ $(r_1,r_2) = (1,2), (1,3), (1,4), (1,5), (1,6), (1,7), (2,3), (2,4), (2,5), (2,6), (3,4), (3,5)$
$m = 3$ $(r_1,r_2,r_3) = (1,2,3), (1,2,4), (1,2,5), (1,3,4)$

The total number of cases is $1+8+12+4 = 25$, and the significance probability is therefore[2]

$$P_H(V_r \leqslant 8) = 25/2^{10} = .0244$$

If instead of the Wilcoxon test the sign test is used, the significance probability is $P_H(S_{10} \leqslant 2) = 56/2^{10} = .0547$, so that the results appear less significant. A detailed comparison of the two tests, given in Chap. 4, Sec. 3, indicates that for the type of data most commonly met in practice the Wilcoxon test is more powerful, that is, has a better chance of detecting a treatment difference when it exists. For this reason, as has already been mentioned, the principal use of the sign test is in situations in which only the signs of the differences are available and not their values. (Another use is described in Chap. 4, Sec. 1.)

To facilitate the use of the Wilcoxon signed-rank test, a table of the null distribution is given as Table H at the end of the book. It is seen from the computation of this distribution in the examples that V_r and V_s have the same null distribution, so that the table applies equally to both. Since the distribution is symmetric about $N(N+1)/4$ (Prob. 61) it is again enough to table one tail. The quantity tabled is

$$P_H(V_s \leqslant v) = P_H(V_r \leqslant v)$$

for $N \leqslant 20$ up to at least the first entry that is $\geqslant .5$. As an illustration, let us check the significance probability $P_H(V_r \leqslant 8)$ when $N = 10$, of Example 3. Entering the table with $N = 10$ and $v = 8$, we find $P_H(V_r \leqslant 8) = .0244$ as before. [For more extensive tables see, for example, McCormack (1965) and Wilcoxon, Katti, and Wilcox (1970).]

[1] Actually, the authors applied a two-sided rather than a one-sided Wilcoxon test.

[2] If the hypothesis of no difference is being tested against the two-sided alternatives that the sutured wounds have either more or less strength than those closed by tape, this significance probability would have to be doubled. (See the discussion in Chap. 1, Sec. 5.)

Beyond the range of the table, one can use the normal approximation (1.26) with $T = V_s$.† The expectation and variance of V_s are shown in formulas (A.28) and (A.29) of the Appendix to be given by

$$E_H(V_s) = \frac{N(N+1)}{4} \tag{3.13}$$

and

$$\mathrm{Var}_H(V_s) = \frac{N(N+1)(2N+1)}{24} \tag{3.14}$$

Let us illustrate the normal approximation by applying it to the significance probability $P_H(V_r \leqslant 8)$ in the case $N = 10$, which we computed earlier by enumeration. Here $E_H(V_r) = 27.5$ and $\mathrm{Var}_H(V_r) = 96.25$. Using the continuity correction, we therefore find for the desired probability the approximation

$$P_H(V_r \leqslant 8) \approx \Phi\left(\frac{8.5-27.5}{9.81}\right) = \Phi(-1.9367) = .0264$$

which is in close agreement with the value .0244 obtained earlier.

A more systematic idea of the accuracy of the approximation can be obtained from Table 3.1, which provides a comparison of the Wilcoxon probability $P(V_s \leqslant v)$ with its normal approximation, both with and without continuity correction. It suggests that at least for the range $\alpha = .01$ to $\alpha = .15$ the approximation with continuity correction will be adequate for most purposes when $N \geqslant 20$, and hence for the sample sizes not covered by Table H.

TABLE 3.1 Exact and approximate values of $P(V_s \leqslant v)$ with and without continuity correction.

v	3	5	11	14	17
Exact	.0049	.0098	.0527	.0967	.1611
Without	.0063	.0109	.0463	.0844	.1423
With	.0072	.0124	.0515	.0926	.1541

$N = 10$

v	37	43	61	70	76
Exact	.0047	.0096	.0527	.1012	.1471
Without	.0055	.0103	.0502	.0957	.1394
With	.0059	.0108	.0522	.0989	.1437

$N = 20$

† For an improvement over this approximation see Fellingham and Stoker (1964), Bickel (1974), and Claypool and Holbert (1974).

The normal approximation is supported by a limit theorem which states that

$$P_H\left[\frac{V_s-\frac{1}{4}N(N+1)}{\sqrt{N(N+1)(2N+1)/24}} \leqslant z\right] \to \Phi(z) \tag{3.15}$$

as $N \to \infty$. A proof is given in Appendix, Sec. 3, Example 13.

The statistic V_s has an interesting alternative interpretation, which will be useful for later applications. Denote the N differences by $Z_1, \ldots, Z_N$ and consider the totality of averages $\frac{1}{2}(Z_i+Z_j)$ with $i \leqslant j$. For $i<j$, these are averages of two of the observed differences; for $i=j$, they reduce to the differences themselves. There are $\binom{N}{2}+N$ such averages, and we shall now show that V_s is the number among them that are positive, that is,

$$V_s = \text{number of positive averages } \tfrac{1}{2}(Z_i+Z_j) \text{ with } i \leqslant j \tag{3.16}$$

To see (3.16), note that $\frac{1}{2}(Z_i+Z_j)$, or equivalently Z_i+Z_j, is positive if and only if the difference with the larger absolute value is positive (or in case $i=j$, if the single difference is positive). Suppose without loss of generality that the differences have been numbered in increasing order of their absolute values, so that the signed rank of Z_j is $+j$ or $-j$ as Z_j is positive or negative. Then for a fixed value of j, the contribution to the right-hand side of (3.16) from pairs (Z_i, Z_j) with $i \leqslant j$ is zero when $Z_j<0$ and is j (and hence equal to the signed rank of Z_j) when $Z_j>0$. Thus the right-hand side of (3.16) is equal to the sum of the positive signed ranks, as was to be proved.

The definition of signed ranks in the preceding section presupposes that no ties occur among the absolute values of the differences and that none of the differences are zero. The modifications required in the presence of zeros or ties are quite analogous to those made in the unpaired case (Chap. 1, Sec. 4).

Suppose, for example, that $N=7$ and that each observation is a score of -2, -1, 0, 1, or 2, corresponding to an assessment of the subject's response as very poor, poor, indifferent, good, or very good. Let the seven pairs of observations be

$$(-1,0) \quad (-2,0) \quad (1,0) \quad (2,2) \quad (0,0) \quad (-1,1) \quad (0,0)$$

with the first member of each pair representing the control and the second the treatment response. The seven differences are $+1$, $+2$, -1, 0, 0, $+2$, 0, their absolute values (in increasing order) 0, 0, 0, 1, 1, 2, 2, and the midranks of these absolute values 2, 2, 2, 4.5, 4.5, 6.5, 6.5. By multiplying each of the midranks by $+1$, -1, or 0 as the corresponding difference is positive, negative, or zero, we obtain the signed midranks shown below:

Difference	-1	0	0	0	$+1$	$+2$	$+2$
Signed midrank	-4.5	0	0	0	$+4.5$	$+6.5$	$+6.5$

The sum of the positive signed midranks is thus

$$V_s^* = 4.5 + 6.5 + 6.5 = 17.5$$

Let us now compute the null distribution of V_s^*. In this process, the zeros can be disregarded and attention restricted to the four nonzero signed midranks -4.5, $+4.5$, $+6.5$, $+6.5$. Note that the zeros are dropped only *after* the midranks have been computed. (For an alternative approach see Sec. 5B.) This corresponds to the fact, noted already in the discussion of the sign test, that the zeros do not involve an element of randomness as do the other signed ranks but will have the same value (zero) regardless of the assignment within the corresponding pairs to treatment and control. In the present case, there are $2^4 = 16$ possible sign combinations, and we list below the different sets of positive signed midranks $(s_1^*, \ldots, s_n^*)$ together with the values of their sum V_s^*, and their probabilities.

$(s_1^*, \ldots, s_n^*)$	None	4.5	6.5	4.5, 4.5	4.5, 6.5	6.5, 6.5
V_s^*	0	4.5	6.5	9	11	13
Probability	$\frac{1}{16}$	$\frac{2}{16}$	$\frac{2}{16}$	$\frac{1}{16}$	$\frac{4}{16}$	$\frac{1}{16}$

$(s_1^*, \ldots, s_n^*)$	4.5, 4.5, 6.5	4.5, 6.5, 6.5	4.5, 4.5, 6.5, 6.5
V_s^*	15.5	17.5	22
Probability	$\frac{2}{16}$	$\frac{2}{16}$	$\frac{1}{16}$

The probability in the second box, for example, is $2/16$ since there are two possible choices of the midrank 4.5. Similarly, the probability $4/16$ in the fifth box reflects the fact that there are two choices each of the midranks 4.5 and 6.5 and hence a total of $2 \times 2 = 4$ choices. Since there are no duplications in the values of V_s^* shown in the second row, the second and third row together give the desired null distribution of V_s^*. If large values of V_s^* are significant, it follows in particular that the significance probability is $P_H(V_s^* \geqslant 17.5) = 3/16$.

Since tables for the null distribution of V_s^* are not practicable and enumeration quickly becomes unmanageable as N increases, we again have recourse to the normal approximation. The required expectation is given by

$$E_H(V_s^*) = \frac{N(N+1) - d_0(d_0+1)}{4} \tag{3.17}$$

where d_0 is the number of zero differences. For $d_0 = 0$, this reduces to the expectation of V_s.

The variance of V_s^* depends not only on d_0 but also on the numbers $d_1, \ldots, d_e$ of ties among the nonzero absolute differences, and is given by the formula

$$\operatorname{Var}_H(V_s^*) = \frac{1}{24}[N(N+1)(2N+1) - d_0(d_0+1)(2d_0+1)] - \frac{1}{48}\sum_{i=1}^{e} d_i(d_i-1)(d_i+1) \tag{3.18}$$

This reduces, as it should, to the variance of V_s when $d_0 = 0$ and the remaining d_i are equal to 1. Proofs of (3.17) and (3.18) are given in formulas (A.30) and (A.31) of the Appendix.

The normal approximation is again supported by a limit theorem, which states that the null distribution of $[V_s^* - E_H(V_s^*)]/\sqrt{\text{Var}_H(V_s^*)}$ tends to the standard normal distribution as the number of nonzero differences tends to infinity. For a proof of this fact see Appendix, Example 14 and Prob. 28.

EXAMPLE 4. *Vitamins and IQ*. In a study of the effect of thiamine (vitamin B) on learning, 74 children living in an orphanage were divided into 37 matched pairs. One child within each pair was selected at random to receive the thiamine, the other received a placebo and served as control. Although the study was primarily concerned with learning, a number of other variables were also observed. The following table shows the gain in IQ during the six weeks of the experiment for 12 of the pairs.[1]

Pair	2	5	8	11	14	17	20	23	26	29	32	35
Treated	14	18	2	4	−5	14	−3	−1	1	6	3	3
Control	8	26	−7	−1	2	9	0	−4	13	3	3	4
Difference	6	−8	9	5	−7	5	−3	3	−12	3	0	−1
Signed midrank	8	−10	11	6.5	−9	6.5	−4	4	−12	4	0	−2

To test the hypothesis of no treatment effect against the alternative that the treated children tend to show a larger increase than those not treated, we can employ the Wilcoxon test. The signed midranks of the differences (treated − control) are shown in the last row of the table. The sum of the positive midranks is 40, and the significance probability is therefore $P(V_s^* \geqslant 40)$. Since $d_0 = 1$ and of the remaining d's, one is equal to 2, one is equal to 3, and six are equal to 1, it is seen from (3.17) and (3.18) that $E(V_s^*) = 38.5$ and $\text{Var}(V_s^*) = 161.625$. This gives the approximate significance probability

$$P(V_s^* \geqslant 40) \approx 1 - \Phi\left(\frac{40 - 38.5}{12.71}\right) = .453$$

The treatment therefore had no significant effect in increasing the IQ scores.

Like V_s, the statistic V_s^* also has an alternative interpretation. Suppose that the N differences are denoted by $Z_1, \ldots, Z_N$ and that each average $\frac{1}{2}(Z_i + Z_j)$, $i \leqslant j$, is scored as 1, $\frac{1}{2}$, or 0 as it is positive, zero, or negative. Then V_s^* differs from the sum of these scores by a constant; more precisely,

$$\begin{aligned} V_s^* = {} & [\text{number of positive averages } \tfrac{1}{2}(Z_i + Z_j) \text{ with } i \leqslant j] \\ & + \tfrac{1}{2}[\text{number of averages } \tfrac{1}{2}(Z_i + Z_j) \text{ with } i \leqslant j \text{ that are zero}] \\ & - \tfrac{1}{4} d_0(d_0 + 1) \end{aligned} \tag{3.19}$$

[1] Data from Ruth Flint Harrell, "Effect of Added Thiamine on Learning," *Contrib. Educ.* no. 877, Teacher's College, Columbia University, 1943, Table 10.

where d_0 is the number of zeros among the Z's. The proof of this formula is analogous to that of formula (1.38), and is indicated in Prob. 71.

3. COMBINING DATA FROM SEVERAL EXPERIMENTS OR BLOCKS

It often happens that after a study has been completed, additional observations become available and the problem then arises how to combine the data from these different sources.

EXAMPLE 5. *Televised vs. live instruction.* To see whether a televised history course is as effective as having the instructor in the classroom, students are asked to volunteer for an experiment. Some of them, chosen at random, will attend the lectures in the classroom in which they are given while the others will watch the lectures on the screen in an adjoining room. Suppose that 12 students agree to participate in the study, and that these students are found to vary considerably in background and ability. It is therefore decided to match them into pairs rather than use a completely randomized comparison. The pairs are formed by matching on grades in previous history courses, number of years in college, and also on sex because it is thought that the psychological effect of having the instructor in the classroom rather than only seeing him on a screen might be different for the two sexes. Suppose that the final examination gives the following six pairs of scores:

TV	70	77	80	80	84	73
Live	73	75	80	83	85	74
Difference	−3	2	0	−3	−1	−1
Signed midrank	−5.5	+4	0	−5.5	−2.5	−2.5

At this point, someone recalls that another instructor had performed a similar experiment the year before. It turns out that the earlier study involved five pairs of students and gave the following results:

TV	85	93	90	91	89
Live	89	92	90	98	87
Difference	−4	1	0	−7	2

These earlier observations may be combined with those of the present study to give the following set of 11 differences:

$$-7 \quad -4 \quad -3 \quad -3 \quad -1 \quad -1 \quad 0 \quad 0 \quad 1 \quad 2 \quad 2$$

Ranking the absolute values of these differences leads to the signed midranks

$$-11 \quad -10 \quad -8.5 \quad -8.5 \quad -4 \quad -4 \quad 0 \quad 0 \quad 4 \quad 6.5 \quad 6.5$$

and hence to the value

$$V_s^* = 4+6.5+6.5 = 17$$

To see how to compute the significance probability $P_H(V_s^* \leqslant 17)$, consider the 2^9 possible sign combinations

$$\pm 1 \quad \pm 1 \quad \pm 1 \quad \pm 2 \quad \pm 2 \quad \pm 3 \quad \pm 3 \quad \pm 4 \quad \pm 7$$

of the nine nonzero differences. The argument of the preceding sections shows that under the hypothesis of no difference between the two types of instruction each sign is plus or minus with probability $^1/_2$ (since the sign of each difference is determined solely by the random assignment of the two subjects to the two methods of instruction) and that the nine signs are independent. Thus, the joint null distribution of the midranks, and hence the null distribution of V_s^*, is exactly the same as if the observations had been obtained on a single occasion rather than from two separate experiments, and the significance probability of the combined observations is .107 (Prob. 44).

Let us now change the conditions of the example. Suppose in the first experiment, without any pairing, 5 of 10 available students are assigned at random to the televised course, the other 5 serve as controls, and the scores on the final examination are

$$68 \quad 69 \quad 74 \quad 82 \quad 93 \qquad \text{and} \qquad 72 \quad 75 \quad 83 \quad 95 \quad 100$$

for the students attending the televised and live lectures, respectively. Let us assume that the comparison is repeated in another semester with eight students participating who are assigned at random, four each, to treatment and control, and that the scores on the final examination are

$$47 \quad 51 \quad 52 \quad 56 \qquad \text{and} \qquad 54 \quad 59 \quad 60 \quad 70$$

for the treatment and control subjects. Can we again proceed as if the observations had been obtained on a single occasion?

To answer this question, note that the combined study involves 18 students, of whom 9 have been assigned to treatment, but that not all $\binom{18}{9} = 48{,}620$ assignments are equally likely. In fact, 5 of the 9 treatment subjects must come from the first group of 10, with the remaining 4 coming from the second group of 8, so that only $\binom{10}{5}\binom{8}{4} = 252 \times 70 = 17{,}640$ of the assignments are possible. These are clearly equally likely and each has probability 1/17,640, and the remaining assignments have probability zero.

If the 18 scores were now ranked, the joint distribution of the 9 treatment ranks would no longer be given by (1.4), and the null distribution of the sum of the treatment ranks would no longer be the Wilcoxon rank-sum distribution but would have to be computed on the basis of the 17,640 equally likely cases. However, there are reasons for not basing a test on these ranks at all. A glance at the two sets of observations indicates that the scores in the second set are significantly lower than those in the first (perhaps the examination was harder, the grading stiffer, or the students less well prepared). This suggests that the comparison of treatment observations of one set with control observations of the other, which is implied by a single ranking of all 18 observations, should be avoided.

Before discussing ways of dealing with this difficulty, let us notice that the same problem arises in the context of the preceding sections where it was assumed that the subjects available for observation are deliberately divided into subgroups or blocks, which may differ widely but each of which is relatively homogeneous.

These may be natural blocks such such as litters of animals, or observations taken on the same day or in the same clinic or school; alternatively, they may be obtained by matching the subjects either by a subjective assessment or by objective criteria such as age, sex, severity of disease, class standing, or income.

Let there be b blocks, with N_i subjects in the ith block from which n_i are selected at random to receive the treatment, and the remaining $m_i = N_i - n_i$ serve as controls. The assignments in the different blocks are assumed to be independent. The number of possible assignments within the ith block is $\binom{N_i}{n_i}$, and the total number of possible assignments is the product

$$\binom{N_1}{n_1} \cdots \binom{N_b}{n_b} \tag{3.20}$$

Under the assumption of independent random assignments, all these assignments are equally likely and the probability of each is therefore the reciprocal of (3.20). As in Chaps. 1 and 2, we shall denote the total number of subjects by N, and the total number of treatment and control subjects by n and m, so that

$$N = N_1 + \cdots + N_b; \qquad n = n_1 + \cdots + n_b; \qquad m = m_1 + \cdots + m_b \tag{3.21}$$

The selection of the n treatment subjects is, however, no longer completely random since not all $\binom{N}{n}$ assignments are possible. Instead, we are dealing with a method of *restricted random assignment*, with only those assignments possible (and equally likely) in which exactly n_i subjects from the ith block receive the treatment. It should be noted that the notation (3.21) does not agree with that of the preceding sections. In the present notation, paired comparisons correspond to the case $N_i = 2$, $m_i = n_i = 1$ for all i. The number of pairs, which was previously denoted by N, in the present notation is b, and N now denotes the total number of subjects in the study.

When the observations are divided into blocks, which presumably vary considerably among each other, observations from different blocks are not directly comparable. For this reason, suppose that the observations are ranked separately in each block.

Let $S_{i1} < \cdots < S_{in_i}$ denote the n_i treatment ranks from the ith block. Then it follows from the independence of the random assignments within the different blocks and from formula (1.4) that the joint distribution of the $n = n_1 + \cdots + n_b$ treatment ranks under the hypothesis of no treatment effect is given by

$$P_H(S_{11} = s_{11}, \ldots, S_{1n_1} = s_{1n_1}; S_{21} = s_{21}, \ldots, S_{2n_2} = s_{2n_2}; \ldots) = \frac{1}{\binom{N_1}{n_1}\binom{N_2}{n_2} \cdots \binom{N_b}{n_b}} \tag{3.22}$$

Let

$$W_s^{(i)} = S_{i1} + \cdots + S_{in_i} \tag{3.23}$$

denote the Wilcoxon rank-sum statistic computed for the ith block, and consider the problem of testing the hypothesis of no treatment effect against the alternatives that the treatment does have an effect and that it is in the same direction for all blocks. Simple test statistics for this problem are provided by the sum of the statistics $W_s^{(i)}$, or more generally by linear combinations $\sum c_i W_s^{(i)}$. When the group sizes m_i and n_i (and hence the block sizes N_i) are the same for all blocks, it seems natural to give the same weight to each of the blocks and hence take as test statistic just the sum of the $W_s^{(i)}$. However, for unequal group and block sizes it turns out that a more powerful test is obtained by using the weights $c_i = 1/(N_i+1)$, and hence by basing the test on the statistic

$$W_s = \sum \frac{W_s^{(i)}}{N_i+1} \tag{3.24}$$

When the block sizes are equal, this statistic is proportional to the sum of the $W_s^{(i)}$, and the test therefore reduces to that based on this sum. Since the null distribution of W_s depends on the group sizes m_i and n_i, a table of this distribution is not practicable. We shall therefore illustrate the computation of significance probabilities by an example and then discuss the normal approximation.

EXAMPLE 6. *Two methods of advertising.* An advertising agency wishes to test the theory that a strikingly unpleasant and obnoxious advertisement will be more effective than a more pleasant one because of its stronger impact. It is thought that the name of the product will then be remembered longer, and the unpleasant impression will quickly be forgotten. To test this theory, the agency persuades an interested client to agree to an experimental advertising campaign for a new product that he is about to introduce in several cities. The advertising in some of the cities will be intentionally strident and unpleasant; in the other cities, it will be equally intensive but will try to please rather than repel.

The subjects in this example are the cities, and they could of course be assigned at random to the two methods, say, half the cities to each. However, the test cities vary considerably in size, and as a result there will also be great variability in the consumption of the product in question. If consumption were proportional to the size of the city, allowance could be made for this factor. Since this assumption is deemed not to be justified, it seems advisable to block the cities according to size. It turns out that they fall into four fairly homogeneous blocks: two rather large cities with about equal population, two groups of three each at intermediate levels, and a group of six smaller towns of about the same size. It is therefore decided to select in these four blocks one, two, one, and three cities, respectively, to be assigned to treatment (the strident advertising) with the remaining ones serving as controls (the more pleasant and conventional advertising). The consumption figures at the end of a month are given in the table below, in which the last two columns show the ranks for each block.

Block	Control	Treatment	Control	Treatment
1	236	255	1	2
2	183	179, 193	2	1, 3
3	115, 128	132	1, 2	3
4	61, 70, 79	67, 84, 88	1, 3, 4	2, 5, 6

To deal with smaller numbers we shall work with the test statistic

$$W_r = \frac{\sum W_r^{(i)}}{N_i+1} \tag{3.25}$$

instead of the equivalent W_s (Prob. 72). The weights $1/(N_i+1)$ for the four blocks are, respectively, $1/3$, $1/4$, $1/4$, and $1/7$. To avoid fractions, it is convenient to multiply the statistic

$$W_r = \tfrac{1}{3}W_r^{(1)} + \tfrac{1}{4}W_r^{(2)} + \tfrac{1}{4}W_r^{(3)} + \tfrac{1}{7}W_r^{(4)}$$

by the smallest common multiple of the denominators, which is 84, and work with the equivalent statistic

$$84W_r = 28W_r^{(1)} + 21W_r^{(2)} + 21W_r^{(3)} + 12W_r^{(4)}$$

The observed value of this statistic is

$$28 \times 1 + 21 \times 2 + 21 \times 3 + 12 \times 8 = 229$$

and the significance probability is $P_H(84W_r \leqslant 229)$. To facilitate the computation of this probability, let us first write down the distributions of the sums $W_r^{(1)}, \ldots, W_r^{(4)}$ of the control ranks in the different blocks, multiplied by 28, 21, 21, and 12, respectively. These can either be obtained through enumeration (they are trivial for the first three blocks) or read from Table B. In the table below, the first line gives the value of the statistic and the second its probability.

$28W_r^{(1)}$		$21W_r^{(2)}$			$21W_r^{(3)}$			$12W_r^{(4)}$					
28	56	21	42	63	63	84	105	72	84	96	108	120	···
$\frac{1}{2}$	$\frac{1}{2}$	$\frac{1}{3}$	$\frac{1}{3}$	$\frac{1}{3}$	$\frac{1}{3}$	$\frac{1}{3}$	$\frac{1}{3}$	$\frac{1}{20}$	$\frac{1}{20}$	$\frac{2}{20}$	$\frac{3}{20}$	$\frac{3}{20}$	···

Let us now fix $28W_r^{(1)}$, $21W_r^{(2)}$, and $21W_r^{(3)}$ at their lowest values 28, 21, and 63; the joint probability of these values is $1/2 \times 1/3 \times 1/3 = 1/18$, and their sum is 112. The sum will remain less than or equal to 229 when $12W_r^{(4)}$ is added provided $12W_r^{(4)} \leqslant 229 - 112 = 117$, an event of probability $7/20$. The total probability of these cases is therefore $7/360$. Fixing $28W_r^{(1)}$, $21W_r^{(2)}$, and $21W_r^{(3)}$ next at 28, 21, and 84, the sum will remain less than or equal to 229 when $12W_r^{(4)}$ is added provided $12W_r^{(4)} \leqslant 96$, and the probability of these cases is $4/360$. Continuing in this manner, we find (Prob. 46)

$$P_H(84W_r \leqslant 229) = \tfrac{20}{360} = .056$$

The normal approximation to the distribution of the statistics W_r and W_s poses no particular problem. The means and variances of these statistics are equal to the sums of the means and variances of the random variables $W_r^{(i)}/(N_i+1)$ and

$W_s^{(i)}/(N_i+1)$, respectively, which by (1.27) to (1.30) are given by

$$E_H \frac{W_r^{(i)}}{N_i+1} = \frac{m_i}{2} \quad ; \quad E_H \frac{W_s^{(i)}}{N_i+1} = \frac{n_i}{2} \tag{3.26}$$

and

$$\mathrm{Var}_H \frac{W_r^{(i)}}{N_i+1} = \mathrm{Var}_H \frac{W_s^{(i)}}{N_i+1} = \frac{m_i n_i}{12(N_i+1)} \tag{3.27}$$

The computation may be illustrated on the data of Example 6. From the expectations and variances for the different blocks listed in the table below, we find

i	m_i	n_i	N_i	$E_H[(W_r^{(i)}/(N_i+1)]$	$\mathrm{Var}_H[W_r^{(i)}/(N_i+1)]$
1	1	1	2	$\frac{1}{2}$	$\frac{1}{36}$
2	1	2	3	$\frac{1}{2}$	$\frac{1}{24}$
3	2	1	3	1	$\frac{1}{24}$
4	3	3	6	$\frac{3}{2}$	$\frac{3}{28}$

that $E_H(W_r) = 3.5$ and $\mathrm{Var}_H(W_r) = .218$, and hence that the normal approximation to the significance probability is

$$\Phi\left(\frac{2.73-3.5}{.467}\right) = \Phi(-1.649) = .050$$

as compared with the value .056 obtained earlier by enumeration.

Asymptotic normality can be proved in the present case under different assumptions representing different types of situations. In particular, $[W_r - E(W_r)]/\sqrt{\mathrm{Var}(W_r)}$ tends to the standard normal distribution when the number b of blocks tends to infinity and the block sizes N_i remain bounded. This is a mathematical representation of the case of many small blocks, and the proof follows easily from Appendix, Prob. 26.

On the other hand, one might be interested in combining the data from a small number of experiments, each of which is relatively large. This could be represented by a sequence in which the number b of blocks is fixed but all the group sizes $m_1, n_1; \ldots; m_b, n_b$ tend to infinity. For such a sequence, asymptotic normality follows from Appendix, Theorem 16 and the asymptotic normality of the individual terms $W_r^{(i)}$.

Since the blocked comparison of two treatments is a generalization of a comparison based on pairs, it is interesting to see what happens to the test statistic W_s in this special case. Each $W_s^{(i)}$ is now 1 or 2 as the control or treated subject in the ith pair comes out ahead; in the notation of Sec. 2, $W_s^{(i)}$ is 1 or 2 as the difference Z_i is negative or positive. Thus $\sum W_s^{(i)}$ is the number of negative differences plus twice the number of positive differences, and hence is the total number of differences plus the number of positive differences. Since the former is constant, the W_s-test in this case reduces to the sign test.

As already mentioned, the sign test has the advantage of requiring only the knowledge of which member of each pair comes out ahead; on the other hand, it typically has relatively low power. The advantage clearly extends to the W_s-test, which requires only a ranking within each block. This is important in situations in which the ranking is not obtained from numerical measurements but from a subjective comparison because it is then usually difficult to compare more than a few subjects at a time. Unfortunately, the disadvantage of low power also extends when the block sizes are small. It is worst for paired comparisons when each block is at its minimum value of two and becomes less serious as the block sizes increase. (See Sec. 5C for references.)

The reason for both the advantage and disadvantage of tests based on separate rankings in the different blocks is the same: Comparisons are made only between subjects that belong to the same block. If the blocks are small, there are then too few comparisons (relative to the total number that are theoretically possible) to permit an effective overall comparison of the two treatments. When numerical responses are available, the Wilcoxon signed-rank test provides a substantial improvement over the sign test in the paired comparison case. We shall now discuss a similar improvement for the general case.

EXAMPLE 6. *Methods of advertising (continued).* Consider once more the consumption figures for the 14 cities of Example 6. These were not directly comparable since they depended heavily on the size of the cities. Let us try to eliminate or at least reduce this size effect by subtracting from each observation the average of the observations for that block. The average of the third block, for example, is $(115+128+132)/3 = 125$, and subtracting this from each observation in the block yields the *aligned observations*

$$115-125 = -10, \qquad 128-125 = 3, \qquad 132-125 = 7$$

Proceeding in this way with each of the four blocks, we replace the original observations with the following aligned observations.

Block	Control	Treatment
1	$-9\frac{1}{2}$	$9\frac{1}{2}$
2	-2	$-6, 8$
3	$-10, 3$	7
4	$-13\frac{5}{6}, -4\frac{5}{6}, 4\frac{1}{6}$	$-7\frac{5}{6}, 9\frac{1}{6}, 13\frac{1}{6}$

Since the aligned observations are of about the same order of magnitude, even observations from different blocks are comparable, and we therefore rank all 14 observations instead of ranking separately the observations within each block. In this new ranking, the smallest observation is $-13\frac{5}{6}$ and therefore is assigned rank 1; next comes -10 with rank 2; and so on. The ranks of the aligned observations are displayed in the table below.

Block	Control	Treatment
1	3	13
2	7	5, 11
3	2, 8	10
4	1, 6, 9	4, 12, 14

To obtain the joint null distribution of the aligned treatment or control ranks, we must now repeat the standard argument (Chap. 1, Secs. 1 and 4 and Chap. 3, Sec. 1). Note first that under the hypothesis H of no treatment effect, and with the method of random assignment that was used, the aligned rank of each subject is independent of which subjects are assigned to treatment and which to control. This follows from the facts that (i) under H the response of each subject is the same whether it is assigned to treatment or control, and (ii) the average response in each block is independent of this assignment. Hence the aligned responses do not depend on the assignment, and neither, therefore, do the aligned ranks. By selecting the subjects that are to receive the treatment, the randomization process also selects their aligned ranks: One from the two aligned ranks 3 and 13 of block one, with probability $\frac{1}{2}$ each; two from the three ranks of block two, each pair with probability $1\Big/\binom{3}{2} = \frac{1}{3}$; one from block three with probability $\frac{1}{3}$ each; and three from block four, each triple with probability $1\Big/\binom{6}{3} = \frac{1}{20}$. Furthermore, the selections from the four blocks are independent.

This distribution provides the basis for computing the null distribution of any test statistic that is defined in terms of the aligned ranks. Consider for example the sum $\hat{W}_r$ of the control ranks of the aligned observations. The observed value is

$$3+7+2+8+1+6+9 = 36$$

so that the significance probability is $P(\hat{W}_r \leqslant 36)$. This is again most conveniently obtained by first considering the corresponding sums $\hat{W}_r^{(1)}, \ldots, \hat{W}_r^{(4)}$ of the aligned control ranks in the different blocks. Since the minimum value of $\hat{W}_r^{(1)} + \hat{W}_r^{(2)} + \hat{W}_r^{(3)}$ is $3+5+10 = 18$, the largest value of $\hat{W}_r^{(4)}$ for which $\hat{W}_r \leqslant 36$ is 18. The null distributions of $\hat{W}_r^{(1)}, \hat{W}_r^{(2)}, \hat{W}_r^{(3)}$ and the relevant part of that of $\hat{W}_r^{(4)}$ are listed below.

i	1		2			3			4			
w	3	13	5	7	11	10	12	18	11	14	16	17
$P_H(\hat{W}_r^{(i)} = w)$	$\frac{1}{2}$	$\frac{1}{2}$	$\frac{1}{3}$	$\frac{1}{3}$	$\frac{1}{3}$	$\frac{1}{3}$	$\frac{1}{3}$	$\frac{1}{3}$	$\frac{1}{20}$	$\frac{1}{20}$	$\frac{1}{20}$	$\frac{1}{20}$

We now proceed as in the first part of Example 6. With $\hat{W}_r^{(1)}, \hat{W}_r^{(2)}, \hat{W}_r^{(3)}$ fixed at their lowest values 3, 5, 10, the sum $\hat{W}_r$ is less than or equal to 36 provided $\hat{W}_r^{(4)} \leqslant 18$, and the total probability of these cases is $\frac{4}{360}$. Continuing in this way, we find (Prob. 51) $P_H(\hat{W}_r \leqslant 36) = \frac{13}{360} = .036$, as compared with the probability .056 obtained from separate rankings.

The above procedure generalizes in the obvious manner. The observations in each block are aligned by subtracting the mean of the block, and the totality of $N = N_1 + \cdots + N_b$ aligned observations is then ranked. If the aligned treatment ranks in the ith block are denoted by $\hat{S}_{i1}, \ldots, \hat{S}_{in_i}$, their joint null distribution is

$$P_H(\hat{S}_{11} = s_{11}, \ldots, \hat{S}_{1n_1} = s_{1n_1}; \hat{S}_{21} = s_{21}, \ldots, \hat{S}_{2n_2} = s_{2n_2}; \ldots) = \frac{1}{\binom{N_1}{n_1}\binom{N_2}{n_2}\cdots\binom{N_b}{n_b}} \tag{3.28}$$

in complete analogy with (3.22).

The method of alignment just described is not the only possible one. Instead of the block mean, some other estimate of location such as, for example, the block median could be subtracted. The estimate must, however, ignore to which of the two treatments the observations have been assigned (thus, the mean of the control observations would not be a permissible estimate) or otherwise the value of the estimate, and hence the aligned observations, would depend on which observations are associated with treatment and which with control. The observations can be aligned with respect to scale as well as with respect to location. If it appears that the observations in some blocks are more spread out than those in others, it may be desirable to adjust for this difference by dividing the observations in each block by the standard deviation of that block (Prob. 52). The observations can be transformed in still other ways, with the transformation varying from one block to another. As long as the same transformation is applied to all observations in the same block, the null distribution (3.28) will remain valid.

From the distribution (3.28), the null distribution of the sum $\hat{W}_s$ of the aligned treatment ranks can now be obtained in the usual way. However, except for small values of N the computation is cumbersome, and one will wish to apply the normal approximation. Since

$$\hat{W}_s = \sum \hat{W}_s^{(i)} \tag{3.29}$$

where $\hat{W}_s^{(i)}$ is the sum of the aligned treatment ranks associated with the ith block and since the $\hat{W}_s^{(i)}$ are independent, the expectation and variance of $\hat{W}_s$ is the sum of the expectations and variances of the variables $\hat{W}_s^{(i)}$. Now $\hat{W}_s^{(i)}$ is the sum of a random sample of size n_i from the N_i aligned ranks, say $k_{i1}, \ldots, k_{iN_i}$ of the ith block. The expectation and variance of $\hat{W}_s^{(i)}$, and also of $\hat{W}_r^{(i)}$, therefore follow from formulas (A.39) and (A.42) of the Appendix and are given by

$$E(\hat{W}_s^{(i)}) = n_i k_i; \qquad E(\hat{W}_r^{(i)}) = m_i k_i \tag{3.30}$$

where

$$k_{i\cdot} = \frac{k_{i1} + \cdots + k_{iN_i}}{N_i}$$

and

$$\mathrm{Var}(\hat{W}_s^{(i)}) = \mathrm{Var}(\hat{W}_r^{(i)}) = \frac{m_i n_i}{N_i(N_i - 1)} \sum_{j=1}^{N_i} (k_{ij} - k_{i\cdot})^2 \tag{3.31}$$

As in the case of the statistic W_s based on separate rankings, asymptotic normality can be proved under a number of different assumptions. The method of aligned ranks is particularly useful when the blocks are small, so we shall here consider only the case that the number b of blocks tends to infinity and the block sizes N_i are bounded. Then it follows from Appendix, Prob. 29 that the null distribution of $[\hat{W}_s - E(\hat{W}_s)]/\sqrt{\mathrm{Var}(\hat{W}_s)}$ tends to the standard normal distribution provided each block contains, after alignment, at least one positive and one negative observation. This condition is always satisfied when the observations are aligned on the block mean or median.

The test based on the sum $\hat{W}_s$ of the aligned treatment ranks tries to accomplish for the general case what was achieved by the paired-comparison Wilcoxon test when all blocks are of size 2. It is therefore interesting to see what happens to the $\hat{W}_s$ test in this case. If the alignment in each block is on the midpoint between the two observations of that block (which is both the mean and the median of the block), it turns out (Prob. 73) that $\hat{W}_s$ and the signed-rank Wilcoxon statistic V_s are related by the equation

(3.32) $$\hat{W}_s = 2V_s - S_N + \tfrac{1}{2}N(N+1)$$

where N now denotes the number of pairs and S_N is the sign test statistic based on N pairs [in the notation of the present section N should be replaced by b in formula (3.32)]. Although the two tests are thus not completely equivalent, it follows from (3.32) that they are asymptotically equivalent in the sense that the probability of their leading to different conclusions (acceptance or rejection of H) tends to zero as N tends to infinity.

4. A BALANCED DESIGN FOR PAIRED COMPARISONS

The paired comparisons design in which one member of each pair is assigned to treatment and the other to control with independent random assignments for each pair can lead to difficulties when it is possible to distinguish the members of each pair, as is often the case. For example, when a pair consists of the two hands of a person, there may well be a systematic difference in the response of the right and left hand. Similarly, in a study of twins, there may be differences between the first and second born; and in a taste experiment in which a subject compares two brands of coffee by tasting one cup of each, it may be important to distinguish which brand was tasted first. Even when the pairs do not occur naturally but are put together by the experimenter through matching on a number of variables such as the sex, age, and severity of disease of a patient, a typing of this kind may be possible. A physician could examine the case histories and determine for each pair which of the two patients has the better prognosis.

Suppose in general that one can distinguish two types of subjects A and B and that one member of each pair is of type A and the other of type B. If the two members of each pair are assigned at random, one to treatment and one to control, the probability is $\frac{1}{2}$ that the treatment is assigned to the member of type A, and the total number, among the N pairs, in which the treatment is assigned to type A is a random variable with binomial distribution corresponding to N trials and success probability $\frac{1}{2}$. It may then happen by chance that most of the pairs have the treatment assigned to type A. In such a case, if the results are significant it is not clear whether this was caused by the treatment being more effective than the control or by type-A subjects outresponding those of type B. Treatment and type are said to be *confounded*.

This difficulty can and should be avoided by employing a different method of random assignment to treatment and control when there exists a typing of subjects that might be related to their responsiveness. Instead of letting the number of treatment cases of type A be random, we can decide that we should like this number to have a fixed value a, with the number of treatment cases of type B then being $b = N - a$. Typically, one will take a and b to be equal or at least nearly so, and then call the assignment *balanced*. Suppose therefore that of the N pairs, we select a at random so that all $\binom{N}{a}$ choices are equally likely, and in these a pairs assign the treatment to the type-A subject, with the type-B subject serving as control. In the remaining $b = N - a$ pairs, the treatment is assigned to the type-B subject and the type-A subject serves as control.

Consider now the N differences $A - B$ for each pair, the response of the A-subject minus the response of the B-subject. The problem is then of the kind discussed in Chap. 1, Secs. 1 and 2. The N pairs constitute the N "subjects"; the differences $A - B$ their "responses." Of these N subjects, a are selected at random and assigned to situation 1 (A-subject treated, B-subject not), the remaining $b = N - a$ are assigned to situation 2 (B-subject treated, A-subject not). The hypothesis being tested, that treatment and control do not differ, implies that there is no difference between situations 1 and 2. Under the alternatives that the response is increased by the treatment, the responses in situation 1 (treatment − control) tend to be larger than those in situation 2 (control − treatment). This is exactly the problem for which the rank-sum Wilcoxon test was developed in Chap. 1, and this test is therefore applicable in the present case.

EXAMPLE 7. *The effect of hypnosis on speech.* In a study of the effect of hypnosis on speech, comparable speech samples were obtained from ten subjects, once under hypnosis and once in a waking state. Among other measures the numbers of words that the subjects produced in the two states were recorded. The hypothesis of no difference was to be tested against the alternative that there would tend to be fewer words under hypnosis. To guard against a possible order effect it was decided to balance the order of presentation: Five subjects were selected at random and were observed first under hypnosis (HW); the speech samples of the remaining five subjects were obtained first while awake (WH). The numbers of words for the two sets were:[1]

[1] From Hunt (1969), "The Speech of the Subject under Hypnosis," *Intern. J. Clin. Exptl. Hypnosis* **17**:209–215.

	WH					HW				
W	255	1250	126	480	371	308	688	345	264	306
H	67	67	89	129	491	304	49	281	131	107
Difference	188	1183	37	251	−120	4	639	64	133	199

Let A and B refer to the first and second presentation, respectively. Then the differences $A-B$ are 188, 1183, 37, 251, -120 for the WH cases, and -4, -639, -64, -133, -199 for the HW cases. Ranking these ten differences we find for the rank-sum of the HW cases $1+2+3+5+6 = 17$. Small values of the sum are significant since under the alternative of disinclination to talk the $W-H$ differences should be larger than the $H-W$ differences. The significance probability is therefore the probability that the rank-sum is $\leqslant 17$. After subtracting $\frac{1}{2}n(n+1) = 15$, we find from Table B that $P(W_{XY} \leqslant 2) = .016$, which is therefore the desired significance probability.

5. FURTHER DEVELOPMENTS

5A. Power of the Sign and Wilcoxon Tests

As we discussed in Chap. 1, Sec. 7D, an important aspect of a test is its power, that is, the probability of rejecting the hypothesis H of no treatment effect when the treatment has in fact a beneficial effect. Again, the framework of the randomization model is not well suited to the assessment of power, and a full discussion of the problem must be postponed to the next chapter. Here we shall only consider very briefly two alternatives to H that can be defined in the present context.

The first of these alternatives is provided by the constant effect model, which is exactly analogous to that discussed in Chap. 1, Sec. 7D. It assumes that the treatment has the effect of adding a fixed amount Δ to the response a subject would have given as a control. The value of Δ is assumed to be the same regardless of which subject of a pair receives the treatment and the same for all N pairs. For some numerical examples, see Hodges and Lehmann (1970).

The second model considered in Chap. 1, Sec. 7D provides an alternative to H for the sign test. Suppose that for each pair of subjects, a (subjective) judgment is made as to which of the two came out ahead. Such judgments involve an element of randomness. Suppose first that the probability p of the treated subject coming out ahead is the same for all N pairs. Then the sign test statistic S_N has the binomial distribution $b(p,N)$ corresponding to N trials and success probability p, and the power can be determined from a binomial table (for details, see Chap. 4, Sec. 1).

Unless the pairs are very homogeneous, one would expect the probability p to have different values for different pairs. The exact evaluation of power is then more laborious. A normal approximation is given by Cramér (1946, pp. 217–218). In addition, it has been pointed out by Hoeffding (1956) [see also David (1960)] that a lower bound for sufficiently large values of the power (roughly, for values exceeding $\frac{1}{2}$) is obtained by computing it under the assumption that the N probabilities are all equal to $\bar{p} = (p_1 + \cdots + p_N)/N$.

5B. Alternative Treatment of Zeros

For a discussion of different ways of handling zeros in the sign test, consider once more the simple version of the second model of the preceding subsection in which there is a common probability p for the treated subject to come out ahead, but suppose that an additional preference statement is permitted, namely, that the treated and control subject have performed equally well, with neither coming out ahead of the other. In generalization of the previous model, let us assume that there are three fixed probabilities: the probability p_+ which was previously denoted by p; the probability p_0 that the two subjects are tied, i.e., that they are judged to have performed equally well; and the probability p_- that the control subject will come out ahead.

In this model, the number N_0 of tied pairs is no longer a constant as it was in the randomization model of Sec. 1. Instead, it is a random variable, distributed according to the binomial distribution $b(p_0,N)$. As a result, the test (*a*) based on the statistic $N_+ + \frac{1}{2}N_0$ and the test (*b*) that discards the N_0 tied pairs are no longer equivalent. It is shown by Hemelrijk (1952) that of these two tests (*b*) has the better power. The problem is considered further by Putter (1955), who shows that (*b*) has the highest power among a large class of tests. This class includes, in particular, the test (*c*), which breaks each tie at random. [For a recent study of this problem, see Krauth (1973).]

In determining whether and how to take account of the zeros, it is essential to be clear of the purpose of the analysis. If a definite choice has to be made between two candidates, two treatments, and so forth, and if no other considerations play a role (such as the price of the treatment or the possibility of toxic side effects), it is intuitively clear that the zeros are of no help in reaching a decision. If, on the other hand, a producer of a brand of coffee wishes to know whether a higher admixture of a more expensive kind of bean would be preferred by sufficiently many customers, it is crucial to know how many persons cannot tell the difference.

An important practical problem arises when one sets up a paired comparisons study based on subjective judgments. Should one insist on a definite decision ($+$ or $-$) for each pair or permit an expression of indecision that could cause a tie? The answer turns out to depend on the circumstances. The problem is discussed by Gridgeman (1959); see also Lehmann (1959).

Let us now turn to the treatment of zeros in the Wilcoxon test. In the definition of V_s^*, all N observations were assigned their midranks and only then were the zeros discarded. An alternative procedure, suggested by Wilcoxon (1949), consists in first discarding the zeros and then ranking the remaining observations. A third possibility, as always, is the breaking of ties at random. A comparison of these and still other possibilities is the subject of a paper by Pratt (1959). For some further developments, see Vorličková (1972) and Conover (1973).

5C. Tests against Omnibus Alternatives

The sign and signed-rank Wilcoxon tests for paired comparisons are designed to test the hypothesis H of no treatment effect against the alternative that the treatment is effective in raising (or lowering) the response. However, as was discussed in Chap. 1, Sec. 6, a treatment may affect the response in many other ways. A test of H against the omnibus alternative of a completely unspecified treatment effect, which is analogous to the Smirnov test of Chap. 1, Sec. 6, is discussed by Smirnov (1947) and Butler (1969).

5D. Efficiency and Generalizations of the Blocked Comparisons Test W_s

In Sec. 3, the test statistic (3.24) was discussed as a means of combining the results from a number of different studies, or from a single study carried out in a number of different blocks. A weighting of the statistics $W_s^{(i)}$ other than with the weights $1/(N_i+1)$ used in (3.24) is considered by van Elteren (1960), who also proves that the weights of (3.24) have certain optimum properties. These comparisons are not discussed in the context of the randomization model of Sec. 3 but in terms of a population model generalizing that of Chap. 2. This model assumes the observations in the ith block or ith study to be random variables, say $X_{i1}, \ldots, X_{im_i}$ (controls) and $Y_{i1}, \ldots, Y_{in_i}$ (treatment), which are independent and distributed according to distributions F_i and G_i, respectively. The hypothesis to be tested is $F_i = G_i$ for all $i = 1, \ldots, b$. In the framework of this model, van Elteren shows that the weights (3.24) maximize the power locally (i.e., near the hypothesis) under conditions which state roughly that the treatment has about the same effect in all blocks.

When the blocking is not the result of different studies being carried out at different times or places but is a deliberate effort to increase the power of the test by dividing nonhomogeneous subjects into more homogeneous blocks, it is important to know just what gain in homogeneity is required of the blocks to make up for the smaller number of comparisons that are a result of blocking. This problem is studied by Noether (1963), who obtains the efficiency of the blocked test (3.24) relative to the unblocked Wilcoxon test.

Generalizations of van Elteren's test for Normal and other scores and the efficiencies of these tests relative to the corresponding linear combination of t-statistics are discussed by Puri (1965).

6. PROBLEMS

(A)

Section 1

1. Use (3.1) to obtain the null distribution of S_N, and draw its histogram, for (i) $N = 4$; (ii) $N = 5$.

2. Use Table G to find $P_H(S_N \leqslant a)$ when (i) $N = 14$, $a = 5$; (ii) $N = 18$, $a = 12$; (iii) $N = 30$, $a = 21$.

3. Use Table G to find $P_H(S_N > a)$ for the three cases of Prob. 2.

4. Use Table G to find $P_H(S_N \geqslant a)$ for the three cases of Prob. 2.

5. Use Table G to find $P_H(S_N = a)$ when (i) $N = 10$, $a = 5$; (ii) $N = 20$, $a = 10$; (iii) $N = 30$, $a = 15$.

6. In a taste test, each of 13 subjects is asked to express a preference between a domestic and a more expensive imported brand of chocolate. If 8 of the 13 prefer the imported brand, find the significance probability of this result.

7. In the preceding problem, suppose that seven subjects prefer the imported brand, four the domestic brand, and two are indifferent. Find the significance probability.

8. In Example 1, suppose that eight of the subjects prefer the new drug, four the standard drug, and three are indifferent. Find the significance probability.

9. Use the normal approximation to compute the probabilities of (i) Prob. 2(i); (ii) Prob. 3(i); (iii) Probs. 5(i) and (ii) both with and without continuity correction, and compare the results with those of the earlier problems.

10. For each value of $N = 6, 8, 10, 12$, use Table G to find the value a for which $P_H(S_N \leqslant a)$ is closest to $\alpha = .07$, and compare these probabilities with their normal approximations obtained (*a*) with and (*b*) without continuity correction.

11. Find the normal approximation to the probability $P_H(S_N \leqslant a)$ both with and without continuity correction and compare it with the value given in Table G when (i) $N = 8$, $a = 1, 2, 3$; (ii) $N = 10$, $a = 1, 2, 3$; (iii) $N = 12$, $a = 2, 3, 4, 5$.

12. In Example 1, use the normal approximation to determine for what values of S_N to reject the hypothesis of no treatment difference at $\alpha = .05$ when (i) $N = 25$; (ii) $N = 50$; (iii) $N = 80$.

13. If $N = 5$, use (3.5) to find the probabilities (i) $P_H(n = 3; S_1 = 1, S_2 = 2, S_3 = 3)$; (ii) $P_H(n = 3; S_1 = 1)$; (iii) $P_H(n \geqslant 1; S_1 = 1)$.

14. Use (3.5) to find the probability $P_H(n = 3; S_1 + S_2 + \cdots + S_n = 10)$ when: (i) $N = 5$; (ii) $N = 6$; (iii) $N = 7$.

15. If $N = 5$, use (3.5) to find the probabilities (i) $P_H(S_1 + \cdots + S_n = 10)$; (ii) $P_H(S_1 + 2S_2 + \cdots + nS_n = 10)$.

16. In a paired comparison study on eight pairs of twins, find the probability that by chance the firstborn of the twins is assigned to treatment (i) in all eight cases; (ii) in at least six of the eight cases; (iii) in exactly half of the eight cases.

Section 2

17. Obtain the null distribution of V_s by the method leading to (3.9) and graph its histogram when (i) $N = 4$; (ii) $N = 5$.

18. Use the results of Probs. 1 and 17 to find all significance levels of the sign and Wilcoxon tests when (i) $N = 4$; (ii) $N = 5$.

19. Use Table H to find $P_H(V_s \leqslant v)$ when (i) $N = 8, v = 12$; (ii) $N = 14, v = 76$; (iii) $N = 16, v = 79$.

20. Use Table H to find $P_H(V \geqslant v)$ when (i) $N = 11, v = 21$; (ii) $N = 15, v = 68$; (iii) $N = 19, v = 116$.

21. Use Table H to find $P_H(V = v)$ when (i) $N = 9, v = 23$; (ii) $N = 16, v = 50$; (iii) $N = 20, v = 129$.

22. Use Tables G and H to find all attainable significance levels $\leqslant .05$ for the sign test and the Wilcoxon test when (i) $N = 8$; (ii) $N = 11$; (iii) $N = 14$.

23. Use Tables G and H to find the closest attainable significance level to .01 for the sign and Wilcoxon test when (i) $N = 6$; (ii) $N = 9$; (iii) $N = 12$.

24. For both the sign and the Wilcoxon test find the significance probability of the data of Example 3 assuming that each of the tensile strengths of the taped wounds is (i) increased by 5; (ii) decreased by 5; (iii) decreased by 10.

25. In a study of the effectiveness of eye exercises in improving near-sightedness, N matched pairs of children are assigned at random to treatment (an hour's supervised exercise daily) or control (no special exercises). State for what values of V_s you would reject the hypothesis of no treatment effect at the closest attainable significance level to .02 if (i) $N = 10$; (ii) $N = 13$; (iii) $N = 16$.

26. Given $N = 21$, use (3.34) of Prob. 66 below and Table H to find (i) $P_H(V_s = 82)$; (ii) $P_H(V_s = 90)$; (iii) $P_H(V_s = 97)$.

27. Find the normal approximation to the probability $P_H(V_s \leqslant v)$ both with and without continuity correction, and compare it with the value given in Table H, when $N = 6$ and (i) $v = 4$; (ii) $v = 6$; (iii) $v = 8$.

28. For each value of (i) $N = 4$; (ii) $N = 5$; (iii) $N = 6$; (iv) $N = 7$, use Table H to find the value of v for which $P_H(V_s \leqslant v)$ is closest to $\alpha = .05$, and compare these probabilities with their normal approximations obtained (*a*) with and (*b*) without continuity correction.

29. In the study of the effect of tape closing on wounds of Example 3, use the normal approximation to determine for what values of V_s to reject the hypothesis of no effect at significance level $\alpha = .05$ when (i) $N = 20$; (ii) $N = 40$; (iii) $N = 60$.

30. Use the normal approximation to obtain the critical values c corresponding to $\alpha = .02$ for the three cases of Prob. 25, and compare (i) the values c with the critical values obtained in Prob. 25; (ii) the probabilities $P_H(V_s \geqslant c)$ as read from Table H with the nominal significance level .02.

31. Find the null distribution of V_s^* and graph its histogram when $N = 5$ and (i) $d_0 = 0$, $d_1 = d_2 = 2$, $d_3 = 1$; (ii) $d_0 = 1$, $d_1 = 2$, $d_2 = d_3 = 1$; (iii) $d_0 = 2$, $d_1 = d_2 = d_3 = 1$; (iv) $d_0 = 1$, $d_1 = d_2 = 2$.

32. If $d_0 = 1, d_1 = 2, d_2 = 1, d_3 = 2$, show that $P_H(V_s^* \leqslant 4) = 4/32$.

33. In Example 3, the results on other sets of 10 rats each were[1]

(i) After 10 days

Tape	63	56	32	56	48	45	45	96	56	71
Sutured	79	143	37	56	40	135	45	96	87	83

(ii) After 150 days

Tape	2334	1228	1596	1798	622	1543	1389	1984	1571	1619
Sutured	1862	1330	1596	1234	692	894	1190	1786	833	1167

Test the hypothesis of no difference (*a*) by a two-sided sign test; (*b*) by a two-sided Wilcoxon test.

34. In 24 pairs of matched patients suffering from chronic schizophrenia, one member of each pair was assigned at random to treatment with a new drug (Stelazine), and the other received a placebo. For 8 of the pairs, the trial was not completed. The behavior ratings of the others after 3 months are shown below, broken down according to the two wards from which the patients came.[2] In each case, the initial value is shown in the first row, the value after 3 months in the second row.

Ward *A*

Pair	1	2	3	4	5	7	9	10	12
Treatment	2.3	2.2	1.9	2.9	2.2	2.3	2.7	1.9	1.1
	3.1	2.6	2.45	2.0	2.54	3.72	3.0	1.61	1.63
Control	2.4	2.0	2.1	3.1	2.2	2.4	2.8	1.9	1.3
	2.0	2.1	2.0	3.7	2.4	3.18	4.54	2.54	1.72

Ward *B*

Pair	2	4	5	6	8	9	11
Treatment	1.9	2.3	2.0	1.6	1.6	2.7	1.7
	1.45	2.45	1.45	1.72	1.63	1.54	1.54
Control	1.9	2.4	2.0	1.5	1.5	2.6	1.7
	1.91	2.54	1.81	1.45	1.54	2.45	2.18

Test the hypothesis that the treatment does not decrease the difference in behavior ratings (after − before) using the combined data from both wards and applying (i) the sign test; (ii) the Wilcoxon test.

35. Work the preceding problem under the assumption that no initial scores are available, by comparing the "after" scores of treatment and control patients.

36. Use the initial scores of the control subjects in Prob. 34 to test the hypothesis of no difference between the two wards.

[1] Data from Ury and Forrester, *loc. cit.*

[2] Data from Stanley and Walton, "Trifluoperazine ('Stelazine'). A Controlled Clinical Trial in Chronic Schizophrenia." *J. Mental Sci.* **107**:250–257 (1961).

37. The initial scores of the eight control patients of Prob. 34 for which the trial was not completed were

2.3 1.5 1.8 (Ward *A*) and 2.6 1.9 2.4 3.2 1.9 (Ward *B*)

Test the hypothesis of no difference in the initial scores of these patients and the control patients for whom the trial was completed.

38. To test the effectiveness of group therapy, an investigator divided 44 boys, institutionalized for delinquency, into 22 matched pairs, with a randomly selected boy from each pair participating in 20 group therapy sessions. The emotional and social adjustment of all 44 boys was rated before and after, and the differences (treated − control) in the gains made in these ratings were as follows:[1] −.6, −.9, −.4, .4, .3, −.4, 1.2, 0, .6, .7, −.2, 0, .5, .5, 0, .4, 0, −1.1, .3, 1.0, .1, .7. To test the hypothesis that the therapy had no effect against the alternative, that it tended to lead to greater gains in the ratings, (i) find the value of V_s^*; (ii) find the approximate significance probability of this value.

39. Suppose that in the comparison of a new with a standard headache remedy (Example 1), the expressions of preference for the new drug by nine subjects are as follows:

Much more	Somewhat more	No difference	Somewhat less	Much less
1	4	2	1	1

Show how the V_s^*-test can be used in this case and find its significance probability.

40. For each of the four cases of Prob. 31 find the critical value that gives the significance probability closest to .1, and compare the actual significance level with its normal approximation.

41. In Prob. 39 suppose that the number of subjects in the study is 30 and that the expressions of preference for the new drug are

Much more	Somewhat more	No difference	Somewhat less	Much less
3	12	8	6	1

Use the normal approximation to find the significance probability.

Section 3

42. In a study of the effect that familiarity with the examiner may have on the performance of a child in an IQ test, 48 first-graders were divided into 24 matched pairs, and one child from each pair was selected at random for the experimental group. Each child in that group spent four 20-min periods with the investigator, who then gave it an IQ test. The corresponding test was also given to the controls, and an initial IQ score was established for all children at the beginning of the study. The differences (treated − control) in the gains achieved in the IQ scores were as follows:[2] −17, −15, −8, −7, −10, 4, −10, 4, 12, 1, 21, 2, −20, 2, 0, 5, −7, 19, −8, 9, −3, −7, −1, 34. Obtain the approximate significance probability of the Wilcoxon statistic for testing the hypothesis of no difference against the alternative that familiarity with the examiner tends to improve the performance on the test.

43. (i) The experiment of Prob. 42 was repeated with 32 children in another school with the following results: −12, 13, −5, 6, −11, 3, 14, −9, −3, 4, 11, −11, −9, −11, 16, 2.

[1] Data from Gerstein, "Group Therapy with Institutionized Juvenile Delinquents," *J. Genetic Psychol.* **80**:35–64 (1952).

[2] Data from Marine, "The Effect of Familiarity with the Examiner upon Stanford-Binet Test Performance," *Contrib. to Educ.*, no. 381, Teacher's College, Columbia University, 1929.

Let V_s^* be the Wilcoxon statistic computed for all 40 differences. Find the approximate significance probability of the observed value of V_s^*.

(ii) For the combined experiment, find the significance probability if the sign test is used instead of the Wilcoxon test.

44. For the combined comparison data of the first part of Example 5, show that the significance probability of V_s^* is .107.

45. For the combined unpaired data of the second part of Example 5, find the significance probability of the test statistic (3.24) or (3.25) modified for ties (i) using the exact method indicated in Example 6; (ii) using the normal approximation.

46. Complete the computation of the significance probability $P_H(84W_r \leqslant 229)$ begun in Example 6.

47. In Example 6, find the complete null distribution of $84W_r$ by obtaining first the distribution of $28W_r^{(1)} + 12W_r^{(4)}$ and of $21(W_r^{(2)} + W_r^{(3)})$, and then the distribution of the sum of these two sums.

48. Suppose that the experiment of Chap. 1, Example 2 on the effect of discouragement is repeated on another occasion where (without any pairing) four of eight available subjects are assigned at random to the treatment (discouragement), the other four serve as controls, and the differences in the scores are -10, -8, -8, -7 and -8, -3, 0, 4 for the treatment subjects and the controls, respectively. Find the significance probability for the combined data using the test statistic (3.24) or (3.25) modified for ties (i) by enumeration; (ii) by means of the normal approximation.

49. Suppose that two studies of the effectiveness of a new medical treatment lead to the data of Probs. 42 and 49 of Chap. 1. Find the significance probability of the combined data using the normal approximation on the test statistic (3.24) or (3.25) modified for ties.

50. In the comparison of two types of paper A and B, eight measurements of smoothness were taken on samples from A and B in each of three different laboratories with the following results:[1]

Laboratory 1	A:	15.7	15.7	13.5	10.3	12.1	16.8	17.1	14.3
	B:	12.9	12.5	12.8	11.0	9.3	13.7	13.1	14.0
Laboratory 2	A:	13.4	12.6	12.0	12.2	12.1	12.0	11.0	12.4
	B:	11.7	9.8	9.6	10.6	8.4	13.2	9.5	9.8
Laboratory 3	A:	10.6	10.9	10.2	10.0	10.6	10.1	12.2	8.8
	B:	11.3	10.5	8.2	11.2	8.9	8.9	7.6	7.2

Use the Wilcoxon statistic (3.24) to test the hypothesis of no difference between A and B against the two-sided alternatives that the measurements of B are systematically either higher or lower than those of A.

51. Show that the significance probability $P_H(\hat{W}_r \leqslant 36)$ of Example 6 (continued) is equal to .036.

52. (i) Divide each aligned observation of Example 6 by the standard deviation of its block, where the standard deviation of the ith block is $\sqrt{\sum (x_{ij} - x_{i\cdot})^2/(N_i - 1)}$ when $x_{i1}, \ldots, x_{iN_i}$ denote the observations of the ith block.

[1] Data studied (but not published in full) by Lashof and Mandel, "Measurement of the Smoothness of Paper," *Tappi* **43**:385–389 (1960).

(ii) Rank the modified observations obtained in (i), and find their significance probability when the $\hat{W}_r$-test is used.

53. Find the significance probability of the combined data of Prob. 48 and of Example 2, Chap. 1 when the test statistic (3.29) is used and when alignment is (i) on the block median; (ii) on the block mean.

54. Apply the normal approximation to the test statistic (3.29) to find the approximate significance probability for Example 6 and compare your result with the value obtained in Prob. 46.

55. In Prob. 49, assigning scores -2, -1, 0, 1, 2 to the five categories, use the normal approximation to find the significance probability when the statistic (3.29) is used in place of (3.24) and alignment is (i) on the block median; (ii) on the block mean.

56. In Prob. 50, use the normal approximation to find the significance probability when the statistic (3.29) is used in place of (3.24) and alignment is (i) on the block mean; (ii) on the block median.

(B)

Section 1

57. Prove formula (3.1). [*Hint:* The probability of any particular set of outcomes for the N pairs (treatment ahead in first pair; control ahead in second pair; and so forth) is $1/2^N$. How many such sets are there for which the treatment comes out ahead in a pairs and the control in the remaining $N-a$ pairs?]

58. Prove that the null distribution of S_N is symmetric about $N/2$. [*Hint:* Note that S_N and $N-S_N$ have the same distribution.]

59. Prove that the probabilities (3.1) increase for $a < \frac{1}{2}N$ (and by symmetry decrease for $a > \frac{1}{2}N$). [*Hint:* It is enough to show that $a \leqslant \frac{1}{2}N$ implies $\binom{N}{a-1} < \binom{N}{a}$, which follows from (1.65).]

60. If $\binom{N}{n}$ is defined by (1.2) for $n = 1, \ldots, N$, show that $\binom{N}{0} = 1$ is the only definition of $\binom{N}{0}$ that is consistent with (i) formula (1.3); (ii) formula (1.67).

Section 2

61. Prove that the null distribution of V_s is symmetric about $N(N+1)/4$. [*Hint:* Use (3.12) and the fact that V_r and V_s have the same null distribution.]

62. Find a constant c' such that the inequalities $V_s \geqslant c$ and $V_r \leqslant c'$ are equivalent.

63. Construct a set of N differences for which S_N is more significant than V_s for the cases (i) $N = 5$; (ii) $N = 6$; (iii) $N = 8$.

64. Show that for $v \leqslant N$, the number $\#(v;N)$ occurring on the right-hand side of (3.10) is independent of N, and evaluate this number for $v = 0, 1, \ldots, 10$ and $N \geqslant v$.

65. For the number $\#(v;N)$ occurring on the right-hand side of (3.10), prove the recursion formula

$$\#(v;N) = \#(v;N-1) + \#(v-N;N-1) \tag{3.33}$$

[*Hint:* See the hint for Chap. 1, Prob. 86.]

66. Use (3.33) to prove the recursion formula

$$P_H(V_s = v|N) = \tfrac{1}{2}[P_H(V_s = v|N-1) + P_H(V_s = v-N|N-1)] \tag{3.34}$$

where the number behind the vertical bar indicates the number of pairs of subjects on which that statistic V_s is based.

67. Suppose that in a paired comparison study each difference (treated − control) is classified as very poor, poor, indifferent, good, or very good. Assign to these five categories the scores -2, -1, 0, 1, 2; let the number of subjects in the five categories be A_2, A_1, d_0, B_1, B_2; and let $d_1 = A_1 + B_1$, $d_2 = A_2 + B_2$.

(i) Express V_s^* in terms of B_1, B_2 and d_0, d_1, d_2.

(ii) Determine the joint null distribution of B_1 and B_2, that is, the probability $P_H(B_1 = b_1, B_2 = b_2)$.

68. For a fixed value of d_0, what are the minimum and maximum possible values of V_s^*?

69. Check relation (3.16) on the data of Example 3.

70. Show that the null distribution of V_s^* is always symmetric. [*Hint:* Use the method of Prob. 61.]

71. Prove the identity (3.19). [*Hint:* Let the number of zero differences be d_0, and suppose that of the d_1 smallest absolute nonzero differences b_1 are positive and a_1 are negative; of the d_2 next smallest b_2 are positive and a_2 negative; and so on. Then the sum V_s^* of the positive midranks is given by

$$V_s^* = \sum_{j \geq 1} b_j[d_0 + \cdots + d_{j-1} + \tfrac{1}{2}(d_j + 1)]$$

where the expression in square brackets is by (1.41) the midrank of the jth group.

On the other hand, the sum of the scores (1, ½, 0, as $Z_i + Z_j$ is positive, zero, or negative) is

$$\sum_{j \geq 1} b_j(d_0 + \cdots + d_{j-1}) + \sum_{j \geq 1} \binom{b_j}{2} + \frac{1}{2}\sum a_j b_j + S_N + \frac{1}{2}\left[d_0 + \binom{d_0}{2}\right]$$

Here the first two terms are the contributions from pairs $i < j$ with $Z_i + Z_j > 0$; the fourth from pairs $i = j$ with $Z_i + Z_j > 0$; and so forth.

The difference between the two displayed expressions is seen to be

$$\frac{1}{2}\left[d_0 + \binom{d_0}{2}\right] = \frac{1}{4}d_0(d_0+1)$$

by using the fact that S_N is the sum of the b_j.]

Section 3

72. Show that the inequality $\sum W_s^{(i)}/(N_i + 1) \geq c$ is equivalent to $\sum W_r^{(i)}/(N_i + 1) \leq c'$ for a suitable c', and determine c'.

73. To prove relation (3.32), let the N pairs of observations be $(X_1, Y_1), \ldots, (X_N, Y_N)$, where the X's denote the control and the Y's the treatment responses, and let the N pairs be numbered in increasing order of the absolute differences $|Z_i| = |Y_i - X_i|$, so that $|Z_1| < \cdots < |Z_N|$. Denote the rank of the treatment observation in the ith pair after alignment by $\hat{S}_i$, so that $\hat{W}_s = \sum \hat{S}_i$. Put $v_i = i$, $u_i = 0$ when $Z_i > 0$ and $v_i = 0$, $u_i = i$ when $Z_i < 0$, and note that $V_s = \sum v_i$.

(i) Show that $\hat{S}_i = N+i = N+v_i$ when $Z_i > 0$ and $\hat{S}_i = N+1-i = N+1-u_i$ when $Z_i < 0$.

(ii) From (i) it follows that

$$\hat{W}_s = \sum{}^{(1)}(N+v_i) + \sum{}^{(2)}(N+1-u_i)$$

where $\sum^{(1)}$ and $\sum^{(2)}$ denote summation over the subscripts i for which $Z_i > 0$ and $Z_i < 0$, respectively. On simplification, the displayed formula reduces to (3.32).

74. If $\hat{W}_s$ is the aligned-rank Wilcoxon statistic and V_s the paired-comparison Wilcoxon statistic, both based on N pairs and related by Eq. (3.32), prove that under the null hypothesis of no treatment difference the probability of the $\hat{W}_s$-test and the V_s-test leading to opposite conclusions tends to zero as N tends to infinity. [*Hint:* Using (3.13), (3.14), and the asymptotic normality of V_s and writing the rejection region of the V_s-test in the form

$$\frac{V_s - \frac{1}{4}N(N+1)}{\sqrt{N(N+1)(2N+1)/24}} \geqslant c_N$$

it follows that $c_N \to u_\alpha$ where α is the significance level and $1-\Phi(u_\alpha) = \alpha$. By (3.32) the $\hat{W}_s$-test can be written in the form

$$\frac{V_s - \frac{1}{4}N(N+1)}{\sqrt{N(N+1)(2N+1)/24}} \geqslant c'_N + \frac{S_N/2}{\sqrt{N(N+1)(2N+1)/24}}$$

Here the last term tends to zero in probability {since $(S_N - \frac{1}{2}N)/\frac{1}{2}\sqrt{N}$ tends to the standard normal distribution} and hence $c'_N \to u_\alpha$.]

7. REFERENCES

Bickel, P. J. (1974): "Edgeworth Expansions in Nonparametric Statistics," *Ann. Statist.* **2**:1–20.

Butler, Calvin, C. (1969): "A Test for Symmetry Using the Sample Distribution Function," *Ann. Math. Statist.* **40**:2209–2210.

Claypool, P. L. and Holbert, D. (1974): "Accuracy of the Normal and Edgeworth Approximations to the Distribution of the Wilcoxon Signed-rank Statistic," *J. Am. Statist. Assoc.* **69**:255–258.

Conover, W. J. (1973): "On Methods of Handling Ties in the Wilcoxon Signed-rank Test," *J. Am. Statist. Assoc.* **68**:985–988.

Cramér, H. (1946): *Mathematical Methods of Statistics*, Princeton University Press, Princeton, N.J.

David, H. A. (1960): "A Conservative Property of Binomial Tests," *Ann. Math. Statist.* **31**:1205–1207. A correction is given in *Ann. Math. Statist.* **32**:1343.

van Elteren, P. H. (1960): "On the Combination of Independent Two-sample Tests of Wilcoxon," *Bull. Inst. Intern. Statist.* **37**:351–361. [Proposes the statistic (3.24).]

Fellingham, S. A. and Stoker, D. J. (1964): "An Approximation for the Exact Distribution of the Wilcoxon Test for Symmetry," *J. Am. Statist. Assoc.* **59**:899–905.

Gridgeman, N. T. (1959): "Pair Comparison, With and Without Ties," *Biometrics* **15**:382–388.

Hemelrijk, J. (1952): "A Theorem on the Sign Test When Ties are Present," *Indagationes Math.* **14**:322–326.

Hodges, J. L., Jr., and Lehmann, E. L. (1962): "Rank Methods for Combination of Independent Experiments in the Analysis of Variance," *Ann. Math. Statist.* **33**:482–497. [Develops the method of aligned ranks of Sec. 3.]

——— and ——— (1970): *Basic Concepts of Probability and Statistics*, 2d ed., Holden-Day, San Francisco, Sec. 13.3, Probs. 10–12.

——— and ——— (1973): "Wilcoxon and t-tests for Matched Pairs of Typed Subject," *J. Am. Statist. Assoc.* **68**:151–158. [Gives a more detailed treatment of the issues of Sec. 4.]

Hoeffding, Wassily (1956): "On the Distribution of the Number of Successes in Independent Trials," *Ann. Math. Statist.* **27**:713–721.

Krauth, J. (1973): "An Asymptotic UMP Sign Test in the Presence of Ties," *Ann. Statist.* **1**:166–169.

Lehmann, E. L. (1959): *Testing Statistical Hypotheses*, John Wiley & Sons, New York, p. 148.

McCormack, Robert L. (1965): "Extended Tables of the Wilcoxon Matched Pair Signed-rank Statistic," *J. Am. Statist. Assoc.* **60**:864–871. [Provides critical values for $N = 4(1)100$.]

Noether, Gottfried E. (1963): "Efficiency of the Wilcoxon Two-sample Statistics for Randomized Blocks," *J. Am. Statist. Assoc.* **58**:894–898.

Osterhoff, J. (1969): "Combination of One-sided Statistical Tests," *Mathematical Centre Tracts 28*, Math. Centrum, Amsterdam. [Discusses the general problem of combining the results from a number of different experiments and surveys the history of this problem.]

Pratt, John W. (1959): "Remarks on Zeros and Ties in the Wilcoxon Signed-rank Procedures," *J. Am. Statist. Assoc.* **54**:655–667.

Puri, M. L. (1965): "On the Combination of Independent Two-sample Tests of a General Class," *Rev. Intern. Statist. Inst.* **33**:229–241.

Putter, Joseph (1955): "The Treatment of Ties in Some Nonparametric Tests," *Ann. Math. Statist.* **26**:368–386.

Smirnov, N. V. (1947): "Sur un Critère de Symmetrie de la Loi de Distribution d'une Variable Aléatore," *Izv. Acad. Sci. USSR* **56**:11–14.

Tables of the Cumulative Binomial Probabilities, Ordnance Corps Pamphlet ORDP 20-1, 1952. [Gives the binomial distribution to seven decimal places for $n = 1(1)150$ and for p at intervals of .01.]

Tables of the Cumulative Binomial Probability Distribution, Harvard University Press, Cambridge, Mass., 1955. [Gives the binomial distribution to five decimal places for selected values of n up to 1,000 and for p at intervals of .01.]

Tukey, John W. (1949): "The Simplest Signed-rank Tests," *Memo Report 17*, Statistical Research Group, Princeton University. [Establishes the equivalence of the statistics (3.6) and (3.16).]

Vorličková, Dana (1972): "Asymptotic Properties of Rank Tests of Symmetry under Discrete Distributions," *Ann. Math. Statist.* **43**:2013–2018.

Wilcoxon, Frank (1945): "Individual Comparisons by Ranking Methods," *Biometrics* **1**:80–83. [Proposes the test statistic (3.6).]

——— (1946): "Individual Comparisons of Grouped Data by Ranking Methods," *J. Econ. Entomol.* **39**:269–270.

——— (1949): Some Rapid Approximate Statistical Procedures, American Cyanamid Co., New York.

———, Katti, S. K., and Wilcox, Roberta A. (1970): "Critical Values and Probability Levels for the Wilcoxon Rank-sum Test and the Wilcoxon Signed-rank Test," *Selected Tables in Mathematical Statistics*, Markam, Chicago, vol. 1, pp. 171–259. [Gives the distribution to four decimals of V_s for $N \leqslant 50$.]

CHAPTER 4
PAIRED COMPARISONS IN A POPULATION MODEL AND THE ONE-SAMPLE PROBLEM

1. POWER AND USES OF THE SIGN TEST

The randomization models of the preceding chapter share the advantages and disadvantages of the randomization model of Chap. 1, Sec. 1. The advantages are the simplicity of the models and the fact that it is easy to carry out the required randomization, thereby ensuring that the assumptions are satisfied. A chief disadvantage is the limitation of the findings to the particular subjects that are included in the study. Unless these bear a known relationship to the population of potential users of the treatment under investigation, no legitimate extension of the results to this population is possible. As in the case of the unblocked comparison of two treatments, this difficulty can be overcome by drawing the subjects for the study from the population in question by means of a well-defined sampling scheme.

A simple scheme for paired comparison situations, in which the population is made up of homogeneous pairs, consists in drawing a simple random sample of such pairs. For example, if the subjects serve as their own controls, the sample of pairs is obtained by drawing a random sample of subjects. Similarly, if each pair consists of the morning and afternoon of the same day, one takes a sample of days. As another illustration, consider the comparison of two newspapers: one might sample news stories to see how the two papers treat the same story. An important example of paired comparisons is provided by the study of twins. However, in this case, even if the twins constitute a random sample from a population of twins the extrapolation of the findings to the general population requires the assumption that the responses are distributed in the same way over the subpopulation of twins as over the population at large.

Suppose now that N pairs of subjects are drawn at random from a population of pairs, that one subject of each pair is assigned to treatment and the other to control, and that the assignments are made independently and at random. If the responses of the control and treatment subject of the ith pair are denoted by X_i and Y_i, the assumption of random sampling implies that each of the N pairs of random variables $(X_1, Y_1), \ldots, (X_N, Y_N)$ has the same bivariate distribution, say

$$P(X_i \leqslant x, Y_i \leqslant y) = M(x, y)$$

In contrast to the independent (unrelated) samples $X_1, \ldots, X_m$ and $Y_1, \ldots, Y_n$ considered in Chap. 2, the above control observations $(X_1, \ldots, X_N)$ and treatment observations $(Y_1, \ldots, Y_N)$ are sometimes described as two *related samples* since X_i and the associated Y_i now typically are dependent.

We shall be concerned only with the differences

$$Z_i = Y_i - X_i$$

which under the above assumptions are also identically distributed, say, with distribution

$$P(Z_i \leqslant z) = L(z) \tag{4.1}$$

Consider now the hypothesis H of no treatment effect. Under H, the probability $M(x,y)$ (that the response of a control is $\leqslant x$ and that of a treatment subject $\leqslant y$) and the probability $M(y,x)$ (that the response of a control is $\leqslant y$ and that of a treatment subject $\leqslant x$) are equal, so that

$$M(x,y) = M(y,x) \tag{4.2}$$

By (4.2) the distribution of the random variables $Z_i = Y_i - X_i$ is the same as that of $-Z_i = X_i - Y_i$ so that

$$P_H(Z_i \leqslant z) = P_H(-Z_i \leqslant z) \tag{4.3}$$

or equivalently

$$L(z) = 1 - L(-z) \qquad \text{under } H \tag{4.4}$$

The hypothesis thus states that the distribution L of the Z_i is symmetric about zero.

For the sake of definiteness, suppose that H is being tested against the alternatives that the treatment has the effect of increasing the response. The Y's will then be larger than the X's and the Z's will therefore be slanted towards positive values. (For a precise definition of this intuitive notion, see Prob. 44.)

An important special case is the shift model, which is the analogue of model

(2.7). This model corresponds to the assumption of additivity, according to which the treatment adds a constant amount Δ to the control response of a subject. Let $E(z)$ denote the null distribution of $Z_i = Y_i - X_i$, which by (4.3) is symmetric about zero, and suppose the treatment effect is Δ. Then the distribution of $(Y_i - \Delta) - X_i$ is E, and the distribution of Z_i is therefore

$$P(Z_i \leqslant z) = E(z - \Delta) \qquad (E \text{ is symmetric about } 0) \tag{4.5}$$

If we now make the additional assumption that the population is large enough for the dependence of different pairs upon each other to be negligible, we are led to the following *population model:*

(4.6) The observations of interest are the random variables $Z_1, \ldots, Z_N$, which are independently, identically distributed with distribution L. The hypothesis to be tested is that L is symmetric about zero against the alternatives that it is slanted toward positive values.

For the more specific shift model (4.5), the problem reduces to that of testing the hypothesis $\Delta = 0$ against the alternatives $\Delta > 0$.

The model (4.6), which we have so far discussed in terms of paired comparisons, arises also in situations in which the observations consist of N independent measurements of a quantity Δ whose value is to be tested. Suppose that the observations are $Z'_1, \ldots, Z'_N$ and that the measurement errors $Z'_i - \Delta$ are symmetrically distributed about zero, and consider the hypothesis $H' : \Delta = \Delta_0$. Here Δ_0 may, for example, be the value of a physical or biological constant, predicted by a theory that is being tested. Or it can be the "normal" or desired value with which the quality of a product or the level of a physiological reaction is being compared. The problem can be reduced to the earlier hypothesis $H : \Delta = 0$ in the model (4.5) by putting $Z_i = Z'_i - \Delta_0$. In this context, the problem of testing or estimating Δ is often called the *one-sample problem.* The term distinguishes the present situation, which is concerned with a sample $Z_1, \ldots, Z_N$ from a single distribution L, from the problem of comparing samples $X_1, \ldots, X_m$ and $Y_1, \ldots, Y_n$ from two distributions F and G, considered in Chap. 2.

The population model not only makes it possible to carry conclusions from the sample to the population but also permits a more useful treatment of power. However, before discussing this aspect, it is necessary to check that the significance level calculated earlier for the randomization model of Chap. 3, Sec. 1 remains valid in the population model. Consider first the sign test, with which we shall be concerned in the present section. The test is based on the number S_N of positive differences $Z_i = Y_i - X_i$ and in the randomization model the null distribution of S_N was seen to be the binomial distribution (3.1), provided none of the Z_i are zero.

Let us now consider the distribution of S_N in the population model (4.6) where, to avoid the possibility of zero values for the Z's, the distribution L will be assumed to be continuous. Under H, the distribution of the Z_i is symmetric about zero, so that the probability of a plus or minus sign is $1/2$ for each i. Since the Z_i are assumed to be independent, so are their signs, and it follows that the null distribution of the number S_N of positive signs is, as in the randomization model, the binomial distribution (3.1) corresponding to N trials and success probability $1/2$. This proves that the significance level is the same in both models.

The power of the sign test in the population model (4.6) is obtained by noting that under an alternative to H, the Z_i and hence their signs continue to be independent, but that the probability of Z_i being positive is no longer $1/2$ but

$$p = 1 - L(0) \tag{4.7}$$

The number S_N of positive observations therefore has a binomial distribution corresponding to N trials and success probability p. The probability of failure will be denoted by $q = 1 - p$.

The power function of the test which rejects when

$$S_N \geqslant c \tag{4.8}$$

depends only on p, and is easily seen to be an increasing function of p (Prob. 39). This shows in particular [in analogy with the power property (i) of the Wilcoxon rank-sum test proved in Chap. 2, Sec. 2] that the test (4.8) is unbiased against all alternatives with $p \geqslant 1/2$. Since

$$p = 1 - E(-\Delta) = E(\Delta) \tag{4.9}$$

is greater than or equal to $1/2$ when $\Delta > 0$, it follows that the sign test is unbiased for testing $H: \Delta = 0$ against the alternatives $\Delta > 0$ in the shift model (4.5). It also follows, as was shown for the Wilcoxon rank-sum test in Chap. 2, Sec. 2, that (4.8) is not only a level α test of $H: \Delta = 0$ but also of the more general hypothesis $H': \Delta \leqslant 0$, which arises when the possibility of a negative treatment effect is admitted.

Since extensive tables of the binomial distribution are readily available,[1] we do not include such a table here but discuss next an approximation to the power function for large N. As is shown in Appendix, Example 10, $(S_N - Np)/\sqrt{Npq}$ is asymptotically normally distributed with zero mean and unit variance. The power of the test (4.8) for the alternative p is

$$\Pi(p) = P(S_N \geqslant c) = P\left(\frac{S_N - Np}{\sqrt{Npq}} \geqslant \frac{c - Np}{\sqrt{Npq}}\right) \tag{4.10}$$

and the normal approximation with continuity correction therefore gives

$$\Pi(p) \approx 1 - \Phi\left(\frac{c - \frac{1}{2} - Np}{\sqrt{Npq}}\right) \tag{4.11}$$

[1] For example, "Table of the Cumulative Binomial Probabilities," Ordnance Corps Pamphlet ORDP 20-1 (1952); and *Tables of the Cumulative Binomial Probability Distribution*, Harvard University Press, Cambridge, Mass. (1955).

EXAMPLE 1. *Gas mileage.* A consumer's organization is investigating a dealer's claim that his cars will average 22 miles/gal under ordinary driving conditions in the city. It is planned to find the average gas consumption for each of a sample of 20 cars and to consider the claim unfounded if too many of these averages fall below 22. Before beginning the investigation, the organization would like to know the probability with which the claim will be rejected if the cars actually average only 21 miles/gal. This probability, which is just the power of the test, depends of course on the significance level and on the shape and variance of the distribution of the observations. It is decided to use a significance level of about .05, and from the binomial tables with $N = 20$ and $p = \frac{1}{2}$ it is found that $P(S_N \geqslant 14) = .0577$ gives the significance level closest to .05.

Suppose that the average mileage Z of a car in the sample can be assumed to be approximately normally distributed with variance $\sigma^2 = 2.25$. Then the probability $p = P(Z < 22)$ is easily seen [Prob. 13(i)] to be equal to

$$p = \Phi\left(\frac{22-21}{1.5}\right) = \Phi(.667) = .7476$$

and the desired power is

$$\Pi = P(S_N \geqslant 14) = P\left(\frac{S_N - Np}{\sqrt{Npq}} \geqslant \frac{13.5 - Np}{\sqrt{Npq}}\right) \approx \Phi(.747) = .80$$

This value of Π may be considered too low, and it may then be of interest to find the sample size N required to bring the power up to some higher level, say, $\Pi = .90$. For this purpose, it is convenient to determine the critical value from the normal approximation rather than the exact null distribution, and write the rejection rule as

$$\frac{S_N - \frac{1}{2}N}{\frac{1}{2}\sqrt{N}} \geqslant u_\alpha \tag{4.12}$$

Here u_α is the upper α-point of the standard normal distribution, defined by Eq. (2.30), which for $\alpha = .05$ gives $u_\alpha = 1.645$. A comparison of (4.12) with (4.8) shows that the critical values c and u_α are related through the equation

$$c = \tfrac{1}{2}N + \tfrac{1}{2}u_\alpha\sqrt{N} \tag{4.13}$$

Substitution in (4.11) and neglecting the continuity correction leads to the approximation of the power of the test by

$$\Pi(p) \approx 1 - \Phi\left[\frac{\frac{1}{2}u_\alpha - \sqrt{N}\,(p - \frac{1}{2})}{\sqrt{pq}}\right] \tag{4.14}$$

If the left-hand side is to be equal to Π, the expression in square brackets should be about u_Π. Solving the resulting equation, we find

$$\tfrac{1}{2}u_\alpha - \sqrt{N}(p - \tfrac{1}{2}) \approx u_\Pi\sqrt{pq}$$

or the approximate equation

$$\sqrt{N} \approx \frac{\frac{1}{2}u_\alpha - \sqrt{pq}\,u_\Pi}{p - \frac{1}{2}} \tag{4.15}$$

Substituting the values $\alpha = .05$, $u_\alpha = 1.645$, $\Pi = .9$, $u_\Pi = -1.282$ gives for N the approximate value $N = 36$.

Let us now consider briefly two extensions of the population model (4.6). It was assumed in (4.6) that the random variables $Z_1, \ldots, Z_N$ are independently and identically distributed. However, in some situations the Z's cannot be assumed to have the same distribution. An example occurs in paired comparisons when the population from which the pairs are to be drawn is made up of relatively homogeneous subpopulations (or strata). A sample of homogeneous pairs can then be obtained by drawing two subjects at random from each of these subpopulations. As an illustration, suppose that in the comparison of two advertising methods of Chap. 3, Example 6, the cities had not yet been decided upon, and it is desired to draw a suitable sample of 12 cities. For this purpose, lists could be obtained of all cities having about 500,000 inhabitants, of those with about 250,000, 100,000, 50,000, 25,000, 10,000, and two cities could be drawn at random from each of these lists. As another illustration, suppose that a paired-comparison study is to be made of school children in which the two children of each pair should be of about the same age and working at about the same level. If, as is the case in some schools, each grade is divided into fairly homogeneous classes, with the level of work varying greatly from one class to another, the sample could be obtained by drawing two children at random from each class.

Let X_i and Y_i denote the responses of a control and treatment subject drawn at random from the ith subpopulation. Then as in (4.2), the joint null distribution M_i of X_i, Y_i satisfies $M_i(x, y) = M_i(y, x)$. If $Z_i = Y_i - X_i$, and if L_i denotes the distribution of Z_i, it follows as before that under the hypothesis of no treatment effect the distributions L_i are symmetric about zero, and under the alternatives they are slanted towards positive values. Since the drawings from the different subpopulations are performed independently, the variables Z_i are independent (without the subpopulations having to be large), and the problem reduces to the following generalization of (4.6):

(4.16) The observations $Z_1, \ldots, Z_N$ are independent with distributions $L_1, \ldots, L_N$. The hypothesis H' that all these distributions are symmetric about zero is to be tested against the alternatives that they are slanted towards positive values.

An important special case is again the shift model, which in generalization of (4.5) is now defined by

$$P(Z_i \leqslant z) = E_i(z - \Delta) \tag{4.17}$$

with the distributions E_i being symmetric about zero. In the paired-comparison situation this corresponds as before to the assumption that the treatment effect is additive. The shift model (4.17) can also arise in a measurement situation, when N measurements $Z_1, \ldots, Z_N$ are taken of a quantity Δ and for some reason the measurement errors $Z_i - \Delta$ do not all follow the same distribution.

Since (4.16) assumes the Z_i to be independent, and since under H' (and for continuous $L_1, \ldots, L_N$)

$$P(Z_i > 0) = \tfrac{1}{2} \qquad \text{for all } i$$

the null distribution of S_N continues to be the binomial distribution corresponding to N trials and success probability $\frac{1}{2}$ which it was under (4.6). The significance level of the sign test, computed under (4.6), therefore is also valid under the broader model (4.16).

The second extension of (4.6) concerns the hypothesis being tested. The hypothesis that L is symmetric about zero can be viewed as a combination of the two hypotheses: (i) that the distribution L of the Z's is symmetric; and (ii) that its center of symmetry is zero (or any other specified value). If (i) is dropped, it may still be of interest to investigate the location of L by testing the value of its center. Here a difficulty arises. The natural way of specifying the location of a symmetric distribution is by its center of symmetry. No such natural measure of location is available for asymmetric distributions. The expectation, which is the most commonly used measure of central tendency, is unsuitable when the distribution is not restricted to a parametric family since arbitrarily large differences in the expectation can result from minute changes in the distribution (Prob. 48). A simple and intuitively meaningful measure of the center of an arbitrary continuous distribution L is its median, that is, the point μ for which

$$P(Z < \mu) = P(Z > \mu) = \tfrac{1}{2} \tag{4.18}$$

when Z has distribution L. (We shall assume here that there is only one point satisfying this equation.)

Consider now the problem of testing the hypothesis $H'': \mu = \mu_0$ against the alternatives that $\mu > \mu_0$. Suppose that $Z_1, \ldots, Z_N$ are independently distributed according to L, and denote by S_N the number of Z's that are greater than μ_0. The events $(Z_i > \mu_0)$ are independent, and under H'' each has probability $\frac{1}{2}$. The null distribution of S_N is therefore, as before, the binomial distribution corresponding to N trials and success probability $p = \frac{1}{2}$, while p is greater than $\frac{1}{2}$ when $\mu > \mu_0$. The sign test is thus appropriate for testing the median of an arbitrary continuous distribution.

As an example, suppose that in a hearing test a patient is exposed to a sound whose pitch is continuously increased up to a level Z at which the patient no longer hears it. Then Z is a random variable, and the median μ of the distribution of Z is a possible measure of the upper range of the patient's hearing. It may be desired to compare his value of μ with some "normal" value μ_0 and in particular to test the hypothesis $\mu = \mu_0$ or $\mu \geqslant \mu_0$ against the alternatives that $\mu < \mu_0$. For this purpose, the test is repeated N times, leading to observed values $Z_1, \ldots, Z_N$, and the hypothesis is rejected according to the sign test if too many of the Z's are less than μ_0.

The sign test is seen to be capable of testing three essentially different hypotheses.

(*a*) If $Z_1, \ldots, Z_N$ are independently distributed according to a common continuous distribution L, it can be used to test that L is symmetric about zero against the alternatives that it is slanted toward positive values. More generally, if the Z's are independently distributed according to continuous distributions $L_1, \ldots, L_N$, it can be used to test that each L_i is symmetric about zero (or any given point).

(*b*) If it is only known whether each Z_i is positive or negative and if $p = P(Z_i > 0)$, the sign test can be used to test the hypothesis $p = \frac{1}{2}$ against the alternatives $p > \frac{1}{2}$. In particular, in a paired comparison of two treatments, the sign test is thus applicable when only qualitative comparisons of the two treatments are available, which determine for each pair the treatment that comes out ahead.

(*c*) In case (*a*) the sign test can also be used to test the broader hypothesis that the median of L (or of $L_1, \ldots, L_N$) has a specified value.

As has already been indicated, and as will be discussed in more detail in Sec. 3, alternative tests are available for case (*a*), which are typically considerably more powerful than the sign test, and the main advantage of the latter in this case is its great simplicity. The principal usefulness of the sign test is in relation to hypotheses (*b*) and (*c*). Of these, we have not emphasized (*b*) here, in spite of the fact that this constitutes the most important area of application of the test, because this is a simple binomial problem, which is essentially parametric rather than nonparametric (although the distinction is not clear cut) and is treated fully in most statistical texts.

In the above cases, the probability of zero observations has so far been assumed to be equal to zero. In (*a*), let us now drop the assumption that $P(Z_i = 0) = 0$ under L and put

$$P(Z_i < 0) = p_- \qquad P(Z_i = 0) = p_0 \qquad P(Z_i > 0) = p_+$$

If, as in Chap. 3, Sec. 1, the number of negative, zero, and positive Z's is denoted by N_-, N_0, and N_+, the joint distribution of these variables is the trinomial distribution

$$P(N_- = a, N_0 = b, N_+ = c) = \frac{N!}{a!b!c!} p_-^{\,a} p_0^{\,b} p_+^{\,c} \tag{4.19}$$

where $a+b+c = N$.

The hypothesis H that L is symmetric about zero asserts that $p_+ = p_-$, but even under H the distribution of the test statistics N_+ or $N_+ + \frac{1}{2}N_0$ depend on the unknown p_0, and the same is true of this hypothesis in case (*c*). The difficulty can be overcome by the device that was used in Chap. 2, Sec. 1, namely, by performing the test conditionally, given the value of N_0. Then the test statistics $N_+ + \frac{1}{2}N_0$ and N_+ become equivalent, as they were in the randomization model, and the conditional distribution of N_+ is the binomial distribution corresponding to $N' = N - N_0$ trials and success probability $p'_+ = p_+/(p_- + p_+)$. In this conditional binomial situation the hypothesis H reduces to $p'_+ = 1/2$, and the test to the sign test in which the zeros are discarded and consideration is restricted to the nonzero observations, and in which the number $N' = N_- + N_+$ of nonzero observations is treated as a constant. In terms of the present model, it is possible to show, although we shall not do so, that the conditional sign test just described is more powerful than the sign test obtained by random breaking of zeros. [See Putter (1955) and Lehmann (1959, p. 147).]

Although a modified sign test thus continues to be applicable to hypotheses (*a*) and (*b*) even when $p_0 > 0$, the situation is less satisfactory with respect to hypothesis (*c*). If L is not necessarily continuous, its median is defined as the point μ such that

$$P(Z < \mu) \quad \text{and} \quad P(Z > \mu) \quad \text{are both less than or equal to } \tfrac{1}{2} \tag{4.20}$$

If p_-, p_0, and p_+ have the same meaning as before, the conditional sign test given N_0 tests the hypothesis that $p_- = p_+$, which is less restrictive than the symmetry assumption of hypothesis (*a*), but it is no longer a test of the median of L. [For a discussion of this problem see Walsh (1951).]

2. POWER OF THE SIGNED-RANK WILCOXON TEST

The considerations of the last section concerning the performance of the sign test in a population model have their analogues for the Wilcoxon signed-rank test with which we shall be concerned in the present section. As in the case of the Wilcoxon rank-sum test (Chap. 2, Theorem 1) and the sign test, it is necessary first to check that the null distribution of the Wilcoxon statistic V_s in the population model is the same as its distribution in the randomization model, which was discussed in Chap. 3, Sec. 2. This result is a special case of the following theorem, which will be proved at the end of the section.

Theorem 1. Let $Z_1, \ldots, Z_N$ be independently distributed according to a common continuous distribution L, and let $S_1, \ldots, S_n$ be the ranks of the positive Z's among the absolute values $|Z_1|, \ldots, |Z_N|$. Then under the hypothesis H that L is symmetric about zero, each of the 2^N possible sets $(n;s_1,\ldots,s_n)$ has probability

(4.21) $$P_H(N_+ = n; S_1 = s_1, \ldots, S_n = s_n) = (1/2)^N$$

where as before N_+ denotes the number of positive Z's.

It follows from Theorem 1 that the null distribution (3.5) remains valid in the population model (4.6), and the same argument shows that this distribution continues to hold if (4.6) is replaced by the more general model (4.16). In particular, the significance level of the Wilcoxon test for testing the hypothesis of symmetry about zero in model (4.6) or (4.16) is computed as before from Table H.

Let us now consider the power

(4.22) $$\Pi_E(\Delta) = P_\Delta(V_s \geqslant c)$$

of the Wilcoxon test (3.7) in the shift model (4.5). Note first that, as for the Wilcoxon rank-sum test, the power function $\Pi_E(\Delta)$ is a nondecreasing function of Δ (Prob. 50). It therefore follows as in the two-sample case that the test is unbiased against the alternatives $\Delta > 0$, and that the level α test of $H: \Delta = 0$ has significance level α also for testing $H': \Delta \leqslant 0$.

The evaluation of the exact power is much more difficult for the Wilcoxon test than for the sign test since it depends on the distribution of the ranks when the distribution of the Z's is not symmetric about zero. We shall not discuss it here (but see Sec. 6C). Instead we shall be content with a large-sample approximation to this power. As is shown in Appendix, Example 21, the distribution of

$$\frac{V_s - E(V_s)}{\sqrt{\operatorname{Var}(V_s)}}$$

tends to the standard normal distribution as N tends to infinity for any distribution L for which

(4.23) $$0 < P(Z < 0) < 1$$

When $P(Z < 0)$ is 0 or 1, the statistic V_s reduces to a constant, and the limit problem does not arise. To apply the normal approximation, it is necessary to determine the expectation and variance of V_s.

The expectation depends on the quantities $p = P(Z > 0)$ and

(4.24) $$p'_1 = P(Z + Z' > 0)$$

where Z and Z' are independent random variables with distribution L, and it is given by

(4.25) $$E(V_s) = \tfrac{1}{2}N(N-1)p'_1 + Np$$

This formula is proved as formula (A.79) of the Appendix. When L is symmetric about zero and in addition is continuous, it is seen that $p = p'_1 = 1/2$ and that the expectation of V_s reduces to $N(N+1)/4$, which agrees with formula (3.13).

The variance of V_s depends on p and p'_1 and also on

$$p'_2 = P(Z+Z' > 0 \text{ and } Z+Z'' > 0) \tag{4.26}$$

where Z, Z', and Z'' are independently distributed according to L, and in Appendix, Example 6, is shown to be

$$\begin{aligned}\operatorname{Var}(V_s) = {} & N(N-1)(N-2)(p'_2 - p'^2_1) + \frac{N(N-1)}{2}[2(p-p'_1)^2 + 3p'_1(1-p'_1)] \\ & + Np(1-p)\end{aligned} \tag{4.27}$$

It is again interesting to check this formula for the case that L is continuous and symmetric about zero. Then $p'_2 = 1/3$, and on substituting this value and those for p and p'_1 into (4.27) one regains formula (3.14) (Prob. 52).

Let us now apply the asymptotic normality of V_s to obtain an approximation to the power of the Wilcoxon test. If the test is expressed in terms of the rejection rule $V_s \geqslant c$, its power $\Pi(L)$ against any fixed alternative L is

$$P(V_s \geqslant c) = P\left[\frac{V_s - E(V_s)}{\sqrt{\operatorname{Var}(V_s)}} \geqslant \frac{c - E(V_s)}{\sqrt{\operatorname{Var}(V_s)}}\right]$$

and the normal approximation with continuity correction therefore gives

$$\Pi(L) \approx 1 - \Phi\left[\frac{c - \frac{1}{2} - E(V_s)}{\sqrt{\operatorname{Var}(V_s)}}\right] \tag{4.28}$$

EXAMPLE 2. *Treatment for anemia.* To test the effectiveness of vitamin B_{12} in treating pernicious anemia, the treatment is applied to 10 patients suffering from this disease. Let X_i and Y_i denote the hemoglobin levels of the ith subject before and after the treatment, and let $Z_i = Y_i - X_i$. Suppose that the Wilcoxon test is to be used at approximate level .05 to test the hypothesis of no effect. Table H shows that the critical value $c = 44$ gives a significance level $\alpha = .0527$. What is the power of the test $V_s \geqslant 44$ against the alternative that the treatment raises the hemoglobin level by about 2 g if the Z's are normally distributed with standard deviation $\tau = 2$?

It is useful to determine the probabilities p, p'_1, p'_2, which are required for this power computation, in the more general case that the Z's are normal with mean Δ and variance τ^2. Then

$$p = P(Z > 0) = P\left(\frac{Z-\Delta}{\tau} > -\frac{\Delta}{\tau}\right)$$

so that

$$p = \Phi\left(\frac{\Delta}{\tau}\right) \tag{4.29}$$

Similarly,

$$p'_1 = P(Z+Z' > 0) = P\left(\frac{Z+Z'-2\Delta}{\sqrt{2}\tau} > \frac{-2\Delta}{\sqrt{2}\tau}\right)$$

and hence

$$p_1' = \Phi\left(\sqrt{2}\frac{\Delta}{\tau}\right) \tag{4.30}$$

Consider finally

$$\begin{aligned} p_2' &= P(Z+Z' > 0 \text{ and } Z+Z'' > 0) \\ &= P\left(\frac{Z+Z'-2\Delta}{\sqrt{2}\tau} > \frac{-\sqrt{2}\Delta}{\tau} \text{ and } \frac{Z+Z''-2\Delta}{\sqrt{2}\tau} > \frac{-\sqrt{2}\Delta}{\tau}\right) \end{aligned} \tag{4.31}$$

Since

$$\frac{Z+Z'-2\Delta}{\sqrt{2}\tau} \quad \text{and} \quad \frac{Z+Z''-2\Delta}{\sqrt{2}\tau}$$

have a bivariate normal distribution with means zero and unit variances and correlation coefficient ½ (by Chap. 2, Prob. 43), p_2' can be obtained from published tables.[1]

For the numerical values $\Delta = 2$ and $\tau = 2$ of the example, one finds (Prob. 9)

$$p = .8413 \qquad p_1' = .9213 \qquad p_2' = .8657$$

and on substitution in (4.28), $\Pi = .90$. The exact value in this case is .89.†

The approximation (4.28) to the power, which in principle can be computed for any alternative, requires the often cumbersome evaluation of the probability p_2'. This difficulty is avoided by an alternative approximation, which is applicable in the shift model (4.5) when Δ is small, and which parallels the approximation (2.29) in the two-sample case. Let the power for the signed-rank Wilcoxon test in the shift model (4.5) be denoted by $\Pi_E(\Delta)$, and let E^* denote the distribution of the sum of two independent random variables each with distribution E. Then if $e(0)$ and $e^*(0)$ denote the densities of E and E^* evaluated at zero, the approximation in question is given by

$$\Pi_E(\Delta) \approx \Phi\left[\frac{N(N-1)\,e^*(0)+Ne(0)}{\sqrt{N(N+1)(2N+1)/24}}\Delta - u_\alpha\right] \tag{4.32}$$

A heuristic derivation of (4.32) is given at the end of the section. A slightly different form of this approximation is derived from a different point of view as formula (A.234) (see Appendix, Prob. 47).

To apply (4.32), it is necessary to find $e^*(0)$ and $e(0)$. Suppose in particular that E is the normal distribution $N(0,\tau^2)$. Then E^* is the distribution of the sum of two independent normal variables with zero mean and variance τ^2, and hence is

[1] National Bureau of Standards, "Tables of the Bivariate Normal Distribution Function and Related Functions," *Appl. Math. Ser.* **50** (1959).

† Given by Klotz (1963).

the normal distribution $N(0,2\tau^2)$. It follows that

$$e(0) = \frac{1}{\sqrt{2\pi}\tau} \qquad \text{and} \qquad e^*(0) = \frac{1}{\sqrt{\pi}2\tau}$$

so that (4.32) becomes

$$\Pi_E(\Delta) \approx \Phi\left[\frac{N(N-1)/2 + N/\sqrt{2}}{\sqrt{N(N+1)(2N+1)/24}}\left(\frac{\Delta}{\tau\sqrt{\pi}}\right) - u_\alpha\right] \tag{4.33}$$

As an illustration, consider once more Example 2, where $\Delta = 2$, $\tau^2 = 4$, $N = 10$, and $\alpha = .05$. Then $u_\alpha = 1.645$ and (4.33) gives $\Pi_E(\Delta) \approx .91$ compared with the value .90 obtained from the earlier approximation and the exact value .89.

For the reasons discussed in Chap. 2, Sec. 3, in the context of the two-sample problem, both approximations (4.28) and (4.32) can be expected to work well for sufficiently small Δ, but particularly the second approximation may become quite unreliable as Δ increases. Table 4.1 shows the two approximations and the true values for a normal shift model with sample size $N = 10$ and level $\alpha = .0244$. The

TABLE 4.1 Power of the Wilcoxon signed-rank test for normal shift alternatives; $N = 10$, $\alpha = .0244$

Δ/τ	0	.25	.50	.75	1.25	1.50
Exact	.0244	.1021	.2822	.5437	.7847	.9276
(4.28)	.0197	.0896	.2420	.4819	.7728	.9958
(4.32)	.0242	.1156	.3354	.6360	.8687	.9708

sample size is too small for either approximation to work very well. However, the approximate power values are at least of the right order of magnitude, and the approximations will presumably be adequate for $N \geqslant 20$. The situation is, however, quite different in the Cauchy case illustrated by Table 4.2. Here, the approximation (4.32) is completely misleading except for very small values of Δ, and it will presumably require quite large sample sizes before this approximation will give usable results. It seems likely that this behavior is a consequence of the unusually heavy tails of the Cauchy distribution (see Appendix, Sec. 2D) and that for a distribution such as the logistic, whose tails are intermediate, the approximations will tend to work better than in the Cauchy case though probably not quite as well as in the normal case.

Before undertaking a study, it is desirable to determine the sample size needed to achieve a given power. An approximation to this sample size is easily obtained from

TABLE 4.2 Power of the Wilcoxon signed-rank test against a shifted Cauchy distribution; $N = 10$, $\alpha = .0244$

Δ	0	.25	.50	1.00	2.00
Exact	.0244	.3725	.5026	.7014	.8297
(4.28)	.0197	.2373	.5056	.8018	.9819
(4.32)	.0244	.3832	.9154	1.0000	1.0000

(4.32). Suppose we wish to find the value of N for which $\Pi_E(\Delta)$ has a specified value Π against a given alternative Δ (and E). If u_Π is defined by $\Phi(-u_\Pi) = \Pi$, it follows from (4.32) that

$$-u_\Pi \approx \frac{N(N-1)\,e^*(0) + Ne(0)}{\sqrt{N(N+1)(2N+1)/24}}\,\Delta - u_\alpha$$

and it is only necessary to solve this equation for N. If N is expected to be large, one can obtain an explicit approximation for N by replacing the numerator on the right-hand side by $N^2\,e^*(0)$ and the denominator by $\sqrt{N^3/12}$. Solving the resulting equation for N one finds the approximation

$$N \approx \frac{(u_\alpha - u_\Pi)^2}{12\Delta^2\,e^{*2}(0)} \tag{4.34}$$

When E is a normal distribution with variance τ^2, (4.34) reduces to

$$N \approx \frac{\pi\tau^2(u_\alpha - u_\Pi)^2}{3\Delta^2} \tag{4.35}$$

As an illustration, suppose that $\alpha = .01$ and that it is desired to achieve power $\Pi = .95$ against the alternatives that the Z's are normally distributed with mean Δ and variance τ^2, where $\Delta/\tau = .5$. Then $u_\alpha = 2.33$ and $u_\Pi = -1.645$, and (4.35) gives the approximate value $N = 67$.

At the beginning of this section, it was stated in Theorem 1 that the joint null distribution of the number n of positive Z's and the ranks $S_1, \ldots, S_n$ of the positive Z's among the absolute values $|Z_1|, \ldots, |Z_N|$ is the same in the population model (4.6) as in the randomization model of Chap. 3, provided the distribution L is continuous. If L is not continuous, so that ties occur with positive probability, the situation is exactly analogous to that discussed in Chap. 2, Sec. 1. In the population model, the distribution of the midranks then depends on the unknown L, and only the conditional distribution given the number of zeros and the configuration of ties is independent of L. For this reason and those discussed in Chaps. 1 and 2 it is desirable to avoid ties. If this is not possible, the tests can be carried out conditionally and then reduce to the tests of Chap. 3. However, the inference then also becomes conditional and the advantages of the population model are lost, exactly as in the case discussed in Chap. 2, Sec. 1.

To conclude the section, we shall now (*A*) prove Theorem 1, and (*B*) give a heuristic derivation of the approximation (4.32).

(A) Proof of Theorem 1. We begin by proving that the signs of the Z's are independent of the absolute values $|Z_1|, \ldots, |Z_N|$. Since the Z's are independent, it is enough to show this for a single Z. A simple computation gives

$$P(Z > 0 \text{ and } |Z| \leqslant z) = P(0 < Z \leqslant z) = \tfrac{1}{2}P(|Z| \leqslant z) = P(Z > 0)P(|Z| \leqslant z)$$

and similarly

$$P(Z < 0 \text{ and } |Z| \leqslant z) = P(Z < 0)P(|Z| \leqslant z)$$

and this proves the desired independence.

It follows from this result that the signs of the Z's are independent also of the ranks of the absolute values of the Z's. This implies that the particular Z, the absolute value of which has rank 1, is positive or negative with probability $\frac{1}{2}$; independently, the same is true of the Z, the absolute value of which has rank 2; and so on. Hence, to each of the ranks $1, \ldots, N$ of the absolute values, the sign $+$ or $-$ is attached independently with probability $\frac{1}{2}$. The probability that any particular ranks $s_1 < \cdots < s_n$ receive the sign $+$ and the remaining ranks the sign $-$ is therefore $(\frac{1}{2})^N = 1/2^N$, as was to be proved.

(B) Heuristic Derivation of the Power Approximation (4.32). Suppose that the sample size is large enough so that the critical value of the Wilcoxon test is determined from the normal approximation rather than the exact null distribution, and write the rejection rule as

$$\frac{V_s - \frac{1}{4}N(N+1)}{\sqrt{N(N+1)(2N+1)/24}} \geqslant u_\alpha \tag{4.36}$$

Comparison of (4.36) with the rule $V_s \geqslant c$ shows the critical values u_α and c to be related through the approximate equation

$$c \approx \tfrac{1}{4}N(N+1) + u_\alpha \sqrt{\frac{N(N+1)(2N+1)}{24}} \tag{4.37}$$

Substituting this value of c into (4.28), neglecting the continuity correction, and using the symmetry of the normal distribution, we can rewrite (4.28) as

$$\Pi \approx \Phi\left[\frac{\frac{1}{2}N(N-1)(p'_1 - \frac{1}{2}) + N(p - \frac{1}{2}) - u_\alpha \sqrt{N(N+1)(2N+1)/24}}{\sqrt{\operatorname{Var}(V_s)}}\right] \tag{4.38}$$

As in the argument leading to (4.29) and (4.30), it is seen in the case of the shift alternative (4.5) that

$$p = P(Z - \Delta > -\Delta) = 1 - E(-\Delta) = E(\Delta) \tag{4.39}$$

and

$$p'_1 = P[(Z - \Delta) + (Z' - \Delta) > -2\Delta] = E^*(2\Delta) \tag{4.40}$$

where E^* is the distribution of the sum of two independent random variables each with distribution E. If E and E^* have probability densities e and e^*, expansion of these two distributions about $\Delta = 0$ gives the approximations

$$p \approx E(0) + \Delta e(0) \qquad \text{and} \qquad p'_1 \approx E^*(0) + 2\Delta e^*(0)$$

and hence, because of the symmetry of E and E^*,

$$p-\tfrac{1}{2} \approx \Delta\, e(0) \qquad \text{and} \qquad p_1' - \tfrac{1}{2} \approx 2\Delta\, e^*(0)$$

The variance of V_s also depends on Δ, but for small Δ it is close to its value for $\Delta = 0$, which is $N(N+1)(2N+1)/24$. Application of these three approximations to (4.38) leads to the desired approximation (4.32).

3. COMPARISON OF SIGN, WILCOXON, AND t-TESTS

On the basis of the results obtained in the last two sections for the power of the sign and Wilcoxon tests, it is now possible to undertake a comparison of these two tests. However, before doing so, we shall briefly discuss Student's t-test to include it in the comparison. The one-sample t-test is derived under the assumption that $Z_1, \ldots, Z_N$ are independently, normally distributed with unknown mean Δ and variance τ^2, to test the hypothesis that the mean has a specified value Δ_0. If Δ_0 is taken (without loss of generality) to be zero, the hypothesis is rejected when

$$\frac{\sqrt{N}\,\bar{Z}}{S} \geqslant c \tag{4.41}$$

where $\bar{Z}$ is the average of the Z's and where

$$S^2 = \frac{\sum (Z_i - \bar{Z})^2}{N-1} \tag{4.42}$$

The critical value c is determined from Student's t-distribution with $N-1$ degrees of freedom, which under the assumption of normality is the null distribution of the t-statistic (4.41). This one-sample (or paired-comparison, when $Z_i = Y_i - X_i$) t-test, like its two-sample analogue, has various optimum properties in the above normal model; it is, for example, uniformly most powerful among all unbiased tests.

In the present section, we shall consider the t-test as a competitor of the sign and Wilcoxon tests and must therefore begin by investigating its behavior under the broader hypothesis that the distribution L of the Z's is symmetric about the origin but no longer necessarily normal. Unlike the sign and Wilcoxon tests, the t-test is not distribution-free in this situation, and when L is not normal, the significance level will typically not be equal to the stated value α. However, like the two-sample t-test, it is asymptotically distribution-free and the significance level is approximately independent of L for large N, provided only the variance τ^2 of L is finite. This follows from the facts that $\sqrt{N}\,\bar{Z}/\tau$ tends to the standard normal distribution and S/τ tends to 1 in probability, which show (by Appendix, Theorem 4) that

$$\frac{\sqrt{N}\,\bar{Z}}{S} = \frac{\sqrt{N}\,\bar{Z}/\tau}{S/\tau}$$

is asymptotically normally distributed with zero mean and unit variance independently of L. If the critical value c on the right-hand side of (4.41) is denoted more precisely by $c_N(\alpha)$, it follows that

$$c_N(\alpha) \to u_\alpha \qquad \text{as } N \to \infty \tag{4.43}$$

and hence that the probability of the inequality (4.41) tends to $1 - \Phi(u_\alpha) = \alpha$, as was to be proved.[1]

The sign and Wilcoxon tests were seen to be distribution-free not only in model (4.6) but also in the more general model (4.16). Under mild additional assumptions, the fact that the t-test is asymptotically distribution-free also extends to (4.16).

To obtain an approximation for the power Π of the t-test against a given alternative L, let $E(Z)$ and τ^2 denote the expectation and variance of the Z's under L, and write Π as

$$\Pi = P\left\{\frac{\sqrt{N}[\bar{Z} - E(Z)]}{\tau} \geqslant \frac{cS}{\tau} - \frac{E(Z)\sqrt{N}}{\tau}\right\}$$

where the Z's are independently distributed, each with distribution L. The asymptotic normality of the left-hand side of the inequality, together with (4.43) and the fact that S/τ tends to 1 in probability, suggests the approximation

$$\Pi \approx 1 - \Phi\left[u_\alpha - \frac{E(Z)\sqrt{N}}{\tau}\right] \tag{4.44}$$

On the basis of (4.14), (4.28) or (4.32), and (4.44), the power of the three tests can now be compared for any sufficiently large N and any distribution L. As an illustration, consider the situation in the paragraph following formula (4.35), where a normal shift model was assumed with $N = 67$, $\alpha = .01$, $\Delta/\tau = .5$, and where the power Π_W of the Wilcoxon test was .95. Substituting the values of N, α, and $E(Z)/\tau = \Delta/\tau$ into (4.14) and (4.44), it is found that the power Π_S of the sign test is .81 and the power Π_t of the t-test is .96, so that

$$\Pi_S = .81 \qquad \Pi_W = .95 \qquad \Pi_t = .96$$

An alternative to the direct comparison of the power of the tests is provided, as in Chap. 2, Sec. 4, by considering the situation with which one is confronted when planning a study. In the above example, suppose that a power of .95 has been prescribed. Then (4.15) and (4.44) can be used to give the sample sizes N_S and N_t required by the sign and t-tests for this purpose, and one finds (Prob. 16)

$$N_S = 101 \qquad N_W = 67 \qquad N_t = 63$$

The corresponding efficiencies are

$$e_{S,t} = 63/101 = .62 \qquad e_{W,t} = 63/67 = .94 \qquad e_{S,W} = 67/101 = .66$$

[1] The effect of dependence of the Z's on the level of the sign, Wilcoxon, and $\bar{Z}$-tests is studied by Gastwirth and Rubin (1971).

These computations suggest that in the normal case the t-test is only slightly more efficient than the Wilcoxon test but that both are considerably more efficient than the sign test. Would the results be similar for smaller sample sizes? Table 4.3 gives some exact values of $e_{S,t}$ (for the two-sided tests) for sample sizes $N = 10$ and 20.† The first line in each case gives the efficiency, the second the power of the sign test, for the given values of N, α, and Δ/τ. A striking feature of the table is the constancy of the efficiencies over a considerable range of Π values. For smaller values of α, the efficiencies are more variable and tend to be somewhat higher. As in the two-sample case discussed in Chap. 2, Sec. 4, the efficiency for any fixed values of α and Π tends to a limit as N tends to infinity, which is known as the Pitman efficiency. [A formula for this efficiency is given in the Appendix as (A.254).] In the normal case, this limit turns out to be $2/\pi = .64$, which is close to the approximate efficiency computed above for $\alpha = .01$, $\Pi = .95$.

TABLE 4.3 Efficiency of sign to t-test; E = normal

Δ/τ	.1257	.2534	.3853	.5244	.6745	.8416	1.0364	1.2816	1.6449
$N = 10$, $\alpha = .1094$	.768	.767	.765	.761	.756	.749	.741	.729	.712
Π	.13	.18	.27	.38	.53	.68	.82	.93	.99
$N = 20$, $\alpha = .1153$	.698	.696	.693	.688	.683	.677	.669		
Π	.15	.26	.42	.61	.79	.91	.98		

It is seen from the numerical results above that the efficiency loss of the sign test relative to the t-test in the normal case is considerable. Normality is of course not the only situation of interest, and distributions exist for which the sign test is much more efficient than the t-test (this tends to be true for distributions with sufficiently heavy tails, see Prob. 61). However, the poor showing of the sign test when the underlying distribution is normal or close to normal limits its usefulness, particularly since the Wilcoxon test does not suffer from the same disadvantage. This naturally does not detract from the other uses of the sign test discussed in Sec. 1.

The efficiency $e_{W,t}$ of the Wilcoxon to the t-test in the normal case was seen to be equal to about .94 when the Wilcoxon test is based on $N = 67$ observations. Again the question arises whether this efficiency has similar values for smaller sample sizes. Table 4.4 shows the results when the Wilcoxon test is based on $N = 10$ observations, for two different significance levels.[1] The first line in each case gives the efficiency, the second the power of the Wilcoxon test.

The constancy of the efficiency for varying Π is even more remarkable here than it was in the preceding case. The results show good agreement with the

† The efficiencies are from Dixon, "Power Functions of the Sign Test and Power Efficiency for Normal Alternatives," *Ann. Math. Statist.* **24**:467–473 (1953), Table III.

[1] From Klotz, "Small Sample Power and Efficiency for the One-sample Wilcoxon and Normal Scores Tests," *Ann. Math. Statist.* **34**:624–632 (1963), Table 1.

approximate value .94 found for $N = 67$ and with the limiting (Pitman) efficiency, which is $3/\pi = .955$. (For a proof see the Appendix, Example 26.) The fact that the Pitman efficiency in the present case is the same as in the corresponding two-sample case is not a special feature of the normal distribution. It turns out quite generally (see the Appendix, Example 26) that the efficiency of the one-sample Wilcoxon test to the t-test in the shift model (4.5) is the same as that of the two-sample Wilcoxon test to the t-test in model (2.7) when $F = E$. Thus the comparison of these tests in Chap. 2, Sec. 4 also applies to the corresponding tests in the one-sample or paired-comparison situation.

TABLE 4.4 Efficiency of Wilcoxon to t-test; E = normal

Δ/τ	.25	.50	.75	1.00	1.25	1.50
$N = 10$, $\alpha = .05237$	.968	.967	.966	.965	.965	.964
Π	.1844	.4274	.7013	.8914	.9734	.9957
$N = 10$, $\alpha = .09668$	.960	.959	.957	.956	.955	
Π	.2862	.5669	.8153	.9476	.9904	.9989

Of the three possible comparisons among the tests under consideration—sign to t, Wilcoxon to t, and sign to Wilcoxon—it remains to comment briefly on the last one. The situation here is very similar to that in the comparison of the sign and t-tests. Quite generally, if three tests T_1, T_2, and T_3 require N_1, N_2, and N_3 observations to achieve the same power, the efficiencies

$$e_{ij} = \frac{N_j}{N_i}$$

satisfy the relations

$$e_{ji} = \frac{1}{e_{ij}} \tag{4.45}$$

and

$$e_{ik} = e_{ij} e_{jk} \tag{4.46}$$

and these relations hold also for the Pitman efficiencies. Since in the normal case $e_{W,t}$ is typically about .95, the efficiency of the sign to the Wilcoxon test is obtained by dividing the efficiency of the sign to the t-test by .95. This increases the latter efficiency slightly but still leaves $e_{S,W}$ considerably below 1 [the Pitman efficiency is $(2/\pi)/(3/\pi) = 2/3$]. The efficiency $e_{S,W}$ is less strongly affected by gross errors than $e_{W,t}$ and $e_{S,t}$ since it tends to depend more on the central portion than on the tails of the distribution of the Z's. In view of the prevalence of distributions whose central part is close to normal, the low value of $e_{S,W}$ in the normal case thus indicates the use of the Wilcoxon rather than the sign test when both are available and applicable.

Much of the present section has closely paralleled the material of Chap. 2, Sec. 4, with Tables 4.3 and 4.4 giving additional support to some of the conclusions. An overall comparison of the Wilcoxon and t-tests would just be a repetition of the corresponding comparisons in the two-sample case, and is therefore omitted. Instead, the reader may wish to reread the relevant parts of the earlier section as applicable to the present situation (with only the obvious changes).

4. ESTIMATION OF A LOCATION PARAMETER OR TREATMENT EFFECT

Averaging a number of observations to obtain a more reliable estimate of the quantity being observed is perhaps the most commonly used of all statistical procedures. To find the market value of a house one may average the appraisals of several realtors; the grade of a student is often determined by averaging the results of his tests or assignments; and to obtain the gas consumption of a car one averages the mileages achieved on a number of occasions with the same amount of gas. The average is not the only possible estimator, however. The median of the observations, for example, is a frequently used alternative.

We shall here compare these and other estimators under the assumption that the observations $Z_1, \ldots, Z_N$ are distributed according to a common symmetric distribution L whose center of symmetry, say, θ, is to be estimated. This problem arises not only when a value (such as the long-term average gas consumption of a car) is being observed with an error that is symmetrically distributed about zero but also in the paired-comparison situations discussed in Sec. 1. The location parameter θ then coincides with the treatment effect, which in (4.5) was denoted by Δ. We shall in the present section adhere to the notation θ even in the latter case.

When the distribution L is normal, the average[1]

$$\bar{\theta} = \frac{Z_1 + \cdots + Z_N}{N} \tag{4.47}$$

has strong optimum properties as an estimator of θ but it is sensitive to gross errors and tends to be inefficient for distributions with heavy tails. An extreme example of this fact is provided by the Cauchy distribution for which the distribution of $\bar{\theta}$ is, for any N, the same as that of Z_1, so that the average is no more accurate than a single observation (Appendix, Prob. 12).[2]

To obtain a more robust estimator, let us proceed as in Chap. 2, Sec. 5. In

[1] The average, which is here denoted by $\bar{\theta}$ to emphasize its being considered as an estimator of θ, will in other contexts be denoted by $\bar{Z}$, and the corresponding convention will also be used for other estimators.

[2] For examples for which the average is even less accurate than a single observation, see Brown and Tukey, "Some Distributions of Sample Means," *Ann. Math. Statist.* **17**:1–12 (1946).

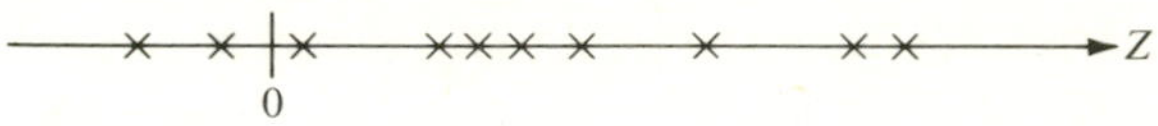

0 Z

FIGURE 4.1
Pattern of $Z_1, \ldots, Z_N$.

view of the fact that the distribution of the variables $Z_1-\theta, \ldots, Z_N-\theta$ is symmetric with respect to zero, take as estimator that value of θ for which the N values $Z_1-\theta, \ldots, Z_N-\theta$ give the best balance relative to the origin. Consider any test statistic for testing the hypothesis that the common distribution of N variables is symmetric with respect to zero. Then the variables $Z_1-\theta, \ldots, Z_N-\theta$ will be best balanced with respect to zero when the test statistic, evaluated for the variables $Z_i-\theta$, takes on the value giving the strongest support to the hypothesis that their distribution is symmetric with respect to zero. This will typically be the case when the statistic takes on its central value. One may expect the resulting estimator to share some of the properties of the test from which it is derived. (See Fig. 4.1.)

As a first example, consider the sign-test statistic. Evaluated for the variables $Z_i-\theta$, it is the number among these variables that are positive. It gives the greatest support to the hypothesis that the distribution of the $Z_i-\theta$ is symmetric about zero if half of the variables $Z_i-\theta$ are positive and half negative. This occurs for the value of θ which is the median of the Z_i, so that the sign test leads to the estimator

$$\tilde{\theta} = \text{med}\,(Z_i) \tag{4.48}$$

(For a more careful argument see Prob. 67.)

Similarly the Wilcoxon statistic based on the variables $Z_i-\theta$ is the number of pairs (i, j) with $i \leqslant j$ for which $(Z_i-\theta)+(Z_j-\theta)$ is positive. This statistic takes on its central value if half of the sums $(Z_i-\theta)+(Z_j-\theta)$ are positive and half negative, that is, if 2θ is the median of the sums Z_i+Z_j with $i \leqslant j$. The resulting estimator is

$$\hat{\theta} = \text{med}_{i \leqslant j}\left[\tfrac{1}{2}(Z_i+Z_j)\right] \tag{4.49}$$

(See again Prob. 67 for a more careful argument.)

EXAMPLE 3. *Weight of one-year-old boys.* From the weights of a sample of 18 one-year-old boys, given below, it is desired to estimate the center θ of the distribution of the weights (assumed to be symmetric about θ) in the population from which the sample is drawn.[1]

12.01	8.99	10.21	12.15	9.54	9.85	10.62	9.52	10.66
9.87	10.44	10.51	10.67	11.16	9.32	9.62	11.11	9.14

On ordering the observations, the median is seen to be equal to

$$\tilde{\theta} = \tfrac{1}{2}(10.21+10.44) = 10.325$$

The computation of the estimator $\hat{\theta}$ can be simplified by first restricting its range. A look at

[1] From Thompson, "Data on the Growth of Children during the First Year after Birth," *Human Biol.* **23**:75–92 (1951), Table 3.

the ordered observations together with the value of the median suggest that one might expect $\hat{\theta}$ to lie between 10 and 10.5. It is simpler to work with the sums $Z_i + Z_j$ than with the averages $(Z_i + Z_j)/2$, and an easy count shows that 59 of these sums are less than 20.0 while 64 are greater than 21.0. Since there are $N(N+1)/2 = 171$ sums with $i \leqslant j$, the median will be the eighty-sixth smallest, and hence the twenty-seventh smallest of those between 20.0 and 21.0, which is 20.48 (Prob. 18). The resulting value 10.24 of $\hat{\theta}$ is in close agreement with both $\tilde{\theta}$ and the mean $\theta = 10.30$.

An alternative method for computing $\hat{\theta}$ is analogous to that of Chap. 2, Example 4. Use the ordered set of Z's (from the smallest to the largest) to head both the N columns and the N rows of an $N \times N$ square. The cell that is in the ith row and jth column has as its entry the sum of the Z's which head that row and column. The only cells of interest are those on or above the main diagonal, corresponding to $i \leqslant j$. To determine the median of these entries, which is equal to $2\hat{\theta}$, compute the entries on the diagonal connecting the SW corner with the NE corner of the square, and proceed as in Chap. 2, Example 4 (Prob. 19). [A graphical method for determining $\hat{\theta}$ is given by Høyland (1964).]

As in the corresponding two-sample problem, a slight difficulty is caused by the ties among the sums $Z_i + Z_j$. These ties occur because the weights are observed only to the nearest hundredth, and the difficulty can be resolved as in Chap. 2, Sec. 5. In the present problem of estimating θ, the difference between the estimator $\hat{\theta}'$ based on the (unavailable) observations Z_i' before rounding, and the estimator $\hat{\theta}$ based on the rounded observations, cannot in absolute value exceed the maximum rounding error, which in the example is 1/200 (Prob. 68). If the rounding error is small, $\hat{\theta}$ will therefore provide an estimated value that is close to what would have been obtained had the observations been exact, and the same is true for the estimators $\tilde{\theta}$ and $\bar{\theta}$ (Prob. 68).

Let us next consider some properties of the estimators $\hat{\theta}$, $\tilde{\theta}$, and $\bar{\theta}$. The proofs of these properties will be given in the text for $\hat{\theta}$ and will be left to the Problems for the other two estimators.

Lemma 1. The distributions of the differences $\hat{\theta} - \theta$, $\tilde{\theta} - \theta$, and $\bar{\theta} - \theta$ are independent of θ.

Proof for $\hat{\theta}$. By definition,

$$\hat{\theta} - \theta = \text{med}\,\{\tfrac{1}{2}(Z_i + Z_j)\} - \theta = \text{med}\,\{\tfrac{1}{2}[(Z_i - \theta) + (Z_j - \theta)]\}$$

Since the distribution of the variables $Z_i - \theta$ and $Z_j - \theta$ is independent of θ, the same is true of $\hat{\theta} - \theta$, as was to be proved.

Theorem 2. If the distribution L of $Z_1, \ldots, Z_N$ is symmetric about θ, the same is true of the distributions of $\hat{\theta}$, $\tilde{\theta}$, and $\bar{\theta}$.

Proof for $\hat{\theta}$. The theorem is concerned with probabilities such as $P_\theta(\hat{\theta} - \theta < a)$, where the subscript θ indicates that the probability is being computed under the assumption that L is symmetric with respect to θ. By Lemma 1, these probabilities are independent of θ, and we can therefore assume without loss of generality that θ is zero. Since L is then symmetric about zero, it follows that $-Z$ has the same distribution as Z, where Z and $-Z$ stand for $(Z_1, \ldots, Z_N)$ and $(-Z_1, \ldots, -Z_N)$, respectively. Thus $\hat{\theta} = \hat{\theta}(Z)$ has the same distribution as $\hat{\theta}(-Z) = -\hat{\theta}(Z) = -\hat{\theta}$, so that $\hat{\theta}$ is symmetrically distributed about zero, as was to be proved.

The theorem shows in particular that when L is symmetric about a point θ, then $\hat{\theta}$ is unbiased for estimating θ; it also satisfies the equation

$$P_\theta(\hat{\theta} < \theta) = P_\theta(\hat{\theta} > \theta) \tag{4.50}$$

so that $\hat{\theta}$ is as likely to overestimate θ as to underestimate it, and these properties also hold for $\tilde{\theta}$ and $\bar{\theta}$.

The estimators $\hat{\theta}$, $\tilde{\theta}$, and $\bar{\theta}$ of the center of symmetry θ of a symmetric distribution L are, by Theorem 2, all centered on θ. Let us next consider their dispersion about this value, and suppose in the following that the distribution L is continuous. Then the distribution of $\hat{\theta}$ is also continuous (and the same is true for $\tilde{\theta}$ and $\bar{\theta}$, Prob. 70), so that

$$P(\hat{\theta} = d) = 0 \quad \text{for any given number } d$$

It is important to keep in mind the distinction between this property of $\hat{\theta}$ and the fact that the distribution of the associated Wilcoxon statistic V_s is not continuous.

For the reasons discussed in Chap. 2, Sec. 5, in relation to $\hat{\Delta}$, we shall measure the dispersion of an estimator in terms of the probability that it falls within a given distance of the true value θ rather than in terms of its variance. The following theorem provides a basis for relating these probabilities to the power of the associated test. To indicate the dependence of the test statistics S_N and V_s on Z, we shall denote them below by $S_N(Z)$ and $V_s(Z)$, respectively.

Theorem 3. (i) If $Z_{(1)} < \cdots < Z_{(N)}$ denote the ordered observations $Z_1, \ldots, Z_N$, then for any integer i between 1 and N and any real number a

$$Z_{(i)} \leqslant a \quad \text{if and only if } S_N(Z-a) \leqslant N-i \tag{4.51}$$

and hence

$$Z_{(i)} > a \quad \text{if and only if } S_N(Z-a) \geqslant N-i+1 \tag{4.52}$$

(ii) Let $M = N(N+1)/2$ and let $A_{(1)} < \cdots < A_{(M)}$ denote the ordered averages $(Z_i + Z_j)/2$ for $i \leqslant j$. Then for any integer i between 1 and M and any real number a

$$A_{(i)} \leqslant a \quad \text{if and only if } V_s(Z-a) \leqslant M-i \tag{4.53}$$

and hence

$$A_{(i)} > a \quad \text{if and only if } V_s(Z-a) \geqslant M-i+1 \tag{4.54}$$

Proof of (ii). The first inequality in (4.53) holds if and only if at least i of the averages

$$\tfrac{1}{2}(Z_i+Z_j)-a = \tfrac{1}{2}[(Z_i-a)+(Z_j-a)]$$

are less than or equal to zero and hence if at most $M-i$ of these differences are greater than zero, that is, if $V_S(Z-a)$ does not exceed $M-i$. [The proof of (i) is exactly analogous.]

By Theorem 2, the distribution of $\tilde{\theta}$ is symmetric about θ, so that

$$P_\theta(|\tilde{\theta}-\theta| \leqslant a) = P_0(|\tilde{\theta}| \leqslant a) = 2P_0(\tilde{\theta} \leqslant a)-1 \tag{4.55}$$

If N is odd, say $N = 2k+1$, then $\tilde{\theta} = Z_{(k+1)}$ and it follows from (4.51) that

$$P_\theta(|\tilde{\theta}-\theta| \leqslant a) = 2P_0[S_N(Z-a) \leqslant N-k-1]-1 \tag{4.56}$$

If N is even, say $N = 2k$,

$$P_0(Z_{(k+1)} \leqslant a) \leqslant P_0(\tilde{\theta} \leqslant a) \leqslant P_0(Z_{(k)} \leqslant a)$$

and it follows from (4.55) and (4.51) that

$$2P_0[S_N(Z-a) \leqslant N-k-1]-1 \leqslant P_\theta(|\tilde{\theta}-\theta| \leqslant a) \leqslant 2P_0[S_N(Z-a) \leqslant N-k]-1 \tag{4.57}$$

Since the distribution of $S_N(Z-a)$ is binomial, with $p \neq 1/2$ when $a \neq 0$, the relations (4.56) and (4.57) show that the probabilities $P_\theta(|\tilde{\theta}-\theta| \leqslant a)$ can be expressed exactly (for N odd) or approximately (for N even) in terms of binomial probabilities with $p \neq 1/2$, which can be obtained from tables of the binomial distribution.

Equation (4.55) remains valid if $\tilde{\theta}$ is replaced by $\hat{\theta}$, and the argument leading to (4.56) and (4.57) shows that the probabilities $P_\theta(|\hat{\theta}-\theta| \leqslant a)$ are related to the non-null distribution of the Wilcoxon statistic. Although this is not tabled and difficult to compute, a large-sample approximation to these probabilities can be obtained from (4.38). To derive this approximation, consider first the case that M is odd, say $M = 2k+1$. Then (4.53) with $i = k+1$ and (4.38) lead to the approximation

$$P_\theta(\hat{\theta}-\theta < a) \approx \Phi\left\{\frac{k+\frac{1}{2}-E[V_s(Z-a)]}{\sqrt{\operatorname{Var}[V_s(Z-a)]}}\right\} = 1-\Phi\left\{\frac{\binom{N}{2}(p'_1-\frac{1}{2})+N(p-\frac{1}{2})}{\sqrt{\operatorname{Var}[V_s(Z-a)]}}\right\}$$

and hence to

$$P_\theta(|\hat{\theta}-\theta| \leqslant a) \approx 1-2\Phi\left\{\frac{\binom{N}{2}(p'_1-\frac{1}{2})+N(p-\frac{1}{2})}{\sqrt{\operatorname{Var}[V_s(Z-a)]}}\right\} \tag{4.58}$$

Here $p = P(Z-a > 0) = P(Z > a)$ and

$$p'_1 = P[(Z-a)+(Z'-a) > 0] = P[\tfrac{1}{2}(Z+Z') > a]$$

Similarly, Var $[V_s(Z-a)]$ is given by (4.27) with

$$p'_2 = P[\tfrac{1}{2}(Z+Z') > a \quad \text{and} \quad \tfrac{1}{2}(Z+Z'') > a]$$

where Z, Z', and Z'' are independently distributed according to L. The argument used for even N and $\hat{\Delta}$ in Chap. 2, Sec. 5, suggests that (4.58) also provides a reasonable approximation for the case of even M in the present context.

The approximation (4.58) can be further simplified for small a by applying the argument that led to (4.32). On using the approximations

$$p - \tfrac{1}{2} \approx -ae(0) \quad \text{and} \quad p'_1 - \tfrac{1}{2} \approx -2ae^*(0)$$

and replacing Var $[V_s(Z-a)]$ by the null variance $N(N+1)(2N+1)/24$, it is seen that (4.58) reduces to

$$\text{(4.59)} \qquad P_\theta(|\hat{\theta}-\theta| \leqslant a) \approx 2\Phi\left\{a\sqrt{\frac{24N}{(N+1)(2N+1)}}[(N-1)e^*(0)+e(0)]\right\}-1$$

When E is a normal distribution with variance τ^2, it was seen earlier that

$$e(0) = \frac{1}{\tau\sqrt{2\pi}} \quad \text{and} \quad e^*(0) = \frac{1}{2\tau\sqrt{\pi}}$$

and (4.58) then reduces to

$$\text{(4.60)} \qquad P_\theta(|\hat{\theta}-\theta| \leqslant a) \approx 2\Phi\left\{a\sqrt{\frac{6N}{(N+1)(2N+1)\pi\tau^2}}[(N-1)+\sqrt{2}]\right\}-1$$

In view of the relations of the probability that $\hat{\theta}$ or $\tilde{\theta}$ falls within a given distance of θ to the power of the Wilcoxon or sign test and the corresponding relation between $\bar{\theta}$ and the asymptotic power of the t-test (Prob. 72), the conclusions concerning the relative efficiencies of these tests to each other carry over to the associated estimators. Here the relative efficiency of two estimators is defined as the ratio of sample sizes required so that they will have the same probability of falling within a given distance of θ. In particular, it follows from the results of Sec. 3 that $\hat{\theta}$ and $\bar{\theta}$ are considerably more efficient than $\tilde{\theta}$ when L is normal, with $\bar{\theta}$ then being slightly more efficient than $\hat{\theta}$. In the presence of gross errors or more generally for distributions with tails heavier than the normal, $\hat{\theta}$ may be considerably more efficient than $\bar{\theta}$ and in general appears to be a reasonable compromise solution. When the observations are easily ordered, the estimator $\tilde{\theta}$ has the advantage of being the easiest to compute, which makes it useful when a quick estimate is required. However, the principal use of $\tilde{\theta}$ occurs when L cannot be assumed to be symmetric, as will be discussed below.

So far we have compared $\hat{\theta}$, $\tilde{\theta}$, and $\bar{\theta}$ as estimators of the center of symmetry of a symmetric distribution L. When the symmetry assumption is not justified, the three estimators are no longer comparable since they then estimate essentially different aspects of L, as is shown by the following theorem.

Theorem 4. (i) If L has a finite expectation λ, the mean $\bar{Z}$ is a consistent estimator of λ.

(ii) If L has a unique median μ, the median $\tilde{Z}$ is a consistent estimator of μ.

(iii) Let ν be such that

$$P[\tfrac{1}{2}(Z+Z') < \nu] = P[\tfrac{1}{2}(Z+Z') > \nu] = \tfrac{1}{2} \tag{4.61}$$

where Z and Z' are independently distributed according to L, and suppose that ν is the only point with this property. Then the estimator $\hat{Z} = \text{med}_{i \leqslant j}[\frac{1}{2}(Z_i+Z_j)]$ is a consistent estimator of ν.

Proof. (i) This is a restatement in statistical language of Khinchine's law of large numbers. (See Appendix, Example 7.)

(ii) This follows from the fact that the frequency of success in N binomial trials with success probability p tends to p in probability, or alternatively, is an easy consequence of the asymptotic normality of the sign-test statistic (Probs. 73 and 74).

(iii) This is proved from the fact that V_s/M is a consistent estimator of p'_1 (Prob. 75), where M is defined in Theorem 3(ii).

Of the three measures of location λ, μ, and ν, the mean λ and the median μ have the advantage of simplicity and intuitive significance. Which measure is most appropriate depends of course on the nature of the problem and the use to be made of the estimator. When nothing is known about the distribution L, it is in some contexts a disadvantage of the mean that a minute change in L can produce an arbitrarily large change in λ (Prob. 76), while μ and ν depend on L in a smoother manner.

The most important application of $\tilde{Z} = \text{med}(Z_i)$ is as an estimator of the median μ of an arbitrary continuous distribution L. (In this connection, it is interesting to note that the derivation of this estimator from the sign test at the beginning of the section, unlike the derivation of $\hat{\theta}$, made no use of the symmetry of L.) For any odd value of N, it is easily seen (Prob. 77) that $\tilde{Z}$ satisfies

$$P(\tilde{Z} < \mu) = P(\tilde{Z} > \mu) \tag{4.62}$$

so that it is as likely to overestimate the true value as to underestimate it. The same relationship holds approximately when N is even (Prob. 77). For large N, of course, Theorem 4 showed $\tilde{Z}$ to be consistent.

5. CONFIDENCE PROCEDURES

Instead of point estimates for μ or θ, one may wish to determine confidence intervals, that is, random intervals which contain the unknown parameter μ or θ with specified probability. The development of such procedures parallels that of Chap. 2, Sec. 6 for the two-sample case and is based on the following result.

Theorem 5. (i) Let $Z_1, \ldots, Z_N$ be independently distributed, each according to the same continuous distribution L. If $Z_{(1)} < \cdots < Z_{(N)}$ denote the ordered Z's, and if μ is any median of L, then

$$(4.63) \qquad P(Z_{(i)} < \mu \leqslant Z_{(i+1)}) = P_{\frac{1}{2}}(S_N = i) = \frac{\binom{N}{i}}{2^N} \qquad \text{for all } i = 0, 1, \ldots, N$$

where $Z_{(0)} = -\infty$ and $Z_{(N+1)} = \infty$.

Here the median μ is no longer assumed to be unique. The subscript $\frac{1}{2}$ on the middle term indicates that the probability is being computed under the assumption that S_N has a binomial distribution with success probability $\frac{1}{2}$.

(ii) Suppose in addition that L is symmetric about θ. Let $M = \frac{1}{2}N(N+1)$, and let $A_{(1)} < \cdots < A_{(M)}$ denote the M averages $\frac{1}{2}(Z_i + Z_j)$ for $i \leqslant j$. Then

$$(4.64) \qquad P_\theta(A_{(i)} < \theta \leqslant A_{(i+1)}) = P_0(V_s = i) \quad \text{for all } i = 0, \ldots, M$$

where $A_{(0)} = -\infty$ and $A_{(M+1)} = \infty$. Here the subscript zero on the right-hand side indicates that the probability is being computed under the null distribution of V_s. The $<$ signs in (4.63) and (4.64) can be replaced by $\leqslant$ signs and vice versa because of the continuity of L.

Proof. (i) The left-hand side of (4.63) is the probability that the number of Z's less than μ is exactly i. Since the Z's are independent and each has probability $\frac{1}{2}$ of being less than μ, the left-hand side is the probability of exactly $N-i$ (or equivalently exactly i) successes in N independent trials with success probability $\frac{1}{2}$, and this proves the result.

(ii) The left-hand side is the probability that exactly i of the averages $\frac{1}{2}(Z_i + Z_j)$ are less than θ, which by the argument of Lemma 1 may be taken to be zero without loss of generality. Since L is then symmetric about zero, the number of negative averages has the same distribution as the number of positive averages, and the latter number is V_s. This completes the proof.

Unless N is very small, it may be desirable to condense the information provided by (4.63) and (4.64), for example, by only giving the values i for which the probabilities

$$(4.65) \qquad \gamma = P_\mu(\mu \leqslant Z_{(i)}) = P_0(S_N \leqslant i-1)$$

or

$$(4.66) \qquad \gamma = P_\theta(\theta \leqslant A_{(i)}) = P_0(V_s \leqslant i-1)$$

have some standard values such as 1, 5, 10, 25, 50, 75, 90, 95, and 99 percent. The associated values $Z_{(i)}$ and $A_{(i)}$ then constitute a set of standard confidence points.

EXAMPLE 4. *Effect of muscle training.* To evaluate the effect of an 8-week muscle training program, the weight was determined that each of a sample of 12 first-graders was able to lift before and after the training program. The differences (after − before) were as follows: 6.0, 7.0, 5.0, 10.5, 8.5, 3.5, 6.1, 4.0, 4.6, 4.5, 5.9, 6.5.† It is seen from (4.66) and Table H that the values of i for which $\gamma = P_\theta(\theta \leqslant A_{(i)})$ is closest to the values 1, 5, ..., 99 percent when $N = 12$ are given by Table 4.5.

TABLE 4.5 Values of i satisfying (4.66) for $N = 12$

γ	.010	.046	.102	.259	.485	.515	.741	.898	.954	.990
i	11	18	23	31	39	40	48	56	61	68

To compute the order statistics $A_{(11)}, A_{(18)}, \ldots, A_{(68)}$ let us exhibit the sums $Z_i + Z_j$ for all $M = \binom{12}{2} + 12 = 78$ combinations of i and j with $i \leqslant j$.

Z_i \ Z_j	3.5	4.0	4.5	4.6	5.0	5.9	6.0	6.1	6.5	7.0	8.5	10.5
3.5	7.0	7.5	8.0	8.1	8.5	9.4	9.5	9.6	10.0	10.5	12.0	14.0
4.0		8.0	8.5	8.6	9.0	9.9	10.0	10.1	10.5	11.0	12.5	14.5
4.5			9.0	9.1	9.5	10.4	10.5	10.6	11.0	11.5	13.0	15.0
4.6				9.2	9.6	10.5	10.6	10.7	11.1	11.6	13.1	15.1
5.0					10.0	10.9	11.0	11.1	11.5	12.0	13.5	15.5
5.9						11.8	11.9	12.0	12.4	12.9	14.4	16.4
6.0							12.0	12.1	12.5	13.0	14.5	16.5
6.1								12.2	12.6	13.1	14.6	16.6
6.5									13.0	13.5	15.0	17.0
7.0										14.0	15.5	17.5
8.5											17.0	19.0
10.5												21.0

There are two sums less than 8, six between 8 and 9 (including 8 but not 9), ten between 9 and 10, and so on; and the desired confidence points are now easily obtained as

$$\begin{array}{lllll} 2A_{(11)} = 9.1 & 2A_{(18)} = 9.9 & 2A_{(23)} = 10.4 & 2A_{(31)} = 10.9 & 2A_{(39)} = 11.6 \\ 2A_{(40)} = 11.8 & 2A_{(49)} = 12.4 & 2A_{(56)} = 13.1 & 2A_{(61)} = 14.0 & 2A_{(68)} = 15.1 \end{array}$$

A graphical method for determining the confidence points $A_{(i)}$ can be found in Moses (1965) and in Noether (1971).

If we were unwilling to assume symmetry for the distribution of the weights, we might instead wish to use (4.63) to determine confidence points for the median μ of the distribution. Unfortunately, for a value of N as small as 12 the possible values of γ are not close to the standard values 1 percent, 5 percent, and so on. The closest that one can come in this case is as shown in Table 4.6.

† From Schweid, Vignos, and Archibald, "Effects of Brief Maximal Exercise on Quadriceps Strength in Children," *Amer. J. Phys. Med.* **41**:189–197 (1962) (Table 1).

TABLE 4.6 Values of i satisfying (4.65) for $N = 12$

γ	.019	.073	.194	.387	.613	.806	.927	.981
i	3	4	5	6	7	8	9	10

For the data of Example 4, the corresponding order statistics are, of course, trivially

$$Z_{(3)} = 4.5 \qquad Z_{(4)} = 4.6 \qquad Z_{(5)} = 5.0 \qquad Z_{(6)} = 5.9$$
$$Z_{(7)} = 6.0 \qquad Z_{(8)} = 6.1 \qquad Z_{(9)} = 6.5 \qquad Z_{(10)} = 7.0$$

In the above discussion, the confidence points $A_{(i)}$ and $Z_{(i)}$ satisfying (4.65) and (4.66) were obtained from Tables H and G. If N is too large for these tables, the points can be determined instead from the appropriate normal approximation. Applying the approximation (3.3) with continuity correction to (4.65), one finds

$$\gamma = P_{\mu}(\mu \leqslant Z_{(i)}) \approx \Phi\left[\frac{2i-(N+1)}{\sqrt{N}}\right] \tag{4.67}$$

and similarly (3.15) with continuity correction gives

$$\gamma = P_{\theta}(\theta \leqslant A_{(i)}) \approx \Phi\left[\frac{2i-1-\frac{1}{2}N(N+1)}{\sqrt{N(N+1)(2N+1)/6}}\right] \tag{4.68}$$

Let us now return to the problem of determining confidence intervals for μ or θ. By (4.63) and (4.64), it is only necessary for this purpose to combine sufficiently many of the intervals $(Z_{(i)}, Z_{(i+1)})$ or $(A_{(i)}, A_{(i+1)})$, so that their probabilities add up to the desired confidence coefficient. Let us illustrate the procedure with the case of confidence intervals $(A_{(i)}, A_{(j)})$ for θ with confidence coefficient .9 when $N = 12$. Typically, it is most natural to choose the intervals symmetrically, i.e., so that the probabilities of the interval $(A_{(i)}, A_{(j)})$ falling entirely to the left or right of θ are equal, or

$$P_{\theta}(\theta < A_{(i)}) = P_{\theta}(A_{(j)} < \theta) \tag{4.69}$$

Each of these probabilities should then be equal to half of $1-.9$ and hence equal to .05. It is seen from Table H that the closest one can come to this for $N = 12$ is to take $i = 18$ and $j = 61$. Then

$$P_{\theta}(\theta < A_{(18)}) = P_{\theta}(A_{(61)} < \theta) = .046$$

and hence

$$A_{(18)} \leqslant \theta \leqslant A_{(61)}$$

with probability $1-.092 = .908$. In Example 4, this interval extends from 9.9 to 14.0.

Quite generally, it follows from (4.66) and from the symmetry of the Wilcoxon distribution that (4.69) will hold when

$$j = \binom{N}{2} + N + 1 - i \tag{4.70}$$

Analogous confidence intervals for μ will satisfy

$$P_\mu(\mu < Z_{(i)}) = P_\mu(Z_{(j)} < \mu) \tag{4.71}$$

when $j = N+1-i$.

Instead of confidence intervals for μ or θ one may wish to determine upper or lower-confidence bounds. Upper-confidence bounds with confidence coefficient γ are given by (4.65) and (4.66). For example, when $N = 12$ it is seen from Tables 4.5 and 4.6 that $A_{(68)}$ is an upper-confidence bound for θ at confidence level .99 and that $Z_{(10)}$ is an upper-confidence bound for μ at level .98. Analogous lower-confidence bounds are easily obtained from (4.69) and (4.71).

If it is desired to compute not a whole set of confidence points but only a confidence bound or interval, one will typically use the shortcut method described in Example 3 of the present chapter and toward the end of Chap. 2, Sec. 6.

The treatment of ties is exactly analogous to that of Chap. 2, Sec. 6. For example, to obtain confidence intervals for θ at confidence level γ from observations rounded to the nearest multiple of ε, one computes the interval $(A_{(i)}, A_{(j)})$ that would be appropriate without ties. Then the interval $A_{(i)} - \varepsilon$ to $A_{(j)} + \varepsilon$ will contain θ with probability greater than or equal to γ.

The confidence statements for μ based on (4.63) can be generalized to quantiles of L other than the median. Let μ_p be a point such that

$$P(Z_i \geqslant \mu_p) = 1 - L(\mu_p) = p \tag{4.72}$$

(The parameter μ_p reduces to the median for the special case $p = \frac{1}{2}$.) Then

$$P(Z_{(i)} < \mu_p \leqslant Z_{(i+1)}) = \binom{N}{i} q^i p^{N-i} \tag{4.73}$$

where $q = 1-p$. To see this, note that the left-hand side is the probability that the number of Z's less than μ_p is exactly i. Since the Z's are independent and each has probability q of being less than μ_p, the left-hand side is the probability of exactly i successes in N independent trials with success probability q, and this completes the proof. Confidence bounds or intervals for μ_p can now be obtained from (4.73) exactly as those for μ followed from (4.65).

6. FURTHER DEVELOPMENTS

6A. Power and Efficiency of the Sign Test

The power of the sign test against normal shift alternatives, for selected sample sizes up to 100, is given by Dixon (1953), and against Cauchy shift alternatives and sample sizes 5 to 10 by Arnold (1965). An Edgeworth expansion of the power function is discussed by Gibbons (1964).

The small-sample efficiency of the sign test relative to Student's t-test is studied by Walsh (1946), Dixon (1953), and Hodges and Lehmann (1956). The latter paper also investigates various properties of the associated Pitman efficiency, which was first derived in the normal case by Cochran (1937) and for more general shift alternatives by Pitman (1948). Alternative notions of asymptotic efficiency are discussed by Blyth (1958), Chernoff (1952), and by Hodges and Lehmann (1956), and the Bahadur efficiency is considered by Bahadur (1960), Klotz (1965), and Woodworth (1970). A comparison of the speed of convergence to the Pitman and Bahadur limits is provided by Hwang and Klotz (1970).

6B. The Absolute Normal Scores Test

An alternative to the Wilcoxon paired-comparison test is provided by the Absolute Normal Scores test, which is the rank analogue of Student's t-test. Suppose for a moment that the variables $Z_i = Y_i - X_i$ are independently normally distributed with zero mean and, say, unit variance. Imagine that the values of the Z's have been lost and that only their signs and the rank of each $|Z_i|$ among $|Z_1|, \ldots, |Z_N|$ are preserved. From this information it is desired to reconstruct the original values as closely as possible. Denote the ordered absolute values by $|Z|_{(1)} < \cdots < |Z|_{(N)}$ and consider one of the Z's, say Z_i, which is positive and the rank of whose absolute value is s. Then its expected value is

$$E[Z_i | Z_i > 0 \text{ and rank of } |Z_i| \text{ is } s] = E[|Z_i| \,|\, Z_i > 0 \text{ and rank of } |Z_i| \text{ is } s]$$

It was shown in the proof of Theorem 1 that $|Z_i|$ is independent of the sign of Z_i. Thus, the desired expectation is equal to the expectation of $|Z_i|$ given that the rank of $|Z_i|$ is s and hence to

$$a_N(s) = E[|Z|_{(s)}] \tag{4.74}$$

the expectation of the sth smallest of the N independent absolute values $|Z_1|, \ldots, |Z_N|$. Now the absolute value of a random variable with distribution $N(0,1)$ has the χ-distribution with 1 degree of freedom and hence (4.74) can be written as

$$a_N(s) = E_{\chi_1}[V_{(s)}] \tag{4.75}$$

where $V_1, \ldots, V_N$ are independently distributed according to χ_1, and $V_{(1)} < \cdots < V_{(N)}$ denote the ordered V's. Tables of the Absolute Normal (or χ_1) Scores (4.75) are provided by Govindarajulu and Eisenstat (1965) for $N \leqslant 100$.

Consider next a negative Z, and suppose that the rank of its absolute value is r. Then the same argument shows that its expected value is $-a_N(r)$. Thus, our attempted reconstruction of the original observations from the ranks $R_1, \ldots, R_m$ and $S_1, \ldots, S_n$ are $a_N(S_1), \ldots, a_N(S_n)$ for the positive Z's and $-a_N(R_1), \ldots, -a_N(R_m)$ for the negative ones. When these reconstructed observations are substituted into the

t-statistic, the resulting test, after some simplification, rejects when

$$K_s \geqslant c \tag{4.76}$$

where

$$K_s = a_N(S_1) + \cdots + a_N(S_n) \tag{4.77}$$

The null distribution of K_s is determined by (4.21). Some critical values are given by Klotz (1963) for $N \leqslant 10$ and by Thompson, Govindarajulu, and Doksum (1967) for $11 \leqslant N \leqslant 17$ (and for very small values of α up to $N = 20$). These latter authors also study the accuracy of the normal approximation and of a more elaborate Edgeworth approximation.

6C. Power and Efficiency of the Wilcoxon and Absolute Normal Scores Test

The power of the Wilcoxon signed-rank and the Absolute Normal Scores tests is given by Klotz (1963) for normal shift alternatives and by Arnold (1965) for shifted t-distributions with $f = \frac{1}{2}$, 1, 2, and 4 degrees of freedom ($f = 1$ is the Cauchy case); in both cases for sample sizes from 5 to 10.

The Pitman efficiency of the Wilcoxon and Absolute Normal Scores tests relative to each other and to Student's t-test is the same in the one- as in the two-sample case, so that the results discussed in Chap. 2, Secs. 4 and 7 continue to apply in the present case. A relationship between Bahadur efficiencies in the one- and two-sample cases is established by Woodworth (1970).

As in the corresponding two-sample case, the Absolute Normal Scores test has the advantage over the Wilcoxon test of taking on a larger number of distinct values. The situation is completely analogous to that discussed in Chap. 2, Sec. 7C.

6D. Tests of Symmetry

The sign and Wilcoxon tests, of the hypothesis H that the common distribution L of $Z_1, \ldots, Z_N$ is symmetric about zero, have obvious extensions to the problem of testing that L is symmetric about a known point μ_0. It is only necessary to subtract μ_0 from each of the Z's to reduce this problem to the earlier one.

A more difficult problem is that of testing the hypothesis H' that L is symmetric but without specifying the point μ of symmetry. The test obtained by estimating μ by the average $\bar{Z}$ and taking as test statistic the number of Z's that are larger than $\bar{Z}$ is studied by Gastwirth (1971).

6E. A Generalized Set of Confidence Points

Tests admitting more significance levels than either the sign or Wilcoxon test and corresponding confidence intervals for θ are described by Walsh [1949(a) and (b)].

Let $Z_{(1)} < \cdots < Z_{(N)}$ denote the ordered Z-values, and consider any subset of the statistics $\frac{1}{2}(Z_{(i)}+Z_{(j)})$, $1 \leqslant i \leqslant j \leqslant N$. For each member of the subset, choose one of the inequalities "less than" or "greater than" and make the associated statement

$$\tfrac{1}{2}(Z_{(i)}+Z_{(j)}) < \theta \qquad \text{or} \qquad \tfrac{1}{2}(Z_{(i)}+Z_{(j)}) > \theta$$

The probability of the chosen inequalities holding for all pairs of the subset is independent of the common distribution L of $Z_1, \ldots, Z_N$ (which is assumed continuous and symmetric about θ), and hence provides a confidence set for θ that, of course, may be empty. In the usual way, the confidence set can be converted into a test of the hypothesis $H: \theta = \theta_0$.

By choosing to make only the single statement $Z_{(i)} < \theta$ or the two statements $Z_{(i)} < \theta < Z_{(j)}$, one obtains the sign test inferences described in earlier sections. [A generalization of these confidence statements is discussed by Noether (1973).] The Wilcoxon confidence points also can be obtained by specializing the Walsh system, as is illustrated by the following example (which is particularly simple and in some ways not quite typical).

Consider the Wilcoxon confidence points $A_1, \ldots, A_M$ defined in Theorem 5 (Sec. 5), for the case $N = 3$. Then $M = \frac{1}{2}N(N+1) = 6$, and the points are $A_1 = Z_{(1)}$, $A_2 = \frac{1}{2}[Z_{(1)}+Z_{(2)}]$, $A_3 = \min[Z_{(2)}, \frac{1}{2}(Z_{(1)}+Z_{(3)})]$, $A_4 = \max[Z_{(2)}, \frac{1}{2}(Z_{(1)}+Z_{(3)})]$, $A_5 = \frac{1}{2}[Z_{(2)}+Z_{(3)}]$, $A_6 = Z_{(3)}$.

The probabilities $P_\theta(\theta < A_i)$ are, respectively, $1/8$, $2/8$, $3/8$, $5/8$, $6/8$, $7/8$. The Walsh system is able to interpolate one further point into the set and thereby provide the missing probability $4/8 = 1/2$. One can choose for it either $Z_{(2)}$ or $\frac{1}{2}[Z_{(1)}+Z_{(3)}]$. It follows that in this case, there are Walsh points yielding all probabilities $i/2^N, i = 1, \ldots, 2^{N-1}$. For larger values of N, the Walsh points also refine the Wilcoxon points, although they are no longer able to realize all probabilities $i/2^N$. [For a different type of extension see Hartigan (1969).]

6F. Bounded-length Sequential Confidence Intervals for θ

Let $(\underline{\theta}_N, \bar{\theta}_N)$ be confidence intervals for θ based on $Z_1, \ldots, Z_N$ such as those discussed in Sec. 5, and let their length be $L_N = \bar{\theta}_N - \underline{\theta}_N$. Then L_N is a random variable that cannot be bounded unless restrictions are placed on the distribution of the Z's since L_N tends in probability to infinity as this distribution becomes more and more spread out. Confidence intervals for θ of length not exceeding any given number can however be obtained by taking observations $Z_1, Z_2, \ldots,$ sequentially as follows. Having observed $Z_1, \ldots, Z_N$, compute $(\underline{\theta}_N, \bar{\theta}_N)$ for $N = 2, \ldots,$ and continue taking observations until $L_N \leqslant l$. The coverage probability of the resulting confidence intervals, their expected length, and their efficiency are investigated by Geertsema (1970) for the case that $(\underline{\theta}_N, \bar{\theta}_N)$ are the intervals derived from the sign or Wilcoxon tests. For further work on this problem, see Sen and Ghosh (1971).

6G. Robust Estimation

In comparison with $\bar{\theta}$, the estimates $\tilde{\theta}$ and $\hat{\theta}$ of Sec. 4 have the advantage of *robustness*, that is, a lack of sensitivity to outlying observations. A general theory of robust estimation is developed by Hampel (1968, 1971, 1974). The robustness properties of a large number of estimates of location are studied in a book-length report by Andrews et al. (1972). Most of these estimates belong to one of the following three classes, whose relationships are investigated by Jaeckel (1971).

(i) *Estimators derived from rank tests.* Consider a signed-rank test for testing $\theta = 0$ such as the sign, Wilcoxon, or Absolute Normal Scores test. From each such test, an estimator can be derived by the procedure which in Sec. 4 led to the estimators $\tilde{\theta}$ and $\hat{\theta}$. It is shown by Jaeckel that these estimators are, in a certain sense, weighted medians of the averages $\frac{1}{2}(Z_i + Z_j)$.

(ii) *Linear combination of order statistics.* These estimators, which are of the form $\sum w_i Z_{(i)}$ (typically with $w_i = w_{N+1-i}$), include in particular the median, the mean, and the trimmed mean. For the asymptotic theory of this class, and references to earlier work, see Chernoff, Gastwirth, and Johns (1967).

(iii) *Maximum-likelihood-type estimators.* This class of estimators is defined by Huber (1964) as solutions of the equations $\sum \psi(Z_i - a) = 0$, where ψ satisfies the condition $\psi(-z) = -\psi(z)$. Huber develops the theory of these estimators, and determines the member of the class possessing a certain minimax property.

Jaeckel establishes a correspondence, which relates each member of one class to a unique member of each of the other two classes. This enables him to conclude that properties proved for some members of these classes also hold for the corresponding members of the other classes.

A survey of the present state of robust estimation is given by Huber (1972); see also Andrews et al. (1972).

6H. Some Optimum Properties of Tests and Estimators

The local optimum properties discussed for the two-sample problem in Chap. 2, Sec. 7G, carry over to the one-sample problem. Thus, for example, the Absolute Normal Scores test maximizes the local power among signed-rank tests against normal shift alternatives. Another such property is possessed by the sign test, which is locally most powerful (even without the restriction to rank tests) against a shifted double exponential distribution [Lehmann (1959)].

Minimax results for nonparametric classes of alternatives are given for the sign test by Hoeffding (1951) and Ruist (1954). A formulation of Doksum and Thompson (1971), which parallels that of Doksum for the two-sample case, leads not to the one-sample Wilcoxon statistic but to a linear combination of the Wilcoxon and sign-test statistics.

An interesting model, which is intermediate between parametric and non-parametric, is considered by Huber (1964) in connection with the problem of estimating θ. He assumes that the common distribution L of the Z's is of the form

$$L(z) = (1-\varepsilon)G(z-\theta)+\varepsilon H(z-\theta) \tag{4.78}$$

where G is a known distribution (or at least known up to a scale factor) and H is unknown, both being symmetric about zero. The first term on the right-hand side represents the cases where observation proceeded normally (with probability $1-\varepsilon$); the second term represents the proportion ε of gross errors, about whose distribution little is usually known. Huber restricts attention to maximum-likelihood-type estimators. To obtain an estimator that is robust to departures from G, he then minimizes the maximum asymptotic variance with respect to the unknown distribution H. His results are related to a corresponding testing problem by Gastwirth (1966), and the restriction to symmetric H is weakened by Jaeckel (1971).

A different approach to robust estimation is developed by Hampel (1968). He gives a definition of the *sensitivity* of an estimator to small changes in the underlying distribution and then minimizes the maximum asymptotic variance in a parametric model among a class of estimators whose sensitivity does not exceed some given bound.

The optimum choice of an estimator within one of the three classes described in Sec. 6G depends on the distribution E of the Z's. If this is unknown, it was first suggested by Stein (1956) that one might use part of the observations to estimate E, and then adapt the estimate of θ to this estimated E, and in this manner obtain an asymptotically efficient estimate of θ. An adaptive estimate of this kind has been proposed and its performance investigated by Takeuchi (1971).

6I. Departures from Assumption

The estimators of the present chapter are intended to estimate the center θ of a symmetric distribution, on the basis of N observations $Z_1, \ldots, Z_N$ from that distribution. What happens if the assumptions of the model are not satisfied?

The performance of $\hat{\theta}$ (and its efficiency relative to $\bar{Z}$) is studied by Høyland (1964) for the case that the observations occur in blocks, with the observations in the same block being dependent. [A more general study of robust estimators under dependence has been undertaken by Gastwirth and Rubin (1970).] Sen (1968) considers the corresponding problems when $Z_1, \ldots, Z_N$ are independent but not necessarily identically distributed, although the N distributions are assumed to be symmetric about a common center θ.

A departure from the last and most central of the three assumptions—independence, common distribution, and symmetry—is studied by Jaeckel (1971). He assumes that L is of the form (4.78) but with H not necessarily symmetric, and with θ continuing to be the quantity one wishes to estimate. In this model, he investigates the asymptotic performance of the maximum-likelihood-type estimators under the additional assumption that the proportion ε of gross errors tends to zero as N tends to infinity at the rate of $1/\sqrt{N}$.

7. PROBLEMS

(A)

Section 1

1. For $n = 16$ and $c = 11$, the following table gives some values of the power function $\Pi(p)$ of the sign test (4.8):

p	.40	.45	.50	.55	.60	.65	.70	.75	.80	.85
$\Pi(p)$	.019	.049	.105	.198	.329	.490	.660	.810	.918	.976

Compare these values with those obtained from the approximation (4.11).

2. In Example 1, find the power of the sign test against the alternative that the gas consumption is (i) 20 miles per gal; (ii) 21 miles per gal when a sample of $N = 25$ cars is taken; (iii) 21 miles per gal when $N = 20$ and the variance is (*a*) $\sigma^2 = 1.00$; (*b*) $\sigma^2 = 1.50$; (*c*) $\sigma^2 = 2.00$.

3. In Chap. 3, Example 1, suppose that the number of patients in the study is 25. Find the value of c for which the significance level of the sign test (4.8) is as close as possible to .02, and use (4.11) to find the power of this test against the alternative that the patients will prefer the new drug with probability (i) $p = .6$; (ii) $p = .7$; (iii) $p = .8$; (iv) $p = .9$. Graph the power function of the test as a function of p.

4. In Chap. 3, Example 1, suppose that the number of patients is 40 and the desired significance level .01. Use (4.15) to graph the power function of the test.

5. In Chap. 3, Example 4, suppose that the number of children is 30 and that the differences of the scores are approximately normally distributed with variance $\sigma^2 = 1.5$. If the sign test is used at level $\alpha = .01$, and if the thiamine treatment adds about Δ to the score, use (4.11) or (4.14) to graph the power function of the test as a function of Δ.

6. In Example 1, against the alternative that the gas consumption is 21 miles per gal find the sample size required to achieve power (i) $\Pi = .90$; (ii) $\Pi = .95$; when the significance level is to be (*a*) $\alpha = .01$; (*b*) $\alpha = .02$.

7. How large a number of children is needed in Chap. 3, Example 4, if the sign test is used at level $\alpha = .02$ and if power .95 is required against a shift of $\Delta = 1$ when the differences of the scores are normally distributed with variance (i) $\sigma^2 = 2$; (ii) $\sigma^2 = 3$; (iii) $\sigma^2 = 4$?

8. In Prob. 5, graph the power function as a function of Δ/σ.

Section 2

9. Verify the numerical values of p and p'_1 given in Example 2.

10. In Example 2, use (4.28) to find the (approximate) power when (i) $\Delta/\tau = .5$; (ii) $\Delta/\tau = .75$. (The exact powers, to two digits, are .43 and .70.†)

11. In Example 2, use (4.28) to find the (approximate) power when $\alpha = .097$ and Δ/τ is equal to (i) .50; (ii) .75; (iii) 1.00. (The exact powers, to two digits, are .57, .82, and .95, respectively.†)

12. In Chap. 3, Example 3, find the probability of detecting at significance level $\alpha = .02$ that tape closing the wounds does increase the tensile strength if in fact it increases it by 10 points, the differences of the scores are assumed to be normally distributed with standard deviation $\tau = 15$, and if the sample sizes are (i) $N = 20$; (ii) $N = 25$; (iii) $N = 40$.

13. Under the assumptions of the first part of Example 1 ($N = 20$, $\alpha = .0577$, $\Delta/\tau = -1/1.5$) (i) verify the value .80 stated in the text for the approximate power; (ii) use (4.28) to find the power if the Wilcoxon test is used instead of the sign test.

14. Under the assumptions of the second part of Example 1 ($\alpha = .05$, $\Delta/\tau = -1/1.5$, $\Pi = .9$), use (4.35) to find the approximate sample size required if the Wilcoxon test is used instead of the sign test.

15. Use (4.33) and (4.35) to solve Probs. 2, 5, 6, and 7 when the Wilcoxon test is used instead of the sign test.

Section 3

16. Use (4.15) and (4.44) to find the approximate sample sizes required by the sign and t-tests at level $\alpha = .01$ to achieve power $\Pi = .95$ in the shift model (4.5) when E is normal and $\Delta/\tau = .5$.

17. Use the approximations of Secs. 1 and 3 to obtain approximate values for $e_{S,t}$ and the power Π of the sign test in the shift model (4.5) when E is normal with unit variance and (i) $N = 10, \alpha = .109, \Delta = .8416$; (ii) $N = 10, \alpha = .109, \Delta = .3853$; (iii) $N = 20, \alpha = .115, \Delta = .8416$; (iv) $N = 20, \alpha = .115, \Delta = .3853$; compare these values with those of Table 4.3.

Section 4

18. Verify the values of $\tilde{\theta}$, $\hat{\theta}$, and $\bar{\theta}$ given in Example 3.

19. Compute the value $\hat{\theta}$ of Example 3 by the alternative method of computation suggested there.

20. The following are measurements (in cm) of the head circumference of 19 boys at the age of 16 weeks[1]

42.8 42.3 40.6 41.7 41.8 40.2 41.1 41.8 42.5 42.2 41.6 41.0 43.0
41.4 41.0 43.4 40.9 38.7 41.8

Determine the estimators $\tilde{\theta}$, $\hat{\theta}$, and $\bar{\theta}$ of the center θ of the distribution of head circumference in the population from which the sample is drawn.

† See Klotz, *loc. cit.*

[1] From Thompson, *loc. cit.*, Table 3.

21. From other data in the study, it appears that the next to last entry (38.7) in the data of the preceding problem was erroneously recorded. Replace this observation by x and determine $\tilde{\theta}$, $\hat{\theta}$, and $\bar{\theta}$ when x has the value (i) $x = 35.0$; (ii) $x = 38.0$; (iii) $x = 42.0$; (iv) $x = 45.0$; (v) $x = 48.0$; and also (vi) when x is deleted from the data.

22. The following figures are behavior ratings before and after 6 weeks of receiving a placebo, for 12 chronic schizophrenics who were used as controls in a clinical trial.[1] Estimate the size θ of the placebo effect by means of $\tilde{\theta}$, $\hat{\theta}$, and $\bar{\theta}$.

Before	2.4	2.2	2.1	2.9	2.2	2.3	2.4	1.5	2.7	1.9	1.8	1.3
After	2.54	3.18	2.54	3.27	2.09	2.45	3.09	1.45	3.45	3.09	1.81	1.45

23. The following data report the weight (in lb) that 12 first-graders were able to lift before and after an 8-week muscle-training program.[2] Determine the estimators $\tilde{\theta}$, $\hat{\theta}$, and $\bar{\theta}$ of the effect θ of training.

Before	14.4	15.9	14.4	13.9	16.6	17.4	18.6	20.4	20.4	15.4	15.4	14.1
After	20.4	22.9	19.4	24.4	25.1	20.9	24.6	24.4	24.9	19.9	21.4	21.4

24. Use a table of the binomial distribution to find $P_\theta(|\tilde{\theta}-\theta| \leqslant a)$ for $a = .05, .10, .20, .25, .30, .40, .50, .75, 1.00$, when $N = 15$ and $Z_1, \ldots, Z_N$ are independently distributed according to the following distributions: (i) normal with unit variance and mean θ; (ii) rectangular over the interval $(\theta - \frac{1}{2}, \theta + \frac{1}{2})$.

25. For $N = 20$, use a table of the binomial distribution to find bounds for $P_\theta(|\tilde{\theta}-\theta| \leqslant a)$ for the values of a and the distributions of the preceding problem. [*Hint:* Note that $P_\theta(|\tilde{\theta}-\theta| \leqslant a) = 2P_\theta(\tilde{\theta} \leqslant a) - 1$ and that $Z_{(10)} \leqslant \tilde{\theta} \leqslant Z_{(11)}$.]

Section 5

26. In analogy with Table 4.6, determine binomial confidence points for the median μ of a continuous distribution L with confidence probabilities as close as possible to 1, 5, 10, 25, 50, 75, 90, 95, and 99 percent for (i) $N = 15$; (ii) $N = 20$.

27. Assuming L in the preceding problem to be symmetric about θ, determine the corresponding Wilcoxon confidence points for θ, in analogy with Table 4.5.

28. Compute the numerical values of the confidence points of Tables 4.5 and 4.6 for the data of Prob. 23, and compare these values with the corresponding ones based on Student's t-test.

29. Determine confidence intervals satisfying (4.69) for the center of a continuous symmetric distribution L with confidence coefficient as close as possible to .95 when (i) $N = 10$; (ii) $N = 14$; (iii) $N = 19$.

30. Determine confidence intervals for the median μ of a continuous (but not necessarily symmetric) distribution L with confidence coefficient as close as possible to .95 for the three sample sizes of the preceding problem.

[1] From Stanley and Walton, "Trifluoperazine ("Stelazine"). A Controlled Clinical Trial in Chronic Schizophrenia," *J. Mental Sci.* **107**:250–257 (1961), Table I.

[2] From Schweid and Archibald, *loc. cit.*, Table 1.

31. Find the numerical values of the endpoints $\underline{\theta}$, $\bar{\theta}$ and $\underline{\mu}$, $\bar{\mu}$ of the confidence intervals determined by the two preceding problems for the data of Prob. 20, and compare them with the corresponding values based on Student's t-test.

32. Solve the preceding problem for the data of Probs. 21(i) to (v).

33. Use (4.68) to determine approximate symmetric confidence intervals for θ when (i) $1-\alpha = .95$, $N = 25$; (ii) $1-\alpha = .95$, $N = 50$; (iii) $1-\alpha = .95$, $N = 100$; (iv) $1-\alpha = .99$, $N = 100$.

34. Use (4.67) to determine approximate symmetric confidence intervals for μ for the four combinations of $1-\alpha$ and N of the preceding problem.

35. Determine an upper-confidence bound for μ_p with confidence coefficient γ as close as possible to .95 when $N = 20$ and (i) $p = .2$; (ii) $p = .3$; (iii) $p = .4$.

36. (i) For the values of N and p of the preceding problem and with confidence coefficient γ as close as possible to .95, determine a lower-confidence bound for μ_p.

(ii) If the lower- and upper-confidence bounds of this and the preceding problem are combined to obtain confidence intervals for μ_p, what is the resulting confidence coefficient?

37. Use a table of the binomial distribution to determine symmetric confidence intervals for μ_p with confidence coefficient γ as close as possible to .95 when $N = 25$ and (i) $p = .2$, (ii) $p = .25$, (iii) $p = .3$.

38. Solve Prob. 26 with the quartile $\mu_{1/4}$ in place of the median μ for $N = 12$ and $N = 15$, and obtain the numerical values of the resulting confidence points for the data of Prob. 23.

(B)

Section 1

39. (i) Show that the power function $\Pi(p)$ of the sign test (4.8) is a strictly increasing function of the probability p defined by (4.7).

(ii) In the shift model (4.5), show that the power function of (4.8) is a nondecreasing function of Δ.

[*Hint* for (i): Let N points be selected at random (i.e., independently, each according to a uniform distribution) on the interval (0,1). Then $\Pi(p)$ is the probability that at least c of the points fall in the interval $(0,p)$.]

40. If $p = P(Z > \Delta_0)$, and the distribution of Z is given by (4.5) with E continuous and symmetric about zero, show that

$$p = E(\Delta - \Delta_0) \tag{4.79}$$

41. If E is the normal distribution with mean zero and variance σ^2, show that (4.79) reduces to

$$p = \Phi[(\Delta - \Delta_0)/\sigma] \tag{4.80}$$

42. Show that if X and Y have a common marginal distribution $F = G$ but are not independent, the distribution of $Y - X$ is not necessarily symmetric about zero. [*Hint:* Let $P(X = i, Y = j) = p_{ij}$ for $i, j = 0, 1, 2$, and write down the conditions for $F = G$.]

43. Show that both the exact power of the sign test (4.8) and its normal approximation are unchanged if the distribution of the Z's is given by (4.5) and if the Z's and Δ are multiplied by the same positive constant.

44. A random variable Z with distribution L is *stochastically positive* if

$$L(z)+L(-z) \leqslant 1 \tag{4.81}$$

for all z, with strict inequality for at least some z.

(i) Explain in what sense (4.81) expresses the fact that L is slanted toward positive values.

(ii) Show that (4.81) holds if and only if there exists a distribution D symmetric about the origin and such that L is stochastically larger than D in the sense of (2.6).

{*Hint* for (i): What does (4.81) say about the tails of the distribution? (ii): If (4.81) holds, let $D(z) = \frac{1}{2}[1+L(z)-L(-z)]$.}

45. Show that (4.5) and $\Delta > 0$ imply (4.81).

46. (i) A random variable Z is said to be *stochastically negative* if $-Z$ is stochastically positive. Give an example of a random variable which is neither stochastically positive nor stochastically negative.

(ii) If Z is stochastically positive, determine whether it is true that aZ is stochastically positive for each $a > 0$.

(iii) Let X and Y be independent. Show that if Y is stochastically larger than X, then $Y-X$ is stochastically positive but that the converse is not necessarily true.

47. If $Z_1, \ldots, Z_N$ are independently distributed according to L, show that the sign test is unbiased against any alternative L satisfying (4.81). [*Hint:* Put $z = 0$ in (4.81).]

48. Show that for any integer N there exist two distributions F and G such that

$$|F(z)-G(z)| \leqslant 1/N \qquad \text{for all } z$$

but the expectations of F and G are arbitrarily far apart. [*Hint:* Let Z be a random variable taking on values a_i with probabilities p_i $(i = 1, \ldots, N)$. Given $a_1, \ldots, a_{N-1}$ and $p_1, \ldots, p_{N-1}$ with $p_1 + \cdots + p_{N-1} = 1-1/N$, there exists a_N such that $E(Z)$ is equal to any preassigned finite number A_N, for example $A_N = 2N$.]

49. (i) Determine the distribution of N_+ in the model (4.19).

(ii) Show that (4.20) is equivalent to the condition that both $P(Z \leqslant \mu)$ and $P(Z \geqslant \mu)$ are greater than or equal to $1/2$.

Section 2

50. Prove that the power function $\Pi_E(\Delta)$ defined in (4.22) (i) is a nondecreasing function of Δ; (ii) tends to 1 or 0 as Δ tends to $+\infty$ or $-\infty$ provided $2^{-N} \leqslant \alpha \leqslant 1-2^{-N}$. [*Hint* for (i): Adapt the method of proof of Chap. 2, Theorem 2 to the present case.]

51. Prove that the Wilcoxon test (3.7) is unbiased against any alternative for which L is stochastically positive.

52. If L is continuous and symmetric about zero, show that

$$p_2' = 1/3$$

and that (4.27) reduces to formula (3.14).

53. If L and L^* denote the distribution of Z and $-Z$, respectively, and if p_1 and p_2 are defined by (2.16) and (2.19), show that

$$p_1 = p_1' \quad \text{and} \quad p_2 = p_2' \quad \text{when} \quad F = L \quad \text{and} \quad G = L^* \tag{4.82}$$

54. Let $p_1(\Delta)$ and $p_2(\Delta)$ denote the probabilities p_1 and p_2 defined by (2.16) and (2.19) when $F(x) = E(x)$ and $G(y) = E(y-\Delta)$, and let $p_1'(\Delta)$ and $p_2'(\Delta)$ denote the probabilities p_1' and p_2' in the shift model (4.5). Show that

$$p_1'(\Delta) = p_1(2\Delta) \quad \text{and} \quad p_2'(\Delta) = p_2(2\Delta) \tag{4.83}$$

55. Show that $p_2' - p_1'^2 \geqslant 0$. [*Hint:* Divide (4.27) by N^3 and let $N \to \infty$.]

56. In the shift model (4.5), prove that p_1' is greater than or less than $\frac{1}{2}$ as Δ is positive or negative. [*Hint:* This follows from Chap. 2, Probs. 54 and 44.]

57. Prove that $V_s' \Big/ \binom{N}{2}$ is consistent for estimating p_1'. [*Hint:* By formula (A.111) of the Appendix it is only necessary to show that the variance of the estimator tends to zero as N tends to infinity.]

58. Show that the statistic V_s' can be expressed as

$$V_s' = \sum_{j=1}^{n} (s_j - 1)$$

59. Show that an alternative to the approximate formula (4.34) for the sample size required by the Wilcoxon test to obtain power Π is given by

$$N \approx 4 \frac{(u_\alpha/\sqrt{12} - u_\Pi \sqrt{p_2' - p_1'^2})^2}{(p_1' - \frac{1}{2})^2} \tag{4.84}$$

[*Hint:* See Chap. 2, Prob. 48.]

Section 3

60. Let $E(z) = (1-\varepsilon)\Phi(z) + \varepsilon\Phi(z/\gamma)$.

(i) Show that the approximate sample size required by the t-test according to (4.44) to achieve a given power Π at a fixed level $\alpha < \Pi$ against a given shift Δ in (4.5) tends to infinity as γ tends to infinity.

(ii) Show that the sample size required by the sign test for the same purpose does not tend to infinity. [*Hint* for (ii): Determine the behavior of p as γ tends to infinity.]

61. If $\alpha < \frac{1}{2} < \Pi$, show that the exact sample size required in part (i) of the preceding problem tends to infinity as γ tends to infinity, and hence that $e_{S,t}$ tends to infinity. [*Hint:* When $\alpha < \frac{1}{2}$, the power of the t-test is less than when $\alpha = \frac{1}{2}$. In the latter case, it is equal to

$$P\left[\frac{\sqrt{N}\bar{Z}}{\sqrt{\text{Var}(Z_1)}} > \frac{-\Delta\sqrt{N}}{\sqrt{\text{Var}(Z_1)}}\right] \tag{4.85}$$

and the result follows from Appendix, Theorems 2 and 3 and the fact that $\text{Var}(Z_1)$ tends to infinity.]

62. For any fixed numbers $a_1, \ldots, a_N$, let $z_i = a_i$ for $i = 1, \ldots, N-1$ and $z_N = a_N + A$. Show that as A tends to infinity, the t-statistic on the left-hand side of (4.41) tends to a finite limit.

63. (i) Show that the Wilcoxon test has the following monotonicity property: if the test rejects for the observations $z_1, \ldots, z_N$, and if $z_i \leqslant z_i'$ for all i, then the test also rejects for $z_1', \ldots, z_N'$.

(ii) Show that this property is not shared by the t-test. [*Hint* for (ii): Use the result of the preceding problem.]

64. (i) Show that the V_s^*-test does not have the monotonicity property of the preceding problem by computing the significance probability of the following two sets of observations: (*a*) 1, 2, 2, −4; (*b*) 2, 2, 2, −4; large values of V_s^* being significant.

(ii) Construct an example analogous to (i) with $N = 5$.

Section 4

65. If $N = 3$ and the three ordered observations are denoted by $Z_{(1)} < Z_{(2)} < Z_{(3)}$ show that

$$\hat{\theta} = \tfrac{1}{4}[Z_{(1)} + 2Z_{(2)} + Z_{(3)}] \tag{4.86}$$

66. Let Z_1, Z_2, and Z_3 be independently normally distributed with mean θ and unit variance. Use a table[1] of the covariances of normal order statistics to determine $\text{Var}(\hat{\theta})$ and $\text{Var}(\tilde{\theta})$, and compare these variances with that of $\bar{\theta}$.

67. Let $T_1, \ldots, T_k$ be a number of random variables the joint distribution of which depends on a location parameter θ, and suppose that the T's are considered most supportive of the hypothesis $H: \theta = \theta_0$ when the number of positive variables $T_i - \theta_0$ is as nearly equal as possible to the number of negative such variables. In analogy with the derivation of the estimator $\hat{\Delta}$ in Chap. 2, Sec. 5, explain how this suggests $\text{med}(T_i)$ as an estimator of θ, distinguishing between the cases that k is odd or even. Show how this argument applies to the derivation of the estimators $\hat{\theta}$ and $\tilde{\theta}$.

68. Let $Z = (Z_1, \ldots, Z_N)$ and $Z + c = (Z_1 + c, \ldots, Z_N + c)$, and let $\delta(Z)$ be an estimator which is nondecreasing in each of the arguments $Z_1, \ldots, Z_N$ and satisfies

$$\delta(Z + c) = \delta(Z) + c \tag{4.87}$$

Show that $|Z_i' - Z_i| \leqslant \varepsilon$ for all $i = 1, \ldots, N$ implies that $|\delta(Z') - \delta(Z)| \leqslant \varepsilon$. [*Hint:* See Chap. 2, Prob. 55.]

69. Let δ satisfy (4.87) and

$$\delta(-Z) = -\delta(Z) \tag{4.88}$$

and suppose that the Z's are independent, and distributed symmetrically about θ. Show that $\delta(Z)$ is distributed symmetrically about θ, and point out where in the proof you are using conditions (4.87) and (4.88).

70. If $Z_1, \ldots, Z_N$ are independently distributed, each according to a continuous distribution, prove that the distributions of (i) $\tilde{Z}$; (ii) $\hat{Z}$ are also continuous. [*Hint:* See Appendix, Prob. 9, and Chap. 2, Prob. 58.]

[1] Such a table is given in, for example, Owen, *Handbook of Statistical Tables*, Addison-Wesley, Reading, Mass., 1962, Sec. 7.3.

71. Obtain an approximation for $P_\theta(|\bar{\theta}-\theta| \leqslant a)$ analogous to (4.58).

72. Show that $\bar{\theta}$ can be obtained from the t-statistic by the same method that led from the Wilcoxon statistic to $\hat{\theta}$.

73. If $Z_1, \ldots, Z_N$ are independently distributed according to a continuous distribution L that has a unique median μ, prove that $\tilde{Z} = \tilde{Z}_N$ provides a consistent sequence of estimators of μ as N tends to infinity through odd values, by the following two methods: (i) from the fact that the frequency of success in N binomial trials with success probability p tends to p in probability; (ii) using the fact that the distribution of $(S_N - Np)/\sqrt{Npq}$ tends to the standard normal distribution. [*Hint:* Let $N = 2k+1$ and show that $P(Z_{(k+1)} < z) \to 0$ or 1 as $z < \mu$ or $z > \mu$. If $T_N(z)$ denotes the number of Z's less than z, then

$$P(Z_{(k+1)} < z) = P[T_N(z) \geqslant k+1]$$

and the result follows by method (i) from the fact that $T_N(z)/N$ tends in probability to $p = P(Z_i < z)$.]

74. (i) Show that if U_N and V_N are two consistent sequences of estimators for estimating a parameter θ, then so is (*a*) $\min(U_N, V_N)$, (*b*) $\max(U_N, V_N)$, (*c*) $\frac{1}{2}(U_N + V_N)$.

(ii) Prove the result of Prob. 73 as N tends to infinity through even values.

[*Hint* for (i): If A_N, B_N are two sequences of events such that $P(A_N)$ and $P(B_N)$ tend to zero as N tends to infinity, then $P(A_N \text{ or } B_N)$ also tends to zero.]

75. Under the assumptions of Theorem 4 (iii), prove that $\hat{Z} = \hat{Z}_N$ provides a consistent sequence of estimators of v as N tends to infinity (i) by showing that $V_s \Big/ \binom{N}{2}$ is consistent for estimating p_1'; (ii) using the fact that the difference between the right- and left-hand sides of (4.58) tends to zero as $N \to \infty$. [*Hint* for (i): Apply the method of the preceding two problems and see Prob. 57.]

76. Let F_k be a sequence of distributions and F_0 a continuous distribution such that $F_k(x) \to F_0(x)$ for all x as $k \to \infty$. Let λ_k denote the mean of F_k and μ_k its median and suppose that the latter is unique.

(i) Prove that $\mu_k \to \mu_0$ as $k \to \infty$.

(ii) Show that it is not necessary for λ_k to tend to λ_0 as $k \to \infty$.

[*Hint* for (i): Suppose that a subsequence of μ_k tends to a limit different from μ_0 and apply Appendix Theorem 3. (ii): Let $F_k(z) = (1-\varepsilon_k)\Phi(z) + \varepsilon_k \Phi(z-\theta_k)$.]

77. Let $Z_1, \ldots, Z_N$ be independently distributed according to a continuous distribution L with median μ.

(i) Show that (4.62) holds when N is odd and holds approximately when N is even.

(ii) Determine bounds for $P(\tilde{Z} < \mu)$ and $P(\tilde{Z} > \mu)$ when $N = 10, 15, 25$. [*Hint:* Use Eq. (4.63).]

Section 5

78. Let $Z_1', \ldots, Z_N'$ be independently distributed according to a continuous distribution L symmetric about θ, and suppose that each observation is rounded to the closest value among $0, \pm\varepsilon, \pm 2\varepsilon, \ldots$. Denote the resulting rounded observations by $Z_1, \ldots, Z_N$. Determine $i < j$ from a table of the signed-rank Wilcoxon distribution so that

$$P_0(V_s < i) + P_0(V_s \geqslant j) = \alpha$$

and let $A_{(i)}$ and $A_{(j)}$ be the ith and jth smallest of the averages $\frac{1}{2}(Z_k + Z_l)$, $k \leqslant l$.

(i) Show that

$$P(A_{(i)} \leqslant \theta \leqslant A_{(j)}) \geqslant 1-\alpha \tag{4.89}$$

for any value of θ which is a multiple of ε.

(ii) Show that

$$P(A_{(i)} - \varepsilon \leqslant \theta \leqslant A_{(j)} + \varepsilon) \geqslant 1-\alpha \tag{4.90}$$

for all θ. [*Hint:* See Chap. 2, Prob. 65.]

8. REFERENCES

Andrews, D. F., Bickel, P. J., Hampel, F. R., Huber, P. J., Rogers, W. H., and Tukey, J. W. (1972): *Robust Estimation of Location: Survey and Advances*, Princeton University Press, Princeton, N.J.

Anscombe, F. J. (1960): "Rejection of Outliers," *Technometrics* **2**:123–147.

——— and Tukey, J. W. (1963): "Examination and Analysis of Residuals," *Technometrics* **5**:141–160.

Arnold, Harvey J. (1965): "Small Sample Power for the One-sample Wilcoxon Test for Non-normal Shift Alternatives," *Ann. Math. Statist.* **36**:1767–1778.

Bahadur, R. R. (1960): "Simultaneous Comparisons of the Optimum and the Sign Tests of a Normal Mean," in *Contributions to Probability and Statistics: Essays in Honor of Harold Hotelling*, Stanford University Press, Stanford, Calif., pp. 79–88.

Blyth, Colin R. (1958): "Note on Relative Efficiency of Tests," *Ann. Math. Statist.* **29**:898–903.

Chernoff, Herman (1952): "A Measure of Asymptotic Efficiency for Tests of a Hypothesis Based on the Sum of Observations," *Ann. Math. Statist.* **23**:493–507.

———, Gastwirth, Joseph L. and Johns, M. V. (1967): "Asymptotic Distribution of Linear Combinations of Functions of Order Statistics with Applications to Estimation," *Ann. Math. Statist.* **38**:52–72.

Cochran, William G. (1937): "The Efficiencies of the Binomial Series Test of Significance of a Mean and of a Correlation Coefficient," *J. Roy. Statist. Soc.* **A100**:69–73.

Dixon, W. J. (1953): "Power Functions of the Sign Test and Power Efficiency for Normal Alternatives," *Ann. Math. Statist.* **24**:467–473.

Doksum, Kjell and Thompson, Rory (1971): "Power Bounds and Asymptotic Minimax Results for One-sample Rank Tests," *Ann. Math. Statist.* **42**:12–34.

Fraser, D. A. S. (1957): "Most Powerful Rank-type Tests," *Ann. Math. Statist.* **28**:1040–1043. [Derives the Absolute Normal Scores test as the locally most powerful rank test in a normal shift model.]

Gastwirth, Joseph L. (1966): "On Robust Procedures," *J. Am. Statist. Assoc.* **61**:929–948.

——— (1971): "On the Sign Test for Symmetry," *J. Am. Statist. Assoc.* **66**:821–823.

——— and Rubin, Herman (1970): "The Behavior of Robust Estimators on Dependent Data," *Purdue Univ. Statist. Mimeo.* ser. no. 197.

——— and ——— (1971): "Effect of Dependence on the Level of Some One-Sample Tests," *J. Am. Statist. Assoc.* **66**:816–820.

Geertsema, J. C. (1970): "Sequential Confidence Intervals Based on Rank Tests," *Ann. Math. Statist.* **41**:1016–1026.

Gibbons, Jean (1964): "Effect of Non-normality on the Power Function of the Sign Test," *J. Am. Statist. Assoc.* **59**:142–148.

Govindarajulu, Zakkula and Eisenstat, Stan (1965): "Best Estimates of Location and Scale Parameters of a Chi Distribution Using Ordered Observations," *Rept. Statist. Appl. Res. Union Japan. Scientists Engrs.* **12**:149–164.

Hajek, Jaroslav and Sidak, Zbynek (1967): *Theory of Rank Tests*, Academic Press, New York.

Hampel, Frank R. (1968): "Contributions to the Theory of Robust Estimation," unpublished Ph.D. thesis, University of California at Berkeley.

——— (1971): "A General Qualitative Definition of Robustness," *Ann. Math. Statist.* **42**:1887–1896.

——— (1974): "The Influence Curve and Its Role in Robust Estimation," *J. Am. Statist. Assoc.* **69**:383–393.

Harter, H. Leon (1969): *Order Statistics and Their Use in Testing and Information*, U.S. Government Printing Office, Washington, D.C., vol. 2.

Hartigan, J. A. (1969): "Using Subsample Values as Typical Values," *J. Am. Statist. Assoc.* **64**:1303–1317.

Hodges, J. L., Jr., and Lehmann, E. L. (1956): "The Efficiency of Some Nonparametric Competitors of the *t*-test," *Ann. Math. Statist.* **27**:324–335.

——— and ——— (1963): "Estimates of Location Based on Ranks," *Ann. Math. Statist.* **34**:598–611. [Proposes the estimate $\hat{\theta}$. See also Sen (1963).]

Hoeffding, Wassily (1951): "'Optimum' Nonparametric Tests," in *Proceedings of the Second Berkeley Symposium on Probability and Statistics*, University of California Press, Berkeley, pp. 83–92.

——— (1968): "Robustness of the Wilcoxon Estimate of Location against a Certain Dependence," *Ann. Math. Statist.* **39**:1196–1201.

Høyland, Arnljot (1964): "Numerical Evaluation of Hodges-Lehmann Estimates," *Norske Vid. Selsk. Forh.* **37**:42–47.

Huber, P. J. (1964): "Robust Estimation of a Location Parameter," *Ann. Math. Statist.* **35**:73–101.

——— (1972): "Robust Statistics: A Review," *Ann. Math. Statist.* **43**:1041–1067.

Hwang, Tea-Yuan and Klotz, Jerome (1970): "On the Approach to Limiting Bahadur Efficiency," Technical Report No. 237, Department of Statistics, University of Wisconsin.

Jaeckel, Louis A. (1971): "Robust Estimates of Location; Symmetry and Asymmetric Contamination," *Ann. Math. Statist.* **42**:1020–1034.

Klotz, Jerome (1963): "Small Sample Power and Efficiency for the One-sample Wilcoxon and Normal Scores Tests," *Ann. Math. Statist.* **34**:624–632.

——— (1965): "Alternative Efficiencies for Signed-rank Tests," *Ann. Math. Statist.* **36**:1759–1766.

Lehmann, E. L. (1959): *Testing Statistical Hypotheses*, John Wiley & Sons, New York.

Moses, Lincoln E. (1965): "Query: Confidence Limits from Rank Tests," *Technometrics* **7**:757–760.

Noether, Gottfried (1971): *Introduction to Statistics*, Houghton Mifflin, Boston.

——— (1973): "Some Simple Distribution-free Confidence Intervals for the Center of a Symmetric Distribution," *J. Am. Statist. Assoc.* **68**:716–719.

Pitman, E. J. G. (1948): *Lecture Notes on Nonparametric Statistics*, Columbia University Press, New York. [Obtains the Pitman efficiency of the Wilcoxon test relative to the *t*-test.]

Putter, Joseph (1955): "The Treatment of Ties in Some Nonparametric Tests," *Ann. Math. Statist.* **26**:368–386.

Ruist, Erik (1954): "Comparison of Tests for Nonparametric Hypotheses," *Arkiv Matematik* **3**:133–163.

Sen, Pranab Kumar (1963): "On the Estimation of Relative Potency in Dilution (-direct) Assays by Distribution-free Methods," *Biometrics* **19**:532–552. [Proposes the estimator $\hat{\theta}$. See also Hodges and Lehmann (1963).]

——— (1968): "On a Further Robustness Property of the Test and Estimator Based on Wilcoxon's Signed Rank Statistic," *Ann. Math. Statist.* **39**:282–285.

——— and Ghosh, Malay (1971): "On Bounded Length Sequential Confidence Intervals Based on One-sample Rank-order Statistics," *Ann. Math. Statist.* **42**:189–203.

Stein, C. (1956): "Efficient Nonparametric Testing and Estimation," in *Proceedings of the Third Berkeley Symposium on Mathematical Statistics and Probability*, University of California Press, Berkeley, vol. 1, pp. 187–195.

Takeuchi, Kei (1971): "A Uniformly Asymptotically Efficient Estimator of a Location Parameter," *J. Am. Statist. Assoc.* **66**:292–301.

Thompson, Rory, Govindarajulu, Z., and Doksum, K. A. (1967): "Distribution and Power of the Absolute Normal Scores Test," *J. Am. Statist. Assoc.* **62**:966–975. [Gives critical values of the Absolute Normal Scores test for $N = 11(1)20$, $\alpha = .001, .005, .01$ and for $N = 11(1)17$, $\alpha = .025, .05, .1$.]

Tukey, John W. (1960): "A Survey of Sampling from Contaminated Distributions," *Contributions to Probability and Statistics: Essays in Honor of Harold Hotelling*, Stanford University Press, Stanford, Calif., pp. 445–485. [A key paper in drawing attention to the influence of gross errors on standard procedures.]

——— (1962): "The Future of Data Analysis," *Ann. Math. Statist.* **33**:1–67. [Continues the discussion of the preceding paper.]

——— and McLaughlin, D. H. (1963): "Less Vulnerable Confidence and Significance Procedures for Location Based on a Single Sample: Trimming/Winsorization I," *Sankhya* **A25**:331–352.

Walsh, John E. (1946): "On the Power Function of the Sign Test for the Slippage of Means," *Ann. Math. Statist.* **17**:358–362.

——— (1949a): "Some Significance Tests for the Median which are Valid under Very General Conditions," *Ann. Math. Statist.* **20**:64–81.

——— (1949b): "Applications of Some Significance Tests for the Median which are Valid under Very General Conditions," *J. Am. Statist. Assoc.* **44**:342–353.

——— (1951): "Some Bounded Significance Level Properties of the Equal-tail Sign Test," *Ann. Math. Statist.* **22**:408–417.

Woodworth, George G. (1970): "Large Deviations and Bahadur Efficiency of Linear Rank Statistics," *Ann. Math. Statist.* **41**:251–283.

CHAPTER 5
THE COMPARISON OF MORE THAN TWO TREATMENTS

1. RANKS IN THE COMPARISON OF SEVERAL TREATMENTS

Comparative studies frequently involve the simultaneous comparison not just of two, but of three or more treatments or conditions. One may, for example, wish to compare the effect on growth of several diets, the mileage obtained from a number of different brands of gasoline, or the weather predictions made by a number of different forecasters. Although somewhat more complex, such comparisons share many of the features found in the comparison of only two treatments. In particular, the development of the present section closely parallels that of Chap. 1, Sec. 1.

EXAMPLE 1. *Three tranquilizers.* To see whether there is any difference in the effect of three brands A, B, and C of tranquilizers, comparable patients in a ward of a mental hospital are assigned at random, two each to brands A and C, and three to brand B. After a month, the effect of the treatment is assessed and produces the following ranking:

$$A: 2, 4 \qquad B: 3, 5, 7 \qquad C: 1, 6$$

with higher ranks corresponding to greater improvement.

Let H denote the hypothesis that there is no difference in the effect of the three brands. Since under H the rank of each patient is determined solely by the state of his health, one may think of each patient's rank as attached to him even before his assignment to one of the drugs.

The division of the patients into three treatment groups then also divides the seven ranks into three groups, and by the randomness of the assignment, each of the possible divisions is equally likely.

How many such divisions are there? The number of possible choices for the ranks in group A is $\binom{7}{2} = 21$; they are displayed below:

12 13 14 15 16 17 23 24 25 26 27 34 35 36 37 45 46 47
56 57 67

For each of these choices for group A, there remain five ranks to choose from for B. For example, when the A-ranks are 13, the B-ranks must be chosen from 24567 and the possible choices are displayed below in the row labeled 13. As further illustrations, the second and third rows show the possible B-ranks when the A-ranks are 16 and 23, respectively.

A-ranks	Possible choices for B-ranks									
13	245	246	247	256	257	267	456	457	467	567
16	234	235	237	245	247	257	345	347	357	457
23	145	146	147	156	157	167	456	457	467	567

Once the A-ranks have been chosen, there are five values available for the three B-ranks, for which there are then $\binom{5}{3} = 10$ possible choices. Correspondingly, each of the three rows in the display has 10 entries. A full table of all possible choices of the A- and B-ranks would thus have $\binom{7}{2} = 21$ rows, each with $\binom{5}{3} = 10$ entries, so that the total number of these choices is 210. After the ranks for both A and B have been determined, the remaining two ranks will automatically constitute the C-group. The total number of possible divisions of the ranks into the three groups is therefore $\binom{7}{2}\binom{5}{3} = 21 \times 10 = 210$. Under the hypothesis H, each of these possible divisions is equally likely and therefore has probability $1/210$.

These considerations are easily generalized. Suppose that N subjects are available for a comparison of s treatments and that it is decided to allocate n_i subjects to the ith treatment, so that

$$n_1 + \cdots + n_s = N \tag{5.1}$$

with the subjects being assigned to the treatments at random. Let the number of possible assignments of the subjects be denoted by $\binom{N}{n_1, \ldots, n_s}$. For example, when $N = 7$, $s = 3$, and $n_1 = 2$, $n_2 = 3$, $n_3 = 2$, we have seen that $\binom{7}{2,3,2} = 210$. The numbers $\binom{N}{n_1, \ldots, n_s}$ are known as *multinomial coefficients* (see Prob. 37), and by the argument of Example 1 are seen to be equal to

$$\binom{N}{n_1, \ldots, n_s} = \binom{N}{n_1}\binom{N-n_1}{n_2}\cdots\binom{N-n_1-\cdots-n_{s-2}}{n_{s-1}} \tag{5.2}$$

(For an alternative formula, see Prob. 33.) For example, if $N = 15$, $s = 4$, and $n_1 = 3$, $n_2 = 6$, $n_3 = 2$, $n_4 = 4$, Table A and (5.2) show that

$$\binom{15}{3,6,2,4} = \binom{15}{3}\binom{12}{6}\binom{6}{2} = 6{,}306{,}720$$

By assumption, the N subjects are assigned to the s treatment groups *at random*; that is, all $\binom{N}{n_1, \ldots, n_s}$ possible assignments are equally likely, so that each has probability $1\Big/\binom{N}{n_1, \ldots, n_s}$. At the termination of the study, the N subjects are ranked, either directly or according to the values of some response. Let the ranks in the 1st, 2nd, ..., sth treatment group be denoted by

$$R_{11}, \ldots, R_{1n_1};\ R_{21}, \ldots, R_{2n_2};\ \ldots;\ R_{s1}, \ldots, R_{sn_s} \tag{5.3}$$

where we shall assume that within each group they are numbered in increasing order so that $R_{11} < \cdots < R_{1n_1}$, $R_{21} < \cdots < R_{2n_2}$, and so on. Then under the hypothesis H of no difference among the treatments, the $\binom{N}{n_1, \ldots, n_s}$ possible divisions of the integers $1, \ldots, N$ into s groups of sizes $n_1, \ldots, n_s$ as treatment ranks each has probability

$$P_H(R_{11} = r_{11}, \ldots, R_{1n_1} = r_{1n_1}; \ldots; R_{s1} = r_{s1}, \ldots, R_{sn_s} = r_{sn_s}) = \frac{1}{\binom{N}{n_1, \ldots, n_s}} \tag{5.4}$$

This is the basic null distribution of the ranks, which generalizes the distribution (1.4) for two treatments to the case of s treatments.

2. THE KRUSKAL-WALLIS TEST

One of the simplest questions that can be asked in a comparison of several treatments is whether there is any difference among the treatments. To obtain an answer, one may wish to test the hypothesis H of no difference, and this is the problem with which we shall be concerned in the present section. Actually, this formulation is frequently an oversimplification. If the hypothesis is rejected, one usually wants to know something about the way in which the treatments differ. Which is the best? Which ones are promising, and which can be discarded? Which are better than the control? And so on. However, the problem of testing H does arise in situations in which it is hoped that the response is unaffected by some factor that can then be disregarded. For example, in a study involving several diet groups, each containing a number of subdiets, one may hope that it is not necessary to distinguish between the diets within each group; or, in a study of tire wear, that it is not necessary to take into account the make of the car on which the tires are mounted.

In deciding on an appropriate test statistic, it is not enough to specify the hypothesis H. One must also be clear about the nature of the alternatives against which H is being tested. As was discussed in Chap. 1 (at the end of Sec. 5 and in Sec. 6), treatment differences may manifest themselves in different ways such as the general level of the responses, the extent to which the responses are spread out, and so on. We shall here suppose that the treatments mainly affect the response level. There is then an order among the treatments: one tends to give the lowest responses; another giving the next lowest responses is second; and so forth. An indication of the position of a treatment in this ordering is provided by its average rank, which for the ith treatment will be denoted by

$$R_{i.} = \frac{R_{i1} + \cdots + R_{in_i}}{n_i}$$

If the treatments differ widely among each other, one would expect big differences among the values of the $R_{i.}$'s. On the other hand, when H is true the $R_{i.}$'s may be expected to be close to each other and hence also to the overall average

$$R_{..} = \frac{(R_{11} + \cdots + R_{1n_1}) + \cdots + (R_{s1} + \cdots + R_{sn_s})}{N}$$

Since the sum of all the ranks is by (1.12) equal to $N(N+1)/2$, it follows that

$$R_{..} = \frac{N+1}{2} \tag{5.5}$$

and an indication of the validity of the hypothesis is obtained by noting how close the averages $R_{i.}$ are to (5.5).

A convenient criterion for measuring the overall closeness of the $R_{i.}$ to $R_{..}$ is a weighted sum of the squared differences $[R_{i.} - \frac{1}{2}(N+1)]^2$, for example, the *Kruskal-Wallis* statistic

$$K = \frac{12}{N(N+1)} \sum_{i=1}^{s} n_i \left(R_{i.} - \frac{N+1}{2} \right)^2 \tag{5.6}$$

The weights in this statistic are chosen to provide a simple approximation to the null distribution when the n_i are large.

Since K is zero when the $R_{i.}$ are all equal and is large when there are substantial differences among the $R_{i.}$, the hypothesis is rejected for large values of K, say, when

$$K \geqslant c \tag{5.7}$$

For $s = 2$, this test reduces to the two-sided Wilcoxon test (1.47) (Prob. 38). By squaring out $[R_{i.} - \frac{1}{2}(N+1)]^2$ and replacing $R_{i.}$ by R_i/n_i where

$$R_i = R_{i1} + \cdots + R_{in_i}$$

is the rank-sum corresponding to the ith treatment, K may be rewritten (Prob. 39) as

$$K = \frac{12}{N(N+1)} \sum_{i=1}^{s} \frac{R_i^2}{n_i} - 3(N+1) \tag{5.8}$$

This expression for K is more convenient to compute since it does not require working with the individual differences $R_{i.} - \frac{1}{2}(N+1)$.

To determine the critical value c in (5.7) that corresponds to a given significance level α, it is necessary to be able to compute the null distribution of K. In principle, this causes no difficulty since the distribution is determined by (5.4). Consider, for example, the case $N = 7$, $s = 3$, and $n_1 = 2$, $n_2 = 3$, $n_3 = 2$ of Example 1. To each of the 210 possible divisions of the $N = 7$ ranks into the three groups corresponds a value of K. Thus, in Example 1,

$$R_{11} = 2 \quad R_{12} = 4; \quad R_{21} = 3 \quad R_{22} = 5 \quad R_{23} = 7; \quad R_{31} = 1 \quad R_{32} = 6$$

so that $R_1 = 6$, $R_2 = 15$, $R_3 = 7$, and hence

$$K = \frac{12}{7 \times 8}\left[\frac{36}{2} + \frac{225}{3} + \frac{49}{2}\right] - 24 = 1.18$$

In this manner, it is possible to compute the value of K corresponding to each of the 210 cases and, since each case has probability $1/210$, to obtain the null distribution of K.

Unfortunately, except for very small values of N, the amount of labor required is prohibitive. When $s = 3$ and none of the group sizes exceeds 5, critical values corresponding to values of α close to 1, 5, and 10 percent have been tabulated by Kruskal and Wallis (1952), and the distribution of K has been tabulated for the same cases by Kraft and van Eeden (1968). Part of the Kruskal and Wallis table is reproduced as Table I at the end of this book. For $n_1 = 2$, $n_2 = 3$, $n_3 = 2$, the table shows that $P_H(K \geqslant 4.4643) = .105$, so that the value $K = 1.18$ of Example 1 is not significant.

A comprehensive table of the null distribution of K is impracticable since it would depend on too many arguments. However, Table I may be supplemented by a simple approximation. As in the normal approximations considered earlier, the probability $P_H(K \geqslant c)$ is approximated by the area to the right of c under a curve representing a probability density function. However, instead of the normal density, it is in the present case the density of the χ^2 (chi-square) distribution with $f = s - 1$ degrees of freedom, which is illustrated in Fig. 5.1 for $f = 3$. The total area under the curve is equal to 1 and the area to the right of c, which provides the desired approximation for $P_H(K \geqslant c)$, will be denoted by $\Psi_{s-1}(c)$.

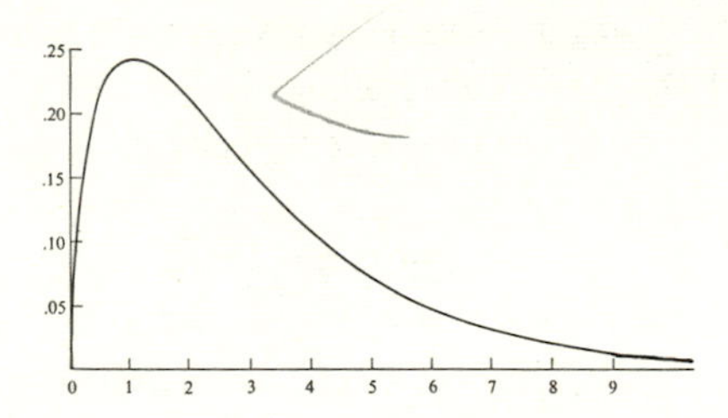

FIGURE 5.1
Graph of χ^2 with 3 degrees of freedom.

The approximation is supported by a limit theorem which states that

$$P_H(K \geqslant c) \to \Psi_{s-1}(c) \tag{5.9}$$

as the group sizes $n_1, \ldots, n_s$ tend to infinity. Exact conditions under which this theorem holds, and a proof, are given in Appendix, Example 31.

A table of the critical values of c for which

$$\Psi_{s-1}(c) = \alpha \tag{5.10}$$

for $\alpha = .001, .005, .01, .025, .05, .10, .20, .30$, and numbers of degrees of freedom $f = 6(1)40(5)100$, and a more complete listing of right-tail probabilities for $f = 2(1)5$ is given in Table J. The approximation typically will be adequate (except for very small significance levels) when either $s = 3$ and the three group sizes are above 5, or when $s > 3$ and all group sizes are above 4.[1] The accuracy of the approximation is illustrated by the following comparison with the exact values for $s = 3$ and $n_1 = n_2 = n_3 = 5$.

c	8.0000	7.9800	5.7800	5.6600	4.5600	4.5000
$P_H(K \geqslant c)$	.009	.010	.049	.051	.100	.102
$\Psi_2(c)$	.018	.018	.056	.059	.103	.106

EXAMPLE 2. *Pooling four subdiets.* In a comparative study of certain diet groups, it is desired to test whether it is necessary to distinguish between the diets within the same group. The following table gives the growth of 25 rats after 12 weeks on four subdiets of diet A together with their ranks.[2]

Diet	Growth Figures	Ranks
A_1	257, 205, 206, 164, 190, 214, 228, 203	1, 4, 8, 10, 11, 14, 18, 22
A_2	201, 231, 197, 185	2, 5, 6, 20
A_3	248, 265, 187, 220, 212, 215, 281	3, 13, 15, 16, 21, 23, 25
A_4	202, 276, 207, 204, 230, 227	7, 9, 12, 17, 19, 24

[1] For a Monte Carlo study of the accuracy of the approximation see Gabriel and Lachenbruch (1969).

[2] From Nieman, Groot, and Jansen, "The Nutritive Value of Butter Fat Compared with That of Vegetable Fats, I.", *Koninkl. Ned. Akad. Wetenschap*, **C55**:588–604 (1952), Table 5. Original data from Boer and Jansen, *Arch. Neerl. Physiol.* **26**:1 (1952).

Here $R_1 = 88$, $R_2 = 33$, $R_3 = 116$, $R_4 = 88$, with a check being provided by the fact that the sum of the R's equals $N(N+1)/2 = 325$. The resulting value of K is 4.2. Since for $f = 3$ the critical value corresponding to $\alpha = .20$ is 4.64 (by Table J), the result is not significant at this level.

Tables I and J do not cover all possible cases. When one or more of the group sizes are too small for the approximation but others are too large to be included in Table I, certain alternative approximations, discussed in Kruskal and Wallis (1952), Rijkoort and Wise (1953), and D. Wallace (1959), are sometimes helpful.

We have assumed so far that there are no ties among the observations. In the presence of ties, we use the method of midranks described in Chap. 1, Sec. 4, with the midrank of an observation defined as the average of the ranks with which it is tied. Suppose as before that the N observations take on e distinct values, and that d_1 of the N observations are equal to the smallest value, d_2 to the next smallest, etc. Let the midranks of the n_i observations on the ith treatment be denoted by $R_{i1}^*, \ldots, R_{in_i}^*$ and their sum by R_i^*. If the statistic K^* is defined by

$$K^* = \frac{[12/N(N+1)]\Sigma R_i^{*2}/n_i - 3(N+1)}{1 - \Sigma(d_i^3 - d_i)/(N^3 - N)} \tag{5.11}$$

the hypothesis H of no difference among the treatments is rejected when

$$K^* \geqslant c \tag{5.12}$$

To illustrate the computation of the null distribution of K^*, suppose that in Example 1 it had not been possible to distinguish between the patients ranked 2, 3, and 4, so that each of these three now receives midrank 3, and similarly for the patients ranked 6 and 7, each of whom now receives midrank 6.5. Treatments A, B, and C of Example 1 would thus have received, respectively, the midranks

$$A\colon 3, 3 \qquad B\colon 3, 5, 6.5 \qquad C\colon 1, 6.5$$

As in Example 1, there are $\binom{7}{2,\,3,\,2} = 210$ equally likely ways of dividing the $N = 7$ patients (and hence their midranks) into three groups of $n_1 = 2$, $n_2 = 3$, $n_3 = 2$, respectively. However, not all these divisions of the midranks are now distinguishable. Let us count the number of choices leading to the arrangement displayed above. There are $\binom{3}{2} = 3$ ways of choosing two of the three patients with midrank 3 to receive treatment A, and $\binom{2}{1} = 2$ ways of choosing one of the two patients with midrank 6.5 to receive treatment B. Once these $3 \times 2 = 6$ choices have been made, there is only one assignment of the remaining patients to their respective treatments that leads to the displayed arrangement. The probability of this arrangement is therefore $6/210 = 1/35$.

As another illustration, let us compute the probability of the arrangement

$$A\colon 1, 3 \qquad B\colon 3, 5, 6.5 \qquad C\colon 3, 6.5$$

Since there are six ways of assigning the three patients with midrank 3, one patient to each of the three treatments, and two ways of assigning the two subjects with midrank 6.5, one patient to B and one to C, the desired probability is $^{12}/_{210}$.

To each of the possible assignments of the seven midranks 1, 3, 3, 3, 5, 6.5, 6.5 to A, B, and C corresponds a value of K^*. (These values are 6.2843 and 2.6373 for the two assignments displayed above.) The null distribution of K^* is obtained by computing for each possible arrangement its probability and the associated value of K^*.

This computation soon becomes cumbersome, and tables are impracticable. However, except for very small values of N, the null distribution of K^* can again be approximated by the χ^2-distribution with $s-1$ degrees of freedom. (For an associated limit theorem, see Appendix, Example 31.)

EXAMPLE 3. *Comparison of four laboratories.* Following are four sets of eight measurements each of the smoothness of a certain type of paper, obtained in four different laboratories:[1]

Laboratory								
A	38.7	41.5	43.8	44.5	45.5	46.0	47.7	58.0
B	39.2	39.3	39.7	41.4	41.8	42.9	43.3	45.8
C	34.0	35.0	39.0	40.0	43.0	43.0	44.0	45.0
D	34.0	34.8	34.8	35.4	37.2	37.8	41.2	42.8

Let us assume that the measurements from the different laboratories are equally variable and test the hypothesis of no systematic differences among the laboratories. As a first step, the 32 measurements are ranked. The following table displays the resulting midranks.

Laboratory									
A	9	17	24	26	28	30	31	32	197
B	11	12	13	16	18	20	23	29	142
C	1.5	5	10	14	21.5	21.5	25	27	125.5
D	1.5	3.5	3.5	6	7	8	15	19	63.5

The sums $R_1^*, \ldots, R_4^*$ of the midranks for the four laboratories are given in the last column. As a check, we note that

$$R_1^* + R_2^* + R_3^* + R_4^* = \tfrac{1}{2}N(N+1) = 528$$

The numerator of K^* is

$$\frac{12}{N(N+1)} \sum_i \frac{R_i^{*2}}{n_i} - 3(N+1) = 12.86861$$

[1] Part of the data from Mandel, *The Statistical Analysis of Experimental Data*, Wiley, Interscience, New York, 1964, Table 13.3.

Of the d's, 26 are equal to 1 and 3 are equal to 2, so that

$$\frac{\sum(d_i^3-d_i)}{N^3-N}=.00055$$

The denominator of K^* is thus .99945, and hence $K^*=12.8757$. Table J shows the significance probability $P(\chi_3^2 \geqslant 12.8757)$ to be approximately equal to .005. Some differences among the laboratories therefore seem to exist.

3. $2 \times t$ CONTINGENCY TABLES

The observable responses in a comparative study need not be numerical but may specify only to which of a number of categories each subject belongs. (Such data are sometimes described as *categorical*.) The subject is observed to be male or female; young, middle-aged, or old; to receive a grade of A, B, C, D, or F in a course; to favor one of several brands of a product or one of several candidates in an election.

EXAMPLE 4. *Live vs. televised instruction.* Consider first the case of just two treatments, and suppose that the comparison is between live and televised instruction in a history course. A number N of students who have volunteered for the study are assigned at random, m to attend the lectures in the room in which they are given; the remaining $n=N-m$ to watch them on a television screen in another room. The performance of each student at the end of the course is measured in terms of his course grade, so that the results can be exhibited as in Table 5.1 below, where, for example, B_3 denotes the number of students who watched the course on TV and received the grade of C, where $d_1=A_1+B_1$, and so on.

The problem of testing the hypothesis of no difference in the effectiveness of the two methods against the alternatives that live instruction is more effective (and hence the grades in the first row of the table will tend to be better than those in the second row) can be viewed as a special case of the comparison of two treatments with ties considered in Chap. 1, Sec. 4. There are d_1 observations tied in the first category, d_2 in the next one, and so on, and the hypothesis may be tested by means of the Wilcoxon statistics W_s^* or W_{XY}^*, the latter given by formula (1.37).

$$W_{XY}^*=B_1(\tfrac{1}{2}A_1)+B_2(A_1+\tfrac{1}{2}A_2)+B_3(A_1+A_2+\tfrac{1}{2}A_3)+\cdots \tag{5.13}$$

The hypothesis is rejected for large values of W_{XY}^*. [For an alternative approach to this problem, see Simon (1973).]

TABLE 5.1 2×5 contingency table

Category	A	B	C	D	F	Total
Treatment						
Live	A_1	A_2	A_3	A_4	A_5	m
TV	B_1	B_2	B_3	B_4	B_5	n
Total	d_1	d_2	d_3	d_4	d_5	N

So far, we have assumed that the effect of following the lectures on TV rather than live, if there is a difference at all, would be in the direction of decreasing the effectiveness of the course. It is not a priori clear that this will be the case. Suppose, for example, that the course is in physics rather than history and that the experiments performed to illustrate the lectures make an important contribution to the understanding of the material. In a large lecture hall, it is frequently difficult to get a clear view of the details of the experiment, and this would give an advantage to the students who are able to watch the experiments on television. Instead of testing the hypothesis of no difference against the *one-sided* alternatives that instruction by TV is less effective than live teaching, one might then wish to consider the *two-sided* alternatives that the televised instruction is either less or more effective. A procedure for dealing with this problem was discussed in Chap. 1, Sec. 5. It consists in accepting the hypothesis of no difference when

$$|W_{XY}^* - \tfrac{1}{2}mn| \tag{5.14}$$

is sufficiently small, and to come to the appropriate alternative conclusion when W_{XY}^* is sufficiently much larger or smaller than its expected value $\frac{1}{2}mn$.

Even the two-sided situation in which the effect of the treatment, if any, is assumed to be either beneficial or deleterious does not exhaust all possibilities. Suppose, for example, that the poorer or less motivated students tend to sit toward the back, and the better students prefer the front seats. In the large room in which the lectures are given, the poorer students may then simply stop listening to an instructor whose words are hard to catch and whose experiments they cannot properly see. On the other hand, the TV screen that they can watch in a small group, perhaps under the supervision of a teaching assistant, may stimulate their interest. In this case, the televised instruction might do more for the poorer students, while the good students (sitting in front) would benefit more from live instruction. This could lead to large values of $A_1, A_2, A_4,$ and A_5 in Table 5.1 (the good students are stimulated and the poor ones give up), and to a relatively large value of B_3 in the second row of the table.

If there are no clear a priori directions for the effect of the treatment, it may be of interest to test the hypothesis of no treatment effect against the *omnibus* alternatives that there is a difference but without specifying in which directions or in which cells of the table to expect an increase or decrease. Let us illustrate this situation with the following example.

EXAMPLE 5. *Mail questionnaires vs. personal interview.* A study of public opinion polls is planned to determine whether polls conducted by mail questionnaires lead to comparable results with those conducted by personal interviews. In an election involving three candidates *A*, *B*, and *C*, it is decided to poll a sample of 100 voters, of whom 40 are assigned at random to be interviewed in person with the remaining 60 receiving a questionnaire by mail. The results are displayed in the following table.

Polling Method	Candidate A	B	C	Total
Questionnaire	25	15	20	60
Interview	16	6	18	40
Total	41	21	38	100

Here a one-sided test would not be appropriate unless it is believed that one of the polling methods will tend to benefit the candidates relative to the other method in some definite order (e.g., A the least, B next, C the most). Similarly, a two-sided test would typically not be appropriate. To see how to apply an omnibus test of the hypothesis H of no difference between the two methods, let us turn the situation around and pretend for a moment that the three candidates represent the three treatments to be compared and "mail questionnaire" and "personal interview" the two possible responses for each subject. Consider the hypothesis H', appropriate to this point of view, that there is no difference between the three candidates with respect to their support by questionnaire or interview, and compare it with the original hypothesis H that the preference for the candidates is independent of the method of polling. These two hypotheses basically make the same statement: each asserts the independence of the preference for a candidate from the polling method.

The hypothesis H' that there is no difference among the three *treatments* A, B, and C with respect to the two possible *responses* can be tested by means of the Kruskal-Wallis test. The *sample sizes* for this test are $n_1 = 41, n_2 = 21, n_3 = 38$. There are only two possible outcomes, one of which, say, *questionnaire*, we may arbitrarily number as 1 and the other, *interview*, as 2. Of the 100 observations $d_1 = 60$ are then equal to 1, and $d_2 = 40$ are equal to 2. The midranks of the 60 1's are equal to

$$\frac{1+\cdots+60}{60} = \frac{61}{2}$$

and those of the 40 2's are equal to

$$\frac{61+\cdots+100}{40} = 60 + \frac{1+\cdots+40}{40} = \frac{161}{2}$$

and the rank sums R_1^*, R_2^*, and R_3^* are

$$R_1^* = 25(\tfrac{61}{2})+16(\tfrac{161}{2}) = 2050.5$$
$$R_2^* = 15(\tfrac{61}{2})+6(\tfrac{161}{2}) = 940.5$$
$$R_3^* = 20(\tfrac{61}{2})+18(\tfrac{161}{2}) = 2059$$

The Kruskal-Wallis statistic defined by (5.11) then has the value 2.00.

Before computing the significance probability of this value, it is necessary to note a difference between the present assumptions and those underlying the original problem. It was assumed (under H) that of the 100 subjects, 41 would have expressed a preference for candidate A, 21 for candidate B, and 38 for candidate C, regardless of which method of polling had been applied to each; also, that of these 100 subjects 60 were assigned at random to receive the questionnaire, with the remaining 40 being interviewed. On the other hand, the present model presupposes the situation to be just the reverse: of the 100 subjects, 60 are marked "Questionnaire" and 40 "Interview"; and the 100 are assigned at random, 41 to A, 21 to B, and 38 to C. To indicate this difference in assumptions we shall denote the statistic (5.11) in the present context by $\bar{K}^*$ rather than by K^*.

As in Table 5.1, let us denote the observations in the first row (labeled "Questionnaire") by A_1, A_2, A_3 and those in the second row (labeled "Interview") by B_1, B_2, B_3. The basic fact, relating the data obtained from the two sets of assumptions, is that the joint null distribution of $A_1, A_2, A_3; B_1, B_2, B_3$ is the same for both. This fact, which will be proved in the next section, implies in particular that the statistic (5.11) has the same null distribution under both sets of assumptions. The significance probability of $\bar{K}^*$ can therefore be computed from the exact distribution or the χ^2-approximation discussed in Sec. 2. In the present example, $s = 3$ and $\bar{K}^* = 2.00$. Since $s-1 = 2$, it is seen from the χ^2-approximation[1] and Table J that the significance probability exceeds .1, so that the result is not significant at the usual levels.

Consider now the general $2 \times t$ table obtained from Table 5.1 by labeling the two treatments A and B and by letting the number of responses be t rather than 5. To find an expression for $\bar{K}^*$ in terms of the entries of the table, let us arbitrarily assign to the "treatments" A and B the scores 1 and 2, respectively [the expression for $\bar{K}^*$ remains unchanged if these values are interchanged or replaced by others; see Prob. 44(ii)].

Since the total number of subjects receiving treatment A is m, the midrank of their scores is

$$\frac{1+\cdots+m}{m} = \frac{m+1}{2}$$

and the midrank of the n 2's receiving treatment B is

$$\frac{(m+1)+\cdots+(m+n)}{n} = m+\tfrac{1}{2}(n+1)$$

In the ith category, there are A_i 1's and B_i 2's, so that the sum of the corresponding midranks is

$$R_i^* = A_i\left(\frac{m+1}{2}\right) + B_i\left(m + \frac{n+1}{2}\right) \tag{5.15}$$

Using the fact that $B_i = d_i - A_i$, it is seen after some calculation [Prob. 44(i)] that

$$\frac{12}{N(N+1)}\sum\frac{R_i^{*2}}{d_i} - 3(N+1) = \frac{3N}{N+1}\left(\sum\frac{A_i^2}{d_i} - \frac{m^2}{N}\right) \tag{5.16}$$

The denominator of $\bar{K}^*$ is

$$1 - \frac{m^3 - m + n^3 - n}{N^3 - N} = \frac{3mn}{N^2-1} \tag{5.17}$$

[1] The accuracy of the χ^2-approximation in a $2 \times t$ contingency table is studied by Lewontin and Felsenstein (1965).

and this finally gives

$$\bar{K}^* = \frac{N(N-1)}{mn}\left(\sum \frac{A_i^2}{d_i} - \frac{m^2}{N}\right) \tag{5.18}$$

The more usual textbook formula replaces the factor $N-1$ by N; in most cases, the difference will be negligible.

There is a slight difference in the properties of the categories required by the one-sided and two-sided tests of Example 4, and the omnibus test of Example 5. In the first case, the categories must be ordered; they and the resulting data may then be described as *ordinal.* Such an order, although it may be present, is not needed in the second case. If there is no order relation between the categories, they and the resulting data are described as *nominal* since they can then be identified only by the name of the category, not by a position such as the smallest, the second smallest, and so forth.

In deciding between an ordered (one- or two-sided) test and an omnibus test, the main consideration is not whether or not the categories are ordered, but the purpose of the inference. If we wish to know whether one treatment has pushed the data in a definite direction (or its opposite), an ordered test is appropriate; if we wish to know only whether or not there is some difference between the treatments, an omnibus test should be applied.

The distinction between an ordered two-sided and an omnibus test disappears in the simplest case, that of a 2×2 table. There are then only two possible alternatives to the hypothesis of no treatment difference: Treatment 2 is either more favorable to category 2 (and hence less favorable to category 1) than treatment 1, or vice versa. Thus, even the most general (omnibus) alternatives are ordered. In terms of the tabular representation, the first alternative means that the responses under treatment 2 tend to be more to the right than those under treatment 1. The one-sided test against this one-sided alternative rejects the null hypothesis when W_{XY}^* is too large, where W_{XY}^* is given by the first two terms of (5.13). Similarly, the two-sided test rejects when $|W_{XY}^* - \frac{1}{2}mn|$ is too large.

In the present context, these tests can be expressed in a simpler form. Since the four entries of Table 5.2 satisfy the relations

$$A_1 + A_2 = m, \qquad B_1 + B_2 = n, \qquad A_1 + B_1 = d_1, \qquad A_2 + B_2 = d_2$$

only one of the four entries is free in the sense that they can all be expressed in

TABLE 5.2 2×2 contingency table

Category	1	2	Total
Treatment			
1	A_1	A_2	m
2	B_1	B_2	n
Total	d_1	d_2	N

terms of any one of them. In terms of A_1, for example, the other entries are given by

(5.19) $$A_2 = m - A_1, \qquad B_1 = d_1 - A_1, \qquad B_2 = d_2 - (m - A_1)$$

Substitution of these values into the statistic (5.13) shows after some simplification that

(5.20) $$W_{XY}^* = \tfrac{1}{2}A_1 N + \tfrac{1}{2}m(d_2 - m)$$

Large or small values of W_{XY}^* therefore correspond, respectively, to large or small values of A_1, which provides an equivalent test statistic.

The null distribution of A_1 is the *hypergeometric* distribution given by the formula

(5.21) $$P_H(A_1 = a_1) = \binom{d_1}{a_1}\binom{d_2}{a_2} \bigg/ \binom{N}{m}$$

where $a_2 = m - a_1$. A proof of this formula is given in Appendix, Sec. 2B. For values of $N \leqslant 25$, the binomial coefficients on the right-hand side of (5.21) can be obtained from Table A.†

EXAMPLE 6. *Anxiety and dependence.* In a psychological investigation, it was desired to compare (for a certain type of person) the wish to be alone or in company when in a state of high and low anxiety. The subjects were divided at random into two groups, and both groups were told that the experiment, to which they had agreed, would involve their receiving some electric shocks. These were ominously described to group 1 as intense and quite painful; group 2 was reassured that the shocks would be mild and painless. Both groups were then told that there would be a 10-min wait before the experiment could be started, and each subject was given the choice of waiting alone or together with some of the other subjects. This, in fact, completed the experiment, which gave the following results:[1]

Number Choosing

	Together	Don't Care and Alone	Total
High anxiety	12	5	$m = 17$
Low anxiety	4	9	$n = 13$
Total	$d_1 = 16$	$d_2 = 14$	$N = 30$

Let us consider first whether a one- or two-sided test is appropriate here. If the study was intended to test the conjecture that high anxiety tends to increase the need to be together with others who share our problems, the hypothesis of no difference in the effect of high and

† Tables from which critical values of the test can be obtained for $3 \leqslant d_1 \leqslant 30$, $d_2 \leqslant d_1$ for $\alpha = .005, .01, .025, .05$ and for $d_2 \leqslant d_1 \leqslant 40$ for $\alpha = .01, .05$ are given in *Tables for Testing Significance in a 2 × 2 Contingency Table*, Cambridge University Press, 1963, compiled by Finney, Latscha, Bennett, and Hsu with an introduction by Pearson.

[1] Part of more extensive data of Schachter, *The Psychology of Affiliation*, Stanford University Press, Stanford, Calif. 1959; taken from Tables 11 and 12.

low anxiety would be rejected if A_1 is sufficiently large. On the other hand, if at the beginning of the study it was felt that the results could go in either direction, a two-sided test would be appropriate, and the hypothesis would be rejected if A_1 were either too large or too small. It follows from (5.21) that

$$P_H(A_1 \geqslant 12) = \frac{\binom{16}{12}\binom{14}{5}+\binom{16}{13}\binom{14}{4}+\binom{16}{14}\binom{14}{3}+\binom{16}{15}\binom{14}{2}+\binom{16}{16}\binom{14}{1}}{\binom{30}{17}}$$

The numerator can be computed from Table A, and turns out to be equal to 4,249,350 [Prob. 15(i)]. The denominator $\binom{30}{17}$ is beyond the range of the table, but can be found in the more extensive table mentioned in Chap. 1, Sec. 1 or from Prob. 15(ii) to be 119,759,850. For the one-sided case, the desired significance probability is therefore $\hat{\alpha} = .035$. For the two-sided case, it is approximately twice this value.

From the limit theorem stated in Chap. 1, Sec. 4, it follows that the limiting null distribution of W_s^*, and hence of W_{XY}^* and of A_1, is normal, provided both m and n tend to infinity and that d_1/N and d_2/N are both bounded away from 1 or equivalently (since $d_1+d_2 = N$) from 0. The normal approximation may therefore be expected to work well if m and n are large and neither d_1/N nor d_2/N are too small. The normal approximation requires the expectation and variance of A_1, which are given by

$$E_H(A_1) = \frac{md_1}{N} \tag{5.22}$$

and

$$\text{Var}_H(A_1) = \frac{mnd_1 d_2}{N^2(N-1)} \tag{5.23}$$

These formulas are proved as (A.43) and (A.44) of the Appendix. Alternatively they can be obtained from (5.20) and formulas (1.39) and (1.40) (Prob. 45).

Consider now the one-sided test that rejects for large values of A_1 and hence when

$$\frac{A_1 - E_H(A_1)}{\sqrt{\text{Var}_H(A_1)}} \tag{5.24}$$

is sufficiently large. The significance probability can then be approximated by the area under the normal curve to the right of

$$\frac{N[a_1-(md_1/N)-\frac{1}{2}]}{\sqrt{mnd_1 d_2/(N-1)}} = \frac{Na_1 - md_1 - \frac{1}{2}N}{\sqrt{mnd_1 d_2/(N-1)}} \tag{5.25}$$

where a_1 is the observed value of A_1 and the terms $\frac{1}{2}$ and $\frac{1}{2}N$ in the numerators represent the continuity correction. Using the relations $m = a_1 + a_2$, $d_1 = a_1 + b_1$, $N = a_1 + a_2 + b_1 + b_2$, it is seen that

$$Na_1 - md_1 = a_1 b_2 - a_2 b_1 \tag{5.26}$$

so that finally the approximate significance probability is obtained as the area under the normal curve to the right of

$$\frac{a_1 b_2 - a_2 b_1 - \frac{1}{2}N}{\sqrt{mnd_1\, d_2/(N-1)}} \tag{5.27}$$

Let us illustrate this procedure on the data of Example 6. Here $a_1 = 12$, $a_2 = 5$, $b_1 = 4$, $b_2 = 9$, so that the value of (5.27) is $73/\sqrt{1707} = 1.77$. This gives the approximation $1 - \Phi(1.77) = .038$ to the significance probability .035 computed earlier.

Consider next the one-sided test that rejects for small values of A_1. An argument completely analogous to that leading to (5.27) shows that the significance probability can be approximated by the area under the normal curve to the left of

$$\frac{a_1 b_2 - a_2 b_1 + \frac{1}{2}N}{\sqrt{mnd_1\, d_2/(N-1)}} \tag{5.28}$$

As pointed out above, in the case of a 2×2 table the two-sided test coincides with the omnibus test. If the normal approximation is used, it is natural—because of the symmetry of the normal curve—to reject in this case when (5.24) is too large in absolute value, or equivalently when

$$\frac{|A_1 B_2 - A_2 B_1|}{\sqrt{mnd_1\, d_2/(N-1)}} \tag{5.29}$$

is too large. The significance probability of the two-sided test is the probability that $A_1 B_2 - A_2 B_1$ is either smaller than $-|a_1 b_2 - a_2 b_1|$ or larger than $|a_1 b_2 - a_2 b_1|$. Using the normal approximation with continuity correction, this probability is approximated by the area under the normal curve to the left of

$$(-|a_1 b_2 - a_2 b_1| + \tfrac{1}{2}N) : \sqrt{\frac{mnd_1 d_2}{N-1}}$$

plus the area to the right of

$$(|a_1 b_2 - a_2 b_1| - \tfrac{1}{2}N) : \sqrt{\frac{mnd_1\, d_2}{N-1}}$$

and therefore by

$$2\left\{1 - \Phi\left[\frac{|a_1 b_2 - a_2 b_1| - \frac{1}{2}N}{\sqrt{mnd_1\, d_2/(N-1)}}\right]\right\} \tag{5.30}$$

This approximation can be expressed in a slightly different way, which is mentioned here only because it is the form frequently found in the literature. For any constant c, the area under the normal curve to the left of $-c$ plus the area to the right of c is equal to $P(|X| \geqslant c)$, where X is a random variable with a standard normal distribution. This probability, in turn, is equal to $P(X^2 \geqslant c^2)$. It follows from Appendix, Sec. 2F that the variable X^2 has a χ^2-distribution with 1 degree of freedom. Hence the significance probability in the two-sided case can also be approximated by the area under the χ^2-curve with 1 degree of freedom, to the right of

$$[|a_1 b_2 - a_2 b_1| - \tfrac{1}{2}N]^2 : [mnd_1 d_2/(N-1)] \tag{5.31}$$

This formula is frequently given without the continuity correction $\frac{1}{2}N$ in the numerator and with the $(N-1)$ in the denominator replaced by N.

The test statistic $\bar{K}^*$ of Example 5 was developed for the comparison of two treatments when the response assigns each subject to one of t categories. If instead there are t treatments and the response is dichotomous (i.e., when it assigns each subject to one of two categories), the same formal expression (5.18) now represents the statistic K^* of the Kruskal-Wallis test. Here the distinction between treatments and categories is meant to indicate that the treatments can be assigned to the subjects at random, but that the category to which a subject belongs is not under the control of the experimenter.

EXAMPLE 7. *Psychological influence of color.* In a study of the psychological effect of different colors, the members of an association were sent cards concerning their membership. Of the 572 members, 147 received white cards, 144 yellow cards, 141 blue cards, and 140 cards of cherry color. Suppose that the assignments were made at random, and that the response to the mailing was as follows:[1]

Response	Yellow	Blue	White	Cherry	Total
Returned	73	65	60	54	252
Not returned	71	76	87	86	320
Total	144	141	147	140	572

Here the four colors represent the treatments, and the two categories are *returned* and *not returned*. The hypothesis of no difference in the effect of the four colors can be tested by means of the Kruskal-Wallis statistic. Using formula (5.18), we find $K^* = 5.1$ and since $s = 4$, the significance probability is seen from Table J to exceed .1.

Under somewhat different circumstances, a one-sided test might have been appropriate; namely, if the experiment had been performed to test a predetermined order (suggested either by some theoretical considerations or by the results of some earlier experiment), say, that yellow may be expected to produce a better return rate than white, white a better rate than blue, and blue a better rate than cherry.

[1] Data from Edwards, *Statistical Methods for the Behavioral Sciences*, Rinehart, New York, 1960, quoting Dunlap (1950), "The Effect of Color in Direct Mail Advertising," *J. Appl. Psych.* **34**:280–281.

4. POPULATION MODELS

The tests of the preceding sections were derived for the randomization model in which the treatments are assigned at random to a given set of N subjects. As was the case in the comparison of two treatments, this framework is somewhat confining because it restricts the inference to the particular N subjects in the study. If the findings of the study are to be applicable to a larger population of potential users of the treatments, a relationship must exist between the subjects of the study and this population. As was discussed in Chap. 2, Sec. 1, a simple relationship can be established by drawing the N subjects at random from the population in question.

Suppose that $N = n_1 + \cdots + n_s$ subjects are drawn from the population by means of simple random sampling and that n_1 of these receive treatment 1, n_2 treatment 2, and so forth. Let the response of a randomly selected subject from the population be denoted by X_i when it receives treatment i. Then X_i is a random variable with distribution, say,

$$F_i(x) = P(X_i \leqslant x) \tag{5.32}$$

and the hypothesis of no difference between the treatments states that the s distributions $F_1, \ldots, F_s$ are the same, that is, that $F_1 = \cdots = F_s$. If the responses of the n_i subjects receiving the ith treatment are denoted by $X_{i1}, \ldots, X_{in_i}$, it follows from the assumption of simple random sampling that each of these n_i variables is distributed according to F_i.

To the assumption of random sampling let us now add the further assumption that the population from which the samples are drawn is large enough so that the dependence of the variables X_{ij} can be neglected and all $N = n_1 + \cdots + n_s$ variables can be treated as independent. These two assumptions together lead to the *population model* according to which $X_{11}, \ldots, X_{1n_1}; X_{21}, \ldots, X_{2n_2}; \ldots; X_{s1}, \ldots, X_{sn_s}$ are independent random variables, the variables in the first group distributed according to F_1, those in the second group according to F_2, and so on. The hypothesis to be tested is the hypothesis of no difference in the effect of the treatments,

$$H: F_1 = F_2 = \cdots = F_s \tag{5.33}$$

The problem of testing H in this population model is frequently called the *s-sample problem* because the observations constitute samples of sizes $n_1, \ldots, n_s$ from the populations characterized by the distributions $F_1, \ldots, F_s$, respectively.

To determine the null distribution of the ranks in this model, let us now introduce the further assumption that the common distribution F of the variables X_{ij} is continuous so that the probability of there not being any ties among the observations is 1. (For a proof of this implication, see Appendix, Probs. 9 and 10 and Theorem 1.) In this case the null distribution of the ranks continues to be given by (5.4). The proof of this fact is completely analogous to the proof of Theorem 1 in Chap. 2, Sec. 1 (Prob. 46). It follows that all results derived from (5.4) for the randomization model continue to hold in the population model. In particular, the null distribution of the Kruskal-Wallis statistic K continues to be the distribution of that statistic discussed in Sec. 2.

Suppose next that F is not continuous, and that ties will therefore occur with positive probability. Then the null distribution of the midranks depends on the common unknown distribution F of the variables X_{ij} (Prob. 47). However, as in the two-sample case (Chap. 2, Sec. 1), the conditional distribution of the midranks given the configuration of the ties, that is, the numbers $d_1, d_2, \ldots, d_e$ of observations tied at the smallest value, the next smallest, and so forth, is independent of F. It is in fact the same as the distribution of the midranks in the randomization model discussed in Sec. 2.

To perform the Kruskal-Wallis test in the population model in the presence of ties, it is therefore only necessary to determine the values of e and $d_1, \ldots, d_e$ and then to carry out the K^*-test for the randomization model corresponding to these values. The significance level is then a conditional probability, valid only for the particular values of e and $d_1, \ldots, d_e$ computed from the data. If all these conditional levels were exactly equal to α, then the unconditional level of the test would also be α. This is, however, impossible to achieve (without randomization) because of the discrete nature of the test statistic. Instead, the conditional levels will be chosen as close to α as possible; as a result, hopefully, the unconditional level will also be close to α. (It can be guaranteed only to lie between the smallest and the largest of the conditional levels.)

The s-sample problem described at the beginning of this section arises not only in the comparison of s treatments but also when the conditions being compared are attributes inseparably attached to the subjects rather than being treatments that can be assigned to the subjects at will. We may, for example, wish to know whether members of different religious denominations are equally regular in church attendance, whether there are personality differences between medical students planning to specialize in different areas of medicine, or whether different weather conditions lead to different crime rates. Here the conditions—the religious affiliation, the student's planned area of specialization, the weather conditions (hot and oppressive, sunny but brisk, cool and overcast, etc.) on a given day—are integral attributes of the subject and therefore cannot be assigned at random.

Suppose that we are interested in comparing these s attributes or conditions for some population of subjects: the medical students of some particular state or country, the days of a given year or season, and so on. They divide the members of this population into s categories or populations. The equality of the distributions of responses for these populations can be tested by drawing a random sample for each. If the sample sizes are $n_1, \ldots, n_s$ and if the n_i responses of the sample from the

ith population are $X_{i1}, \ldots, X_{in_i}$, the problem reduces to the s-sample problem described above. The distributions $F_1, \ldots, F_s$ of the s-sample model in the present context represent the distributions of the responses over the s populations being compared, and the hypothesis of no difference in these response distributions is given by (5.33).

As was already pointed out in Chap. 2, Sec. 1, there is an important difference between the comparison of treatments that can be assigned to the subjects at will and attributes that are inseparably attached to the subjects. In the former case, a significant result indicates a differential *effect* of the treatments; in the latter case, it can only indicate an *association* between attribute and response. Correspondingly, the hypothesis being tested is that of the absence, respectively, of treatment differences or of an association.

EXAMPLE 8. *IQ scores at four universities.* The following table gives the IQ scores of 100 stage I students at each of four New Zealand universities: Auckland, Wellington, Canterbury, and Otago.[1]

IQ	90–94	95–99	100–104	105–109	110–114	115–119	120–124	125–129	130–134	135–139	140–	Total
A	1	2	9	13	16	16	14	13	9	5	2	100
W	1	2	9	9	12	15	18	14	9	6	5	100
C	3	5	7	13	15	14	12	12	9	6	4	100
O	2	3	7	13	17	15	11	14	8	5	5	100
Total	7	12	32	48	60	60	55	53	35	22	16	400

Although they were not obtained by random sampling, but are the scores of students for whom they were available, let us suppose that the four sets of responses constitute random samples from the student populations of the four universities.

Ranking all $N = 400$ scores, the midranks of the 11 columns are found to be, respectively, 4, 13.5, 35.5, 75.5, 129.5, 189.5, 247, 301, 345, 373.5, 392.5. The sums of the midranks for the four universities are

$$R_1^* = 19{,}564.5 \qquad R_2^* = 21{,}395 \qquad R_3^* = 19{,}397 \qquad R_4^* = 19{,}843.5$$

and from formula (5.11) with $n_1 = n_2 = n_3 = n_4 = 100$ and $d_1 = 7, d_2 = 12, \ldots, d_{11} = 16$ given by the last row of the table, it is found that $K^* = 1.909$. Referring this value to the χ^2-distribution with $s-1 = 3$ degrees of freedom, we find the (approximate) significance probability to be $\alpha^* = .592$. There is therefore no indication of a significant difference in the IQ distributions for the four universities.

Instead of drawing a sample from each of the s populations being compared, it is sometimes considerably easier to draw a sample from the combined population.

[1] Data from Parkyn, *Success and Failure at the University*, vol. 1, New Zealand Council for Educational Research, Wellington, 1959, Table 2.10.

This would not be the case in the example above, but might be so, for example, if the samples were to be taken of medical students planning to work in different specialties. Suppose that for the sake of convenience a sample of size N is drawn at random from the combined population. The numbers $n_1, \ldots, n_s$ of subjects in this sample belonging to the s subpopulations will then be random variables. However, conditionally, given the values of $n_1, \ldots, n_s$, the distribution of the responses X_{ij} will be exactly as if samples of these fixed sizes had been drawn from the s subpopulations. The test discussed above for the comparison of s populations is therefore applicable without any change, if it is interpreted as a conditional test.

It follows from the above discussion that the null distribution (5.4) of Sec. 1 is applicable to four different comparison problems. These can be summarized in four models that are the obvious extensions of Models 1 to 4 described at the end of Chap. 2, Sec. 1 for the comparison of two treatments or populations. Various aspects of these models, and also of a measurement model, are discussed there, and this discussion carries over to the extensions.

In the presence of ties, the null distribution of the midranks for the case of s samples under Model 1 is given in Sec. 2. The same distribution obtains in Models 2 and 3 as the conditional distribution of the midranks given the configuration of ties, and in Model 4 given $n_1, \ldots, n_s$ and the configuration of ties.

An explicit expression is available for this distribution in the particular case of an $s \times 2$ or $2 \times t$ contingency table.[1] As in the general case, this distribution applies both directly in the randomization Model 1 and as a conditional distribution in the

TABLE 5.3

(*a*) $s \times 2$ table

Treatment	Response A	B	Total
1	A_1	B_1	n_1
2	A_2	B_2	n_2
...	...	...	...
s	A_s	B_s	n_s
Total	d_A	d_B	N

(*b*) $2 \times t$ table

	1	2	...	t	Total
A	A_1	A_2	...	A_t	m
B	B_1	B_2	...	B_t	n
Total	d_1	d_2	...	d_t	N

[1] For extensions to general $s \times t$ contingency tables, see Chap. 7, Sec. 4.

population Models 2 to 4. The data can be represented as in Tables 5.3 (*a*) and (*b*) and can arise as special cases of Models 1 to 4, with the first three models each occurring in two different types of situations, corresponding to parts (*a*) and (*b*), respectively. The resulting different distributions may be summarized as follows.

Model 1. (*a*) In a comparison of s treatments, the N available subjects are assigned at random, n_1 to treatment 1, n_2 to treatment 2, ..., n_s to treatment s. There are only two possible responses A and B, and it is observed how many subjects in each treatment group give response A and how many B (or how many fall into the categories A and B). Then the joint null distribution of the A's and B's is the multiple hypergeometric distribution

$$(5.34)\qquad P_H(A_1 = a_1, \ldots, A_s = a_s; B_1 = b_1, \ldots, B_s = b_s) = \binom{d_A}{a_1, \ldots, a_s}\binom{d_B}{b_1, \ldots, b_s}\Big/\binom{N}{n_1, \ldots, n_s}$$

[For a proof see Eq. (A.260) in the Appendix, Sec. 7B.]

(*b*) In a comparison of two treatments A and B, the N available subjects are assigned at random, m to A and n to B, and it is observed how many of the subjects receiving each treatment give response (or fall into category) $1, \ldots, t$. In this case the natural expression for the null distribution of the A's and B's is

$$(5.35)\qquad P_H(A_1 = a_1, \ldots, A_t = a_t; B_1 = b_1, \ldots, B_t = b_t) = \binom{d_1}{a_1}\cdots\binom{d_t}{a_t}\Big/\binom{N}{m}$$

[For a proof see Eq. (A.259) in the Appendix, Sec. 7B.]

In the course of Example 5 (in Sec. 3), it was asserted that this distribution is the same as that given by (5.34) when

$$(5.36)\qquad t = s, \qquad m = d_A, \qquad n = d_B, \qquad d_1 = n_1, \ldots, d_s = n_s$$

This is easily seen by expressing the coefficients on the right-hand sides of (5.34) and (5.35) in terms of factorials. [See Appendix, Sec. 7B, after Eq. (A.259).]

Models 1(*a*) and (*b*) are randomization models. In the corresponding population models to be considered below some or all of the totals $m, n; d_1, \ldots, d_s$; etc., all of which are given constants in Model 1, will instead be random variables. In the population models it will turn out that the joint distribution of the A's and B's continues to be the distribution (5.34) or (5.35) provided this distribution is interpreted as being conditional on those totals that are random.

Model 2. (*a*) Samples of size $n_1, n_2, \ldots, n_s$ $(n_1 + n_2 + \cdots + n_s = N)$ are drawn at random from the population in question and are assigned to treatments $1, 2, \ldots, s$, respectively, and it is observed how many subjects in each treatment group give response A and how many B.

The distribution F_i [defined by (5.32)] of the response under the ith treatment in the present case specifies the probabilities p_i and q_i that a randomly selected subject from the population, when receiving the ith treatment, would give response A or B, respectively. If the population is sufficiently large so that the N responses can be considered as independent, the responses of the n_i subjects receiving treatment i constitute n_i binomial trials with probability p_i of success (response A) and q_i of failure (response B). The distribution of A_i is thus binomial,

$$(5.37) \qquad P(A_i = a_i) = \binom{n_i}{a_i} p_i^{a_i} q_i^{b_i}$$

The hypothesis being tested states that $p_1 = \cdots = p_s$. If this common probability is denoted by p, the joint null distribution of $A_1, \ldots, A_s$ and hence of the A's and the B's is given by

$$(5.38) \qquad P(A_1 = a_1, \ldots, A_s = a_s) = \binom{n_1}{a_1} \cdots \binom{n_s}{a_s} p^{d_A} q^{d_B}$$

where

$$(5.39) \qquad d_A = a_1 + \cdots + a_s \qquad \text{and} \qquad d_B = b_1 + \cdots + b_s$$

As was mentioned in Sec. 3 and as follows from the result for general F_i stated earlier in the present section, the distribution (5.38) becomes independent of p if it is conditioned on d_A and d_B, and this conditional distribution is the same as the joint distribution of A_i in the randomization model, which is given by (5.34). A direct proof is given in Appendix, Theorem 13(i).

(*b*) Samples of size m and n ($m+n = N$) are drawn from the population in question and are assigned to treatments A and B, respectively, and it is observed how many subjects in each of the two treatment groups give responses $1, \ldots, t$.

Let the response distributions under treatments A and B be denoted by F_A and F_B. These distributions specify the probabilities, say $\pi_1, \ldots, \pi_t$, that a subject drawn at random from the population and receiving treatment A will give response $1, \ldots, t$, and the probabilities $\pi'_1, \ldots, \pi'_t$ of these responses for a randomly drawn subject receiving treatment B. If the population is sufficiently large that the dependence of the subjects in the sample can be neglected, the m subjects receiving treatment A constitute m multinomial trials, and the joint distribution of $A_1, \ldots, A_t$ is given by the *multinomial distribution*

$$(5.40) \qquad P(A_1 = a_1, \ldots, A_t = a_t) = \binom{m}{a_1, \ldots, a_t} \pi_1^{a_1} \ldots \pi_t^{a_t}$$

The joint distribution of the B's is given by the corresponding formula with a_i replaced by b_i, π_i by π'_i, and m by n. Since the two samples are independent, the joint distribution of the A's and B's is the product of these two distributions.

The hypothesis being tested states that $\pi_1' = \pi_1, \ldots, \pi_t' = \pi_t$, so that the joint null distribution of the A's and B's is

$$(5.41) \qquad P_H(A_1 = a_1, \ldots, A_t = a_t; B_1 = b_1, \ldots, B_t = b_t)$$
$$= \binom{m}{a_1, \ldots, a_t}\binom{n}{b_1, \ldots, b_t}\pi_1^{a_1+b_1} \cdots \pi_t^{a_t+b_t}$$

Again, the distribution becomes independent of the unknown probabilities $\pi_1, \ldots, \pi_t$ through suitable conditioning, this time on $d_1, \ldots, d_t$. In Appendix, Theorem 13(i) the conditional distribution of $A_1, \ldots, A_t; B_1, \ldots, B_t$ given $d_1, \ldots, d_t$ is in fact shown to be the distribution (5.35), or equivalently, (5.34) with the notational changes (5.36).

Models 1 and 2 were concerned with the comparison of treatments; Models 3 and 4 represent situations in which the conditions being compared are attributes of the subject or categories to which they belong.

Model 3. (*a*) In this model s such categories are being compared and the response is dichotomous. Samples of size $n_1, \ldots, n_s$ are drawn at random from category $1, \ldots, s$ and it is observed how many subjects from each category give response A and how many B. The distribution F_i of the responses in category i specifies the probabilities p_i and q_i that a subject selected at random from the ith category will give response A or B. If the number of subjects in the ith category is sufficiently large so that the n_i responses in the sample can be considered independent, the joint null distribution of $A_1, \ldots, A_s$ is given by (5.38) as before. This shows that (5.34) continues to be the conditional null distribution of the A's given the values of d_A and d_B.

(*b*) Here two categories A and B are being compared and there are t possible responses. Samples of size m and n are drawn at random from A and B, respectively, and the same argument as before shows that the conditional distribution of $A_1, \ldots, A_t; B_1, \ldots, B_t$ given $d_1, \ldots, d_t$ is the distribution (5.35).

Model 4. Here a sample of size N is drawn at random from the whole population, and it is observed how many of the N responses fall into each of the $2t$ cells of Table 5.3(*b*).† While in Model 1 all the totals m, n, and $d_1, \ldots, d_t$ were given constants, and in Models 2 and 3 one of the sets m, n, and $d_1, \ldots, d_t$ was given and the other random, in the present case all the totals are random, with only their sum fixed:

$$(5.42) \qquad m+n = d_1 + \cdots + d_t = N$$

† All except notational distinctions between Tables 5.3(*a*) and (*b*) disappear in the present case. The discussion could equally well be given in terms of Table 5.3(*a*).

If the population is again large enough so that the dependence of the subjects can be neglected, the N responses constitute N multinomial trials, each of which has $2t$ possible outcomes. If the probabilities of these outcomes are denoted by $p_{A1}, \ldots, p_{At}$ for the cells of the first row and $p_{B1}, \ldots, p_{Bt}$ for the cells of the second row of Table 5.3(*b*), the joint distribution of the A's and B's is

$$\binom{N}{a_1, \ldots, a_t; b_1, \ldots, b_t} p_{A1}^{a_1} \ldots p_{At}^{a_t} p_{B1}^{b_1} \ldots p_{Bt}^{b_t} \tag{5.43}$$

The hypothesis being tested states that there is no association between the row and the column category to which a subject belongs. If the marginal row and column probabilities are

$$p_A = p_{A1} + \cdots + p_{At} \qquad p_B = p_{B1} + \cdots + p_{Bt}$$

and

$$\pi_1 = p_{A1} + p_{B1}, \; \ldots, \; \pi_t = p_{At} + p_{Bt}$$

the hypothesis is formalized by the equations

$$p_{A1} = p_A \pi_1, \ldots, p_{At} = p_A \pi_t; \qquad p_{B1} = p_B \pi_1, \ldots, p_{Bt} = p_B \pi_t \tag{5.44}$$

Under H, the distribution (5.43) therefore reduces to

$$P_H(A_1 = a_1, \ldots, A_t = a_t; B_1 = b_1, \ldots, B_t = b_t) \tag{5.45}$$
$$= \binom{N}{a_1, \ldots, a_t; b_1, \ldots, b_t} p_A^{a_1 + \cdots + a_t} p_B^{b_1 + \cdots + b_t} \pi_1^{a_1 + b_1} \ldots \pi_t^{a_t + b_t}$$

This distribution, of course, depends on the unknown probabilities p_A, p_B, $\pi_1, \ldots, \pi_t$. However, the conditional distribution of the A's and B's given all the marginal totals (since all are now random)

$$A_1 + \cdots + A_t = m, \qquad B_1 + \cdots + B_t = n; \qquad A_1 + B_1 = d_1, \ldots, A_t + B_t = d_t \tag{5.46}$$

is again independent of the unknown probabilities and is, in fact, given by (5.35). This is proved in Appendix, Theorem 13(ii).

5. ONE-SIDED PROCEDURES

The Kruskal-Wallis test for comparing s treatments or populations reduces to the two-sided Wilcoxon test when $s = 2$ (Prob. 38). It is then natural to ask what procedures for comparing s treatments generalize the one-sided Wilcoxon test. In the present section, we shall sketch two such generalizations.

5A. Comparing Several Treatments with a Control

Suppose that s treatments are under investigation as possible improvements over a standard treatment (which may consist of doing nothing). These may, for example, be drugs for combating depression, techniques for soothing babies, or diets for reducing weight. The question then arises which of these treatments (if any) really have the desired effect.

A natural procedure here is to apply a separate Wilcoxon test to the comparison of each of the new treatments with the standard or control. For each i, the n_i observations on the ith treatment and the n_0 control observations are combined and ranked, and the ith treatment is "selected" (i.e., judged to be superior to the control) if its rank-sum R_i is sufficiently large, say,

$$R_i \geqslant c_i \tag{5.47}$$

Note that R_i does not have the same meaning here as in Sec. 2. There it denoted the rank-sum of the ith treatment in a simultaneous ranking of all N subjects in the study; in the present case, it is the rank-sum of the ith treatment when only the n_i+n_0 subjects are ranked that are either in the ith treatment group or are controls.

To determine the procedure completely, it is necessary to specify the critical values c_i. If H_i denotes the hypothesis of no effect of the ith treatment (i.e., no difference between the ith treatment and the control), it seems natural to specify the critical values in terms of significance levels α_i through the relations

$$P_{H_i}(R_i \geqslant c_i) = \alpha_i \tag{5.48}$$

Suppose in particular that $n_1 = \cdots = n_s$ (although these need not necessarily $= n_0$) and that also in other respects the s new treatments play a sufficiently symmetric role, so that one would want to choose α_i independent of i, say, $\alpha_i = \alpha$. The c_i then also become independent of i, and procedure (5.47) reduces to

$$R_i \geqslant c \tag{5.49}$$

with c determined by

$$P_{H_i}(R_i \geqslant c) = \alpha \tag{5.50}$$

Under H_i, the statistic R_i has the distribution of the Wilcoxon statistic W_s with group sizes $m = n_0$ and $n = n_i$, so that the critical values c_i or c can be determined from Table B or for large values of n_0 and n_i from the normal approximation, as discussed in Chap. 1, Secs. 2 and 3.

The error probabilities α_i and α refer to the individual comparisons of each treatment with the control. For the procedure (5.49), α is the *error rate per comparison* since it indicates the frequency of false significance statements in a large number of such comparisons for which H_i is true. If the ultimate decisions or recommendations based on the s rejections or acceptances are unrelated, it seems reasonable to determine c from (5.50) and to think of α as one would if only a single test were applied to a given body of data. (A slight complication in this interpretation is introduced by the dependence of the statistics R_i that result from the use of a common set of controls.) It is, however, important to realize a consequence of this point of view. If α is fixed at a conventional level, regardless of the value of s, then for large s there will be substantial chance for some rejections in any given study, even if none of the new treatments has any effect.

EXAMPLE 9. *Some birth conditions and IQ.* To test whether certain conditions are associated with a lowering of IQ, an IQ score is obtained for 24 girls of which 6 each are, respectively, normal, anoxic, premature, and Rh negative, with the following results[1]

(0) Normal:	103	111	136	106	122	114
(1) Anoxic:	119	100	97	89	112	86
(2) Rh negative:	89	132	86	114	114	125
(3) Premature:	92	114	86	119	131	94

As an illustration, let us take $\alpha = .05$. To obtain a comparison of rows (1) and (0), the 12 scores from these two rows are ranked, and the resulting rank-sum for row (1) is

$$1+2+3+4+8+10 = 28$$

A similar ranking of the entries of rows (2) and (0) shows the sum of the midranks of row (2) to be 38, and a ranking of rows (3) and (0) gives a midrank sum of 33.5 to row (3). It is seen from Table B that $P(W_s \leqslant 28) = .0465$, and the probabilities $P(W_s{}^* \leqslant 38)$ and $P(W_s{}^* \leqslant 33.5)$ are substantially larger (of the order of .4 and .2, respectively). The condition of anoxia thus seems barely significant, and the other two seem to have no effect.

As an alternative to the above approach, it is sometimes suggested that the s comparisons should be viewed as a package. In particular, one would then specify the critical value c in (5.49) in terms of the ability of the overall procedure to control false rejections of the hypotheses H_i. One measure of this ability is the probability of at least one false rejection among the s comparisons. From this point of view, the whole experiment is faulted if even a single false rejection occurs. The probability of at least one false rejection indicates the frequency with which an experiment or study is spoiled by a faulty analysis in a long sequence of such studies with the same value of s and such that the number of hypotheses H_i that are true is the same for all experiments in the sequence. This probability is therefore the *error rate per experiment.*

It is intuitively plausible and will be proved below under certain assumptions that the probability of at least one false rejection takes on its maximum value when all the hypotheses H_i are true. Let H denote this overall null hypothesis that none of the treatments has an effect. Under H, the probability of at least one false rejection with procedure (5.49) is

$$\alpha' = P_H(\max R_i \geqslant c) \tag{5.51}$$

where $\max R_i$ denotes the largest of the rank-sums $R_1, \ldots, R_s$. From the present point

[1] Data of Graham presented by Steel (1959).

of view, it appears natural to determine c so that the probability (5.51) has a specified value. Unfortunately, even when $n_1 = \cdots = n_s$, the value of c then depends on the common value of $n_1, \ldots, n_s$, on the size n_0 of the control group, on the number s of new treatments, and on α'. For a fixed value of α', it thus depends on three arguments. Since the required table would be very extensive, critical values have been tabulated only for the case in which all group sizes are equal:

$$n_0 = n_1 = \cdots = n_s \tag{5.52}$$

Some critical values for $s = 2, 3$ and the common group size $n = 3, 4, 5$ are given by Steel (1959), who first proposed the procedure. A more extensive table for $\alpha' = .01$ and $\alpha' = .05$ is provided by Miller (1966, Table VIII). This latter table is actually based on an approximation to the null distribution of $\max R_i$. This approximation, or rather the associated limit theorem, parallels the χ^2-approximation of Sec. 2. For details, the reader should consult the book by Miller (1966, Chap. 4, Sec. 3).

Determination of c by setting the experimentwise error rate (5.51) equal to a conventional level α', regardless of the value of s, has the opposite disadvantage from that found when the error rate per comparison was used for large s. The earlier error rate led to a liberal selection policy, and, as a result, to a substantial probability of following some red herrings when H holds or when at least a large number of the hypotheses H_i are true. Instead, the error rate per experiment α' sets the value of c so high when s is large that the selection procedure becomes very exclusive and provides little chance for an effective new treatment to prove itself unless it is superior to the control by a very wide margin.

As an illustration of this procedure consider once more the data of Example 9. The tables of Steel and Miller show that for $\alpha' = .05$, a rank-sum must be less than or equal to 26 to be significant. At this level of the error rate per experiment, therefore, none of the three conditions would be declared significant.

An idea of the relationship between α and α' may be obtained from the two tables below, which for fixed values of one of the error rates show how the other changes with s and n. Table 5.4(a) shows how α' increases with the number of treatments when α is fixed. (Note that the value of α is given by the first column of the table.) Thus, if $n = 3$ and c is determined from (5.50) with $\alpha = .05$, the value of α' is .119 when three treatments are being compared with the control. Similarly, Table 5.4(b) shows, for a number of values of s and n, how small α has to be if α' is to be equal to .05.

The probability α' arises also in a context different from the one described above. Upon looking over the rank averages (in advance of any formal analysis of the data), one may be struck by a large value for a particular treatment and wish to know whether it is significantly better than the control at a given level α. A significance test such as the Wilcoxon test carried out at level α is not appropriate because one

TABLE 5.4

(a) Values of α' for various s and n ($s = 1$ gives α)

n	s = 1	2	3
3	.05	.088	.119
4	.014	.027	.037
4	.029	.052	.072
4	.057	.100	.144
5	.016	.030	.041

(b) Values of α for various s and n when $\alpha' = .05$

n	s = 1	2	3	4	5	6	7	8
8	.05	.025	.014	.010	.010	.007	.007	.005
10	.05	.026	.018	.012	.009	.009	.007	.007

must take account of the fact that this particular treatment was suggested by the data. Instead, the appropriate level and critical value are just α' and c, given by (5.51). The argument is exactly analogous to that given below in Sec. 6 for the probability α' defined by (5.74).

So far, we have discussed the procedure (5.49) in terms of fixed levels α and α' defined by (5.50) and (5.51). These levels play a role analogous to the significance level α for testing a single hypothesis. As was discussed in Chap. 1, Sec. 2, the significance probability $\hat{\alpha}$ in this case provides more information and insight than a simple decision to accept or reject the hypothesis at the given level. The analogous remark applies to a multiple testing procedure such as (5.49). For such a procedure, it is useful to compute the significance probabilities

$$\hat{\alpha}_i = P_{H_i}(R_i \geqslant r_i) \qquad i = 1, \ldots, s$$

where r_i is the observed value of R_i, as was done in fact in Example 9 above.

Related to the error rates α and α' is a third measure of the ability of procedure (5.47) to control false significance statements. This is the expected number η of false significance statements when in fact all the hypotheses H_i and hence the overall null hypothesis H are true. It follows from the addition law for expectation and more particularly from Eq. (A.4) of the Appendix that

$$\eta = \alpha_1 + \cdots + \alpha_s$$

when the procedure is specified by (5.48), and hence that

$$\eta = s\alpha \tag{5.53}$$

when it is specified by (5.50). Typically, η provides a better intuitive feeling for the significance of a set of comparisons than does α.

The quantity η also bears an interesting relation to α' in that it always satisfies

$$\alpha' \leqslant \eta \tag{5.54}$$

To see this, let S denote the number of hypotheses H_i that are rejected, and let

$$p_k = P_H(S = k) \qquad k = 0, 1, \ldots, s$$

Then clearly

$$\alpha' = p_1 + p_2 + \cdots + p_s$$

while

$$\eta = p_1 + 2p_2 + \cdots + sp_s$$

A comparison of these two expressions shows (5.54) to be true.

If α' is sufficiently small and H is true, it is relatively unlikely that more than one of the hypotheses H_i will be rejected; that is, $p_2, \ldots, p_s$ will be small compared to p_1. One would then expect the upper bound η to be fairly close to α'. To give an idea of the closeness of the bound for small s, Table 5.5 shows the values of η corresponding to the values of α' given in columns 2 and 3 of Table 5.4(a).

TABLE 5.5 Relation of η and α'

	$s = 2$					$s = 3$				
α'	.088	.027	.052	.100	.030	.119	.037	.072	.144	.041
η	.100	.028	.058	.114	.032	.150	.042	.087	.171	.048

Of course, for large s the values can no longer be close as is seen from the fact that η increases indefinitely with s, and α' can never exceed 1.

Additional insight into the procedure (5.47) is obtained by assuming a population model. As in Sec. 4, the responses $X_{i1}, \ldots, X_{in_i}$ of the n_i subjects receiving the ith treatment are then independent random variables with a common distribution F_i. The hypothesis H_i states that there is no difference between the distributions F_i and F_0 corresponding, respectively, to the ith treatment and the control

$$H_i\colon \quad F_i = F_0$$

Similarly, the overall null hypothesis H states that none of the treatments have an effect; that is,

$$H\colon \quad F_0 = F_1 = \cdots = F_s$$

We shall now make the further assumption of additivity introduced in Chap. 2, which supposes that each treatment adds a constant amount to the control response of a subject, independent of the value of that response. If this amount is denoted by Δ_i for the ith treatment, so that Δ_i measures the effect of the ith treatment, it follows in generalization of Eq. (2.7) that

$$F_i(x) = F_0(x - \Delta_i) \tag{5.55}$$

In terms of this *shift model*, the aim of the selection procedure is to select those of the new treatments (if any) for which $\Delta_i > 0$. The procedure (5.49) attempts this task by performing a set of tests of the hypotheses $H_i'\colon \Delta_i \leqslant 0$ against the alternatives $\Delta_i > 0$. We can now make precise a result that was indicated earlier.

Theorem 1. For the procedure (5.47), [and hence for (5.49)], the maximum probability of falsely rejecting at least one of the hypotheses H_i' occurs when $\Delta_1 = \cdots = \Delta_s = 0$.

Proof. Of the Δ's, some will typically be less than or equal to zero and some greater than zero. Suppose that

$$\Delta_{i_1}, \ldots, \Delta_{i_k} \leqslant 0 < \Delta_{j_1}, \ldots, \Delta_{j_l} \tag{5.56}$$

where $k+l = s$, and for simplicity consider the case (5.49). A false rejection then occurs if one of the rank-sums $R_{i_1}, \ldots, R_{i_k}$ exceeds c, so that the probability of at least one false rejection is

$$P[\max(R_{i_1}, \ldots, R_{i_k}) \geqslant c] \tag{5.57}$$

Now the probability (5.57) increases as the values of any of the parameters $\Delta_{i_1}, \ldots, \Delta_{i_k}$ increase. This is seen by applying the proof of Chap. 2, Theorem 2, with only the obvious modifications. It follows that (5.57) attains its maximum, subject to (5.56), when $\Delta_{i_1} = \cdots = \Delta_{i_k} = 0$. To complete the proof, it is only necessary to notice that (5.57), evaluated at $\Delta_{i_1} = \cdots = \Delta_{i_k} = 0$, increases with k and hence takes on its maximum value when $k = s$ (and hence $l = 0$). The argument for (5.47) is exactly analogous.

For the sake of simplicity, Theorem 1 was stated under the assumption of a shift model. However, the corresponding result is easily seen to be true in the more general case that the distributions F_i are only assumed to be stochastically smaller or larger than F_0. This result is an extension of the statements (i′) to (iii′) at the end of Chap. 2, Sec. 2.

In the shift model (5.55), the set of tests (5.49) can be used to obtain simultaneous lower-confidence bounds for $\Delta_1, \ldots, \Delta_s$ in generalization of the confidence bounds for Δ of Chap. 2, Sec. 6. Similarly, one can obtain upper-confidence bounds and hence simultaneous confidence intervals for the shift parameters $\Delta_1, \ldots, \Delta_s$. Another use of (5.49) is the assessment of the significance of one or more values of Δ_i suggested by the data. These applications are discussed in a similar context on pages 242–244.

5B. Testing Equality against Ordered Alternatives

The one-sided Wilcoxon test provides a comparison of a new with a standard treatment, and this aspect of the test was generalized in the procedures discussed above. Let us now instead focus attention on the fact that in the one-sided test, the hypothesis of equality of two treatments is tested against the alternatives that they are in a definite order, say, that the second treatment is more effective than the first. Generalized to s treatments, this suggests the problem of testing the hypothesis of equality against the alternatives of a definite order of these treatments: that the second is better than the first, the third better than the second, and so on. In testing the effectiveness of a headache remedy, one may, for example, compare a control group with two treatment groups, one of which receives a heavier dose than the other. In a study of the effect of class size on learning, one may wish to test the hypothesis of no effect against the alternatives that the effectiveness decreases as the size of the class increases successively from fewer than 30 students, to 30–60, 60–100, and more than 100. Similarly, one may compare groups with different degrees of psychiatric training in their ability to evaluate a diagnostic test that supposedly can be read fairly mechanically after a few hours of training.

Suppose that N subjects are assigned to the s treatments and their responses ranked, as in Sec. 1. Under the hypothesis H of no difference among the treatments, the null distribution of the ranks is then given by (5.4). Suppose now that the treatments are ordered (of course, before the responses have been observed) in such a way that under the alternatives to H one would expect larger responses under treatment 2 than under treatment 1 (which may be the control), under treatment 3 than under treatment 2, and so on. The Kruskal-Wallis test is then no longer appropriate since it rejects H whenever the rank averages $R_{i.}$ are sufficiently different, regardless of their order, while in the present case only an increasing trend in the sequence $R_{1.}, \ldots, R_{s.}$ supports the alternatives over the hypothesis.

To obtain a test that is suitable against such ordered alternatives, let us denote by $X_{i1}, \ldots, X_{in_i}$ and $X_{j1}, \ldots, X_{jn_j}$ the observations on the ith and jth treatments, and by W_{ij} the Mann-Whitney statistic (defined in Chap. 1, Sec. 2) for these two sets; that is, W_{ij} is the number of pairs (α,β) for which $X_{i\alpha} < X_{j\beta}$. A test statistic proposed by Terpstra (1952) and Jonckheere (1954) is then the sum of the W_{ij}

$$W = \sum_{i<j} W_{ij} \tag{5.58}$$

the hypothesis being rejected for large values of W. A table of the null distribution of W or rather of

$$J = 2W - \sum_{i<j} n_i n_j \tag{5.59}$$

for equal group sizes $n_1 = \cdots = n_s = n$ for several combinations of small values of s and n is given as Table K. Only the probabilities up to .5 are tabled. The remaining values can be obtained from the fact that J is symmetrically distributed about zero (Prob. 49).

For large values of N, the statistic W is approximately normally distributed. To apply this approximation, one requires the expectation and variance of W. It follows from formula (1.29) that the expectation of W_{ij} is $\frac{1}{2}n_i n_j$ and hence (Prob. 50) that

$$E_H(W) = \tfrac{1}{2} \sum_{i<j} n_i n_j = \frac{(N^2 - \Sigma n_i^2)}{4} \tag{5.60}$$

The variance of W is given by

$$(5.61) \qquad \mathrm{Var}_H(W) = \frac{N^2(2N+3) - \Sigma n_i^2(2n_i+3)}{72}$$

(see Appendix, Prob. 5). A limit theorem supporting the normal approximation is given in the Appendix, Prob. 44.

EXAMPLE 10. *Three diets.* In a comparison of three basic diets A, B, and C, 15 rats were assigned at random, 5 to each of the three diets. After 7 weeks, the following growth rates were recorded:[1]

A					B					C				
133	139	149	160	184	111	125	143	148	157	99	114	116	127	146

Suppose that from past experience and for theoretical reasons it is believed that, if there is a differential effect, the growth rate would be highest for A, intermediate for B, and lowest for C. It is then of interest to test the hypothesis of no differences among the three diets against the ordered alternatives that the growth rate decreases from A to C.

To compute W, let us first compare the observations from A and B. It is seen that 133 exceeds 111 and 125, that 139 exceeds 111 and 125, that 149 exceeds 111, 125, 143, 148, and so on. Thus,

$$W_{BA} = 2+2+4+5+5 = 18$$

and similarly

$$W_{CB} = 1+3+4+5+5 = 18 \qquad \text{and} \qquad W_{CA} = 4+4+5+5+5 = 23$$

It is seen that $W = 18+18+23 = 59$ and $J = 118-75 = 43$. Large values of the W's and hence of J are significant, and Table K shows that for $s = 3$ treatments and common group size $n_1 = n_2 = n_3 = 5$, $P(J \geqslant 43) = .0120$.

For comparison, let us compute the same significance probability by means of the normal approximation. We find

$$E(W) = 37.5 \qquad \text{and} \qquad SD(W) = 9.465$$

and hence

$$P(W \geqslant 59) \approx 1 - \Phi\left(\frac{59-37.5}{9.465}\right) = .0116$$

This agrees quite well with the exact value.

It should be emphasized that the use of Jonckheere's test in this example is justified only if the ordered alternatives $A > B > C$ and the decision to use this test were specified before the experiment was performed, or at least before the observations were obtained. If this was not the case, but the particular order was suggested by the data, the hypothesis would also have been rejected had the data been equally significant for some other order such as $A < B < C$ or

[1] The data are taken from Tables 3 and 4 of the study quoted in Example 2. They are the observations on diet groups A_{VIII}, B_{II}, and C_{III}, with one or two randomly selected observations deleted from each group to bring the data within the scope of Table K.

$A > C > B$. (A post hoc theoretical explanation can usually be found for any particular order.) The computed significance probability for the order $A > B > C$ would then be the probability of only part of the rejection rule. The true significance probability would be much larger, and the claimed significance of the result not justified. The issue is exactly the same as that discussed in Chap. 1, Sec. 5, regarding the choice between a one- or two-sided test in the comparison of two treatments.

When no particular order is determined a priori as the only possible one (or the only one of interest), the appropriate test is not Jonckheere's but that of Kruskal and Wallis discussed in Sec. 2. In the present example, the significance probability with this test turns out to be .072 instead of the value .012 found with Jonckheere's test. As expected, the results appear considerably less significant if no prior order is assumed.

In the definition (5.58) of W it was assumed that there are no ties. When some of the observations are tied, the ranks are replaced by the corresponding midranks. Equivalently, W_{ij} can be replaced by W_{ij}^*, the Mann-Whitney statistic in the presence of ties [defined by Eq. (1.37)] for the ith and jth group, so that W_{ij}^* is the number of pairs (α,β) for which $X_{i\alpha} < X_{j\beta}$ plus half the number of pairs (α,β) for which $X_{i\alpha} = X_{j\beta}$. Then the proposed test statistic is

$$W^* = \sum_{i<j} W_{ij}^* \tag{5.62}$$

There are no tables for the null distribution of W^*, but a normal approximation can be applied. For an associated limit theorem as the group sizes $n_i \to \infty$ see D'Abrera (1973). Since it was shown in Chap. 1, Sec. 4, that $E(W_{ij}^*) = E(W_{ij})$, it follows that

$$E(W^*) = E(W) \tag{5.63}$$

The variance of W^* is more complicated. Suppose as in Sec. 2 that the N observations take on e distinct values, and that d_1 of the N observations are equal to the smallest value, d_2 to the next smallest, etc. Then the variance of W^* is given by

$$\begin{aligned}\operatorname{Var}(W^*) = \frac{1}{72}\Bigg[N(N-1)(2N+5) - \sum_{i=1}^{s} n_i(n_i-1)(2n_i+5) - \sum_{j=1}^{e} d_j(d_j-1)(2d_j+5)\Bigg] \\ + \frac{1}{36N(N-1)(N-2)}\Big[\sum n_i(n_i-1)(n_i-2)\Big]\Big[\sum d_j(d_j-1)(d_j-2)\Big] \\ + \frac{1}{8N(N-1)}\Big[\sum n_i(n_i-1)\Big]\Big[\sum d_j(d_j-1)\Big]\end{aligned} \tag{5.64}$$

This formula will not be proved here; a proof can be found in Kendall (1970, p. 72). Although this is not apparent at first sight, (5.64) reduces to (5.61) when there are no ties, as it should (see Prob. 51).

Let us now briefly mention an alternative test for testing H against an upward (or downward) trend in the treatments. If the increasing trend in the alternatives is sufficiently strong, it will typically be reflected in the rank-averages, and one would ideally expect that $R_{1.} \leqslant \cdots \leqslant R_{s.}$. Actually, this order will be violated at least in some places of the sequence. Consider any pair of consecutive treatments, say i and $i+1$, for which this is the case, so that $R_{i+1.} < R_{i.}$. Taking this as an indication of equality of the ith and $(i+1)$st treatments, combine the n_i+n_{i+1} ranks from these two groups into a single group, that is, pretend that they constitute n_i+n_{i+1} observations from a single treatment. With the situation now reduced to $s-1$ groups consisting of $n_1, \ldots, n_{i-1}$, n_i+n_{i+1}, $n_{i+2}, \ldots, n_s$ members, respectively, repeat this amalgamation process by again combining two neighboring groups among the $s-1$, whose average ranks are in the "wrong" order. This process is repeated, reducing the number of groups by one at each step until all the averages of the consolidated groups are in the "right" order. When there are several inversions, it is immaterial in which order they are resolved; the final result will always be the same. [For a proof of this result, see Miles (1959).]

In the final grouping, after the amalgamation process is completed, let the number of groups be t; let $R'_{i.}$ denote the average of the n'_i ranks in the ith amalgamated group, and consider the test statistic

$$K' = \frac{12}{N(N+1)} \sum_{i=1}^{t} n'_i[R'_{i.} - \tfrac{1}{2}(N+1)]^2 \tag{5.65}$$

where large values are significant. This test was proposed for the case of equal group sizes by Chacko (1963) and extended to the general case by Shorack (1967). At present, the test suffers from the disadvantage that no tables of the null distribution exist for small group sizes and even large-sample tables are available only for the case $n_1 = \cdots = n_s$. In this case, the probability that K' exceeds a constant c is approximately equal to

$$P(K' \geqslant c) \approx \sum_{i=2}^{s} p_{i,s} P(\chi^2_{i-1} \geqslant c) \tag{5.66}$$

where χ^2_{i-1} indicates a χ^2 variable with $i-1$ degrees of freedom and the weights $p_{i,s}$ are shown in Table L for $3 \leqslant s \leqslant 10$. The theory behind this approximation will not be given here; it can be found in the papers by Chacko and Shorack, and in the book by Barlow et al. (1972).

EXAMPLE 11. *Effect of added information on the ranking of candidates.* Twelve psychologists were divided at random into three groups of four each, and presented with information about 15 (hypothetical) candidates for graduate work in psychology. The first group was given the verbal and quantitative scores on the candidate's Graduate Record Exam; the second group was furnished in addition with the grade point average, and the status of the candidate's undergraduate school; finally, the third group was also given the number of undergraduate science courses taken by each candidate, and the score of the psychology portion of his Graduate Record Exam. On this basis, each member was asked to rank the 15 candidates.

The corresponding information was then provided for a second set of 15 candidates. These were ranked twice for each of the 12 psychologists: once directly, and once by a prediction formula on the basis of his ranking of the first set of candidates. For each member, a measure of the agreement of these two rankings (the Spearman rank correlation coefficient, which will be discussed in Chap. 7) was then computed with the following results:[1]

Group 1				Group 2				Group 3			
.97	.84	.98	.92	.46	.82	.67	.81	.65	.69	.83	.95

There are reasons for believing that an increase in information, if it has an effect, would tend to decrease the closeness of the agreement. Let us therefore test the hypothesis of no difference among the three groups against the ordered alternatives that the correlation coefficients tend to decrease from the first to the third group. A ranking of all 12 observations gives $R_{1.} = 10$, $R_{2.} = 3.75$, $R_{3.} = 5.75$. Since the predicted order is $R_{1.} \geqslant R_{2.} \geqslant R_{3.}$, the observed rank averages of the second and third groups are in the "wrong" order, and these two groups are therefore amalgamated into a single group of eight observations. In this new grouping, the rank average of the first group continues to be 10, and that of the amalgamated second group is 4.75. Since these two averages are in the "right" order, the amalgamation process is complete, so that $t = 2$; $n'_1 = 4$, $R'_{1.} = 10$; $n'_2 = 8$, $R'_{2.} = 4.75$. The statistic K' has the value

$$\frac{12[4(10-6.5)^2+8(4.75-6.5)^2]}{12.13} = 5.65$$

and the significance probability is approximately

$$P(K' \geqslant 5.65) \approx p_{23}P(\chi_1^2 \geqslant 5.65)+p_{33}P(\chi_2^2 \geqslant 5.65)$$

From Tables J and L we find

$$p_{23} = \tfrac{1}{2}, \qquad p_{33} = \tfrac{1}{6}; \qquad P(\chi_1^2 \geqslant 5.65) = .018, \qquad P(\chi_2^2 \geqslant 5.65) = .059$$

so that the significance probability is approximately

$$.009+.0098 = .019$$

The question naturally arises whether Jonckheere's test or that of Chacko and Shorack is preferable. A tentative answer can be given in terms of the population model that assumes the observations $X_{i1}, \ldots, X_{in_i}$ to be distributed independently according to

$$P(X_{i\alpha} \leqslant x) = F(x-\theta_i) \qquad \text{for } \alpha = 1, \ldots, n_i \tag{5.67}$$

with the θ's satisfying

$$\theta_1 \leqslant \theta_2 \leqslant \cdots \leqslant \theta_s \tag{5.68}$$

Work of Bartholomew (1961) suggests that if the θ's are about equally spaced, Jonckheere's test appears to be more powerful; if the spacings are widely different, the K' test seems to be preferable. [In this connection, see also Tryon and Hettmansperger (1973) and D'Abrera (1973). The D'Abrera paper also proposes a further test which possesses certain optimal properties when the spacings are unknown.]

[1] Part of the data of Einhorn (1971), "Use of Nonlinear, Noncompensatory Models as a Function of Task and Amount of Information," *Org. Behav. Human Perform.* **6**:1–27.

This conclusion is illustrated by Examples 10 and 11. In Example 10, the difference between treatments A and B appears comparable to that between B and C; in Example 11 there seems to be a substantial decrease from group 1 to group 2 but no appreciable further decrease from group 2 to group 3. The significance probabilities for these examples under the two tests in question are shown in Table 5.6; the third column shows the corresponding significance probabilities under the Kruskal-Wallis test.

TABLE 5.6 Significance probabilities for three s-sample tests

	Jonckheere's Test	Chacko-Shorack Test	Kruskal-Wallis Test
Example 10	.012	.025	.072
Example 11	.063	.019	.054

The remark made earlier regarding the choice between the tests of Jonckheere and Kruskal-Wallis apply equally to the problem of choosing between all three tests: the choice must be made before the results of the study become known. Unfortunately, even when a particular order of the treatments is indicated, a prior knowledge of the spacing of the θ's often will not be available. In this case, the K' test should be used in preference to Jonckheere's test. It is hoped that tables allowing the use of this test for unequal group sizes will become available before too long.

6. SELECTION AND RANKING PROCEDURES

In the preceding sections, we have discussed generalizations of the one- and two-sided Wilcoxon test to the case of more than two treatments. We shall now give some generalizations of the Wilcoxon procedure for determining the better of two treatments, described in Chap. 1, Sec. 5. [A different approach has been developed by Gabriel and Sen (1968) and Peritz (1971) and for contingency tables by Gabriel (1966).]

6A. Ranking Several Treatments

The procedure (1.54) can be viewed as a ranking procedure. It attempts to rank two treatments A and B, permitting also the possibility of remaining in doubt if the order cannot be established with enough conviction. Let us now consider how to order s treatments, again with the possibility of remaining in doubt concerning the relative order of some of the treatments. Such an ordering can be obtained by reaching a decision for each of the $\binom{s}{2}$ pairs of treatments, for example, according to (1.54). If $R_{i.}$ denotes the average rank of the ith treatment, the jth treatment is declared superior to the ith if $R_{j.}-R_{i.}$ exceeds a critical value c_{ij},

$$R_{j.}-R_{i.} \geqslant c_{ij} \tag{5.69}$$

where $c_{ji} = c_{ij}$. The order of the two treatments is left in doubt if $|R_{j.} - R_{i.}| < c_{ij}$.

The results of such a set of comparisons can be presented visually by marking the numbers of the treatments on a horizontal line. The number of a treatment is displayed to the right of another to indicate that it has been found superior; a set of treatments is underlined to show that it contains no significant pair. Suppose for example that $s = 4$. Then $3\ \underline{2\ 1\ 4}$ indicates that treatment 3 has been found to be inferior to 2, 1, and 4 but that none of the pairs (2, 1), (2, 4), and (1, 4) have been found to differ significantly. Similarly,

$$3\ \underline{2\ 1}\ 4 \quad (\text{with } \underline{1\ 4} \text{ also underlined})$$

shows that treatment 3 has been found inferior to 2, 1, and 4, and 2 to 4 but that the pairs (2, 1) and (1, 4) have not been found to differ significantly.

The overall comparison will be quite clear-cut if the treatments are divided into a number of groups without overlap as in the cases $3\ \underline{2\ 1}\ 4$, $3\ 2\ \underline{1\ 4}$, or $\underline{3\ 2\ 1}\ 4$. Alternatively, there may be considerable overlap between the groups, in which case no simple summary statement is possible. The following three cases with $s = 6$ illustrate some of the possibilities, where for the sake of simplicity the treatments are assumed to be numbered in the "correct order."

1 2 3 4 5 6	1 2 3 4 5 6	1 2 3 4 5 6
(a)	(b)	(c)

An alternative method of presentation is obtained by using < to indicate that treatment i has been found inferior to treatment j and $\approx$ to indicate that the difference between the two has not been found to be significant. The information summarized in diagrams (a) to (c) can then also be shown in the following tabular form. Here, for example, the < sign in row 2 and column 5 of (a) indicates that treatment 5 has been found to be superior to treatment 2.

	2	3	4	5	6	2	3	4	5	6	2	3	4	5	6
1	<	<	<	<	<	<	<	<	<	<	≈	≈	<	<	<
2		≈	≈	<	<		≈	<	<	<		≈	≈	<	<
3			≈	≈	<			≈	<	<			≈	≈	<
4				≈	≈				≈	≈				≈	≈
5					≈					≈					≈
			(a)					(b)					(c)		

In determining the critical level c_{ij} of (5.69), one must decide as with procedure (5.47) of the preceding section whether to control the error rate per comparison or per experiment (or whether to adopt a compromise between these two extremes). However, in addition a new choice arises. When comparing the ith and jth treatment, it is possible to rank only the $n_i + n_j$ observations in the ith and jth group (from 1 to $n_i + n_j$), which requires reranking as the comparison goes from one pair to the next.

Alternatively, one can perform a single ranking (from 1 to N) of all $N = n_1 + \cdots + n_s$ observations and base the comparison of the ith and jth treatment on the average ranks $R_{i.}$ and $R_{j.}$ resulting from this combined ranking.

From the point of view of available tables, the simplest procedure is that based on separate ranking of each pair with control of the error rate per comparison. For a pairwise ranking procedure, the notation (5.69) is inadequate since the average rank of the ith treatment depends also on the treatment with which it is being compared. Let us denote the average rank of the ith treatment when it is being compared with the jth treatment by $R_{i.}^{(j)}$ and the corresponding rank sum by $R_i^{(j)} = n_i R_{i.}^{(j)}$. The procedure then simply consists in applying the three-decision Wilcoxon procedure (1.54) to all pairs (i,j) and declaring treatment j superior to treatment i if

$$R_j^{(i)} \geqslant c_{ij} \tag{5.70}$$

If the error rate per comparison is to be α, the critical values c_{ij} are determined by

$$P_{H_{ij}}(R_j^{(i)} \geqslant c_{ij}) = P_{H_{ij}}(R_i^{(j)} \geqslant c_{ji}) = \tfrac{1}{2}\alpha \tag{5.71}$$

where H_{ij} denotes the hypothesis of no difference between treatments i and j. The sum of the two probabilities (5.71), which is the probability of falsely declaring the difference between treatments i and j to be significant in either direction, is α. If all the group sizes n_i are equal, the constant c_{ij} is independent of i and j, and treatment j is declared superior to treatment i if

$$R_j^{(i)} \geqslant c \tag{5.72}$$

By (1.13) the above procedures can be restated in terms of the variables W_{ij} of the preceding section. Treatment j is then declared superior to treatment i if W_{ij} is sufficiently large. For the computations, it is useful to recall that W_{ij} is distributed symmetrically about $\frac{1}{2}n_i n_j$, and that $W_{ij} + W_{ji} = n_i n_j$.

EXAMPLE 12. *Comparison of four laboratories (continued).* In comparing measurements of the smoothness of paper from four different laboratories in Example 3, we found that the laboratories differed significantly ($\hat{\alpha} = .005$). Let us now try to obtain a more detailed analysis of the situation through a pairwise comparison of the laboratories. At significance level $\alpha = .03$, it is seen from Table B that treatment j will be declared superior to treatment i if $W_{ji} \leqslant 11$. A comparison of the first two rows of the data of Example 3 shows that $W_{AB} = 15$ while $W_{BA} = 64 - 15 = 49$. The difference between laboratories A and B is thus not significant. If one neglects the few ties within and between laboratories C and D, the corresponding comparisons between all laboratories show that the measurements from laboratory D are significantly smaller than those from laboratories A and B, and that none of the other differences are significant. The conclusions can be summarized diagramatically as

$$\underline{D \quad C} \quad B \quad A$$
$$ \underline{C \quad B \quad A}$$

The problem of finding a suitable grouping for a number of treatments arises also in the one-sided case in which there is a prior order of the treatments. Let us suppose, as in the preceding section, that the second treatment is known to be at least as good as the first, the third at least as good as the second, and so on. Then if $i < j$, the jth treatment will be declared superior to the ith when (5.70) holds, where c_{ij} is now determined by

$$P_{H_{ij}}(R_j^{(i)} \geqslant c_{ij}) = \alpha \tag{5.73}$$

rather than by (5.71), since in the present case (5.70) represents the only way of falsely declaring the difference between treatments i and j to be significant.

To illustrate this procedure, consider once more the data of Example 12, but suppose it is known, on the basis of extensive past experience, that the measurements of laboratory A tend to run higher than those of B, which in turn tend to run higher than those of C, and so on. If we try to set $\alpha = .03$ as before, Table B shows that the measurements of laboratory j will be declared larger than those of i if $W_{ji} \leqslant 14$ ($\alpha = .032$). The statistics W_{ij} have the values $W_{AB} = 15$, $W_{AC}^* = 13$, $W_{AD}^* = 3$, $W_{BC}^* = 29$, $W_{BD}^* = 8$, $W_{CD}^* = 16.5$, and the conclusions can be summarized in the following diagram:

$$\underline{D \quad C} \quad B \quad A$$
$$\quad\; \underline{C \quad B}$$
$$\qquad\quad \underline{B \quad A}$$

Let us now return to the case in which no prior order of the treatments is available and consider the error rate per experiment, restricting attention to the case of equal group sizes $n_i = n$, so that the procedure is given by (5.72). A traditional approach considers this error rate for the overall hypothesis H that there exist no treatment differences and determines c so that the probability is α' of declaring at least one difference significant when H is true. With this procedure, c is given by the equation

$$\alpha' = P_H(\max R_j^{(i)} \geqslant c) \tag{5.74}$$

where the maximum is taken over all $s(s-1)$ pairs $i \neq j$.

Tables of the critical values c for $\alpha' = .01$ and $\alpha' = .05$ (based on an approximation) are given by Miller (1966, Table IX) and Steel (1961). In computing $\max R_j^{(i)}$, it is actually necessary to determine only half the $s(s-1)$ sums $R_j^{(i)}$, say, those for $i < j$, since

$$R_i^{(j)} + R_j^{(i)} = n(2n+1)$$

This relation follows from the fact that the left-hand side is the sum of all the integers from 1 to $2n$.

The comparison of the error rates α and α' made in the preceding section applies qualitatively also to the present problem. The quantitative relationship may be illustrated as in Table 5.7, which shows the value of α required to achieve

$\alpha' = .05$ for various values of s and n. (The column $s = 2$ shows in parentheses the exact values of α' for $s = 2$ corresponding to the asymptotic value $\alpha' = .05$.) A comparison with Table 5.4(b) shows the values of α to be considerably smaller in the present case; this is because the number of comparisons is now $\frac{1}{2}s(s-1)$ instead of the previous smaller number s.

TABLE 5.7 Values of α for various s and n when $\alpha' = .05$

n \ s	2		3	4	5	6
6	.05	(.042)	.008	.004	.002	.002
8	.05	(.038)	.014	.006	.002	.002
10	.05	(.044)	.014	.006	.004	.002

The definition of error in the discussion above does not correspond to the definition given in Chap. 1, Sec. 5 and (implicitly) at the beginning of the present section. There, the problem was viewed as that of determining for each pair of treatments (i, j) which of the two treatments is superior, with a claimed superiority of j over i being considered in error only if treatment j is in fact inferior to treatment i but not in case of equality.

The probability (5.74) arises, however, in two other contexts, which we shall now consider. The first refers to a standard significance test of the hypothesis of no difference between i and j against the alternative that j is better than i but where the particular values of i and j being tested have been suggested by the data. When looking over the differences $R_j^{(i)} - R_i^{(j)}$ between the jth and ith treatments, one may be struck by a large value for a particular pair and wish to know whether it is significant at a given level α. A significance test such as the Wilcoxon test carried out at level α is not appropriate, since one must take account of the fact that it had not been planned a priori to compare the treatments of this particular pair but that this comparison was suggested by the data.

Suppose now that a pair (i,j) is selected in any way whatever (possibly in the light of the observations), and consider the maximum probability of falsely declaring the difference between treatments i and j to be significant under the assumptions of model (5.67). Here the maximum is taken over all possible choices and all values of $\theta_1, \ldots, \theta_s$. It can be shown that this maximum probability is the value α' given by (5.74). The proof is based on the fact that the probability of falsely declaring the difference between the treatments of the selected pair significant is less than or equal to the probability of at least one false significance statement when all pairs are being tested. By an argument similar to that of Chap. 2, Sec. 2, which we shall omit here, this latter probability can be shown to take on its maximum value when $\theta_1 = \cdots = \theta_s$, and the desired result then easily follows.

Closely related to the problem of deciding on the significance of a pair of treatments suggested by the data is that of setting simultaneous confidence limits for all differences $\theta_j - \theta_i$ in the model (5.67). For a single difference, confidence intervals were derived in Chap. 2, Sec. 6. In generalization of the notation used there, let

$$D_{(1)}^{(ij)} < D_{(2)}^{(ij)} < \cdots < D_{(n_i n_j)}^{(ij)}$$

denote the $n_i n_j$ ordered differences $X_{j\beta} - X_{i\alpha}$ ($\alpha = 1, \ldots, n_i; \beta = 1, \ldots, n_j$) from the ith and jth samples, and let $W_{ij}(\Delta)$ denote the number of pairs (α, β) for which

$$X_{i\alpha} < X_{j\beta} - \Delta$$

Then it was seen in Eq. (2.58) that for any fixed integer a

$$D_{(a)}^{(ij)} \leqslant \theta_j - \theta_i < D_{(a+1)}^{(ij)} \qquad \text{if and only if } W_{ij}(\theta_j - \theta_i) = n_i n_j - a$$

The argument that led to (2.60) now shows that

$$P[\theta_j - \theta_i \leqslant D_{(a)}^{(ij)} \quad \text{for all } i,j] = P_0[W_{ij} \leqslant a - 1 \quad \text{for all } i,j] \tag{5.75}$$

where actually a could also depend on i and j. Now the differences $X_{j\beta} - X_{i\alpha}$ are just the negatives of the differences $X_{i\alpha} - X_{j\beta}$ so that

$$-D_{(a)}^{(ji)} = D_{(n_i n_j - a + 1)}^{(ij)}$$

Therefore

$$\theta_i - \theta_j \leqslant D_{(a)}^{(ji)}$$

if and only if

$$\theta_j - \theta_i \geqslant D_{(n_i n_j - a + 1)}^{(ij)}$$

and the left-hand side of (5.75) is equal to

$$P[D_{(n_i n_j - a + 1)}^{(ij)} \leqslant \theta_j - \theta_i \leqslant D_{(a)}^{(ij)} \text{ for all } i,j] \tag{5.76}$$

To relate the probability (5.76) to α' defined by (5.74), note that by (1.13) the right-hand side of (5.75) is equal to

$$P_0[R_j^{(i)} \leqslant \tfrac{1}{2}n_j(n_j + 1) + a - 1 \quad \text{for all } i,j] \tag{5.77}$$

Suppose now that $n_1 = \cdots = n_s = n$ and that α' is determined by (5.74) with $c = a + \frac{1}{2}n(n+1)$. Then the probability of (5.77) and hence of (5.75) is equal to $1 - \alpha'$.

Let us illustrate the computation of the simultaneous confidence intervals (5.76) for a given confidence level $\gamma = 1 - \alpha'$ on the data of Example 3, where for the sake of simplicity we shall consider only laboratories A, B, and D and disregard the single tie in the measurements of laboratory D. To determine simultaneous confidence intervals for $\theta_A - \theta_B$, $\theta_A - \theta_D$, and $\theta_B - \theta_D$ at confidence level .95, note from Table 5.7 with $s = 3, n = 8$ that $\alpha' = .05$ corresponds to $\alpha = .014$. As was pointed out in relation to (5.71), α is the error probability for a two-sided Wilcoxon test, and the corresponding one-sided error probability is $\frac{1}{2}\alpha = .007$. Table B shows that for $m = n = 8$

$$P_0(W_{ij} \leqslant 8) = P_0(W_{ij} \geqslant 56) = .007$$

so that

$$P_0(W_{ij} \geqslant 56 \text{ for at least one pair } i,j) = .05$$

and

$$P_0(W_{ij} \leqslant 55 \text{ for all } i,j) = .95$$

Thus, finally, by (5.75) and (5.76)

$$P[D_{(9)}^{(ij)} \leqslant \theta_j - \theta_i \leqslant D_{(56)}^{(ij)} \text{ for all } i,j] = .95$$

For the data of Example 3, an easy computation shows the numerical values for the three intervals to be

$$-1.4 \leqslant \theta_A - \theta_B \leqslant 8.5$$
$$2.6 \leqslant \theta_A - \theta_D \leqslant 13.7$$
$$.2 \leqslant \theta_B - \theta_D \leqslant 8.5$$

To obtain a grouping of s treatments, we have so far considered procedures based on separately ranking (from 1 to $n_i + n_j$) the $n_i + n_j$ observations from each of the possible pairs (i,j). As mentioned earlier, an alternative is provided by the joint ranking (from 1 to $N = n_1 + \cdots + n_s$) of all N observations. Changing notation, let us denote by $R_{i\cdot}$ the average rank of the ith treatment under this new ranking. A grouping of the s treatments can then again be obtained in terms of the inequalities (5.69). If the sample sizes are equal, the constants c_{ij} will be taken to be equal, and treatment j will be judged to be superior to treatment i when

$$R_{j\cdot} - R_{i\cdot} \geqslant c \tag{5.78}$$

with the order of the treatments left in doubt when $|R_{j\cdot} - R_{i\cdot}| < c$.

No tables of c are available for the error rate per comparison with this procedure. For a simple normal approximation when the common sample size n is not too small, see Miller (1966, Chap. 4, Sec. 6.1). Tables of c for the error rate per experiment under the overall null hypothesis $\theta_1 = \cdots = \theta_s$,

$$\alpha' = P_H[\max(R_{j\cdot} - R_{i\cdot}) \geqslant c] \tag{5.79}$$

are given by McDonald and Thompson (1967) and by Tobach, Smith, Rose, and Richter (1967). It is, however, doubtful whether this error rate is again equal to the maximum probability of a false significance statement suggested by the data.

Of the two possibilities, pairwise or joint ranking, the pairwise ranking approach seems to have a number of advantages: (1) the availability of tables for the error rate per comparison; (2) the fact that α' controls the probability of significance statements suggested by the data; (3) the fact, not yet mentioned, that it is much more difficult to obtain simultaneous confidence intervals for the differences $\theta_j - \theta_i$ on the basis of joint than of pairwise ranking.

Another feature that has been mentioned as a drawback of the joint ranking procedure leading to (5.78) is that the comparison of a given pair of treatments depends not only on the observations from these treatments but also on those from all the other treatments. It is, however, not clear that this phenomenon is as unreasonable as it may appear at first sight. Suppose that we are willing, once more, to assume the shift model (5.67), which implies a definite order among the treatments. Such an order carries with it at least the possibility that the other samples may provide some useful information for the comparison of a given pair.

Unfortunately, pairwise ranking also has a rather undesirable feature: it may lead to inconsistencies. More specifically, it can happen that procedure (5.69) will declare treatment j superior to treatment i and treatment k superior to j, without declaring treatment k superior to i. To see this, consider the case of three treatments with three observations on each. Suppose that the order of the observations is

$$ZXXYYYZZX$$

where the X's, Y's, and Z's denote the observations on the first, second, and third treatment, respectively. Then in a ranking of the X's and Y's, the rank sum of the Y's is $3+4+5 = 12$; in a ranking of the Y's and Z's, the rank sum of the Z's is $1+5+6 = 12$; but in a ranking of the X's and Z's, the rank sum of the Z's is only $1+4+5 = 10$. The procedure would therefore be afflicted with the inconsistency in question if the critical value were 11.

This difficulty, the probability of which tends to be small, does not arise in joint ranking, at least in the case of equal group sizes. For let $\sum R_i$, $\sum S_j$, $\sum T_k$ denote the rank sums of three of the treatments in the joint ranking of s treatments, and suppose that the difference in a treatment pair is declared significant if the difference in their rank sums exceeds c. Then the second treatment will be declared superior to the first if $\sum S_j > \sum R_i + c$ and the third to the second if $\sum T_k > \sum S_j + c$. These two inequalities together imply that $\sum T_k > \sum R_i + 2c$ and hence that the third treatment will be declared superior to the first. [Issues related to the inconsistency problem are discussed by Gabriel (1969). The whole subject of simultaneous tests and confidence sets is surveyed by Miller (1966) and by O'Neill and Wetherill (1971).]

6B. Selecting the Best of Several Treatments

The problem of selecting the better one of two treatments has generalizations other than that considered in Sec. 6A. In particular, there is the obvious one to the problem of selecting the best of a number of treatments, for example, the brand of gasoline giving the greatest mileage per gallon or the most promising of a number of candidates for a position. A simple and natural procedure, based on the joint ranking of the $N = n_1 + \cdots + n_s$ observations, consists in selecting as best the treatment with the largest average rank. (To complete the description, it is necessary to specify how to act in case of a tie for first place. The choice in that case can be made at random, or preferably it can sometimes be based on the performance of the tied treatments with respect to an auxiliary criterion.)

The chief problem that has been considered in connection with this procedure is the determination of appropriate sample sizes. One would typically want to have equal sample sizes, say $n_1 = \cdots = n_s = n$, and the problem is that of determining the value of n required to guarantee some desirable property. One such requirement is

$$P(\text{selected treatment is good}) \geqslant \gamma$$

where γ is a specified probability. Here the ith treatment might for example be considered good under the assumption of model (5.67), if θ_i does not fall too far below the largest θ, say,

$$\theta_i \geqslant \theta_{\max} - \Delta \tag{5.80}$$

where Δ is a given constant. For a discussion of this problem see Lehmann (1963a), Rizvi and Woodworth (1970), Ghosh (1973), and for certain extensions Puri and Puri (1969). A sequential procedure is proposed by Geertsema (1972).

Let us now consider another way of approaching this problem. When trying to determine the best of a number of treatments, it may be useful as a first stage to select a subgroup that is reasonably certain to contain the best treatment. Having narrowed the choice, one could then as a second stage perform more extensive experimentation on the selected subgroup. Consider now only the first of these stages, and as before let $R_{i.}$ denote the average rank of the ith treatment in the joint ranking of all N observations. Gupta and McDonald (1970) discuss the following three selection rules: include the ith treatment in the selected subgroup if $R_{i.}$ is greater than or equal to $\max_j R_{j.} - c$; $\max_j R_{j.}/c$; c. Here the constants c are of course different in the three cases.

In view of the ultimate aim of selecting the best treatment, it is desirable to control the probability of a *correct selection*, i.e., the probability that the selected subgroup will include the best population. (For the sake of simplicity, suppose that the treatments are distinct, so that the best treatment is uniquely defined.) This can be achieved through appropriate choice of c, which will therefore be determined so that

$$P(\text{correct selection}) \geqslant \gamma \tag{5.81}$$

Subject to this condition, one would wish the subgroup to be as small as possible. It is not difficult to see that this means choosing c so that the minimum value with respect to the θ's of the left-hand side is equal to γ.

Consider now once more the shift model (5.67) and suppose that $n_1 = \cdots = n_s = n$. Without loss of generality the treatments can then be numbered in such a way that

$$\theta_1 < \theta_2 < \cdots < \theta_s$$

Under these assumptions Gupta and McDonald show that for any fixed value of

θ_s, the probability of a correct selection with the third procedure decreases as $\theta_1, \theta_2, \ldots, \theta_{s-1}$ increase toward θ_s. (This is intuitively plausible since it is then increasingly more difficult to identify θ_s as the best population.) It follows that the probability of a correct selection approaches its minimum value as $\theta_1, \ldots, \theta_{s-1}$ approach θ_s, and hence that for the third procedure the critical value c is determined by the condition

$$P_{\theta_1 = \cdots = \theta_s}(R_{s\cdot} \geqslant c) = \gamma \tag{5.82}$$

Now $nR_{s\cdot}$ is the rank sum of n X's (namely, of $X_{s1}, \ldots, X_{sn}$) in the joint ranking of $N = sn$ X's, which in (5.82) are assumed to be independently and identically distributed. Hence, $nR_{s\cdot}$ has the distribution of the Wilcoxon statistic W_s tabled in Table B, with group sizes $m = (s-1)n$ and n. The critical value c can therefore be obtained from this table.

The determination of c for the first two procedures is considerably more complicated. For details, the reader is referred to the paper by Gupta and McDonald (1970).

7. FURTHER DEVELOPMENTS

7A. Power and Efficiency

An important aspect of any test is its power and, related to it, the determination of the sample size necessary to achieve a given power. For the two-sample Wilcoxon test, these problems were treated in Chap. 2, Secs. 2 and 3, and for the sign test and paired-comparison Wilcoxon tests in Chap. 4. The reason for not providing here an analogous discussion of the power of the Kruskal-Wallis test is the greater complication of the problem in this case. It is not difficult to determine the joint large-sample distribution of the rank sums $(R_1, \ldots, R_s)$ in generalization of the asymptotic distribution of W_s given in Chap. 2, but unfortunately the resulting distribution of K is too complicated to be very useful.

There is available an alternative that parallels the approximation (2.29) the idea of which is due to Neyman (1937). As in the definition of Pitman efficiency, consider not a fixed alternative but rather a sequence of alternatives for which the power has a fixed value. For the case of the location model (5.67), the associated limit distribution of K is obtained by Andrews (1954), and turns out to be a noncentral χ^2-distribution with a relatively simple noncentrality parameter. Approximate values of the power can therefore be found from tables of the noncentral χ^2-distribution such as the tables of Fix (1949), Fix, Hodges, and Lehmann (1959), Haynam, Govindarajulu, and Leone (1970), and Patnaik (1949). This approximation may be expected to give good results for the normal shift model but to be less reliable in some other cases.

The approach in which the power is kept fixed while the alternative is permitted to vary also leads to the Pitman efficiency of the K test. Andrews (1954) shows that this efficiency relative to the standard F-test in the shift model (5.67) has the same value $e_{W,t}(F)$ as that of the two-sample Wilcoxon test relative to the t test, which was discussed in Chap. 2, Sec. 4.

7B. Estimation of Several Differences in Location

One of the central problems concerning the location model (5.67) is the estimation of the treatment differences $\theta_i - \theta_j$. Simultaneous confidence intervals for these differences were discussed in Sec. 6; let us now turn to the problem of point estimation. Application of the estimate $\hat{\Delta}$ of Chap. 2, Sec. 5, suggests

$$\hat{\Delta}_{ij} = \text{med}\,(X_{i\alpha} - X_{j\beta}) \tag{5.83}$$

the median of the $n_i n_j$ differences $X_{i\alpha} - X_{j\beta}$ ($\alpha = 1, \ldots, n_i$; $\beta = 1, \ldots, n_j$) as an estimate of the location difference $\Delta_{ij} = \theta_i - \theta_j$ between the distribution of the $X_{i\alpha}$ and the $X_{j\beta}$.

The estimates $\hat{\Delta}_{ij}$ suffer from the disadvantage that $\hat{\Delta}_{ik}$ (the estimate of $\theta_i - \theta_k$) will typically not be equal to $\hat{\Delta}_{ij} + \hat{\Delta}_{jk}$ [the sum of the estimates of $(\theta_i - \theta_j)$ and $(\theta_j - \theta_k)$]. To avoid this awkwardness, Lehmann (1963b) proposes to replace (5.83) by the estimate

$$\hat{\Delta}_{i\cdot} - \hat{\Delta}_{j\cdot} \tag{5.84}$$

where $\hat{\Delta}_{i\cdot} = \sum_{k=1}^{s} \hat{\Delta}_{ik}/s$, and then shows that these adjusted estimates of $\theta_i - \theta_j$ have essentially the same properties as those established for the estimate $\hat{\Delta}$ in Chap. 2, Sec. 5.

If the sample sizes are unequal, an improvement over the estimates (5.84) is suggested by Spjøtvoll (1968). Instead of the unweighted averages $\hat{\Delta}_{i\cdot}$ he proposes using the weighted averages

$$\hat{\Delta}_{i*} = \frac{\sum n_k \hat{\Delta}_{ik}}{\sum n_k} \tag{5.85}$$

and hence as an estimate of $\theta_i - \theta_j$ the difference

$$\hat{\Delta}_{i*} - \hat{\Delta}_{j*} \tag{5.86}$$

7C. The Estimation of Contrasts

The differences $\theta_i - \theta_j$ considered in the last subsection are not the only functions that one may wish to estimate. For comparative studies, an important class are the so-called *contrasts*. These are linear functions of the θ's, say,

$$\gamma = \sum c_k \theta_k \tag{5.87}$$

with the property that the sum of the coefficients is zero:

(5.88) $$\sum c_k = 0$$

The differences $\theta_i - \theta_j$ themselves are of course of this form ($c_i = 1$, $c_j = -1$, the remaining c's $= 0$). Another example is the difference between the averages of two groups of nonoverlapping θ's such as $\frac{1}{2}(\theta_1 + \theta_4) - \frac{1}{3}(\theta_2 + \theta_3 + \theta_5)$. To see how such differences arise, suppose that $\theta_1, \ldots, \theta_{12}$ indicate the level of some characteristic such as height or IQ, of children born in January, ..., December, respectively. Then one might wish to estimate the difference in the average θ-values for the summer and winter months, say, $\frac{1}{3}(\theta_7 + \theta_8 + \theta_9) - \frac{1}{3}(\theta_1 + \theta_2 + \theta_3)$. Still another example of a contrast is $\theta_i - \bar{\theta}$, the amount by which the ith treatment exceeds the average of all treatments.

It follows from (5.88) that the contrast γ can be expressed in terms of the differences $\theta_i - \theta_j$, say, as

(5.89) $$\gamma = \sum\sum d_{ij}(\theta_i - \theta_j)$$

An estimate of γ is then obtained by substituting for $\theta_i - \theta_j$ its estimate $\hat{\Delta}_{i\cdot} - \hat{\Delta}_{j\cdot}$. The resulting estimate

(5.90) $$\hat{\gamma} = \sum\sum d_{ij}(\hat{\Delta}_{i\cdot} - \hat{\Delta}_{j\cdot})$$

is discussed by Lehmann (1963b), where it is shown to be uniquely defined (although the d_{ij} are not) and where the asymptotic efficiency of $\hat{\gamma}$ relative to the classical estimate $\sum\sum d_{ij}(X_{i\cdot} - X_{j\cdot})$ is proved to be the same as that of $\hat{\Delta}_{ij}$ relative to $X_{i\cdot} - X_{j\cdot}$. Again, in the case of unequal sample sizes it may be preferable to replace $\hat{\gamma}$ by

(5.91) $$\gamma^* = \sum\sum d_{ij}(\hat{\Delta}_{i*} - \hat{\Delta}_{j*})$$

A problem of considerable interest is that of finding simultaneous confidence intervals for all contrasts since this makes it possible to assess any particular contrasts whose significance is suggested by the data. Large-sample solutions of this problem are discussed by Sen (1966) and Marascuilo (1966); for some related work see Crouse (1969).

7D. Normal Scores and Smirnov Tests for the s-sample Problem

As an alternative to the Kruskal-Wallis statistic K, consider the test statistic

(5.92) $$\frac{N-1}{\sum_{i=1}^{N} [a_N(i)]^2} \sum_{j=1}^{s} \frac{1}{n_j} \left[\sum_{k=1}^{n_j} a_N(R_{jk}) \right]^2$$

where the $a_N(i)$ are the Normal Scores defined by (2.67). (This can be obtained by replacing the variables X_{ij} in the classical F-test by their Normal Scores.) It was proved by Puri (1964) that the null distribution of (5.92), under the hypothesis $F_1 = \cdots = F_s$, tends to the χ^2-distribution with $s-1$ degrees of freedom as the sample sizes tend to infinity. [For a more precise statement see Hajek and Sidak (1967).] He also showed that the asymptotic efficiency of the test based on (5.92) relative to the Kruskal-Wallis test or to the classical F-test is the same as that of the Normal Scores test relative to the Wilcoxon two-sample test or the t-test, respectively. No tables of the null distribution of (5.92) are available. However, for the reasons discussed in Chap. 2, Sec. 7, one would expect the χ^2-approximation to work well even for relatively small values of N.

Of the many different possible generalizations of the Smirnov two-sample test to the problem of s samples [see, for example, Kiefer (1959)], we mention only the following. If $F_{n_i}^*$ denotes the sample cumulative distribution function of the n_i observations on the ith treatment, and $\bar{F}_N^*$ denotes the sample cumulative distribution function of all $N = \Sigma\, n_i$ observations, consider

$$\max_x \sum n_i [F_{n_i}^*(x) - \bar{F}_N^*(x)]^2$$

For this statistic Kiefer obtains and tabulates the limit of the null distribution. As in the preceding case, no small-sample tables are available. In view of the slow convergence of the null distribution of the Smirnov test, it seems advisable also in the present case not to trust the limit distribution unless the sample sizes are fairly large.

8. PROBLEMS

(A)

Section 1

1. If $N = 12, s = 4$, and $n_1 = 5, n_2 = 2, n_3 = 3, n_4 = 2$, display all possible choices for the second group of two ranks when the five ranks of the first group are (i) 2, 3, 5, 10, 11; (ii) 1, 4, 9, 10, 12.

2. In the preceding problem, display all possible choices for the third group of three ranks when the ranks of the first two groups are, respectively, (i) 2, 3, 5, 10, 11 and 7, 9; (ii) 2, 3, 5, 10, 11 and 1, 8; (iii) 1, 4, 9, 10, 12 and 3, 6; (iv) 1, 2, 4, 10, 12 and 5, 11.

3. Use (5.2) and Table A to compute the multinomial coefficients (i) $\binom{12}{5,2,3,2}$; (ii) $\binom{12}{2,4,2,4}$; (iii) $\binom{12}{3,3,3,3}$; (iv) $\binom{12}{8,1,2,1}$.

4. Use (5.2) and Table A to compute the multinomial coefficients (i) $\binom{15}{4,4,4,3}$; (ii) $\binom{15}{2,3,4,6}$; (iii) $\binom{15}{1,1,4,9}$; (iv) $\binom{15}{6,6,2,1}$.

5. If $n_1 = 2, n_2 = 3, n_3 = 3$, find the probabilities [defined by (5.4)]

(i) $P_H(R_{11} = 1, R_{12} = 2; R_{21} = 3, R_{22} = 4, R_{23} = 5)$

(ii) $P_H(R_{11} = 1, R_{12} = 3; R_{21} = 5, R_{22} = 6, R_{23} = 7)$

(iii) $P_H(R_{11} = 1, R_{21} = 2)$ (iv) $P_H(R_{11} = 1, R_{12} = 2)$

(v) $P_H(R_{11} = 1, R_{21} = 4)$ (vi) $P_H(R_{11} = 1, R_{21} = 2, R_{31} = 3)$

6. If $n_1 = 3, n_2 = 2, n_3 = 5$, find the probabilities [defined by (5.4)]

(i) $P_H(R_{11} = 1, R_{12} = 2, R_{13} = 3; R_{21} = 5, R_{22} = 7)$

(ii) $P_H(R_{11} = 2, R_{12} = 4, R_{13} = 6; R_{21} = 1, R_{22} = 5)$

(iii) $P_H(R_{11} = 1, R_{12} = 3, R_{13} = 5; R_{21}+R_{22} = 10)$

(iv) $P_H(R_{21} = 3, R_{22} = 6; R_{11}+R_{12}+R_{13} = 11)$

Section 2

7. Obtain the null distribution of K for the cases (i) $s = 3$, $n_1 = n_2 = 1$, $n_3 = 6$; (ii) $s = 3$, $n_1 = n_2 = n_3 = 2$.

8. Fourteen rats were divided into three groups of sizes $n_1 = n_2 = 5$, $n_3 = 4$, and assigned to three different kinds of food. Their weights in grams after 2 months were[1] 123, 121, 159, 138, 178; 144, 172, 165, 143, 179; 139, 146, 161, 149. Use Table I to determine whether the data are significant at level $\alpha = .05$.

9. In the preceding problem apply the χ^2-approximation to determine an approximate significance probability.

10. Compute the significance probability of the data of Example 3 under the assumption that the measurement 58.0 of laboratory A had been copied incorrectly and really should have been 38.0.

11. The following table shows measurements of the smoothness of a certain kind of paper, five measurements each by 15 different laboratories.[2] Test the hypothesis that there is no difference between the laboratories.

1	2	3	4	5	6	7	8	9	10	11	12	13	14	15
173	135	165	185	108	120	125	135	98	150	136	104	127	120	125
185	165	175	137	112	135	138	155	93	140	169	132	145	125	178
141	145	155	162	113	140	125	140	142	130	165	113	142	185	143
133	160	135	125	137	135	155	155	106	155	146	77	145	130	118
160	120	160	157	146	130	173	140	114	150	132	134	148	120	148

Section 3

12. A group of 59 kittens was divided into three subgroups of 18, 20, and 21, respectively. Those in the first group were reared together with a mouse or rat from the age of about one week; those in the second group were reared in isolation from any rodent; finally, the kittens in the third group saw their mothers kill a rat or mouse every 4 days outside the cage. In the test

[1] Part of the data of Petitpierre quoted by Linder, *Statistische Methoden*, 2d ed., Birkhaüser, Basel, p. 12.

[2] Data from Mandel, *The Statistical Analysis of Experimental Data*, Interscience Publishers, New York, 1964, Table 3.5.

situation, each kitten was put together with a rat, with the following results.[1]

	Reared with Rodent	Reared in Isolation	Watched Mothers
Killed	3	9	18
Did not kill	15	11	3

Compute the significance probability using (i) the one-sided Wilcoxon test of Chap. 1, Sec. 4; (ii) the two-sided Wilcoxon test of Chap. 1, Sec. 5; (iii) the Kruskal-Wallis K^*-test. Which of the three do you feel is most appropriate?

13. For the data of Prob. 49 of Chap. 1, suppose that it is not clear a priori what effect the treatment would have (if any), so that it is desired to test the hypothesis of no treatment effect against the general alternative of an unspecified effect. Find the approximate significance probability for the K^*-test, and compare the result with that of the earlier problem.

14. The following table[2] shows the number of male patients who were free of postoperative nausea (after ether anesthesia) under four different drugs. Compute the approximate significance probability using the K^*-test for testing the hypothesis of no difference in the effectiveness of the drugs.

	Chlorpromazine 50 mg	Dimenhydrinate 100 mg	Pentobarbital 100 mg	Pentobarbital 150 mg
Number of patients	64	39	22	30
Free of nausea	33	14	8	18

15. (i) Verify the value given in Example 6 for the numerator of $P_H(A_1 \geqslant 12)$; (ii) Compute the binomial coefficient $\binom{30}{13}$ using formula (1.66) and Table A.

16. The following table[3] shows the incidence of colds among 354 men, of whom 265 received an infectious secretion while the other 89 served as controls.

	Developed Cold	Did not Develop Cold	Total
Treated	68	197	265
Control	12	77	89

Find the approximate significance probability of these results.

Section 4

17. The following table gives the units in the third decimal place for determinations of the gravitational constant G (for example, 83 corresponds to an observation of 6.683), using three different substances.[4]

Gold:	83	81	76	79	76
Platinum:	61	61	67	67	64
Glass:	78	71	75	72	74

Determine the approximate significance probability of K^*.

[1] Data from (1930), "The Genesis of the Cat's Response to the Rat," *J. Comp. Psych.* **11**:1–33, as reported by Edwards, *Statistical Methods for the Behavioral Sciences*, Rinehart, New York, 1960.

[2] From Beecher, *Measurement of Subjective Response*, Oxford University Press, London, 1959, part of Table 17.12.

[3] Part of data from Dowling, Jackson, and Inouye (1957), "Transmission of the Experimental Common Cold in Volunteers," *J. Lab. Clin. Med.* **50**:516–525.

[4] Data from Heyl (1930), "A Determination of the Constant of Gravitation," *J. Res. Natl. Bureau Std.* **5**:1243–1250, as quoted by Brownlee, *Statistical Theory and Methodology in Science and Engineering*, John Wiley & Sons, New York, 1960.

18. Among appendectomies of 103 female patients, age ten to twenty-nine years, the appendix of 60 patients was found to be pathological at operation while 43 appendixes were found to be normal. Sixty-eight of the patients were located for follow-up studies nine to twelve years after the operation; of these 39 had had a pathological appendix and 26 a normal one.

The following table[1] shows the number among these patients who did and did not have additional operations during the intervening period. Determine whether there is a significant difference between the two categories in this respect.

	No Additional Operations	One or More Additional Operations
Pathological	26	13
Normal	10	16

19. In a survey of Scottish school children, aged approximately ten to twelve years, the teacher observed whether the pupil wrote with the left or right hand, with the following results:[2]

	Left Hand	Number Observed	Percentage
Boys	991	12,629	7.84
Girls	1,478	25,045	5.89

Is the difference in the proportions significant?

20. The following table[3] gives the number of convictions and the number of dismissals among criminal cases in three California counties during 1931–1932. Test whether the proportions differ significantly in the three counties.

	Alameda	Los Angeles	San Francisco
No. convicted	482	3,672	804
No. dismissed	76	1,268	405

21. The following results were obtained in a study (conducted in 1948) of the 1944 presidential election in Elmira, N.Y.[4]

	Individual Interviewed on		
1944 Presidential Vote	First Call	Second or Later Call	Total
Roosevelt	138	217	355
Dewey	124	200	324
Did not vote	90	142	232
Other or too young	39	78	117
Total	391	637	1,028

Test the hypothesis that the distribution of responses is the same for individuals reached on the first call as for those interviewed on the second or later calls.

Section 5

22. The following table[5] shows the number of female patients who were free of postoperative

[1] Data from Meyer, Unger, and Slaughter (1964), "Investigation of a Psychological Hypothesis in Appendectomies," *Psychosomat. Med.* **26**:671–681.

[2] From Clark, *Left Handedness*, University of London Press, London, 1957.

[3] Beattie, *A System of Criminal Judicial Statistics for California*, University of California Press, Berkeley, 1936, Table 12.

[4] Taken from McCarthy, *Introduction to Statistical Reasoning*, McGraw-Hill Book Company, New York, 1957, p. 330.

[5] Data from the same table as those of Prob. 14.

nausea under four different drugs and in a control situation where only a placebo was given. Determine which of the treatments provide a significant improvement over the control at an error rate per comparison of $\alpha = .005$.

	Chlorpromazine 50 mg	Dimenhydrinate 100 mg	Pentobarbital 100 mg	Pentobarbital 150 mg	Placebo
No. of patients	88	46	45	55	111
Free of nausea	29	7	9	10	12

23. Suppose that $s = 5$, $n_1 = n_3 = n_4 = 2$, $n_2 = n_5 = 3$, and that the ranks of the five groups are 2, 7; 4, 6, 11; 5, 8; 3, 12; 1, 9, 10.

Find the amalgamated rank averages $R'_{i.}$, and the values of t, and $n'_1, n'_2, \ldots$.

24. Obtain the null distribution of the statistic W defined by (5.58) for the cases of Prob. 7.

25. In the study of Example 11 there were actually 10 psychologists in each of the three groups, and the 30 correlation coefficients were as follows:

Group 1	.97	.70	.84	.92	.98	.86	.92	.82	.76	.73
Group 2	.80	.69	.46	.53	.82	.92	.67	.68	.81	.82
Group 3	.89	.84	.65	.23	.69	.80	.83	.47	.95	.55

Find the significance probability of these data using the J-test.

26. In the preceding problem, find the significance probability if the test used is that of (i) Chacko-Shorack; (ii) Kruskal-Wallis.

27. To determine whether a certain diagnostic test can be interpreted successfully without much psychological training, 72 judges were presented with the results of the test on 200 carefully matched patients, half of whom were psychiatrically hospitalized and half medically hospitalized without apparent psychiatric disturbance. Of the judges, 21 were staff members and 23 trainees at Veterans Administration hospitals; the remaining 28 were undergraduate psychology majors who were given only brief instructions on the interpretation of the test. The following were the accuracies of the judges in terms of percent correctly identified:[1]

Staff 74.5 75.0 75.0 73.5 74.0 68.5 70.5 70.5 72.0 73.5 73.0 70.5
76.5 69.0 75.5 69.0 74.0 71.5 75.0 76.5 76.0

Trainees 70.5 73.5 71.5 74.0 70.0 68.5 74.0 71.0 70.0 74.5 69.0
71.5 62.5 74.5 69.5 66.0 70.0 69.5 70.5 63.0 74.0 71.5 73.0

Undergraduates 74.0 74.5 72.0 72.0 71.5 71.0 72.5 69.0 69.0 69.0
68.5 64.5 65.5 58.0 74.0 74.0 73.0 72.0 71.5 71.5
71.5 70.5 69.5 70.0 68.5 66.0 66.5 60.0

If training and experience have an effect, the staff members could be expected to be most accurate, the trainees next, and the undergraduates least. Under this assumption, compute the significance probability of the data using Jonckheere's test.

28. In the situation of the preceding problem, an investigator wishes to determine whether either the trainees or the students are significantly less accurate than the staff members. Use

[1] Data from Oskamp (1962), "The Relationship of Clinical Experience and Training Methods to Several Criteria of Clinical Prediction," *Psychol. Monographs* **47,** no. 547.

the procedure defined by (5.47) and (5.48) at levels $\alpha_1 = \alpha_2 = .01$ to answer this question, treating the staff members as the controls.

29. Obtain the null distribution of $\bar{K}'$ for the cases of Prob. 7.

30. Verify the values of Table 5.6 not worked out in the text.

Section 6

31. Suppose that in Example 11 it was not known a priori how the addition of information would affect the correlation coefficients. Apply a pairwise comparison procedure based on separate ranking for each pair and with an error rate per comparison of $\alpha = .01$ to the data of (i) Example 11; (ii) Prob. 25.

32. The data of Example 10 with the deleted observations restored are

A	*B*	*C*
133 139 149 160 162 184	111 125 137 143 148 153 157	99 100 114 116 127 146

Assuming no prior knowledge regarding the likely effect of the three diets, apply a pairwise comparison procedure based on separate ranking for each pair and with an error rate per comparison of $\alpha = .02$.

(B)

Section 1

33. Show that an alternative formula for the multinomial coefficients (5.2) is

$$\binom{N}{n_1, n_2, \ldots, n_s} = \frac{N!}{n_1! n_2! \cdots n_s!} \tag{5.93}$$

where $n!$ is defined in Chap. 1, Prob. 69. [*Hint:* In (5.2), express each of the binomial coefficients by means of (1.65).]

34. It follows from (5.93) that the coefficients (5.2) are unchanged if $n_1, n_2, \ldots, n_s$ are arranged in a different order. Check this result by using (5.2) to compute (i) $\binom{8}{1,1,2,4}$; (ii) $\binom{8}{1,2,1,4}$; (iii) $\binom{8}{2,1,1,4}$; (iv) $\binom{8}{4,1,1,2}$.

35. Let $N = sn$. For any integers $n_1, \ldots, n_s$ not all equal and satisfying $n_1 + \cdots + n_s = N$, show that

$$\binom{N}{n_1, \ldots, n_s} < \binom{N}{n, \ldots, n}$$

[*Hint:* It is enough to show that $n! \cdots n! < n_1! \cdots n_s!$. To see this, cancel each $n!$ against one $n_i!$ leaving $n(n-1) \cdots (n_i+1)$ on the left-hand side when $n_i < n$ and $n_i(n_i-1) \cdots (n+1)$ on the right-hand side when $n_i > n$. Compare the number and sizes of factors on the two sides after cancellation.]

36. Let $n_1 + \cdots + n_s = N$ and let $n'_1 = \cdots = n'_{s-1} = 1$, $n'_s = N - s + 1$. Show that

$$\binom{N}{n_1, \ldots, n_s} \geqslant \binom{N}{n'_1, \ldots, n'_s}$$

[*Hint:* Cancel n'_s against the largest of $n_1, \ldots, n_s$ and apply the argument of the preceding problem.]

37. In the expansion of $(a_1 + \cdots + a_s)^N$, show that the coefficient of $a_1^{n_1} \cdots a_s^{n_s}$ is $\binom{N}{n_1, \ldots, n_s}$. (This explains the term *multinomial coefficient.*)

Section 2

38. Show that the Kruskal-Wallis test (5.7) reduces to the two-sided Wilcoxon test when $s = 2$.

39. Show that the expressions for K given by (5.6) and (5.8) agree.

40. When $s = 3$, $n_1 = n_2 = 2$, $n_3 = 3$, find (i) the smallest value that K can take on; (ii) the probability of K taking on this value.

41. (i) Show that zero is a possible value of K when $n_1, \ldots, n_s$ are all even.
(ii) Find a combination of group sizes for which K cannot take on the value zero.

42. Generalize the scale problem discussed in Chap. 1, Sec. 6, to the case of s treatments, and discuss whether the statistic K, based on a Siegel-Tukey ranking, would provide a suitable test statistic for this problem.

43. Show that the maximum value of K is $(N^3 - \sum n_i^3)/N(N+1)$. [*Hint:* From the identity

$$\sum\sum [R_{ij} - \tfrac{1}{2}(N+1)]^2 = \frac{N(N+1)}{12} K + \sum\sum (R_{ij} - R_{i\cdot})^2$$

it follows that K is maximized by minimizing the second term on the right-hand side. To complete the proof, note that $\sum_{j=1}^{n_i} (R_{ij} - R_{i\cdot})^2$ is minimized when the R_{ij} take on the values of n_i consecutive integers and then is equal to $(n_i^3 - n_i)/12$.]

Section 3

44. (i) Verify Eq. (5.16).
(ii) In the derivation of the expression (5.18) for K^*, replace the scores 1 and 2 assigned to treatments A and B by scores a and b. Show that this does not change the expression for K^*.

45. Use (5.20) and the expectation (1.39) and variance (1.40) of W^*_{XY} given in Chap. 1 to prove formulas (5.22) and (5.23).

Section 4

46. If the variables $X_{ij}(j = 1, \ldots, n_i; i = 1, \ldots, s)$ have a common continuous distribution F, show that the distribution of the ranks continues to be given by (5.4).

47. Let F be a discrete distribution that assigns probabilities p and $1-p$ to the two values a and b $(a < b)$. If $n_1 = n_2 = n_3 = 1$, find the joint distribution of the three midranks.

Section 5

48. For the comparison of two treatments with a control, find the null distribution [under H of (5.51)] of $\max(R_1, R_2)$ when (i) $n_0 = n_1 = n_2 = 2$; (ii) $n_0 = n_1 = 1, n_2 = 2$.

49. Prove that the null distribution of J is symmetric about zero.

50. Prove formula (5.60).

51. Show that the expression (5.64) for the variance of W^* reduces to (5.61) when there are no ties present.

9. REFERENCES

Andrews, Fred C. (1954): "Asymptotic Behavior of Some Rank Tests for Analysis of Variance," *Ann. Math. Statist.* **25**:724–736.

Barlow, R. E., Bartholomew, D. J., Bremmer, J. M., and Brunk, H. D. (1972): *Statistical Inference under Order Restrictions*, John Wiley & Sons, New York.

Bartholomew, D. J. (1959): "A Test of Homogeneity of Ordered Alternatives," *Biometrika* **46**:36–48, 328–335.

——— (1961): "A Test of Homogeneity of Means under Restricted Alternatives," *J. Roy. Statist. Soc.* **B23**:239–281.

Chacko, V. J. (1963): "Testing Homogeneity against Ordered Alternatives," *Ann. Math. Statist.* **34**:945–956.

Crouse, C. F. (1969): "A Multiple Comparison Rank Procedure for a One-way Analysis of Variance," *S. African Statist. J.* **3**:35–48.

D'Abrera, Howard (1973): "Some Rank Tests for Ordered Alternatives," unpublished Ph.D. thesis, University of California at Berkeley.

Dunn, O. J. (1964): "Multiple Comparisons Using Rank Sums," *Technometrics* **6**:241–257. [See also Nemenyi (1963).]

Fix, Evelyn (1949): "Tables of Noncentral χ^2," *Univ. Calif. (Berkeley) Publ. Statist.* **1**:15–19.

———, Hodges, J. L., Jr., and Lehmann, E. L. (1959): "The Restricted Chi-square Test," in *Probability and Statistics* (The Harold Cramer volume) (Ulf Grenander, ed.), Almquist and Wiksell, Stockholm, John Wiley & Sons, New York, pp. 92–107.

Gabriel, K. R. (1966): "Simultaneous Test Procedures for Multiple Comparisons on Categorical Data," *J. Am. Statist. Assoc.* **61**:1081–1096.

——— (1969): "Simultaneous Test Procedures—Some Theory of Multiple Comparisons," *Ann. Math. Statist.* **40**:224–250.

——— and Lachenbruch, P. A. (1969): "Non-parametric ANOVA in Small Samples: A Monte Carlo Study of the Adequacy of the Asymptotic Approximation," *Biometrics* **25:**593–596.

——— and Sen, P. K. (1968): "Simultaneous Test Procedures for One Way ANOVA and MANOVA Based on Rank Scores," *Sankhya* **A30:**303–312.

Geertsema, J. C. (1972): "Nonparametric Sequential Procedures for Selecting the Best of k Populations," *J. Am. Statist. Assoc.* **67:**614–616.

Ghosh, Malay (1973): "Nonparametric Selection Procedures for Symmetric Location Parameter Populations," *Ann. Statist.* **1:**773–779.

Gupta, Shanti and McDonald, Gary C. (1970): "On Some Classes of Selection Procedures Based on Ranks," in *Nonparametric Techniques in Statistical Inference*, M. L. Puri (ed.), Cambridge University Press, London, pp. 491–514.

Hajek, J. and Sidak, Z. (1967): *Theory of Rank Tests*, Academic Press, New York.

Haynam, G. E., Govindarajulu, Z., and Leone, F. C. (1970): "Tables of the Cumulative Non-central Chi-square Distribution," in *Selected Tables in Mathematical Statistics*, Harter and Owen (eds.), Markham, Chicago, vol. 1, pp. 1–78.

Jonckheere, A. R. (1954): "A Distribution Free k-sample Test against Ordered Alternatives," *Biometrika* **41:**133–145.

Kendall, M. G. (1970): *Rank Correlation Methods*, 4th ed., Griffin, London.

Kiefer, J. (1959): "K-sample Analogues of the Kolmogorov-Smirnov and Cramér-von Mises Tests," *Ann. Math. Statist.* **30:**420–447.

Kraft, Charles H. and van Eeden, Constance (1968): *A Nonparametric Introduction to Statistics*, Macmillan, New York.

Kruskal, William H. (1952): "A Nonparametric Test for the Several Sample Problem," *Ann. Math. Statist.* **23:**525–540.

——— and Wallis, W. Allen (1952, 1953): "Use of Ranks in One-criterion Variance Analysis," *J. Am. Statist. Assoc.* **47:**583–612 (1952), **48:**907–911. [These papers and the preceding one propose the test statistics K and K^*, discuss some of their properties, and prove the asymptotic null distributions.]

Lehmann, E. L. (1963a): "A Class of Selection Procedures Based on Ranks," *Math. Ann.* **150:**268–275.

——— (1963b): "Robust Estimation in Analysis of Variance," *Ann. Math. Statist.* **34:**957–966.

Lewontin, R. C. and Felsenstein, J. (1965): "The Robustness of Homogeneity Tests in $2 \times n$ Tables," *Biometrics* **21:**19–33.

Marascuilo, L. A. (1966): "Large-sample Multiple Comparisons," *Psychol. Bull.* **64:**280–290.

——— and McSweeney, Maryellen (1967): "Nonparametric Post Hoc Comparisons for Trend," *Psychol. Bull.* **67:**401–412.

McDonald, B. J. and Thompson, W. A. (1967): "Rank Sum Multiple Comparisons in One- and Two-way Classifications," *Biometrika* **54:**487–497.

Miles, R. E. (1959): "The Complete Amalgamation into Blocks by Weighted Means of a Finite Set of Real Numbers," *Biometrika* **46:**317–327.

Miller, Rupert G., Jr. (1966): *Simultaneous Statistical Inference*, McGraw-Hill, New York.

Nemenyi, Peter (1963): "Distribution-free Multiple Comparisons," unpublished Ph.D. thesis,

Princeton University. [Proposes the joint ranking procedure (5.78), which is also discussed by Dunn (1964).]

Neyman, J. (1937): "Smooth Tests of Goodness of Fit," *Skand. Aktuarietidskr.* **20**:149–199.

O'Neill, R. and Wetherill, G. B. (1971): "The Present State of Multiple Comparison Methods," *J. Roy. Statist. Soc.* **B33**:218–250.

Patnaik, P. B. (1949): "The Non-central χ^2- and F-distributions and Their Applications," *Biometrika* **36**:202–232.

Peritz, E. (1971): "On a Statistic for Rank Analysis of Variance," *J. Roy. Statist. Soc.* **B33**:137–139.

Puri, Madan Lal (1964): "Asymptotic Efficiency of a Class of c-sample Tests," *Ann. Math. Statist.* **35**:102–121.

——— and Puri, Prem S. (1969): "Multiple Decision Procedures Based on Ranks for Certain Problems in Analysis of Variance," *Ann. Math. Statist.* **40**:619–632.

Rijkoort, P. J. and Wise, M. E. (1953): "Simple Approximations and Monograms for Two Ranking Tests," *Indagationes Math.* **15**:294–302, 407.

Rizvi, M. Haseeb and Woodworth, George G. (1970): "On Selection Procedures Based on Ranks: Counter-examples Concerning Least Favorable Configurations," *Ann. Math. Statist.* **40**:1942–1951.

Sen, P. K. (1966): "On Nonparametric Simultaneous Confidence Regions and Tests for the One Criterion Analysis of Variance Problem," *Ann. Inst. Statist. Math.* **18**:319–336.

Sherman, Ellen (1965): "A Note on Multiple Comparisons Using Rank Sums," *Technometrics* **7**:255–256. [Compares the joint and separate ranking procedures of Chap. 5, Sec. 6 and obtains their asymptotic efficiency relative to the corresponding normal theory procedures.]

Shorack, Galen R. (1967): "Testing against Ordered Alternatives in Model I Analysis of Variance; Normal Theory and Nonparametric," *Ann. Math. Statist.* **38**:1740–1752.

Simon, Gary (1973): "Analysis of the Ordered $2 \times k$ Contingency Table," Technical Report 32, Series 2, Department of Statistics, Princeton University.

Spjøtvoll, Emil (1968): "A Note on Robust Estimation in Analysis of Variance," *Ann. Math. Statist.* **39**:1486–1492.

Steel, Robert G. D. (1959): "A Multiple Comparison Rank Sum Test: Treatment Versus Control," *Biometrics* **15**:560–572. [Proposes the procedure (5.47) and provides some tables.]

——— (1960): "A Rank Sum Test for Comparing All Pairs of Treatments," *Technometrics* **2**:197–207. [Proposes the pairwise ranking procedure (5.70) and provides some tables.]

——— (1961): "Some Rank Sum Multiple Comparison Tests," *Biometrics* **17**:539–552. [Further discussion and additional tables for the procedures (5.47) and (5.70).]

Terpstra, T. J. (1952): "The Asymptotic Normality and Consistency of Kendall's Test against Trend, When Ties Are Present in One Ranking," *Indagationes Math.* **14**:327–333.

Tobach, Ethel; Smith, Mark; Rose, George; and Richter, Donald (1967): "A Table for Rank Sum Multiple Paired Comparisons," *Technometrics* **9**:561–567.

Tryon, Peter V. and Hettmansperger, Thomas P. (1973): "A Class of Nonparametric Tests for Homogeneity against Ordered Alternatives," *Ann. Statist.* **1**:1061–1070.

Wallace, David L. (1959): "Simplified Beta Approximations to the Kruskal-Wallis H Test," *J. Am. Stat. Assoc.* **54**:225–230.

CHAPTER 6
RANDOMIZED COMPLETE BLOCKS

1. RANKS IN RANDOMIZED COMPLETE BLOCKS

Chapter 5 was concerned with the comparison of s treatments when the N available subjects are assigned at random, n_1 to treatment 1, n_2 to treatment 2, and so on. As pointed out earlier in the case of two treatments, this type of comparison is likely to be ineffective when the subjects are highly variable, since existing differences between the treatments are then blurred by the variation among the subjects receiving the same treatment. In such cases, a sharper comparison can often be obtained by dividing the subjects into more homogeneous subgroups or *blocks* and by comparing the subjects receiving different treatments only within these blocks. This chapter will be concerned with such blocked comparisons and thus bears the same relationship to Chap. 5 as did Chap. 3 to Chap. 1.

A particularly simple blocked design is obtained by generalizing the paired-comparison design for the comparison of two treatments discussed in Chap. 3, Sec. 1. Such a *randomized complete block design* for the comparison of s treatments requires blocks of size s, with the s subjects in each block being assigned to the s treatments at random. The blocks may consist of animals from the same litter, of agricultural plots that are close together, of students that are matched by class and past performance, or patients matched on such variables as age, sex, and severity of disease.

Important examples are also provided by situations in which the s treatments can be compared on a single subject, for example, in the comparison of a number of different headache remedies or sleeping pills, or in taste tests for comparing a number of different products. In such cases, in which each subject constitutes a block, it is often the order of application of the s streatments that is randomized.

EXAMPLE 1. *Three tranquilizers.* Suppose that three brands of tranquilizers, A, B, and C, are to be compared as in Chap. 5, Example 1 and that 12 patients are available for the study. In the earlier consideration of such a comparison, the patients were assumed to be "comparable," and it was therefore thought appropriate to assign them to the three brands completely at random. Suppose, however, that the patients vary widely in their response to such medication. Then the random assignment will tend to balance the subjects, so that to each treatment are assigned some subjects who react strongly and some who do not. This balance, which is a principal purpose of the random assignment, will lead to a corresponding balancing of the ranks and hence to acceptance of the null hypothesis even when the treatments differ in their effectiveness unless, of course, the treatment differences are so pronounced that they can overwhelm the differences among the subjects.

For this reason it is decided to restrict the comparisons to patients whose responses may be expected to be comparable. The possibility is considered of having each patient try all three brands, in random order. However, this proposal is found to entail a number of difficulties. In particular, it is feared that there may be a carry-over effect of unknown duration, which would confound the effects of the different treatments. Also, if each subject was given all three drugs, the study would take three times as long, which is felt to be undesirable. As a compromise, it is therefore decided to divide the 12 subjects into four groups of 3 each, in such a way that the three patients in any one group are fairly homogeneous in their responsiveness to tranquilizers. The three patients within each group are assigned at random to the three treatments, with the assignments in different groups being made independently of each other.

At the end of the study, some measure of the effectiveness of the treatment is obtained for each of the 12 patients, and on this basis the 3 patients in each group are ranked (from 1 to 3) with the following results:

Treatment \ Group	1	2	3	4	Total
A	3	2	3	3	11
B	2	3	1	1	7
C	1	1	2	2	6

Let H denote the hypothesis that there is no difference in the effect of the three brands, and consider one of the groups, say the third. The usual argument (see, for example, Chap. 5, Sec. 1, Example 1) shows that each of the following $3! = 6$ possible orders of the three ranks

1 2 3 | 1 3 2 | 2 1 3 | 2 3 1 | 3 1 2 | 3 2 1

is equally likely under H, and therefore has probability $1/6$. Since the assignments in the different groups are assumed to be independent, the probability of any of the 6^4 possible arrangements of the four columns of ranks is $(1/6)^4$.

These considerations easily generalize. In the comparison of s treatments, the responses of the subjects in each block are ranked separately from 1 to s. Thus, if R_{ij} denotes the rank of the ith treatment in the jth block, the ranks $(R_{1j}, \ldots, R_{sj})$ for each j are the integers $(1, \ldots, s)$ in some order, that is, a permutation of the integers $(1, \ldots, s)$. Under the hypothesis of no difference among the treatments, the random assignment of the s treatments to the subjects of each block implies that all

$$s! = 1 \cdot 2 \cdot \cdots \cdot s$$

possible orders of $(R_{1j}, \ldots, R_{sj})$ are equally likely, so that each has probability $1/s!$. Since the assignments, and hence the ranks, in different blocks are independent, the overall null distribution of the ranks is given by

$$P_H(R_{11} = r_{11}, \ldots, R_{s1} = r_{s1}; \ldots; R_{1N} = r_{1N}, \ldots, R_{sN} = r_{sN}) = \left(\frac{1}{s!}\right)^N \tag{6.1}$$

where N is the number of blocks.

2. THE TESTS OF FRIEDMAN, COCHRAN, AND McNEMAR

Many different alternatives can be envisaged to the hypothesis H of no difference among s treatments. As in Chap. 5, Sec. 2, we shall here consider the problem of testing H against the alternatives that the treatments affect mainly the level of the response (rather than, for example, its variability). There is then an ordering of the treatments, one tending to produce the lowest responses, another the next lowest, and so on. An indication of the position of the ith treatment in this ordering is provided by the average rank

$$R_{i.} = \frac{R_{i1} + \cdots + R_{iN}}{N}$$

which it achieves over the N blocks. Many questions concerning the ordering pattern of the treatments may be of interest (some of which will be discussed in Secs. 5B and 5C), but for the moment we shall consider only the simple problem of testing H.

If the treatments differ widely among each other, this will typically be reflected in large differences among the $R_{i.}$'s. On the other hand, when H is true, chance sees to it that the $R_{i.}$'s tend to be close to each other and hence also to the overall average

$$R_{..} = \frac{(R_{11} + \cdots + R_{s1}) + \cdots + (R_{1N} + \cdots + R_{sN})}{sN}$$

Since the numerator is equal to $N\frac{1}{2}s(s+1)$, it is seen that

$$R_{..} = \tfrac{1}{2}(s+1) \tag{6.2}$$

A convenient criterion for measuring the overall closeness of the $R_{i.}$ to $\frac{1}{2}(s+1)$ is the sum of the squared differences $[R_{i.}-\frac{1}{2}(s+1)]^2$ or the *Friedman* statistic

$$Q = \frac{12N}{s(s+1)}\sum_{i=1}^{s}[R_{i.}-\tfrac{1}{2}(s+1)]^2 \tag{6.3}$$

where the factor is chosen so as to provide a simple approximation to the null distribution when N is large. The statistic Q is zero when the $R_{i.}$ are all equal and is large when there are substantial differences among the $R_{i.}$. The hypothesis is therefore rejected when

$$Q \geqslant c \tag{6.4}$$

An alternative expression for Q that is computationally more convenient is obtained by squaring the terms on the right-hand side of (6.3) and replacing $R_{i.}$ by R_i/N, where

$$R_i = R_{i1} + \cdots + R_{iN} \tag{6.5}$$

is the rank sum associated with the ith treatment. It is then easy to see (Prob. 16) that

$$Q = \frac{12}{Ns(s+1)}\sum_{i=1}^{s} R_i^2 - 3N(s+1) \tag{6.6}$$

The critical value c in (6.4) is obtained from the null distribution of Q. To illustrate the computation, consider the case $s = 3$, $N = 4$ of Example 1. The 12 ranks

$$R_{11} \quad R_{12} \quad R_{13} \quad R_{14} \qquad R_{21} \quad R_{22} \quad R_{23} \quad R_{24} \qquad R_{31} \quad R_{32} \quad R_{33} \quad R_{34}$$

can then take on $(3!)^4 = 1{,}296$ different sets of values, each of which determines a value of Q. Thus, in Example 1,

$$\begin{array}{llll} R_{11} = 3 & R_{12} = 2 & R_{13} = 3 & R_{14} = 3 \\ R_{21} = 2 & R_{22} = 3 & R_{23} = 1 & R_{24} = 1 \\ R_{31} = 1 & R_{32} = 1 & R_{33} = 2 & R_{34} = 2 \end{array}$$

so that $R_1 = 11$, $R_2 = 7$, $R_3 = 6$. [A check is obtained by noting that $R_1 + \cdots + R_s$ is the sum of all the ranks and hence by (6.2)

$$R_1 + \cdots + R_s = \tfrac{1}{2}Ns(s+1) \tag{6.7}$$

For $s = 3$, $N = 4$, the right-hand side of (6.7) is 24, which agrees with $11+7+6$.] Substituting the values of the R_i in (6.6) shows that

$$Q = \frac{12}{4 \times 3 \times 4}(121+49+36) - 3 \times 4 \times 4 = 3.5$$

In this way, the value of Q could be computed for each of the 1,296 cases. Since each has probability 1/1,296 this would provide the null distribution of Q.

Tables of this distribution are given by Friedman (1937), who first proposed the test, and by Owen (1962), Kraft and van Eeden (1968), and Kendall (1970). For the case $s = 3$, $N = 4$ they show that $P_H(Q \geqslant 3.5) = .273$. In spite of the rather striking appearance of superiority of treatment A over B and C, the test therefore does not establish significance of the results at the usual levels. [A table of the probabilities less than or equal to .15 for $s = 3$, $N = 2(1)15$ and $s = 4$, $N = 2(1)8$ is given as Table M at the end of the book.]

For values of s and N beyond the range of the tables, there is again available an approximation. As in the case of the Kruskal-Wallis statistic (Chap. 5, Sec. 2), the probability $P_H(Q \geqslant c)$ can be approximated by the area to the right of c under the χ^2-curve with $s-1$ degrees of freedom, so that

$$P_H(Q \geqslant c) \approx \Psi_{s-1}(c) \tag{6.8}$$

Tables and critical values of Ψ are given in Table J. The approximation (6.8) is supported by a limit theorem, which states that the left-hand side tends to the right-hand side as $N \to \infty$. A proof of this theorem is given in the Appendix, Example 29 (Sec. 8).

EXAMPLE 2. *Effectiveness of hypnosis.* In a study of hypnosis, the emotions of fear, happiness, depression, and calmness were requested (in random order) from each of eight subjects during hypnosis. The following table gives the resulting measurements of skin potential (adjusted for initial level) in millivolts:[1]

Subject	1	2	3	4	5	6	7	8
Fear	23.1	57.6	10.5	23.6	11.9	54.6	21.0	20.3
Happiness	22.7	53.2	9.7	19.6	13.8	47.1	13.6	23.6
Depression	22.5	53.7	10.8	21.1	13.7	39.2	13.7	16.3
Calmness	22.6	53.1	8.3	21.6	13.3	37.0	14.8	14.8

Ranking the observations in each of the eight blocks yields the following table of ranks:

Subject	1	2	3	4	5	6	7	8	Total
Fear	4	4	3	4	1	4	4	3	27
Happiness	3	2	2	1	4	3	1	4	20
Depression	1	3	4	2	3	2	2	2	19
Calmness	2	1	1	3	2	1	3	1	14

As a check, note that the sum of all the ranks is $\frac{1}{2}Ns(s+1) = 80$ which agrees with the sum of totals $27+20+19+14 = 80$. The values $N = 8$, $s = 4$, $R_1 = 27$, $R_2 = 20$, $R_3 = 19$, $R_4 = 14$ give $\sum R_i^2 = 1{,}686$ and $Q = 6.45$. From Table J, the probability $P(Q \geqslant 6.45)$ that a χ^2-variable with three degrees of freedom exceeds 6.45 is seen to be .09. The results are thus

[1] From Damaser, Shore, and Orne, "Physiological Effects during Hypnotically Requested Emotions," *Psychosomatic Medicine* **25**:334–343 (1963), Table 2. The data have been rounded for the sake of simplicity.

not significant at the 5 percent level. (Table M shows the exact significance probability to be .091.)

The accuracy of the χ^2-approximation, which was seen to be quite adequate in the example, tends to improve as s and N increase. Certain alternative approximations are discussed by Kendall and Smith (1939), Friedman (1940), Moses (1953), and Rijkoort and Wise (1953).

So far, we have assumed that there are no ties among the observations within any one block. (Ties among observations in different blocks cause no difficulty since only observations in the same block are being ranked.) In the presence of ties, we can use the method of midranks described in Chap. 1, Sec. 4. In the present case of randomized blocks, the midrank of an observation is the average of the ranks *in the same block* with which the observation is tied. Suppose that the s observations in the first block take on e_1 distinct values, with d_{11} of the observations being equal to the smallest value, d_{21} to the next smallest, and so forth. Similarly, the s observations in the second block take on e_2 distinct values, with d_{12} of the observations being equal to the smallest value, d_{22} to the next smallest, etc. As an example, suppose that $s = 5$ and that the observations in the first two blocks are 1.3, 1.1, 1.1, 1.6, 1.1 and 1.9, 1.7, 1.9, 1.9, 1.7. Then $e_1 = 3$ with $d_{11} = 3$, $d_{21} = 1$, $d_{31} = 1$, and $e_2 = 2$ with $d_{12} = 2$, $d_{22} = 3$.

Let the midrank of the subject receiving the ith treatment in the jth block be denoted by R_{ij}^* and the sum of the midranks associated with the ith treatment by

$$R_i^* = R_{i1}^* + \cdots + R_{iN}^*$$

If the statistic Q^* is defined by

(6.9)
$$Q^* = \frac{[12/Ns(s+1)]\sum R_i^{*2} - 3N(s+1)}{1 - \sum_{j=1}^{N} \sum_{i=1}^{e_j} (d_{ij}{}^3 - d_{ij})/Ns(s^2-1)}$$

the hypothesis of no difference among the treatments is rejected if

(6.10)
$$Q^* \geqslant c$$

The null distribution, which in principle can be obtained by enumeration (Prob. 11), will in practice again be approximated by the χ^2-distribution with $s-1$ degrees of freedom. Typically, the approximation will work reasonably well provided $sN \geqslant 30$, with the accuracy also depending on the configuration of ties. A limit theorem supporting the approximation is given in the Appendix, Example 30 (Sec. 8).

EXAMPLE 3. *Association and memory.* In a study of the effect of emotional associations on memory, each of 15 subjects was asked to recall the titles of 18 stories, which he had written earlier. Of these stories, six each had been given a positive rating (+), six a negative rating

(−), and six had been left unmarked (0). The ratings were believed by the subjects to be meaningful, but they had in fact been distributed at random. The following table gives the number of forgotten titles in each category:[1]

	Subject														
	1	2	3	4	5	6	7	8	9	10	11	12	13	14	15
+	3	5	3	2	1	1	2	3	1	1	3	2	0	1	1
0	3	1	0	4	0	4	2	1	4	5	3	1	3	1	3
−	4	3	3	5	1	3	3	5	1	4	4	3	2	2	0

Let us test the hypothesis that the rating has no significant effect on the ability to recall the title against the alternatives that it has an unspecified effect. It is, for example, conceivable that the stories with positive and negative ratings are recalled more easily than those left unmarked, or that those with positive rating are most easily recalled and those with negative rating least easily, and so on.

For this purpose, the three scores for each subject are ranked. The resulting midranks are shown in the following table:

	Subject															
	1	2	3	4	5	6	7	8	9	10	11	12	13	14	15	R^*
+	1.5	3	2.5	1	2.5	1	1.5	2	1.5	1	1.5	2	1	1.5	2	25.5
0	1.5	1	1	2	1	3	1.5	1	3	3	1.5	1	3	1.5	3	28
−	3	2	2.5	3	2.5	2	3	3	1.5	2	3	3	2	3	1	36.5

As a check, note that the sum of all the midranks is $\frac{1}{2}Ns(s+1) = 90$, which agrees with $R_+^* + R_0^* + R_-^* = 25.5 + 28 + 36.5$. Substituting the values $N = 15$, $s = 3$, and the values for the R^*'s in (6.9), the numerator of Q^* is found to be equal to 4.433. To compute the denominator, recall that the d's are the numbers of observations tied in the various blocks. Seven of the d's are equal to 2 and 31 are equal to 1, so that

$$\sum\sum(d_{ij}^3 - d_{ij}) = 7 \times 6 = 42$$

It follows that the denominator of Q^* is equal to .883 and $Q^* = 5.021$. Since $P(\chi_2^2 \geqslant 5.021) = .082$, the results are not significant at the 5 percent level.

What happens to the Friedman statistic Q when the number s of treatments is 2, so that the complete randomized block design reduces to the paired-comparisons design considered in Chaps. 3 and 4? In this case, the two ranks in each block are 1 and 2. Suppose that in A of the blocks treatment 1 receives rank 1 and treatment 2 rank 2, with the situation reversed in the remaining $N - A$ blocks. Then

$$R_1 = A + 2(N - A) = 2N - A \quad ; \quad R_2 = N - A + 2A = N + A$$

Using the expression (6.3) for Q and the fact that $s = 2$, it is seen that

$$Q = 4N\left(\frac{A}{N} - \frac{1}{2}\right)^2$$

[1] Data from Korner (1950), "Experimental Investigation of Some Aspects of the Problem of Repression: Repressive Forgetting," *Contributions to Education*, no. 970, Teacher's College, Columbia University, Table III.

The hypothesis is rejected when Q, or equivalently $|A/N-\frac{1}{2}|$, is sufficiently large. Since A is the number of pairs in which the second treatment comes out ahead of the first, this is seen to be the two-sided sign test, the one-sided version of which was discussed in Chaps. 3 and 4.

As was pointed out in these chapters (particularly in Chap. 4, Sec. 1), the sign test finds its principal usefulness in cases where it is known for each pair which of the two treatments came out ahead, but not by how much. If this additional information is available, there exist tests such as the paired-comparison Wilcoxon test, which are typically more effective. This remark generalizes to Friedman's Q-test. For s greater than 2, this test continues to be best suited to situations in which only the ranks of the s treatments in each block are available, but no more detailed numerical observations. We shall discuss an alternative, more effective procedure when such additional information is available in the next section. The disadvantage of the Q-test is most serious for small s and decreases as the number of treatments becomes larger. (For details see Sec. 4.) Besides this drawback, the Friedman test also shares with the sign test the advantage of simplicity and ease of computation. In particular, the fact that the observations need to be ranked only within each block makes the test convenient if ease of computation is a consideration.

The test statistic Q^* defined by (6.9) can be given a somewhat simpler form in the important special case that the responses are dichotomous: yes or no, cure or no cure, success or failure. If a success is then represented by a one and a failure by a zero, the responses in each block constitute a set of s zeros and ones. Let L_j denote the number of successes in the jth block and B_i the number of successes for the ith treatment. Then we shall show below that Q^* reduces to the statistic proposed by Cochran (1950):

$$Q^* = \frac{s(s-1)\sum_{i=1}^{s}(B_i-\bar{B})^2}{s\sum_{j=1}^{N}L_j-\sum_{j=1}^{N}L_j^2} \tag{6.11}$$

where $\bar{B}$ denotes the average $(B_1+\cdots+B_s)/s$. The computation is simplified if the numerator is replaced by the equivalent expression

$$(s-1)[s\sum B_i^2-(\sum B_i)^2] \tag{6.12}$$

EXAMPLE 4. *Different methods of soothing babies.* In a comparison of different methods of soothing young babies, four methods (warm water, rocking, pacifier, sound) were applied to each of 12 neonates. Each of the 48 combinations was classified as successful (+) or unsuccessful (−), with the following results:[1]

[1] Adapted from the data for the first 12 subjects reported in Prob. 10.

	Subject 1	2	3	4	5	6	7	8	9	10	11	12	Total +
W	−	−	−	−	+	−	+	−	−	−	+	+	4
R	−	+	−	−	+	+	+	+	+	−	+	+	8
P	−	−	+	−	+	+	−	−	+	+	−	+	6
S	−	−	−	−	+	+	−	+	+	−	+	+	6
	0	1	1	0	4	3	2	2	3	1	3	4	24

Here $N = 12$ and $s = 4$. The subjects form the blocks, and we suppose that the four treatments were applied to each subject in random order. The L's are the block totals: $L_1 = 0, L_2 = 1, \ldots$; and the B's are the treatment totals: $B_1 = 4$, $B_2 = 8, \ldots$. Substituting these values in (6.11) and (6.12), it is seen that $Q^* = 8/17 = .47$, and Table J of the χ^2-distribution with $s-1 = 3$ degrees of freedom shows the significance probability to be $P(\chi_3^2 \geqslant .47) = .93$. The test therefore indicates no significant differences among the treatments.

A case of the Q^*-test that is of special interest is the case $s = 2$, where the blocks reduce to pairs and there are four possible responses for each block: (0,0), (0,1), (1,0), and (1,1). The numbers A, B, C, and D of these four types constitute the results of the experiment and can be shown as in Table 6.1.

TABLE 6.1 Matched pairs experiment with dichotomous response

(0,0)	(0,1)	(1,0)	(1,1)	Total
A	B	C	D	N

More commonly, although perhaps less conveniently, they are displayed in a 2×2 table as in Table 6.2.

TABLE 6.2 2×2 table for matched pairs

		Treatment 2 F	S
Treatment 1	F	A	B
	S	C	D

It is seen that $B_1 = C+D$ and $B_2 = B+D$. On the other hand, A of the L_j's are equal to 0, $B+C$ equal to 1, and D equal to 2, so that $\sum L_j = B+C+2D$ and $\sum L_j{}^2 = B+C+4D$. On substituting these expressions into (6.11) and (6.12) and putting $s = 2$, one finds that (Prob. 20)

$$Q^* = \frac{(B-C)^2}{B+C} \tag{6.13}$$

The statistic (6.13), which was introduced by McNemar (1947), therefore has a limiting χ^2-distribution with $s-1 = 1$ degree of freedom. For recent discussions of the above and related tests, and references to the literature, see Altham (1971) and Nam (1971).

There is a simpler way of looking at McNemar's test, which gives an exact null distribution and provides the basis for a one-sided test. Note that (6.13) does not involve A and D. In fact, under the usual hypothesis of random assignment within each pair, the values of A and D are fixed rather than random, and so therefore is $B+C=k$. Each of the k pairs in which there is one success and one failure has probability $1/2$ of being assigned to either (1,0) or (0,1). Since the assignments in the different pairs are independent, the number B of cases (0,1) is thus a binomial random variable corresponding to k trials and success probability $1/2$.

The hypothesis is rejected against the one-sided alternatives that treatment 2 tends to increase the number of success if $B \geqslant c$ and against two-sided alternatives if $|B-\frac{1}{2}k| \geqslant c$. Here the critical value c is determined for small values of k from the binomial distribution with k trials and $p = 1/2$ and for large values of k from the normal approximation to this distribution discussed in Chap. 3, Sec. 1. This latter approximation is equivalent to (6.13).

The above tests for matched pairs correspond to the tests for the 2×2 table, Table 5.2, where there is no matching. It is important to note the different organization of this earlier table from the present Table 6.2. For a more detailed discussion of the differences between the two tables, see Mosteller (1952).

To complete the section, let us now prove formula (6.11). To this end, we must relate the quantities $L_1, \ldots, L_N$ and $B_1, \ldots, B_s$ to those used in defining Q^*. In the general discussion leading up to this definition, it was assumed that the observations are ranked separately in each block, and the quantities d_{ij} denote the numbers of observations tied at the ith value in the jth block. Of the s observations in the first block, for instance (in Example 4, this is the first column), d_{11} are tied at the smallest value, d_{21} at the next smallest, etc. In the present case, therefore, d_{11} is the number of zeros in the first block and d_{21} the number of ones, so that $d_{21} = L_1$ and $d_{11} = s-L_1$ (in Example 4, $s = 4$, $d_{21} = 0$, and $d_{11} = 4$). Similarly, the jth block consists of $d_{2j} = L_j$ ones and $d_{1j} = s-L_j$ zeros. The midranks of the $s-L_j$ zeros are

$$\frac{1+2+\cdots+(s-L_j)}{s-L_j} = \tfrac{1}{2}(s-L_j+1)$$

and the midranks of the L_j ones are

$$s-L_j+\tfrac{1}{2}(L_j+1) = s-\tfrac{1}{2}L_j+\tfrac{1}{2}$$

[see (1.41)].

Let us now convert the observations of the original table, which are all zeros and ones, to midranks. Then each zero or one in the jth block is replaced, respectively, by $\frac{1}{2}(s-L_j+1)$ or by $s-\frac{1}{2}L_j+\frac{1}{2}$. The sum of the midranks for the ith treatment is then

$$R_i^* = \textstyle\sum^{(1)}\frac{1}{2}(s-L_j+1)+\sum^{(2)}(s-\frac{1}{2}L_j+\frac{1}{2})$$

where $\Sigma^{(1)}$ extends over the $N-B_i$ blocks in which the ith treatment is a failure and $\Sigma^{(2)}$ over the B_i blocks in which it is a success. Combining the terms in L_j from the two sums, it follows that

$$R_i^* = (N-B_i)\tfrac{1}{2}(s+1)+B_i(s+\tfrac{1}{2})-\tfrac{1}{2}\sum_{j=1}^{N} L_j$$

If one now squares the right-hand side and sums over i, it turns out after some simplification and use of the fact that $\Sigma B_i = \Sigma L_j$ that (Prob. 19)

$$\sum_{i=1}^{s} R_i^{*2} = \frac{s^2}{4}\sum_{i=1}^{s}(B_i-\bar{B})^2+\frac{s(s+1)^2N^2}{4}$$

Hence the numerator of Q^* reduces to $3s\Sigma(B_i-\bar{B})^2/N(s+1)$.

Consider now the denominator. Since $d_{1j} = s-L_j$ and $d_{2j} = L_j$,

$$\sum_i\sum_j(d_{ij}^3-d_{ij}) = \sum_j[(s-L_j)^3-(s-L_j)+(L_j^3-L_j)]$$

The terms in L_j^3 cancel, and after some simplification (Prob. 19), the denominator is seen to be equal to

$$\frac{3(s\Sigma L_j-\Sigma L_j^2)}{N(s^2-1)}$$

Together with the expression for the numerator this proves formula (6.11).

3. ALIGNED RANKS

The Friedman test for the comparison of s treatments was shown in the preceding section to reduce to the sign test when $s = 2$, that is, for the problem of paired comparisons. For this case, the Wilcoxon signed-rank test provides a more effective alternative, and it is therefore natural to seek a similar improvement of the Friedman test for other values of s. The relatively low sensitivity of the Friedman test, which is most pronounced for small values of s and becomes less serious as s increases, is a result of the fact that separate ranking in each block provides comparisons only of the responses within each block. A direct comparison of responses in different blocks is not meaningful because of the variation between blocks, of which some may give consistently high and others consistently low responses. The different blocks can, however, be made more comparable by subtracting from the observations in each block some estimate of the location of the block, such as the average of the observations in the block or the median of these observations. This method of "aligning" the observations was already discussed for blocked comparison of two treatments in Chap. 3, Sec. 3.

EXAMPLE 5. *Effect of hypnosis* (*continued*). Consider once more the data on the effectiveness of hypnotic suggestion given in Example 2. Here there are eight blocks corresponding to the eight different subjects. From the observations in the jth block, let us subtract the average of this block. The averages are

22.725 54.4 9.825 21.475 13.175 44.475 15.775 18.75

and the aligned observations are

Subject	1	2	3	4	5	6	7	8
Fear	.375	3.2	.675	2.125	−1.275	10.125	5.225	1.55
Happiness	−.025	−1.2	−.125	−1.875	.625	2.625	−2.175	4.85
Depression	−.225	−.7	.975	−.375	.525	−5.275	−2.075	−2.45
Calm	−.125	−1.3	−1.525	.125	.125	−7.475	−.975	−3.95

The ranks (or midranks in the case of ties) of these aligned observations are found to be

Subject	1	2	3	4	5	6	7	8	Total
Fear	21	29	24	27	10	32	31	26	200
Happiness	18	11	16.5	7	23	28	5	30	138.5
Depression	15	13	25	14	22	2	6	4	101
Calm	16.5	9	8	19.5	19.5	1	12	3	88.5
Total	70.5	62	73.5	67.5	74.5	63	54	63	528

Here the last row shows the rank sums for the eight blocks and the last column the rank sums for the four treatments. A check is obtained by adding the entries in the last row, and separately those in the last column. They must of course agree and must be equal to the sum of the integers from 1 to sN, which is $\frac{1}{2}sN(sN+1)$, or in our case $\frac{1}{2}\times 4\times 8\times 33 = 528$.

In general, denote by $\hat{R}_{ij}$ the aligned rank (or midrank) of the observation in the jth block that corresponds to the ith treatment. Suppose now that there is no treatment effect, so that the s treatments are in fact identical. Then each block would have received the same set of aligned ranks regardless of which subjects (or positions in the block) were assigned to which treatment. Thus, in the example, the four responses in the first block would in any case have been 23.1, 22.7, 22.5, 22.6; therefore, the aligned observations would have been .375, −.025, −.225, −.125; and the aligned ranks 21, 18, 15, 16.5. The $s!$ possible assignments (in the example, 4! possible orders of the "treatments") in any given block would simply result in $s!$ permutations of the aligned ranks in this block. The total number of equally likely permutations within the N sets of aligned ranks in all the blocks is $(s!)^N$ and the null distribution of the aligned ranks is therefore

$$P_H(\hat{R}_{11} = \hat{r}_{11}, \ldots, \hat{R}_{s1} = \hat{r}_{s1}; \ldots; \hat{R}_{1N} = \hat{r}_{1N}, \ldots, \hat{R}_{sN} = \hat{r}_{sN}) = \frac{1}{(s!)^N} \tag{6.14}$$

It should be emphasized once more that in this distribution not all $(sN)!$ permutations of the $\hat{r}_{ij}$ are permissible but only permutations among the aligned ranks within each block. Thus, (6.14) is completely analogous to the distribution (6.1), with the N sets of aligned ranks of the N blocks now playing the role of the N s-tuples $(1, \ldots, s)$ assigned to the N blocks in (6.1).

Under the hypothesis, one would expect the aligned ranks achieved by the different treatments to average out over the N blocks, so that the average ranks

$$\hat{R}_{i.} = \frac{\hat{R}_{i1} + \cdots + \hat{R}_{iN}}{N}$$

would be close to each other. Since the sum of all sN aligned ranks is $1+\cdots+sN=\frac{1}{2}sN(sN+1)$, it follows that

$$\sum \hat{R}_{i.} = \tfrac{1}{2}s(sN+1)$$

If the $\hat{R}_{i.}$ are close to each other, each must therefore be close to $\frac{1}{2}(sN+1)$. By analogy with (6.3) this suggests as a test statistic a suitable multiple of

$$\sum [\hat{R}_{i.} - \tfrac{1}{2}(sN+1)]^2$$

Since this choice leads to a convenient limit distribution, we shall in fact use

$$\hat{Q} = \frac{s-1}{\sum\sum (\hat{R}_{ij} - \hat{R}_{.j})^2} \sum_{i=1}^{s} [N\hat{R}_{i.} - \tfrac{1}{2}N(sN+1)]^2 \tag{6.15}$$

where

$$\hat{R}_{.j} = \frac{\hat{R}_{1j} + \cdots + \hat{R}_{sj}}{s}$$

At first sight, the factor may be surprising, since it appears to be a random variable rather than a constant. This impression is, however, misleading. For consider

$$\sum\sum (\hat{R}_{ij} - \hat{R}_{.j})^2 = \sum_{i=1}^{s} \sum_{j=1}^{N} \hat{R}_{ij}{}^2 - \frac{1}{s} \sum_{j=1}^{N} (s\hat{R}_{.j})^2 \tag{6.16}$$

The first term on the right-hand side is a constant because it is simply the sum of squares of all the ranks or midranks. Moreover, $s\hat{R}_{.j}$ is the sum of the ranks in the jth block, and since this set of s ranks is fixed with only the order being random, the quantities $s\hat{R}_{.j}$ are also constants and so therefore is the second term on the right-hand side of (6.16).

For computational purposes, it is typically more convenient to square out the terms in both the numerator and denominator sum of squares of $\hat{Q}$. After slight simplification one finds for the numerator

$$\sum_{i=1}^{s} [N\hat{R}_{i.} - \tfrac{1}{2}N(sN+1)]^2 = \sum_{i=1}^{s} (N\hat{R}_{i.})^2 - \tfrac{1}{4}sN^2(sN+1)^2 \tag{6.17}$$

If there are no ties, the expression (6.16) for the denominator can be simplified further since then $\sum\sum \hat{R}_{ij}{}^2$ is the sum of squared integers from 1 to sN and hence

$$\sum\sum \hat{R}_{ij}{}^2 = \frac{sN(sN+1)(2sN+1)}{6}$$

Except for very small values of s and N, the distribution of $\hat{Q}$ is too cumbersome to compute, and tables are not feasible since the distribution (and even the values of $\hat{Q}$) depend on the way the aligned ranks are distributed over the N blocks. However, for large values of sN, the statistic $\hat{Q}$ is approximately distributed as χ^2 with $s-1$ degrees of freedom (see Appendix, Prob. 75). Some numerical values of the accuracy of this approximation are given by Gilbert (1972).

As an illustration let us now compute $\hat{Q}$ for the data of Example 5. We find the expressions (6.16) and (6.17) to be equal to 2644.75 (Prob. 14) and 7915.5, respectively, and hence $\hat{Q} = 3 \times 7519.5/2644.75 = 8.53$. From Table J, it is seen that $P(\chi_3^2 \geqslant 8.53) = .036$, compared with a significance probability of .09 when the Friedman test was used. With the additional information obtained through alignment, the data have become significant at the 5 percent level.

Alignment was described at the beginning of the section as "subtracting from the observations in each block some estimate of the location of the block." It should have been mentioned that this estimate must be a symmetric function of the observations of the block (this condition is of course satisfied by both the block mean and the block median). On the other hand, it is not necessary to restrict alignment to location. If the blocks show substantial differences in scale, for example, it may be desirable to adjust the original observations for both location and scale and use as the aligned observations variables of the form

$$\hat{X}_{ij} = \frac{(X_{ij} - \hat{b}_i)}{\hat{a}_i}$$

where $\hat{a}_i$ and $\hat{b}_i$ are symmetric functions of the observations in the ith block. The null distribution of the corresponding aligned ranks will then continue to be given by (6.14), so that the null distribution of $\hat{Q}$ will be as before. Similar adjustments are possible also to compensate for other differences among the blocks.

4. POPULATION MODELS AND EFFICIENCY

The randomization model of Sec. 1, which has constituted the framework for the preceding sections, has the usual disadvantage of restricting the inference to the subjects in the study. We shall now consider population models that allow extrapolating from these subjects to populations of potential users of the treatments under investigation. At the same time, such models provide the basis for determining the efficiencies of the tests based on separate and aligned ranking.

(i) Suppose first that the blocks represent different laboratories, hospitals, farms, or different teachers or doctors, to which sN subjects, drawn at random from a large population Π, are assigned at random, s to each block. Let the response X_{ij} of a random subject from Π be distributed according to

(6.18) $$P(X_{ij} \leqslant x) = F_{ij}(x) \qquad (i = 1, \ldots, s; j = 1, \ldots, N)$$

if it is assigned to block j and receives treatment i. The hypothesis of no differences among the treatments then states that

(6.19) $$H: \quad F_{1j} = \cdots = F_{sj} \qquad \text{for each } j = 1, \ldots, N$$

Let us now add the assumption that the population Π is large enough so that the subjects in the sample, and hence the variables X_{ij}, can be taken to be independent. Then we have the population model:

(6.20) The random variables X_{ij} are independent with distributions $F_{ij}(i = 1, \ldots, s;$ $j = 1, \ldots, N)$; the hypothesis to be tested is that these distributions satisfy (6.19).

(ii) Actually, somewhat different assumptions frequently are more realistic. The patients at different hospitals or of different doctors, or the students in different schools, will often come from different localities with different distributions of religion, education, income, and so forth. It is then more reasonable to assume that one is dealing with random samples from different populations or strata, each corresponding to a different hospital, doctor, or school. The reason for taking such a stratified sample (i.e., a random sample of given size from each stratum) is of course the same as that discussed for blocking at the beginning of the chapter. The total population is assumed to be much less homogeneous than the individual strata, and the comparisons can then be sharpened by taking the differences between the strata into account. This can be done either through separate ranking and by considering only comparisons within each stratum, or by trying to eliminate the strata differences through alignment.

Let $\Pi_1, \ldots, \Pi_N$ denote the different strata, and suppose that a sample of size s is drawn from each. If X_{ij} denotes the response of a subject drawn at random from Π_j and receiving treatment i, and if F_{ij} is the distribution of X_{ij}, the hypothesis of no treatment differences is again represented by (6.19). If in addition the subpopulations $\Pi_1, \ldots, \Pi_N$ are large enough so that the subjects in the sample can be taken to be independent, we are thus again led to Model (6.20).

(iii) In Model (i), the blocks were assumed to be fixed and subjects were assigned to them at random. Instead, the blocks themselves may be drawn at random from a population of blocks; for example, in taste tests where the block is a subject comparing a number of flavors, or in the examples of Model (i) when the hospitals, teachers, doctors, and so on, which constitute the blocks, have been drawn at random from some population. The s responses in a block are then dependent since they are all influenced by the particular block produced by the sampling process. In this case, the N s-tuples $(X_{1j}, \ldots, X_{sj})$ are independently, identically distributed according to an s-variate distribution M. The hypothesis of no treatment effect finds its mathematical expression in the assumption that M is symmetric in its s arguments, i.e., that

$$M(x_{i_1}, \ldots, x_{i_s}) = M(x_1, \ldots, x_s)$$

for all permutations $(i_1, \ldots, i_s)$ of $(1, \ldots, s)$.

In Models (i) and (ii), the random variables $X_{1j}, \ldots, X_{sj}$ in the jth block are—under H—independent and identically distributed. If this common distribution is continuous, it follows from Chap. 5, Prob. 46 that for separate ranking all $s!$ permutations of the ranks $1, \ldots, s$ in the jth block are equally likely so that the null distribution (6.1) continues to hold. The null distribution of these ranks is given by (6.1) also in Model (iii), as is clear from symmetry.

It follows from these considerations that the Friedman statistic Q will have the same null distribution in Models (i) to (iii) as in the randomization model of Sec. 1.

When the joint distribution of the X's is not continuous, so that ties can occur, the distribution of the midranks depends on this underlying distribution. However, as in the population models considered in Chaps. 2, 4, and 5, the conditional distribution of the midranks given the numbers d_{ij} of ties is again distribution-free and coincides with the distribution of the midranks in the randomization model. Thus, in particular, the conditional null distribution of the variable B defined in Tables 6.1 and 6.2, given that $B+C=k$, continues to be the binomial distribution corresponding to k trials and success probability $\frac{1}{2}$.

The above results for Q are not true for $\hat{Q}$, whose null distribution depends on the joint distribution of the X's, even when the latter is continuous. The situation is quite analogous to that caused by the presence of ties. A distribution-free test can again be obtained by conditioning; this time, on the configuration of the aligned ranks in the various blocks, that is, the values of the aligned ranks in block 1, block 2, ..., block N. The conditional distribution (given the configuration) of the aligned ranks associated with the various treatments is in fact the same as the distribution of these ranks discussed for the randomization model in the preceding section. An analysis based on this conditional distribution of course loses the advantages of the population model. It turns out, however, that for large N the conditional null distribution of $\hat{Q}$ tends to the χ^2-distribution with $s-1$ degrees of freedom for all configurations (Appendix, Prob. 75) so that asymptotically the $\hat{Q}$-test can be interpreted as unconditional.

The $\hat{Q}$-test is considerably more laborious than the Friedman test, and to decide whether this extra effort is warranted one may wish to know by how much it will increase the efficiency of the procedure. As a basis for such an efficiency comparison, let us specialize the above population model. In analogy with the assumptions made for such models earlier, let us assume that F_{ij} is of the form

$$F_{ij}(x) = F_j(x-a_i) \tag{6.21}$$

We shall make the additional assumption that

$$F_j(x) = F(x-b_j)$$

so that

$$P(X_{ij} \leqslant x) = F(x-a_i-b_j)$$

where the a's and b's are unknown constants.

In the terminology of Chap. 2, Sec. 1, we are assuming that the effects of block and treatment are additive, i.e., that the size of one does not depend on the value of the other. (This assumption would not be satisfied, for example, if the blocks were teachers and the treatments textbooks, and if different teachers found that they could be more effective with different texts.)

A more useful representation of the model is obtained by introducing the "effect" of treatment i as the difference

$$\alpha_i = a_i - \bar{a} \qquad \left(\bar{a} = \frac{\sum a_i}{s}\right)$$

which compares the ith treatment with the average of all the treatments; and the "effect" of block j as the difference

$$B_j = b_j - \bar{b} \qquad \left(\bar{b} = \frac{\sum b_j}{N}\right)$$

Writing $\delta = \bar{a} + \bar{b}$, we then have $a_i + b_j = \delta + \alpha_i + \beta_j$, and hence

$$P(X_{ij} \leqslant x) = F(x - \delta - \alpha_i - \beta_j) \tag{6.22}$$

where both the α's and the β's add up to zero. The representation (6.22) avoids the difficulty that the parameters a_i and b_j are unidentifiable in the sense that the X_{ij} will have the same distribution for a_i, b_j and a_i', b_j' provided $a_i + b_j = a_i' + b_j'$. In model (6.22), the hypothesis of no treatment difference becomes

$$H\colon \quad \alpha_1 = \cdots = \alpha_s = 0 \tag{6.23}$$

That the common value is zero follows from the fact that the sum of the α's is zero.

To complete the model (6.22) it is still necessary to specify F. Classical statistical theory assumes F to be normal, and under this assumption derives the rejection rule

$$\frac{N \sum_{i=1}^{s} (X_{i.} - X_{..})^2/(s-1)}{\sum\sum (X_{ij} - X_{i.})^2/(N-1)(s-1)} \geqslant c \tag{6.24}$$

as possessing various optimum properties. Here the critical value c is determined from the F-distribution with $s-1$ and $(s-1)(N-1)$ degrees of freedom.

The reason for the interest in alternatives to (6.24) (such as Q or $\hat{Q}$) is, of course, that one does not trust the assumptions under which (6.24) is derived. In particular, suppose that we are satisfied with the model (6.22) but have doubts about the assumption of normality like those discussed in Chap. 2, Sec. 4. As we saw above, such doubts do not affect the significance level of the Friedman test, which is distribution-free in the sense that the null distribution of the test statistic, and hence the significance level, is independent of F. This is not true of the test (6.24), whose significance level does depend on F. It can, however, be shown, quite analogously to the corresponding proof for the t-test (Chap. 2, Sec. 4 and Appendix Example 11), that the test (6.24) is asymptotically distribution-free in the sense that, as $N \to \infty$, the null distribution of the statistic (6.24) tends to the χ^2-distribution with $s-1$ degrees of freedom irrespective of F, provided only the variance σ^2 of F is finite.

Let us now compare the Friedman test with (6.24) in terms of their asymptotic (or Pitman) efficiency, a concept introduced in Chap. 2, Sec. 4. In the present context, this efficiency is the limiting value, as $N \to \infty$, of the ratio of the numbers of observations (or equivalently of the numbers of blocks) required by the two tests to achieve the same power against the same alternatives when both are carried out at the same significance level. It turns out [see van Elteren and Noether (1959)] that the asymptotic efficiency of the Friedman test relative to (6.24) is[1]

$$\frac{s}{s+1} e_{W,t}(F) \tag{6.25}$$

where $e_{W,t}(F)$ is the corresponding efficiency of the Wilcoxon test to Student's t-test discussed in Chap. 2, Sec. 4. The efficiency is therefore lowest when $s = 2$, in which case it is only two-thirds of $e_{W,t}(F)$, and increases with s, tending to the full value $e_{W,t}(F)$ as s tends to infinity. When F is normal, Table 6.3 shows the value of (6.25) for a number of values of s:

TABLE 6.3 Asymptotic efficiency of Friedman test vs. test (6.24)

s	2	3	4	5	10	15	∞
Asymptotic efficiency	.637	.716	.764	.796	.868	.895	.955

It is seen from (6.25) that the need for improvement of the Friedman test through alignment is greatest when the block size s is small and becomes less important as s increases. Let us now see how much the test actually is improved by alignment. The improvement can be measured by the asymptotic efficiency $e_{\hat{Q},Q}(F)$ of the aligned test $\hat{Q}$ to the Friedman test Q. A general expression for this efficiency was obtained by Mehra and Sarangi (1967), who show that it decreases as s increases and who give the values shown in Table 6.4 when F is normal. Some numerical values of the efficiencies of Q and $\hat{Q}$ relative to (6.24) for somewhat different alternatives are given by Gilbert (1972).

TABLE 6.4 Asymptotic efficiency $e_{\hat{Q},Q}$ (Normal)

s	2	3	4	5	∞
$e_{\hat{Q},Q}$ (Normal)	1.5	1.355	1.263	1.210	1

In Sec. 2, the Friedman test was seen to reduce to the sign test when $s = 2$. We conclude this section by showing that in this case the $\hat{Q}$-test is for large N essentially equivalent to the paired-comparison Wilcoxon test. To see this, denote the difference

[1] For some efficiencies in more general models see Sen (1968a and 1968b).

between the observation for the second and first treatment in block j by

$$Z_j = X_{2j} - X_{1j} \qquad j = 1, \ldots, N$$

If the observations in the jth block are aligned on the block mean $\frac{1}{2}(X_{1j}+X_{2j})$, the aligned observation $\hat{X}_{2j}$ in the jth block becomes

$$\hat{X}_{2j} = X_{2j} - \tfrac{1}{2}(X_{1j}+X_{2j}) = \tfrac{1}{2}Z_j$$

and the aligned observation $\hat{X}_{1j}$ becomes $-\frac{1}{2}Z_j$. Consider now the absolute values $|Z_1|, \ldots, |Z_N|$, and let R_j denote the rank of $|Z_j|$ among these N absolute values. Then the aligned rank $\hat{R}_{2j}$ (that is, the rank of $\hat{X}_{2j}$ among the $2N$ aligned observations $\hat{X}_{11}, \ldots, \hat{X}_{1N}$; $\hat{X}_{21}, \ldots, \hat{X}_{2N}$ and hence of Z_j among the $2N$ variables $-Z_1, \ldots, -Z_N$; $Z_1, \ldots, Z_N$) is equal to

$$\hat{R}_{2j} = \begin{cases} R_j + N & \text{if } Z_j > 0 \\ N+1-R_j & \text{if } Z_j < 0 \end{cases} \tag{6.26}$$

and similarly

$$\hat{R}_{1j} = \begin{cases} N+1-R_j & \text{if } Z_j > 0 \\ R_j + N & \text{if } Z_j < 0 \end{cases} \tag{6.27}$$

The expression for $\hat{Q}$ will involve two quantities that are familiar from Chap. 3. The sum of the ranks R_j summed over those values of j for which $Z_j > 0$ is the paired-comparison Wilcoxon statistic which in Chap. 3 was denoted by V_s, while the number of positive Z's is the sign statistic S_N. Summing (6.26) over j therefore gives

$$\sum \hat{R}_{2j} = N^2 + V_s + (N - S_N) - [\tfrac{1}{2}N(N+1) - V_s]$$

and hence

$$\sum \hat{R}_{2j} - \tfrac{1}{2}N(2N+1) = 2V_s - S_N - \tfrac{1}{2}N^2$$

Analogously,

$$\sum \hat{R}_{1j} - \tfrac{1}{2}N(2N+1) = S_N - 2V_s + \tfrac{1}{2}N^2$$

Substituting these expressions into (6.17), we find that $\hat{Q}$ is proportional to

$$(V_s - \tfrac{1}{2}S_N - \tfrac{1}{4}N^2)^2 \tag{6.28}$$

It is easy to show (see Appendix, Example 21) that $[V_s - \frac{1}{4}N(N+1)] - \frac{1}{2}[S_N - \frac{1}{2}N]$ is asymptotically equivalent to $V_s - \frac{1}{4}N(N+1)$ and hence that (6.28) is asymptotically equivalent to $[V_s - \frac{1}{4}N(N+1)]^2$. This latter statistic is the square of the test statistic for the two-sided Wilcoxon test. Thus, in the case $s = 2$, the $\hat{Q}$-test is asymptotically equivalent to the paired-comparison Wilcoxon test.

5. FURTHER DEVELOPMENTS

5A. More General Blocks

It was assumed in Sec. 1 that the size of each block is equal to the number s of treatments being compared. However, it is frequently convenient to have blocks of size less than s, and not every treatment can then be represented in each block (the blocks are said to be *incomplete*). Conversely, it may happen (as in Chap. 3, Sec. 3) that the blocks are larger than s; this might be the case, for example, if it is desired to combine a number of studies of s treatments, performed under different conditions, but each of the type considered in Chap. 5. The Friedman test for the absence of a treatment effect is generalized to cases in which the block sizes may differ from s, and in particular to the case of balanced incomplete blocks, by Durbin (1951) and Benard and van Elteren (1953). The asymptotic efficiency of Durbin's procedure is studied by van Elteren and Noether (1959). A generalization of the $\hat{Q}$-test to the situation in which the number n_{ij} of observations on the ith treatment in the jth block is not necessarily equal to 1 is developed by Mehra and Sarangi (1967) and by Sarangi and Mehra (1969); some efficiency results are given by Sen (1971). An extension of the $\hat{Q}$-test to Normal and other scores is given by Sen (1968a).

5B. One-sided Tests and Ranking Procedures

Most of the procedures of Chap. 5, Secs. 5 and 6 have been adapted to the case of randomized complete blocks. Following are references to some of this work.

(i) The problem of comparing several treatments with a control is considered by Hollander (1967) following an earlier discussion by Nemenyi (1963).

(ii) A modification of Friedman's test to the case that there is an a priori ordering of the treatment effects, say, $\alpha_1 \leqslant \cdots \leqslant \alpha_s$ is proposed by Shorack (1967). It consists simply in applying the amalgamation process described in Chap. 5, Sec. 5, to the rank sums R_i of the present chapter. The method is extended to the case of aligned ranks by Sen (1968a). For other approaches see Pirie (1974).

(iii) Paired comparisons leading to a partial ranking of the s treatments are discussed for the case of ranking within each block by Nemenyi (1963), Miller (1966), and McDonald and Thompson (1967), and for the case of aligned ranking by Sen (1968a).

(iv) A method for determining the significance of a treatment suggested by the data is given by Thompson and Wilkie (1963).

5C. Estimation of Treatment Differences and Other Contrasts

Consider the problem of estimating the treatment difference $\alpha_i - \alpha_j$ or, more generally, the contrast

$$\theta = \sum_{i=1}^{s} c_i \alpha_i \qquad (\sum c_i = 0) \tag{6.29}$$

in the model (6.22), under the additional assumption that the distribution F is symmetric. To obtain an estimate of θ, form the N linear functions

$$V_\mu = \sum_{i=1}^{s} c_i X_{i\mu} \qquad \mu = 1, \ldots, N$$

Then $V_1, \ldots, V_N$ are independent, identically distributed with a common distribution G_c, which is symmetric about θ. A possible estimate of θ is therefore given by (4.49), which in the present case becomes

$$\hat{\theta} = \text{med}_{\mu \leqslant \nu} \left[\tfrac{1}{2}(V_\mu + V_\nu)\right] \tag{6.30}$$

As in Chap. 5, Secs. 7B and 7C, the estimates (6.30) for different contrasts may be inconsistent. For example, the estimate of $\alpha_1 + \alpha_2 - 2\alpha_3$ does not coincide with the sum of the estimates of $\alpha_1 - \alpha_3$ and $\alpha_2 - \alpha_3$. This difficulty can be overcome as in the earlier section. Let $\hat{\theta}_{ij}$ denote the estimate (6.30) of the difference $\alpha_i - \alpha_j$, so that it is given by

$$\hat{\theta}_{ij} = \text{med}_{\mu \leqslant \nu} \left[\tfrac{1}{2}(X_{i\mu} - X_{j\mu} + X_{i\nu} - X_{j\nu})\right] \tag{6.31}$$

Then it is proposed by Lehmann (1964) to replace (6.31) by

$$\hat{\theta}_{i.} - \hat{\theta}_{j.} \tag{6.32}$$

and to define the estimate of any contrast in analogy with (5.90).

The development up to this point quite closely parallels that of the earlier case, but this is not true of the properties of the estimates (6.32). The change from the estimates $\hat{\Delta}_{ij}$ given by (5.83) to $\hat{\Delta}_{i.} - \hat{\Delta}_{j.}$ can be shown to involve only a minor adjustment, with the two estimates being asymptotically equivalent. In contrast, the estimates (6.31) and (6.32) have different properties, even in large samples. Fortunately, it turns out that the change from (6.31) to (6.32) represents an improvement since the asymptotic efficiency of (6.32) relative to (6.31) is always greater than or equal to 1.

A different approach to estimating θ is proposed by Doksum (1967), who takes as his starting point the estimates

$$V_{ij} = \text{med}_\mu (X_{i\mu} - X_{j\mu}) \tag{6.33}$$

of $\alpha_i - \alpha_j$. To avoid inconsistencies, these estimates are replaced by

$$V_{i.} - V_{j.} \tag{6.34}$$

and the estimate of any contrast is defined in analogy with (5.90). The resulting estimates are simpler to compute than those based on (6.32) but in typical cases tend to be less efficient.

5D. Combination of Independent Tests

The problem of combining data from several experiments is essentially the same as that of combining data from different blocks, but it is treated in a separate literature and frequently under different assumptions. Let us here consider how to combine the data on the same s treatments from a number of different experiments, and for this purpose distinguish three situations. First, there is a natural prior order of the treatments as in Chap. 5, Sec. 5. The same order would then typically prevail in all the experiments. For $s = 2$, this is the problem considered in Chap. 3, Sec. 3. The approach of Jonckheere discussed there is adapted to general s by D'Abrera (1974). Second, the order of the treatments is not assumed known a priori, but if there are treatment differences, the treatments are assumed to be in the same order in all the experiments being combined. This is just the problem considered in Sec. 5A. Third, it may not be realistic in the second situation to assume a common order for the different experiments (or blocks): the best fertilizer will depend on the type of soil; which of a number of textbooks is most effective may depend on the type of school; and so on. In this case, a simple procedure for testing the overall hypothesis of no treatment differences consists in adding the Kruskal-Wallis statistics $K_1, \ldots, K_N$ for the N separate experiments. Since each K_i has approximately a χ^2-distribution with $s-1$ degrees of freedom, the approximate distribution of $K_1 + \cdots + K_N$ will be χ^2 with $N(s-1)$ degrees of freedom, and large values of the sum are significant.

This test is carried out in the hope that even small effects in the separate experiments, when combined, will yield a total effect which can be detected. However, the detection of such a total effect would typically seem to be of little value. One would want to know instead for which of the different experimental conditions, and in which direction, any given treatment difference is significant. This would require noticeable differences in the separate experiments and a multiple testing procedure such as those described in Chap. 5, Secs. 5 and 6. An interesting discussion of a number of different tests for the three situations when the data for each separate experiment consist of a 2×2 table is given by Cochran (1954). For more recent work on this problem, see Radhakrishna (1965) and Zelen (1971).

Much of the literature about the combination of independent tests is concerned with the case that only significance probabilities are available for the separate experiments but not the complete data. An account of this work and some recent results are provided by Oosterhoff (1969) and Littell and Folks (1971).

6. PROBLEMS

(A)

Section 1

1. In Example 1 find the probability that the ranks of treatment A are

(i) Equal to 3 in all four groups

(ii) Equal to 3 in the first three groups and equal to 2 in the last

(iii) Equal to 3 in some three of the groups and equal to 2 in the remaining one

2. In Example 1 find the probability that

(i) The sum of the ranks of treatment A is equal to 10

(ii) The sum of the ranks of treatment A is 10 and that of treatment B is 4

3. (i) If $N = 5$ and $s = 3$, and if the rank sum of treatments A and B are 10 and 8, respectively, what is the rank sum for treatment C?

(ii) What is the answer to the question of part (i) if $N = 5$ and $s = 4$?

4. If the distribution of the ranks is given by (6.1), what is the probability that

(i) The first treatment receives rank 1 in all N blocks

(ii) The first treatment receives rank 1 and the second rank 2 in all N blocks

Section 2

5. Obtain the null distribution of Q for the cases (i) $s = 3$, $N = 2$; (ii) $s = 2$, $N = 4$.

6. The following table presents the total number of coughs per day of seven patients, under three different medications and a placebo administered in random order over a number of days:[1]

	Subject						
	1	2	3	4	5	6	7
Heroin, 5 mg	251	126	49	45	233	291	1385
Dextromethorphan, 10 mg	207	180	123	85	232	208	1204
Codeine, 10 mg	167	104	63	147	233	158	1611
Placebo	301	120	186	100	250	183	1913

Determine whether there is a significant difference between the four treatments.

7. Compute the approximate significance probability of the data of Example 2 if the response of subject 1 under fear is recorded as 13.1 instead of 23.1.

8. Compute the approximate significance probability of the data of Example 2 if instead of eight subjects there are nine, with the responses of the ninth under fear, happiness, etc., being 43.5, 47.4, 40.1, 42.0.

9. Ten subjects were told under hypnosis to regress to ages fifteen, ten, and eight. At each stage, an Otis test of Mental Ability was administered. The following table shows the IQ scores, adjusted to the age requested under hypnosis:[2]

Subject	*A*	*B*	*C*	*D*	*E*	*F*	*G*	*H*	*I*	*J*
Waking	113	118	119	110	123	111	110	118	126	124
Fifteen yr	112	119	121	109	126	113	114	118	127	123
Ten yr	110	122	123	107	127	112	112	118	127	126
Eight yr	112	120	119	107	125	111	112	119	125	125

Test the hypothesis of no difference in the four rows.

10. The following table shows the behavior level (1 indicates quiet, ..., 6 indicates extreme agitation) summed over three trials, of 35 newborns (2 to 3 days) under four different soothing

[1] Part of the data given in Beecher, *Measurement of Subjective Responses*, Oxford University Press, London, 1959, Table 19.4.

[2] Data from Kline (1950), "Hypnotic Age Regression and Intelligence," *J. Genet. Psychol.* **77**:129–132.

conditions.[1] Test the hypothesis that the four methods are equally effective.

Neonate no.	Stimulus			
	Warm water	Rocking	Pacifier	Sound
1	12.1	12.5	13.3	11.5
2	11.3	11.0	13.1	13.7
3	13.8	15.3	8.0	18.0
4	13.1	12.2	11.8	11.8
5	9.3	9.1	10.7	8.4
6	11.3	9.2	10.0	8.3
7	7.2	7.2	11.3	12.4
8	16.8	10.6	13.8	10.9
9	14.3	10.8	9.6	10.4
10	12.3	11.7	10.0	13.5
11	8.8	10.3	11.3	10.9
12	5.7	8.3	8.1	4.9
13	11.2	13.3	10.8	10.6
14	12.3	11.9	11.3	10.2
15	10.9	11.3	13.0	11.8
16	8.3	8.5	8.1	9.0
17	13.0	9.1	11.5	13.3
18	8.3	10.4	10.9	10.0
19	9.0	11.9	10.6	9.8
20	13.3	10.0	11.7	9.8
21	6.6	10.7	9.0	9.9
22	11.7	10.0	12.3	10.7
23	11.0	9.6	8.9	9.7
24	9.7	11.0	11.7	13.7
25	10.9	10.3	10.6	15.0
26	11.7	11.1	13.0	13.5
27	10.4	16.1	9.8	11.6
28	6.9	16.0	10.4	10.8
29	10.6	12.0	11.1	12.9
30	7.8	15.9	9.8	11.7
31	7.0	11.8	10.7	9.8
32	11.4	11.6	12.3	12.5
33	12.1	10.8	12.4	11.8
34	14.3	8.4	17.1	12.8
35	10.8	12.7	10.6	12.0

11. Find the null distribution of Q^* if $s = 3$, $N = 2$, and the midranks of the two blocks are (i) 1.5, 1.5, 3 and 1, 2, 3, respectively; (ii) 1.5, 1.5, 3 and 1, 2.5, 2.5, respectively.

12. The experiment of Example 3 was repeated on another set of 23 subjects with the following results:[2]

[1] Data from Birns, Blank, and Bridger (1966), "The Effectiveness of Various Soothing Techniques on Human Neonates," *Psychosomat. Med.* **28**:316–322.

[2] Data from Korner, "Experimental Investigation of Some Aspects of the Problem of Repression: Repressive Forgetting," *Contributions to Education* no. 970, Teacher's College, Columbia University, New York,

Subject	1	2	3	4	5	6	7	8	9	10	11	12	13	14	15	16	17	18	19	20	21	22	23
+	1	0	1	2	1	2	3	2	2	0	0	2	1	2	4	3	2	1	1	3	3	0	1
0	2	0	3	2	0	1	3	1	1	1	1	3	3	3	3	4	0	1	0	2	3	1	0
–	2	1	2	4	0	2	4	0	0	1	1	4	3	5	5	4	2	1	2	3	4	0	0

Obtain the significance probability under the hypothesis of Example 3 (i) on the basis of the present data; (ii) on the basis of the present data combined with those of Example 3.

13. In an investigation of the effect of four different media A, B, C, D for growing diphtheria bacilli, specimens were taken from the throats of 69 suspected cases. Each specimen was grown on each of the four media, with the following results:

Media	A	B	C	D	No. of Cases
	1	1	1	1	4
	1	1	0	1	2
	0	1	1	1	3
	0	1	0	1	1
	0	0	0	0	59
Totals	6	10	7	10	

Thus, for example, four specimens exhibited growth on all four media; two specimens showed growth in media A, B, and D but not in C; and so forth. Find the approximate significance probability of the Cochran statistic (6.11). (The exact significance probability is .045.†)

Section 3

14. For the data of Example 5 show that (6.16) has the value 2644.75.

15. Find the significance probability of $\hat{Q}$ for the data of (i) Prob. 6; (ii) Prob. 10; (iii) Prob. 12.

(B)

Section 2

16. Show that the expressions for Q given by (6.3) and (6.6) agree.

17. (i) Show that zero is a possible value of Q when N is even.
(ii) Find a combination of s and N for which zero is not a possible value of Q.

18. Show that the maximum value of Q is $(s-1)N$. {*Hint*: From the identity

$$\sum\sum[R_{ij}-\tfrac{1}{2}(s+1)]^2 = \frac{s(s+1)}{12}Q+\sum\sum(R_{ij}-R_{i.})^2$$

it follows that Q is maximized by minimizing the second term on the right-hand side. This term can be made equal to zero by having $R_{ij} = R_{i.}$ for all i and j.}

19. Verify the numerator and denominator of (6.9) given at the end of Sec. 2 for the case of dichotomous response.

20. Show that (6.11) reduces to (6.13) when $s = 2$.

† This example is taken from Cochran (1950).

Section 4

21. If

$$\xi_{ij} = \mu + \alpha_i + \beta_j \qquad (\sum \alpha_i = \sum \beta_j = 0) \tag{6.35}$$

show that μ, the α's, and the β's are uniquely determined. [*Hint:* Sum the equations (6.35) over i, over j, and over both i and j, and use the resulting equations to express μ, α_i, and β_j in terms of the ξ's.]

7. REFERENCES

Altham, P. M. E. (1971): "The Analysis of Matched Proportions," *Biometrika* **58**:561–576.

Benard, A. and van Elteren, P. (1953): "A Generalization of the Method of *m* Rankings," *Indagationes Math.* **15**:358–369.

Cochran, William G. (1950): "The Comparison of Percentages in Matched Samples," *Biometrika* **37**:256–266. [Proposes and discusses the test statistic (6.11).]

——— (1954): "Some Methods for Strengthening the Common χ^2-test," *Biometrics* **10**:417–451.

D'Abrera, Howard (1974): "Rank Tests for Ordered Alternatives in Randomized Blocks" (to be published).

Doksum, Kjell (1967): "Robust Procedures for Some Linear Models with One Observation Per Cell," *Ann. Math. Statist.* **38**:878–883.

Durbin, J. (1951): "Incomplete Blocks in Ranking Experiments," *Brit. J. Psychol. (Statist. Section)* **4**:85–90.

van Elteren, P. and Noether, G. E. (1959): "The Asymptotic Efficiency of the χ^2-test for a Balanced Incomplete Block Design," *Biometrika* **46**:475–477.

Friedman, Milton (1937): "The Use of Ranks to Avoid the Assumption of Normality Implicit in the Analysis of Variance," *J. Am. Statist. Assoc.* **32**:675–701. [Develops the test (6.4).]

——— (1940): "A Comparison of Alternative Tests of Significance for the Problem of *m* Rankings," *Ann. Math. Statist.* **11**:86–92.

Gilbert, Richard O. (1972): "A Monte Carlo Study of Analysis of Variance and Competing Rank Tests for Scheffe's Mixed Model," *J. Am. Statist. Assoc.* **67**:71–75.

Hollander, Myles (1967): "Rank Tests for Randomized Blocks When the Alternatives Have an A Priori Ordering," *Ann. Math. Statist.* **38**:867–877.

Kendall, Maurice G. (1970): *Rank Correlation Methods*, 4th ed. (1st ed., 1948), Charles Griffin, London.

——— and Babington Smith, B. (1939): "The Problem of *m* Rankings," *Ann. Math. Statist.* **10**:275–287.

Kraft, Charles H. and van Eeden, Constance (1968): *A Nonparametric Introduction to Statistics*, Macmillan, New York.

Lehmann, E. L. (1964): "Asymptotically Nonparametric Inference in Some Linear Models with One Observation Per Cell," *Ann. Math. Statist.* **35**:726–734.

Littell, Raman C. and Folks, J. Leroy (1971): "Asymptotic Optimality of Fisher's Method of Combining Independent Tests," *J. Am. Statist. Assoc.* **66**:802–806.

McDonald, B. J. and Thompson, W. A. (1967): "Rank Sum Multiple Comparisons in One- and Two-way Classifications," *Biometrika* **54:**487–497.

McNemar, I. (1947): "Note on the Sampling Error of the Difference between Correlated Proportions or Percentages," *Psychometrika* **12:**153–157.

Mehra, K. L. and Sarangi, J. (1967): "Asymptotic Efficiency of Certain Rank Tests for Comparative Experiments," *Ann. Math. Statist.* **38:**90–107. [Develops the test based on (6.15).]

Miller, Rupert G., Jr. (1966): *Simultaneous Statistical Inference*, McGraw-Hill Book Company, New York.

Moses, Lincoln E. (1953): "Nonparametric Methods," in Walker and Lev, *Statistical Inference*, Henry Holt, New York, chap. 18, pp. 426–450.

Mosteller, F. (1952): "Some Statistical Problems in Measuring the Subjective Response to Drugs," *Biometrics* **8:**220–226.

Nam, Jun-Mo (1971): "On Two Tests for Comparing Matched Proportions," *Biometrics* **24:**945–959.

Nemenyi, Peter (1963): "Distribution-free Multiple Comparison Procedures," unpublished Ph.D. thesis, Princeton University.

Oosterhoff, J. (1969): "Combination of One-sided Statistical Tests," *Mathematical Centre Tracts 28*, Math. Centrum, Amsterdam.

Owen, Donald B. (1962): *Handbook of Statistical Tables*, Addison-Wesley, Reading, Mass.

Pirie, Walter R. (1974): "Comparing Rank Tests for Ordered Alternatives in Randomized Blocks," *Ann. Statist.* **2:**374–382.

Radhakrishna, S. (1965): Combination of Results from Several 2 × 2 Contingency Tables," *Biometrics* **21:**86–98.

Rijkoort, P. G. and Wise, M. E. (1953): "Simple Approximations and Monograms for Two Ranking Tests," *Indagationes Math.* **15:**294–302, 407.

Sarangi, J. and Mehra, K. L. (1969): "Some Further Results on Hodges-Lehmann Conditional Rank Tests," *Bull. Calcutta Statist. Assoc.* **18:**25–41.

Sen, Pranab Kumar (1968a): "On a Class of Aligned Rank Order Tests in Two-way Layouts," *Ann. Math. Statist.* **39:**1115–1124.

——— (1968b): "Robustness of Some Nonparametric Procedures in Linear Models," *Ann. Math. Statist.* **39:**1913–1933.

——— (1971): "Asymptotic Efficiency of a Class of Aligned Rank Order Tests for Multiresponse Experiments in Some Incomplete Block Designs," *Ann. Math. Statist.* **42:**1104–1112.

Shorack, Galen R. (1967): "Testing against Ordered Alternatives in Model I Analysis of Variance; Normal Theory and Nonparametric," *Ann. Math. Statist.* **38:** 1740–1752.

Tate, Merle W. and Brown, Sara M. (1970): "Note on the Cochran *Q* Test," *J. Am. Statist. Assoc.* **65:**155–160.

Thompson, W. A., Jr., and Wilkie, T. A. (1963): "On an Extreme Rank Sum Test for Outliers," *Biometrika* **50:**375–383.

Walker, Helen M. and Lev, Joseph (1953): *Statistical Inference*, Henry Holt, New York.

Zelen, M. (1971): "The Analysis of Several 2 × 2 Contingency Tables," *Biometrika* **58:**129–137.

CHAPTER 7
TESTS OF RANDOMNESS AND INDEPENDENCE

1. THE HYPOTHESIS OF RANDOMNESS

In investigations of the effect of a treatment or some other factor on an observable characteristic of interest (the response), it often happens that the treatment is not simply either present or absent, as was assumed in the treatment versus control comparisons of Chaps. 1 to 4, but that it can be present to different degrees or at different levels. In the study of the effect of anxiety on dependence (Chap. 5, Example 6), for instance, anxiety could be provoked at a number of different levels ranging from quite light to very severe. As another example, consider the effect of the size of a hospital on its efficiency. Here the factor is size, which can take on a large number of different values.

In this chapter we shall be concerned primarily with situations in which the treatment or condition can occur at a number of levels and one observation is taken at each level. In particular, this is frequently the case when the condition in question is time; for example, when one wishes to determine whether winters are getting colder, or whether medical costs are increasing more rapidly than the general cost of living.

EXAMPLE 1. *Time of day and quality.* Suppose a manufacturer wishes to know whether the quality of the items coming off a production line tends to deteriorate in the course of the day.

Let us assume that only one day is available for taking the necessary observations. The raw material, which is fed into the process in batches, is assigned at random to the first, second, . . . batch and a finished item is taken off the line for inspection after $1\frac{1}{2}$, 3, $4\frac{1}{2}$, 6, and $7\frac{1}{2}$ hr. The quality of the five items is ranked (rank 1 for the best, ..., rank 5 for the worst), with the following results:

Time period	1	2	3	4	5
Rank of quality	2	3	5	1	4

Let H denote the hypothesis that the time of day has no effect so that the quality of an item is determined by the quality of the raw material (if variations from other sources can be assumed to be sufficiently small). The rank of the item (and hence of the time period) is then determined by the rank of the batch of raw material from which it came. Therefore, the assignment of the raw material also assigns to each of the five periods listed in the first row of the table one of the five ranks as shown in the second row. By the randomness of these assignments, they are all equally likely.

To determine how many such assignments there are, note that each possible second row corresponds to an arrangement of the numbers 1, 2, 3, 4, 5. The number of such arrangements is easily obtained by an argument similar to that used in Chap. 5, Sec. 1. There are five possible choices for the first position of the second row, namely, 1, 2, 3, 4, 5; for each of these choices, there are four numbers left to choose from for the second position. For example, when the first number chosen is 2, the second number can be 1, 3, 4, 5. The first two positions can therefore be filled in

$$4+4+4+4+4 = 5 \times 4 = 20$$

ways, namely,

12 13 14 15 21 23 24 25 31 32 34 35 41 42 43 45
51 52 53 54

For each of the 20 choices of the first two numbers, three numbers are left to choose from for the third position, so that the total number of choices for the first three numbers is $5 \times 4 \times 3 = 60$. Once the first three numbers are chosen, only two numbers are left for the fourth position. The number of possible arrangements of the first four numbers is thus $5 \times 4 \times 3 \times 2 = 120$, and this is the total number of possible arrangements since the choice for the first four positions determines the arrangement completely. (If, for example, the first four numbers are 2531, the arrangement is 25314.) The total number of equally likely arrangements is therefore $1 \times 2 \times 3 \times 4 \times 5 = 120$, so that the probability of any one of them is 1/120.

These considerations easily generalize. Suppose that the effect of a treatment is being investigated by applying it at N levels, to N subjects which are assigned to the levels at random. Let the responses be ranked and let T_i denote the response-rank of the subject assigned to the ith lowest level. The results can then be displayed in the following table, the second row of which constitutes an arrangement (permutation) of the integers $1, \ldots, N$.

(7.1)

Rank of level	1	2	. . .	N
Rank of response	T_1	T_2	. . .	T_N

If the treatment has no effect, it follows from the random assignment of the subjects that all possible arrangements of the second row of (7.1) are equally likely. How many such arrangements are there? The argument of the example immediately extends. There are N possible choices for the first position; for each of these, there are $N-1$ choices for the second position; and so on. The total number of arrangements is therefore $N(N-1)\cdots 2\cdot 1$. This product of the first N integers is denoted by

$$N! = 1 \times 2 \times \cdots \times N \qquad \text{(called } N \text{ factorial)} \tag{7.2}$$

Under the hypothesis of no treatment effect, the probability that the ranks $T_1, \ldots, T_N$ take on any particular set of values $t_1, \ldots, t_N$ is therefore equal to

$$P_H(T_1 = t_1, \ldots, T_N = t_N) = \frac{1}{N!} \tag{7.3}$$

which is the desired null distribution of these ranks. The hypothesis expressed by (7.3), that all $N!$ orderings of the N responses are equally likely, is sometimes called the *hypothesis of randomness.*

Instead of dealing with N fixed subjects, which are assigned to the N treatment levels at random, we may be concerned with a sample of size N drawn at random from a population of such subjects. Let the responses of the N subjects be denoted by $Z_1, \ldots, Z_N$, and let the distribution of a subject's response under treatment i be

$$P(Z_i \leqslant z) = F_i(z) \tag{7.4}$$

The hypothesis of no treatment effect then states that $F_1 = \cdots = F_N$. If the population is sufficiently large, so that the dependence of the subjects can be neglected, the mathematical assumptions can be summarized in the following *population model.* The responses $Z_1, \ldots, Z_N$ are independent random variables with distributions $F_1, \ldots, F_N$ respectively. The hypothesis to be tested is

$$H: \quad F_1 = \cdots = F_N \tag{7.5}$$

This is completely analogous to the population model of Chap. 2, Sec. 1 and is a special case of the population model described in Chap. 5, Sec. 4. From the corresponding fact stated there, it follows that in the present case, the null distribution of $T_1, \ldots, T_N$ continues to be given by (7.3).

Unfortunately, the problems to be considered in the present chapter typically do not permit random assignment or sampling of subjects. They are much closer instead to the measurement situation represented by Model 5 in Chap. 2, Sec. 1. Consider for example the question: Are winters getting colder? The measurements here might consist of an average winter temperature, obtained under conditions that are as similar as possible for a number of consecutive years. Even if there is no trend or other systematic change of temperature, the observed temperature will vary from year to year. The fluctuations are due to the cumulative effect of a large number of factors affecting the weather from day to day, and the resulting observations will have the character of random variables. It may be reasonable to assume that these are independent (obviously, this would not be the case if we were studying the weather on successive days instead of successive years), and the hypothesis of no systematic change would find its expression in (7.5). The mathematical model would therefore coincide with the population model considered above.

2. TESTING AGAINST TREND

To test the hypothesis H of the preceding section, that a treatment or factor does not affect the variable being observed, it is necessary to decide for which arrangements $(t_1, \ldots, t_N)$ to reject H. This in turn depends on what type of arrangement one would expect when the hypothesis is false.

EXAMPLE 2. *Pollution of Lake Michigan.* In a study of the pollution of Lake Michigan, the number of "odor periods" were observed for each year, with the following results for the period 1950–1964.[1]

Year	50	51	52	53	54	55	56	57	58	59	60	61	62	63	64
Number of odor periods	10	20	17	16	12	15	13	18	17	19	21	23	23	28	28

Here H denotes the hypothesis that the degree of pollution has not changed with time. Under H, the measurement model of the preceding section seems appropriate, according to which the numbers $Z_1, \ldots, Z_{15}$ of odor periods in successive years are independent random variables with the same distribution, which reflects the effects of wind, temperature, and other factors affecting the intensity of the odor.

Suppose it is believed that the only likely change in pollution (if any) would be a gradual increase over the period in question, so that the hypothesis is to be tested against the alternative of an *upward trend*. The existence of such a trend would increase the probability of arrangements of the ranks $T_1, T_2, \ldots$, in which large t-values occur late in the series and small values early. If the effect were very marked, it could even happen that the T's would be perfectly ordered, i.e., that $T_1 = 1$, $T_2 = 2$, …, $T_N = N$, or at least nearly so.

Let us now consider the general problem of testing the hypothesis H of no change against the alternative of an upward trend. Analogy with the Mann-Whitney form of the Wilcoxon test statistic suggests as a possible test statistic the number of

[1] Data from Gerstein (1965), "Lake Michigan Pollution and Chicago's Supply," *Am. Water Works Assoc. J.* **57**:841–857.

pairs (T_i, T_j) with $i < j$ for which $T_i < T_j$. If the observations tend to increase with time, and hence the T's with their subscripts, the number of such pairs will be large. Large values of the test statistic will therefore be significant. The statistic can be represented formally as $\sum_{i<j} U_{ij}$, where

$$U_{ij} = 1 \text{ or } 0 \qquad \text{as } T_i < T_j \text{ or } T_i > T_j \tag{7.6}$$

Instead of this statistic [proposed by Mann (1945)] we shall consider the closely related statistic

$$D' = \sum_{i<j} (j-i) U_{ij} \tag{7.7}$$

which weights more heavily the pairs $T_i < T_j$ for which i and j are further apart. Again, H is rejected for large values of D' when one is testing against an upward trend (and for small values of D' when the alternatives specify a downward trend).

The statistic D' has a simple expression in terms of the ranks $(T_1, \ldots, T_N)$. As will be shown at the end of the section, it is equal to

$$D' = \tfrac{1}{6}N(N^2-1) - \tfrac{1}{2}\sum_{i=1}^{N} (T_i - i)^2 \tag{7.8}$$

Thus, instead of rejecting for large values of D', it is equivalent to reject for small values of

$$D = (T_1-1)^2 + (T_2-2)^2 + \cdots + (T_N-N)^2 \tag{7.9}$$

which is more convenient to compute than D'. The statistic D is an intuitively reasonable test statistic since under the alternatives large values of T_i will tend to occur for large values of i and small values of T_i for small values of i, so that the differences $(T_i - i)^2$ will tend to be small.

For computing the null distribution of D, it is convenient to realize that D can take on only even values. To see this, expand the squares in (7.9) to find that

$$D = \sum T_i^2 - 2\sum iT_i + \sum i^2$$

Since $(T_1, \ldots, T_N)$ is a rearrangement of the integers $(1, \ldots, N)$, it follows that $\sum T_i^2 = \sum i^2$, so that the sum of the first and last terms on the right-hand side of the displayed equation is twice an integer and hence even. The middle term is also even, and the result follows. Substitution of the expression for the sum of squares given as formula (1.68) permits rewriting the displayed equation as

$$D = \tfrac{1}{3}N(N+1)(2N+1) - 2\sum iT_i \tag{7.10}$$

This representation will prove useful later in the section.

Let us illustrate the computation of the significance probability of an observed value of D on the data of Example 1. The ranking there was $(T_1, \ldots, T_5) = (2,3,5,1,4)$, so that $D = 1+1+4+9+1 = 16$, and the significance probability is $P_H(D \leqslant 16)$. The smallest possible value of D is zero, and this is taken on only when $(T_1, ..., T_5) = (1,2,3,4,5)$; thus, $P(D = 0) = 1/5! = 1/120$. The next smallest value, 2, occurs when $(T_1, ..., T_5)$ is either (1,2,3,5,4), (1,2,4,3,5), (1,3,2,4,5), or (2,1,3,4,5), so that $P(D = 2) = 4/120$. Similar counts show that $P(D = 4) = 3/120$ and $P(D = 6) = 6/120$. We could continue in this way but shall be content with noting that the significance probability $P(D \leqslant 16)$ will certainly be larger than $P(D \leqslant 6) = 14/120 > .1$. The data are thus not significant at the usual levels.

As N increases, $N!$ grows extremely fast, and even rudimentary computations quickly become prohibitive. However, the null distribution of D has been tabled for $N \leqslant 11$, and the table is reproduced as Table N.† For $N = 5$, it is seen from this table that $P_H(D \leqslant 16) = 47/120 = .392$, which is therefore the significance probability sought in the preceding paragraph. The table gives only the probabilities up to the smallest probability greater than or equal to ½. The rest of the distribution can be obtained from the fact that it is symmetric about $(N^3 - N)/6$ [Prob. 34(ii)].

For sufficiently large N, the null distribution is approximately normal. The expectation and variance required for the approximation are (Appendix, Prob. 3)

$$E_H(D) = \frac{N^3 - N}{6} \tag{7.11}$$

and

$$\text{Var}_H(D) = \frac{N^2(N+1)^2(N-1)}{36} \tag{7.12}$$

Asymptotic normality is proved in Example 18 of the Appendix.

To illustrate the normal approximation, consider once more the probability $P(D \leqslant 16)$ when $N = 5$. Then $E(D) = 20$ and $\text{Var}(D) = 100$, and $P(D \leqslant 16)$ is approximated by

$$\Phi\left(\frac{16-20}{10}\right) = \Phi(-.4) = .345$$

A comparison with the tabular value .392 shows the error to be substantial. The approximation is improved, at least in this case, by use of the continuity correction. Since the values taken on by D are ..., 14, 16, 18, ..., the discussion of the continuity correction in Chap. 1, Sec. 3 suggests that in the computation of $P(D \leqslant 16)$ the approximating area should extend halfway to the next value and thus be $\Phi[(17-20)/10] = \Phi(-.3) = .382$.

An idea of the accuracy of the approximation in the area of greatest interest

† Taken from Owen, *Handbook of Statistical Tables*. For tables of critical values, see Glasser and Winter (1961), "Critical Values of the Coefficient of Rank Correlation for Testing the Hypothesis of Independence," *Biometrika* **48**:444–448 for $n = 11(1)30$ and Zar (1972), "Significance Testing of the Spearman Rank Correlation Coefficient," *J. Am. Statist. Assoc.* **67**:578–580 for $n = 4(1)50(2)100$.

may be obtained by inspecting some examples at the edge of the table. In Table 7.1, in each case the first line gives the exact value of $P(D \leqslant d)$; the second and third lines give the normal approximation without and with continuity correction, respectively.

TABLE 7.1 Exact and approximate values of $P(D \leqslant d)$

	$N = 10$			$N = 11$		
d	44	74	92	66	104	128
Exact	.010	.052	.102	.010	.050	.102
Approximation without continuity correction	.014	.049	.092	.013	.048	.093
Approximation with continuity correction	.015	.051	.095	.014	.049	.095

Let us consider next the modification of the D-test required in the presence of ties. As in earlier chapters, we shall replace each rank T_i by its midrank T_i^* (defined in Chap. 1, Sec. 4). Suppose that N fixed subjects are assigned at random to the N treatment levels, so that we are dealing with a randomization model. The null distribution of $(T_1^*, \ldots, T_N^*)$ is then no longer given by (7.3), but it can be obtained by the same argument since it follows from the random assignment that all possible (distinct) arrangements of $(T_1^*, \ldots, T_N^*)$ are equally likely.

To illustrate, suppose that $N = 5$ and that the second and third smallest observations are tied. The five midranks are then 1, 2.5, 2.5, 4, 5. If we artificially distinguish between the two equal values, there are as before $5! = 120$ possible orderings, all of which are equally likely. However, there are only half as many distinct orderings since each arrangement is equal to that obtained by interchanging the two equal values. The number of distinct orderings is thus 60, and the 60 possibilities are equally likely, so that each has probability 1/60.

In general, suppose that the N midranks take on e distinct values, with d_1 equal to the smallest, d_2 to the next smallest, and so on. If we again artificially distinguish between equal values, there are $N!$ equally likely orderings of the N midranks. However, the d_1 values equal to the smallest can be ordered in $d_1!$ ways, all of which are really indistinguishable. Similarly, the next set of d_2 equal values can be ordered in $d_2!$ ways, and so on. For any possible arrangement, there are thus $d_1! \cdots d_e!$ arrangements that are equal to it. This shows that the number of distinct arrangements is $N!/d_1! \cdots d_e!$, each of which has probability

$$P_H(T_1^* = t_1^*, \ldots, T_N^* = t_N^*) = \frac{d_1! \cdots d_e!}{N!} \tag{7.13}$$

A possible modification of the test statistic D in the presence of ties is

$$D^* = \sum (T_i^* - i)^2 \tag{7.14}$$

The null distribution is determined by (7.13), and for large N may again be approximated by the normal distribution. This approximation is supported by a limit theorem (proved in the Appendix, Example 19), which states that the null distribution of $[D^* - E_H(D^*)]/\sqrt{\mathrm{Var}_H(D^*)}$ tends to the standard normal distribution as N tends to infinity provided

$$\max \frac{d_i}{N} \text{ is bounded away from 1 as } N \to \infty \tag{7.15}$$

Thus, the normal approximation should only be used if the largest of the ratios $d_1/N, \ldots, d_e/N$ is not too close to 1. The expectation and variance required for the approximation are (see Appendix, Prob. 3)

$$E_H(D^*) = \tfrac{1}{6}(N^3 - N) - \tfrac{1}{12}\sum (d_i^3 - d_i) \tag{7.16}$$

and

$$\mathrm{Var}_H(D^*) = \frac{(N-1)N^2(N+1)^2}{36}\left[1 - \frac{\sum (d_i^3 - d_i)}{N^3 - N}\right] \tag{7.17}$$

As an illustration of the normal approximation, let us find the approximate significance probability of the pollution data of Example 2. The following table shows in the second row the midranks of the number of odor periods for the first

1	2	3	4	5	6	7	8	9	10	11	12	13	14	15
1	10	6.5	5	2	4	3	8	6.5	9	11	12.5	12.5	14.5	14.5
0	8	3.5	1	3	2	4	0	2.5	1	0	.5	.5	.5	.5

through the fifteenth year, and in the third row the absolute difference between each midrank and the rank of the corresponding year. By squaring the entries in the third row and adding we find $D^* = 114.5$. From (7.16) and (7.17), the expectation and standard deviation of D^* are seen to be $E(D^*) = 558.5$ and $SD(D^*) = 149.26$. The significance probability $P(D^* \leqslant 114.5)$ is therefore approximately equal to

$$\Phi\left(\frac{114.5 - 558.5}{149.26}\right) = \Phi(-2.97) = .0015$$

An interesting special case of tied observations occurs when each subject falls into one of two categories, so that there are only two possible responses. This would be the case, for instance, in Example 1 if each item were only classified as satisfactory or defective. Let us call the two categories success and failure, and arbitrarily associate with each a value, say 1 and 0. If there are m zeros and n ones, it follows from (7.13) with $d_1 = m$ and $d_2 = n$ that under the hypothesis each of the $\binom{m+n}{n}$ possible positions of the n ones are equally likely. On the other hand, under the alternative of an upward trend, the pattern of zeros and ones will tend to have the zeros occurring toward the left and the ones toward the right; if the factor under investigation is time, this means that the zeros tend to occur early and the ones late.

To see what happens to the test in this case, let us expand the squares in the expression (7.14) for D^*, which then becomes

$$D^* = \sum T_i^{*2} - 2\sum iT_i^* + \sum i^2 \tag{7.18}$$

Now the m zeros share the ranks $1, \ldots, m$, so their midrank is $(1+\cdots+m)/m = \frac{1}{2}(m+1)$. Similarly, the midrank of the n ones is $[(m+1)+\cdots+(m+n)]/n = m+\frac{1}{2}(n+1)$. The first term on the right-hand side of (7.18) is therefore a constant depending only on m and n. Since the third term is also a constant, rejecting for small values of D^* is equivalent to rejecting for large values of $\sum iT_i^*$. Let us denote, for the present discussion only, the positions of the m zeros by $R_1, \ldots, R_m$ and those of the n ones by $S_1, \ldots, S_n$. The statistic $\sum iT_i^*$ then becomes

$$\tfrac{1}{2}(m+1) \sum_{i=1}^{m} R_i + [m+\tfrac{1}{2}(n+1)] \sum_{j=1}^{n} S_j$$

Since $\sum R_i + \sum S_j$ is a constant, rejection for large values of $\sum iT_i^*$ is thus equivalent to rejection for large values of $W_s = S_1 + \cdots + S_n$. In the case of a dichotomous response, the test is therefore seen to be expressible in terms of a test statistic that has the structure and null distribution of the Wilcoxon statistic.

In this context, the test was proposed by Haldane and Smith (1948) as a procedure for testing whether some disease or abnormality is more likely to occur in later children of a family than in the earlier ones. Typically in such a situation the data would consist of the records not of a single family but of many families. For this problem Haldane and Smith proposed as test statistic the sum of the rank-sums W_s obtained for the various families. The corresponding test for the two-sample problem was discussed in Chap. 3, Sec. 3. The results obtained there apply without change in the present situation.

That the data for testing the hypothesis of randomness against an upward trend consist not of a single series of observations but of a number of such series occurs not only when the response is dichotomous. In the situation of Example 1, for instance, the test would typically be based on observations taken on a number of different days, each day providing such a series. A natural procedure is then obtained by ranking separately the observations in each series and basing the test on $\sum D_k$, where D_k denotes the statistic D computed for the kth series. The distribution of this sum of the independent statistics D_k will typically be close to normal provided the total number of observations (the sum of the numbers in the individual series) is not too small. The expectation and variance of $\sum D_k$ required for the approximation are of course $\sum E(D_k)$ and $\sum \text{Var}(D_k)$, where the expectation and variance of D_k are given by (7.11) and (7.12) with the size N_k of the kth series in place of N.

EXAMPLE 3. *Trend in illegitimate births.* The following table shows the number of illegitimate births during the years 1940, 1945, 1947, 1950, 1955, and 1957 for a number of states.[1] Do the data support the statement that during this period the number has tended to increase in these states?

	1940	1945	1947	1950	1955	1957
Alabama	660	745	724	642	661	740
Delaware	113	127	155	150	164	176
District of Columbia	254	379	457	352	409	450
Kansas	318	593	564	530	694	722
Lousiana	520	517	598	560	691	726

Ranking the numbers for each state leads to the following table of ranks:

	1940	1945	1947	1950	1955	1957	D
Alabama	2	6	4	1	3	5	32
Delaware	1	2	4	3	5	6	2
District of Columbia	1	3	6	2	4	5	16
Kansas	1	4	3	2	5	6	8
Lousiana	2	1	4	3	5	6	4

The D-values for the five states are shown in the last column, and it is seen that $\sum D_k = 62$. From (7.11) and (7.12) it follows that $E(\sum D_k) = 175$ and $SD(\sum D_k) = 35$. Hence $P(\sum D_k \leqslant 62)$ is approximately equal to $\Phi(-3.2) = .0007$ (but see Prob. 6).

Let us now return to the problem of ties and the distribution of the midranks, which were considered above in the framework of the randomization model. Ties occur not only when it is possible to assign subjects at random to the different treatment levels but also in the population and measurement stituations discussed at the end of Sec. 1. The modifications required in these situations are completely analogous to those discussed in Chap. 2, Sec. 1, and it is therefore enough to summarize them briefly. Ties may occur if the distributions $F_1, \ldots, F_N$ are not continuous, and it then follows that the distribution of the midranks under the hypothesis $F_1 = \cdots = F_N$ depends on the common distribution F. What is independent of F is the *conditional* distribution of the midranks given the configuration of the ties, that is, the numbers e and $d_1, \ldots, d_e$. In fact, this conditional distribution is just the distribution (7.13). Thus, the computation of the null distribution of D^* discussed above for the randomization model remains valid for the population and measurement models with the proviso that the resulting significance level or probability must be interpreted as being conditional.

To conclude the section, we now prove formula (7.8). To this end we shall utilize (7.10) and express $\sum iT_i$ in terms of the variables U_{ij} defined by (7.6). Since T_j is the rank of the jth observation, $T_j - 1$ is the number of observations less than the jth and hence is the sum of the variables U_{ij} summed over all values of i different

[1] Data from *Vital Statistics—Special Reports 47(8)* (1960), "Illegitimate Births in the United States 1938–1957." These numbers may be of great interest to the agencies concerned with the fate of such children, but other considerations might emphasize rates (rather than numbers) of illegitimate births. For the corresponding data, see Prob. 6.

from j. Thus

$$T_j - 1 = \sum_{i<j} U_{ij} + \sum_{i>j} U_{ij}$$

and

$$\sum j(T_j - 1) = \sum\sum_{i<j} jU_{ij} + \sum\sum_{i>j} jU_{ij} \tag{7.19}$$

Here the single summation sign in the first and the double summation sign in the second equation indicate that in the first case j is fixed so that the summation is only with respect to i, while in the second case the summation extends over both i and j.

In the second sum on the right-hand side of (7.19), let us interchange the summation variable i and j so that this sum becomes $\sum\sum_{i<j} iU_{ji}$. This is permissible since these variables can be replaced by arbitrary letters without changing the meaning of the sum. Now

$$U_{ji} = 1 - U_{ij} \tag{7.20}$$

since $U_{ji} = 1$ when $U_{ij} = 0$ and $U_{ji} = 0$ when $U_{ij} = 1$. Using the fact that by (1.12)

$$\sum j = \tfrac{1}{2}N(N+1)$$

and also interchanging i and j on the left-hand side of (7.19), we thus have

$$\sum iT_i = \tfrac{1}{2}N(N+1) + \sum\sum_{i<j} (j-i)U_{ij} + \sum\sum_{i<j} i$$

The last sum on the right-hand side is equal to $\sum i(N-i) = N\sum i - \sum i^2$, and by (1.12) and (1.68) this reduces to $N(N^2-1)/6$. Consolidation of the first and last term on the right-hand side now shows that

$$\sum iT_i = D' + \tfrac{1}{6}N(N+1)(N+2)$$

and substitution of this expression for $\sum iT_i$ in (7.10) completes the proof of (7.8).

3. TESTING FOR INDEPENDENCE

Some of the most important applications of the trend test of the preceding section occur in situations in which the factors being studied are not treatments that the investigator can assign to his subjects but conditions or attributes which are inseparably attached to these subjects. To illustrate the distinction between the two types of situations, consider a study of the effects of anxiety on dependence. One possibility, a modification of the study described in Chap. 5, Example 6, would be to place the subjects (at random) in situations designed to produce anxiety to varying degrees and then test for the strength of their feeling of dependence. A quite different study would determine for each subject measures of his tendency both to anxiety (as a personality trait) and to dependence.

Often, the second kind of investigation is the only one available. Thus, it is, for example, not possible to assign people at an early age to various smoking habits in order to study the effect of smoking on a person's health. The impossibility of assigning subjects at random to different levels of the factor whose influence is being investigated has serious consequences for the conclusions that can be drawn from such a study. As was already discussed in Chap. 2, Sec. 1, it is then no longer possible to infer a causal relationship. One cannot conclude that anxiety brings forth a feeling of dependence, that smoking causes lung cancer, but only that an association exists between the two factors. This association might be the result of a causal effect of either factor on the other, or it might be the result of certain common influences acting on both. The hypothesis to be tested thus no longer asserts the absence of an effect of the first on the second factor but the absence of a relationship between the two.

When both factors are attached to their subjects so that random assignment becomes impossible, sampling of subjects at predetermined levels of one of the factors is often impracticable. Instead, the experimental material may consist of a sample of N subjects from the population of interest. In such situations, the levels of both factors will of course be random.

To summarize, the present section will be concerned with testing whether there exists a relationship between two factors in a population of subjects. These may be people, manufactured items, institutions, and so on. With each subject are associated two characteristics: a person's ability for mathematics and music; the color of a pineapple and its ripeness; the size and crime rate of a city; and so forth. To test the *hypothesis of independence*, that there is no relationship between the two characteristics, a sample of N subjects is drawn from the population, and the values of the two characteristics are obtained for each member of the sample. Let us now illustrate how these data can be used to test the hypothesis of independence.

EXAMPLE 4. *Performance in language and arithmetic.* From a group of 98 students enrolled for a statistics course, nine are selected at random and given a simple arithmetic test and an artificial language test with the following results:[1]

Code no. of student	74	91	33	27	76	29	09	25	67
Language score	50	23	28	34	14	54	46	52	53
Arithmetic score	38	28	14	26	18	40	23	30	27

[1] Taken from Walker and Lev, *Statistical Inference*, Henry Holt, New York, 1953, Table 13.1, where these and other test scores are given for 98 students.

Let us rank separately the scores on each of the two tests to produce the following table:

Code no. of student	74	91	33	27	76	29	09	25	67
Language rank	6	2	3	4	1	9	5	7	8
Arithmetic rank	8	6	1	4	2	9	3	7	5

A clearer view of the relationship of the two sets of ranks is obtained by arranging one of them, say, the first, in its natural order. By discarding the code numbers of the students, we obtain the following presentation of the data that, at least formally, is seen to be of the kind considered in the preceding section:

Language rank	1	2	3	4	5	6	7	8	9
Arithmetic rank	2	6	1	4	3	8	7	5	9

If we denote the ranks in the second row by $(T_1, \ldots, T_N)$, with $N = 9$ in the present case, what can we say about the null distribution of $(T_1, \ldots, T_N)$? The hypothesis asserts that there is no relationship between the language and arithmetic scores, or more formally, that the arithmetic score of each subject is independent of his language score. It seems intuitively clear that under this assumption all $N!$ orderings of $(T_1, \ldots, T_N)$ are equally likely, so that the null distribution (7.3) continues to apply. A formal proof of this fact is given at the end of the section.

For what values of $(T_1, \ldots, T_N)$ should the hypothesis of independence be rejected? Suppose we are willing to assume that the two scores are either unrelated or that there is a *positive association* in the sense that high scores on one test tend to be associated with high scores on the other, and similarly for low scores. Under the alternative, the T's would then behave as in the case of an upward trend: large values would tend to occur to the right of the second row and low values to the left. The test statistic $D = \sum (T_i - i)^2$ of the preceding section would therefore be appropriate also in the present situation, with small values of D being significant. Since the null distribution of the T's is given by (7.3), the statistic D has the same null distribution as in the preceding section. In particular, Table N continues to be applicable, as well as the normal approximation with expectation and variance given by (7.11) and (7.12).

In our example, D has the value

$$1+16+4+0+4+4+0+9+0 = 38$$

and from Table N we find $P_H(D \leqslant 38) = .025$. The hypothesis of independence would therefore be rejected in favor of the alternative of a positive association at the 5 percent level but not at 1 percent.

An alternative expression is available for the statistic D. Let us denote the ranks of the two characteristics associated with the first subject by (R_1, S_1), those associated with the second subject by (R_2, S_2), and so on. In the example, for instance,

$$\begin{pmatrix} R_1 \ldots R_N \\ S_1 \ldots S_N \end{pmatrix} = \begin{pmatrix} 6 & 2 & 3 & 4 & 1 & 9 & 5 & 7 & 8 \\ 8 & 6 & 1 & 4 & 2 & 9 & 3 & 7 & 5 \end{pmatrix}$$

Since this is just a rearrangement of

$$\begin{pmatrix} 1 \ldots N \\ T_1 \ldots T_N \end{pmatrix} = \begin{pmatrix} 1 & 2 & 3 & 4 & 5 & 6 & 7 & 8 & 9 \\ 2 & 6 & 1 & 4 & 3 & 8 & 7 & 5 & 9 \end{pmatrix}$$

it follows that the differences $(T_i - i)^2$ and $(S_i - R_i)^2$ are the same N numbers in a different order, and hence that

$$D = \sum (S_i - R_i)^2 \tag{7.21}$$

In computing D, it is thus not necessary to arrange one of the rows in its natural order.

The statistic D is related to the correlation coefficient of the ranks $(R_1, \ldots, R_N)$ and $(S_1, \ldots, S_N)$, which we shall denote by r_s and which is defined to be equal to

$$r_s = \frac{\sum (R_i - \bar{R})(S_i - \bar{S})}{\sqrt{\sum (R_i - \bar{R})^2 \sum (S_i - \bar{S})^2}} \tag{7.22}$$

where $\bar{R} = \sum R_i/N$ and $\bar{S} = \sum S_i/N$. Known as *Spearman's rank correlation coefficient*, r_s was proposed by Spearman (1904) as a measure of the strength of the association between the two characteristics. By (A.12) and (A.13) of the Appendix, $\bar{R} = \bar{S} = \frac{1}{2}(N+1)$ and

$$\sum (R_i - \bar{R})^2 = \sum (S_i - \bar{S})^2 = (N^3 - N)/12$$

Since the numerator of r_s is equal to

$$\sum R_i S_i - \bar{S} \sum R_i - \bar{R} \sum S_i + N\bar{R}\bar{S} = \sum R_i S_i - N\bar{R}\bar{S} = \sum i T_i - \tfrac{1}{4}N(N+1)^2$$

it follows from (7.10) that

$$r_s = 1 - \frac{6D}{N^3 - N} \tag{7.23}$$

Thus, r_s is an equivalent test statistic to D, with large values of r_s significant against the alternatives of positive association.

It is seen from (7.23) that r_s takes on its maximum value if and only if D takes on its minimum value 0. In this case, which occurs only when $T_1 = 1, \ldots, T_N = N$ and hence $R_1 = S_1, \ldots, R_N = S_N$, the rank correlation coefficient has the value $r_s = 1$. Similarly, r_s takes on its minimum value -1 if and only if $T_1 = N, \ldots, T_N = 1$ so that the two characteristics for the N subjects are ranked in exactly opposite order. In view of the symmetry of the null distribution of D about $(N^3 - N)/6$ [Prob. 34(ii)], it follows from (7.23) that the null distribution of r_s is symmetric about zero.

The correlation coefficient r_s is a popular estimate of the strength of the association between the two characteristics in the population from which the sample is drawn. Unfortunately, the precise measure of strength being estimated is somewhat complicated, and difficult to interpret, and we shall not consider this aspect any further here. For a detailed discussion of this and related measures see, for example, Kruskal (1958) and Kendall (1970).

If ties occur in the ranks $(T_1, \ldots, T_N)$, the test statistic D can be replaced by the statistic D^* defined in the context of the trend problem by (7.14). The null distribution of D^* discussed in the preceding section also applies in the present situation, with the understanding that it is a conditional distribution, given the configuration of ties. Here the conditioning is made necessary by the fact that we are dealing with a population model. The test based on D^* can be used when ties occur in either one of the two characteristics but not in both.

Let us now consider the situation in which ties occur in the ranks of both characteristics. If R_i^* and S_i^* denote the midranks of the ith subject that result from ranking each of the two sets of N responses separately, a natural generalization of D and D^* is the statistic

$$D^{**} = \sum (S_i^* - R_i^*)^2 \tag{7.24}$$

To see how to compute the null distribution of D^{**}, consider a simple example. Suppose that $N = 5$ and that the two sets of midranks are 1, 2.5, 2.5, 4, 5 and 1, 3, 3, 3, 5. Under the hypothesis of independence of the two sets, it was seen on page 293 that for the first set there are 60 equally likely orderings. Similarly, for the second set there are five possible positions for rank 1; for each of these, there are four possibilities for rank 5; and once the positions of these two ranks are fixed, the remaining three positions are filled with the three 3's, so that altogether there are $5 \times 4 = 20$ equally likely possibilities. [This is a special case of formula (7.13) with $e = 3$, $d_1 = d_3 = 1$, $d_2 = 3$, and hence $N!/d_1!\,d_2!\,d_3! = 120/6 = 20$.] Putting these two results together, we see that the total number of distinct, equally likely, arrangements of $(R_1^*, \ldots, R_N^*)$ and $(S_1^*, \ldots, S_N^*)$ is $60 \times 20 = 1{,}200$.

In computing the null distribution of D^{**}, a great simplification is possible, however, for if both the first and second row of

$$\begin{pmatrix} R_1^* \ldots R_N^* \\ S_1^* \ldots S_N^* \end{pmatrix}$$

are subjected to the same rearrangement, the statistic D^{**} remains unchanged. We can therefore take a fixed arrangement (say, the natural order) of one of the rows, and then need to consider only the possible arrangements of the other row. In the present example, the greater saving is achieved by fixing the first row since there are then only 20 possible arrangements for the second row; this shows that D^{**} has at most 20 distinct values. These 20 arrangements, which are equally likely, are easily displayed. Computation of the corresponding values of D^{**} provides the null distribution of D^{**} (Prob. 13).

For large N, we again have a normal approximation (see the Appendix, Example 19) with the required expectation and variance given by (Appendix, Prob. 4)

$$E_H(D^{**}) = \tfrac{1}{6}(N^3 - N) - \tfrac{1}{12}\sum (d_i^3 - d_i) - \tfrac{1}{12}\sum (f_j^3 - f_j) \tag{7.25}$$

and

$$(7.26)\quad \mathrm{Var}_H(D^{**}) = \frac{(N-1)N^2(N+1)^2}{36}\left[1 - \frac{\sum(d_i^3 - d_i)}{N^3 - N}\right]\left[1 - \frac{\sum(f_j^3 - f_j)}{N^3 - N}\right]$$

where $d_1, \ldots, d_e$ and $f_1, \ldots, f_g$ denote for the first and second series, respectively, the numbers of observations tied at the smallest value, the next smallest, and so on.

EXAMPLE 5. *The crying of babies and their IQ.* To test whether children who cry more actively as babies later tend to have higher IQs, a cry count was taken for a sample of 22 children aged five days and was later compared with their Stanford–Binet IQ scores at age three with the results shown below:[1]

Cry count	20	17	15	19	23	14	27	17	18	15	15
IQ score	90	94	100	103	103	106	108	109	109	112	112
Cry count	23	21	16	12	19	18	19	16	17	26	21
IQ score	113	114	118	119	120	124	132	133	141	155	157

As a first step, the midranks are computed for both series; they are shown in the first two rows of the following table, with the third row displaying the absolute difference of the two midranks:

IQ score	1	2	3	4.5	4.5	6	7	8.5	8.5	10.5	10.5
Cry count	16	9	4	14	19.5	2	22	9	11.5	4	4
	15	7	1	9.5	15	4	15	.5	3	6.5	6.5
IQ score	12	13	14	15	16	17	18	19	20	21	22
Cry count	19.5	17.5	6.5	1	14	11.5	14	6.5	9	21	17.5
	7.5	4.5	7.5	14	2	5.5	4	12.5	11	0	4.5

Squaring the absolute differences and adding, we find $D^{**} = 1601.5$. For the calculation of the expectation and variance of D^{**}, we note that there are three pairs of ties among the IQ scores, so that $e = 19$ and $d_4 = d_7 = d_8 = 2$ and the remaining d's are equal to 1. Similarly, ties among the cry counts occur at seven values with $f_4 = f_6 = f_9 = f_{10} = 2$ and $f_3 = f_5 = f_7 = 3$, the remaining f's being equal to 1. Since $N = 22$, the expectation and standard deviation are found to be $E_H(D^{**}) = 1761.5$ and $SD(D^{**}) = 384.4$. The significance probability is thus approximated by $\Phi(-160/384.4) = \Phi(-.4162) = .3386$, confirming the visual impression of a lack of association.

Let us finally give a more formal justification of the null distribution (7.3) for the model of the present section. Recall that this is a population model, in which the N subjects are assumed to have been drawn at random from a population of such subjects. With each subject are associated two characteristics, and the hypothesis asserts that these two characteristics are unrelated. Let the two characteristics of a randomly drawn subject be denoted by X and Y. Then X and Y are random variables with a joint distribution

$$(7.27)\qquad P(X \leqslant x, Y \leqslant y) = F(x,y)$$

[1] Part of the data given by Karelitz, Fisichelli, Costa, Karelitz, and Rosenfeld, "Relation of Crying Activity in Early Infancy to Speech and Intellectual Development at Age Three Years," *Child Develop.* **35**:769–777 (1964).

The formal statement of the hypothesis of independence is that X and Y are independent, so that $F(x,y)$ factors, say,

$$F(x,y) = F_1(x)F_2(y) \tag{7.28}$$

where F_1 and F_2 are the marginal distributions of X and Y, respectively. If the characteristics of the N subjects in the sample are $(X_1, Y_1), \ldots, (X_N, Y_N)$, it follows from the assumption of simple random sampling that under the hypothesis each of the X's is distributed according to F_1 and each of Y's according to F_2.

To the assumption of random sampling let us now add a second assumption. The population from which the sample is drawn will be assumed to be sufficiently large so that the characteristics of the N subjects can be assumed to be independent. Then $X_1, \ldots, X_N$ are independently distributed according to F_1, and it follows from the discussion of the hypothesis (7.5) in Sec. 1 that all $N!$ permutations of $R_1, \ldots, R_N$ are equally likely, and similarly for $S_1, \ldots, S_N$.

The argument for $(T_1, \ldots, T_N)$ is slightly more complicated in the present model. Let $X_{i_1}, \ldots, X_{i_N}$ be the X's with ranks $1, \ldots, N$ in this order, so that $Y_{i_1}, \ldots, Y_{i_N}$ have ranks $T_1, \ldots, T_N$. Then $(i_1, \ldots, i_N)$ is a random permutation of $(1, \ldots, N)$, which under H is independent of the values of the Y's. Now for any fixed permutation $(j_1, \ldots, j_N)$, the variables $Y_{j_1}, \ldots, Y_{j_N}$ are again independently distributed according to F_2. Hence this is also true of $Y_{i_1}, \ldots, Y_{i_N}$, and the result now follows from that of the preceding paragraph.

4. $s \times t$ CONTINGENCY TABLES

In Chap. 5, Secs. 3 and 4, we considered $2 \times t$ contingency tables as a means of representing the data when two treatments are being compared and there are t possible responses. It was also seen there that the resulting test can be adapted to cover the comparison of t treatments when the response is dichotomous, that is, when there are only two possible responses. Further, the test was seen to be applicable to $2 \times t$ tables resulting from the division of the subjects according to two attributes (one with two categories, the other with t), neither of which is a treatment in the sense that it can be assigned to a subject at random. In the present section we shall drop the restriction that one of the dimensions of the table is equal to 2 and consider the general case of an $s \times t$ table.

EXAMPLE 6. *Interviewer effect.* A study of possible interviewer effects in a political poll employs N subjects and s interviewers. Of the subjects, n_1 are assigned at random to interviewer 1, n_2 to interviewer 2, and so on. Suppose that the response is an expression of preference for one of t candidates $C_1, \ldots, C_t$ running for a political office. Here the interviewers represent the treatments and the candidates the possible responses. The results of the poll can be displayed in the $s \times t$ Table 7.2 below, in which, for example, A_2 denotes the number of subjects interviewed by interviewer 1 who express a preference for candidate 2:

TABLE 7.2 $s \times t$ contingency table

Interviewer	Response 1	2	...	t	Total
1	A_1	A_2	...	A_t	n_1
2	B_1	B_2	...	B_t	n_2
...	...	...	...	...	...
s					n_s
Total	d_1	d_2	...	d_t	N

The hypothesis H to be tested is that the interviewer has no effect on the resulting response. Since under H the preferred choice of each subject is unaffected by the interviewer, let us assume that it was decided upon before each subject was assigned to an interviewer. There are then N subjects: d_1 favoring candidate 1, d_2 favoring candidate 2, and so on. The probability that of the n_1 subjects assigned to the first interviewer, a_1 favor candidate 1, a_2 favor candidate 2, and so on, is then

$$P_H(A_1 = a_1, A_2 = a_2, \ldots; B_1 = b_1, B_2 = b_2, \ldots; \ldots) = \frac{\binom{d_1}{a_1, b_1, \ldots}\binom{d_2}{a_2, b_2, \ldots}\cdots\binom{d_t}{a_t, b_t, \ldots}}{\binom{N}{n_1, n_2, \ldots}} \tag{7.29}$$

For a proof, see formula (A.263) of the Appendix, which agrees with (7.29) except for a change in notation.

To decide upon an appropriate test, it is necessary to be clear about the alternatives against which the hypothesis is to be tested. An important distinction was made in Chap. 5, in the $2 \times t$ case, between one-sided, two-sided, and omnibus alternatives. The same distinction applies in the present situation.

(i) As a first question one could ask: Are the differences in the responses for the different interviewers more marked than would be expected by chance, so that there is an indication of an interviewer effect? This is the generalization of Chap. 5, Sec. 3, Example 5, with omnibus alternatives and with s instead of two treatments and t instead of three responses.

(ii) As a second possibility, suppose that the candidates can be arranged from 1 to t in a way that is relevant to the election. In a local election there may, for example, exist one overriding issue, so that the candidates can be ordered according to their stand on this issue. Alternatively, an ordering according to a liberal-conservative spectrum may be reasonable. In such a situation, it is sometimes possible (although this is not necessarily the case) to assume that any differences would take the form of an ordering also of the interviewers, who might tend to favor the candidates toward the left or the right of the table. Suppose that such an assumption is warranted and that the resulting order of the interviewers is unknown, as would typically be the case. It would then be appropriate to test the hypothesis of no difference among the rows of the $s \times t$ table against the alternatives that some rows are shifted toward the right and some toward the left, with the pattern of these shifts being unknown. This is a generalization of the two-sided case of Chap. 5, Example 4 (the two-sided rather than the one-sided case since it is not assumed known for any pair of rows which is to the left or to the right of the other).

(iii) It remains to consider the generalization of the one-sided alternatives of the $2 \times t$ table. These alternatives are appropriate in situation (ii), if the order of the interviewers can be assumed to be known. Such knowledge might have been obtained, for example, by asking for their ranking of the candidates or through an initial interview.

Of the three cases described above, clearly situation (i) requires the fewest assumptions and situation (iii) the most. Why then not play it safe and apply the test appropriate to situation (i)? The reason is, of course, that in that case the test must divide its power, that is, its ability to detect treatment differences when they exist, among alternatives in all directions, and thereby has less to spare against the particular alternatives (ii) and (iii). Put somewhat differently, in deciding between situations (i) and, say, (ii), the investigator must decide between (*a*) a more modest chance of detecting any kind of interviewer effect or (*b*) a better chance of detecting the more special effects (ii) but very little chance of detecting an effect if it has a different form.

In discussing the tests appropriate to these three situations, it will be convenient to begin with case (ii), continue with (iii), and leave (i) to the end.

Case (ii). Consider the problem of testing the hypothesis of no difference among the treatments in an $s \times t$ table against the alternatives that there is an unknown order among the treatments: one tends to give the lowest (leftmost) responses; another gives the next lowest responses; and so on. This situation can be viewed not only as a generalization of a $2 \times t$ table but also as a special case of the comparison of several treatments considered in Chap. 5, Sec. 2. The specialization consists in the fact that the responses can take on only t different values so that there are many ties. These values have no numerical meaning, but since they are ordered this does not interfere with the application of a rank test statistic such as K^*, which is thus appropriate in the present situation.

This statistic was defined by Eq. (5.11) as

$$K^* = \frac{[12/N(N+1)] \sum R_i^{*2}/n_i - 3(N+1)}{1 - \sum (d_i^3 - d_i)/(N^3 - N)} \tag{7.30}$$

where R_i^* is the sum of the midranks of the ith treatment and large values of K^* are significant. In terms of the entries of the original $s \times t$ table, we have

$$R_1^* = A_1[\tfrac{1}{2}(d_1+1)] + A_2[d_1 + \tfrac{1}{2}(d_2+1)] + \cdots + A_t[d_1 + \cdots + d_{t-1} + \tfrac{1}{2}(d_t+1)] \tag{7.31}$$

The rank sum R_2^* is given by the same expression with A_1, A_2, and so on replaced by B_1, B_2, and so on, and similarly for $R_3^*, \ldots, R_s^*$. The exact null distribution of K^*

can thus in principle be computed from (7.29) (Prob. 22). For large values of N the distribution can be approximated by χ^2 with $s-1$ degrees of freedom provided none of the n_i are too small and none of the ratios n_i/N and d_j/N are too close to 1. (See Appendix, Example 31.)

EXAMPLE 7. *Drugs for nausea.* In one phase of a study regarding the effectiveness of several drugs on postoperative nausea, 167 patients were assigned at random, 30 to drug P, and 67 to drug C; the remaining 70 received a placebo. The numbers of patients suffering from severe, moderate, slight, or no nausea are shown below:[1]

	Severe	Moderate	Slight	None	Total
Placebo	8	8	19	35	70
Drug P	2	3	5	20	30
Drug C	3	4	15	45	67
Total	13	15	39	100	167

The midranks of the 13 patients with severe nausea is the average of the ranks 1 to 13, which is $14/2 = 7$; the midrank of the 15 patients with moderate symptoms is $13 + \frac{1}{2} \times 16 = 21$; similarly, the midranks of the third and fourth columns are, respectively, 48 and 117.5. With these values, the sum of the midranks of the three treatments becomes

$$R_1^* = 8(7)+8(21)+19(48)+35(117.5) = 5248.5$$
$$R_2^* = 2(7)+3(21)+\ \ 5(48)+20(117.5) = 2667$$
$$R_3^* = 3(7)+4(21)+15(48)+45(117.5) = 6112.5$$

With $d_1 = 13, d_2 = 15, d_3 = 39, d_4 = 100, n_1 = 70, n_2 = 30, n_3 = 67$, and $N = 167$, it is then seen (Prob. 23) that $K^* = 5.5009$. Table J shows that a χ^2-variable with two degrees of freedom (one less than the number of treatments) exceeds the value 5.5009 with probability .064, which is therefore the approximate significance probability of the observed value of K^*.

Case (iii). The one-sided version of case (ii), in which a *known* order is assumed for the treatments, is of course a special case of the one-sided version of the s-treatment problem considered in Chap. 5, Sec. 4, and the test discussed there is applicable also in the present situation. Alternatively, the present one-sided problem can also be viewed as a generalization of the trend problem treated in Sec. 2 of the present chapter. It differs from that problem in that there are several observations at each level (i.e., for each treatment) instead of just one.

The test statistic proposed for the trend problem was $D^* = \sum(T_i^* - i)^2$, where T_i^* is the midrank of the observation on the ith treatment. When there are several observations on each treatment, it is natural to consider the n_i observations on treatment i as observations with tied treatment ranks and to assign to them the associated midrank i^*. The proposed test statistic is then

$$D^{**} = \sum(T_i^* - i^*)^2 \tag{7.32}$$

[1] From Beecher, *Measurement of Subjective Responses*, Oxford University Press, London, 1959, Table 17.8.

Let us illustrate the computation of D^{**} on the data of the preceding example.

EXAMPLE 7. *Drugs for nausea (continued).* Suppose that in the situation of Example 7 there is a conviction a priori (that is, before the start of the experiment) that if the drugs have any effect, this can only be in the direction of decreasing the nausea, and if there is a difference in the effectiveness of the two drugs, then drug C will be more effective than P.

To compute D^{**}, note that 70 subjects receive treatment 1 (placebo) and are therefore assigned midrank

$$1^* = \frac{1+\cdots+70}{70} = \frac{71}{2} = 35.5$$

Similarly, there are 30 applications of treatment 2 (drug P) and 67 of treatment 3 (drug C), with midranks

$$2^* = 70+\tfrac{1}{2}\times 31 = 85.5 \qquad \text{and} \qquad 3^* = 70+30+\tfrac{1}{2}\times 68 = 134$$

Consider now the eight observations receiving treatment 1 and giving response 1 (severe nausea). Their treatment midrank is 1^* and their response midrank is $T_1^* = 7$. Their contribution to D^{**} is therefore

$$8(T_1^*-1^*)^2 = 8(7-35.5)^2$$

Analogously, the contribution of the eight observations corresponding to treatment 1 and response 2 is

$$8(T_2^*-1^*)^2 = 8(21-35.5)^2$$

Continuing in this way with all 12 cells of the 3×4 table, we find (Prob. 24) that $D^{**} = 521{,}459.5$.

The general expression for D^{**} in terms of the original $s \times t$ table is (Prob. 39)

$$(7.33)\quad D^{**} = \tfrac{1}{3}N(N+1)(2N+1)-\tfrac{1}{12}\sum(n_i^3-n_i)-\tfrac{1}{12}\sum(d_j^3-d_j)$$
$$-2\{\tfrac{1}{2}(n_1+1)R_1^*+[n_1+\tfrac{1}{2}(n_2+1)]R_2^*+\cdots+[n_1+\cdots+n_{s-1}+\tfrac{1}{2}(n_s+1)]R_s^*\}$$

where R_1^* is given by (7.31) and the remaining R_i^* by analogous expressions. The null distribution of D^{**} can then in principle be obtained from (7.29) (Prob. 25). For large N, this distribution is approximately normal, with the necessary expectation and variance given by (7.25) and (7.26). Although these formulas were derived for a different model (to which we shall return below), the assumed null distribution is the same, and the formulas are therefore also applicable in the present situation.

To illustrate the normal approximation, consider once more the situation of Example 7 (continued). Under the assumptions of this example, one finds (Prob. 26)

$$E_H(D^{**}) = 631{,}606 \qquad \sqrt{\text{Var}_H(D^{**})} = 48{,}955.79$$

and hence

$$\frac{D^{**}-E_H(D^{**})}{\sqrt{\text{Var}_H(D^{**})}} = -2.2499$$

Since small values of D^{**} are significant, the approximate significance probability is therefore of the order .0123; but the data seemed much less significant in Example 7, where the test statistic was K^*.

The normal approximation is supported by a limit theorem which states that D^{**} has a normal limit distribution as $n_1, \ldots, n_s$ tend to infinity (with s remaining fixed) provided all the ratios n_i/N and d_j/N are bounded away from 1. This result is proved in the Appendix, Sec. 4 (Example 19).

Case (i). Let us finally consider the case of omnibus alternatives to the hypothesis of no treatment effect. These do not give any indications about the response categories that are increased or decreased by the different treatments. Under such alternatives any ordering of the responses is irrelevant, and accordingly an appropriate test statistic will not involve any ranking of the observations. Although the problem is thus outside the scope of this book, we shall for the sake of completeness briefly describe the standard χ^2-test for this case.

This test has a somewhat similar structure to the Kruskal-Wallis test of Chap. 5 or the Q- and $\hat{Q}$-tests of Chap. 6. In these earlier cases, the test statistics were based on the sums of the treatment ranks or, more precisely, on the differences of these sums from their expectation under the hypothesis. The test statistic in each case was a weighted sum of the squares of these differences. In the present case, the entries from the different cells of the $s \times t$ table should be neither ranked nor should the different entries from the same treatment be combined into a sum because otherwise an increase in one cell might be balanced by a decrease in another and hence go undetected. Instead, we consider the difference of each individual cell entry from its expectation and use as test statistic a weighted sum of squares of these differences. The expectation of A_j (under the hypothesis) is $n_1 d_j/N$, that of B_j is $n_2 d_j/N$, and so on (see, for example, Appendix, Sec. 7), and the proposed test statistic is

$$L = N\left[\sum_{j=1}^{t} \frac{1}{n_1 d_j}\left(A_j - \frac{n_1 d_j}{N}\right)^2 + \sum_{j=1}^{t} \frac{1}{n_2 d_j}\left(B_j - \frac{n_2 d_j}{N}\right)^2 + \cdots\right] \tag{7.34}$$

The null distribution of L, as was the case with K^* and D^{**}, is determined by (7.29). For large values of N, the statistic L has approximately a χ^2-distribution with $(s-1)(t-1)$ degrees of freedom. [Proofs can be found, for example, in Cramér (1946), Rao (1965), and Lancaster (1969).] The conditions for this distribution to provide a good approximation are different from the corresponding conditions for the statistics K^* and D^{**} defined by (7.30) and (7.32). These latter conditions involve only the marginal totals n_i and d_j of the contingency table given by Table 7.2, but the conditions for L require that the expected number of observations in none of the st cells be too small. A detailed discussion of these conditions is given by Cochran (1954).

EXAMPLE 7. *Drugs for nausea (continued).* As an illustration of the L-test, let us apply it to the data of Example 7 to test the hypothesis of no difference among the three treatments against the omnibus alternatives of a completely unspecified difference. The value of L for the data of Example 7 turns out to be $L = 6.358$ (Prob. 27). Table J shows that a χ^2-variable with $(3-1)(4-1) = 6$ degrees of freedom exceeds the value 6.358 with probability greater than .3. This probability turns out to be .384, which is therefore the approximate significance probability of the observed value of L.

The issue raised by Example 7 is so basic that it may be worthwhile to return to it once more. The three tests described above give quite different results: the data appear highly significant under the one-sided test, case (iii); barely significant under test (ii); not significant under test (i). Test (iii) thus seems to be the most desirable of the three. The choice of test (iii) is justified if a priori considerations narrow the possible alternatives to the one-sided class for which this test is appropriate or if there is no interest in detecting the existence of treatment differences (by rejecting the hypothesis) if these take other forms.

If such justification on prior grounds is not available, one must guard against the temptation to choose test (iii) because the particular order of the alternatives is suggested by the data. If the choice is made under the influence of the data, the highly significant result becomes illusory, the significance probability meaningless. Since the hypothesis would then also have been rejected if the data had been equally significant in some other direction, the computed significance probability is the probability of only part of the rejection rule. The true significance probability, which has to take account of all the different arrangements for which the hypothesis would also have been rejected, is much larger, and the claimed significance is a fraud. To make sure that a look at the data does not in some way influence the choice between the three tests, this choice should always be made before the results of the study are available.

The tests (i) to (iii) were described above in terms of a randomization model, in which the s treatments are assigned at random to a given set of N subjects. Instead, it is often more appropriate to assume a population model in which the subjects are drawn at random from a larger population of potential users of the treatments. The discussion of this and other population models so closely parallels that of Chap. 2, Sec. 1 and of Chap. 5, Sec. 4 that it will only be sketched here rather briefly.

Let $N = n_1 + \cdots + n_s$ subjects be drawn from a population by simple random sampling, and let n_1 be assigned to treatment 1, n_2 to treatment 2, and so on. Suppose that the population is large enough so that the responses of the different subjects can be assumed to be independent. If p_j denotes the probability that a subject receiving treatment 1 will give response j, the joint distribution of $A_1, \ldots, A_t$ is the multinomial distribution

$$P(A_1 = a_1, \ldots, A_t = a_t) = \binom{n_1}{a_1, \ldots, a_t} p_1^{a_1} \ldots p_t^{a_t}$$

If, similarly, $q_1, \ldots, q_t$ denote the probabilities of the different responses for a subject receiving treatment 2, and so on, it follows from the assumed independence of the N subjects that the joint distribution of the A's, B's, and so on is the product

$$(7.35) \quad P(A_1 = a_1, \ldots, A_t = a_t; B_1 = b_1, \ldots, B_t = b_t; \ldots)$$
$$= \binom{n_1}{a_1, \ldots, a_t} p_1^{a_1} \ldots p_t^{a_t} \binom{n_2}{b_1, \ldots, b_t} q_1^{b_1} \ldots q_t^{b_t} \ldots$$

The hypothesis of no treatment effect now takes the form

$$(7.36) \qquad H: \quad p_1 = q_1 = \cdots; p_2 = q_2 = \cdots; \ldots$$

Under H, the joint distribution of the A's, B's, and so forth still depends on the common values $p_1, \ldots, p_t$, which are unknown. However, a test statistic with known distribution can be obtained by considering the conditional distribution of the A's, B's, and so forth given the values of $d_1, \ldots, d_t$. In fact, this conditional null distribution is the distribution (7.29), which was introduced as the unconditional distribution of the same variables in the earlier randomization model. (For a proof, see Appendix, Sec. 7.) The three tests described earlier in the section can therefore be applied as before, with the understanding that the resulting significance probabilities are now conditional.

Let us next suppose that the comparison is not between s treatments but between s attributes which cannot be separated from their subjects, such as the size of the family into which an individual is born, his religious affiliation, or his profession. These attributes divide the population being studied into s subpopulations. If samples of sizes $n_1, \ldots, n_s$ are drawn from these, it is seen that the distribution of the A's, B's, and so on continues to be given by (7.35), where $p_1, \ldots, p_t$ are the probabilities of the t responses for the first subpopulation, $q_1, \ldots, q_t$ for the second, and so on.

Often, it is difficult to sample the s subpopulations defined by s categories such as different smoking habits, different abilities, or medical conditions. If then, instead, a sample of size N is drawn from the population as a whole, the numbers $n_1, \ldots, n_s$ of subjects from the different subpopulations included in the sample become random variables. Thus, both the row and column totals of the $s \times t$ table are then random. The model for this case is obtained by generalizing (5.35), which defines the model in question for $s = 2$. As is shown in the Appendix, Sec. 7, the null distribution (7.29) retains its validity also in this case, if it is taken as the conditional distribution given the values of both the n_i and the d_j. With this understanding, the three tests can be applied as before.

A comparison of attributes rather than treatments requires another important change in interpretation, which has already been discussed in Chap. 2, Sec. 1. Suppose that the investigation of Example 6 had been concerned with the influence on voter preferences not of the interviewer but, say, of religious affiliation. Then random assignment is not possible and, as a result, a small significance probability would no longer indicate a causal relationship but only an association, about the cause of which the data provide no information.

When a sample of size N is drawn at random from the combined population and the comparison is between attributes rather than treatments, the asymmetry between rows and columns of the $s \times t$ table disappears. Rows and columns then simply represent two different categorizations (in the study of voter preference, for example, these would be, respectively, the different religious affiliations of the voters and the different candidates they might prefer). It is seen that the situation is that of Sec. 3. Two characteristics are attached to each subject of the population (e.g., his religion and a preferred candidate). These can be represented by the numbers $1, \ldots, s$ and $1, \ldots, t$ as in Table 7.2. The hypothesis being tested is the independence of the two characteristics. The alternatives can be one-sided, two-sided, or omnibus. Only the first of these possibilities was considered in Sec. 3, and the D^{**}-test proposed there for this case is identical with the D^{**}-test of the present section.

5. FURTHER DEVELOPMENTS

5A. Pitman Efficiency of D

As an alternative to the hypothesis of randomness, suppose that the distribution of Z_i is

$$F_i(z) = F(z - \alpha - i\beta) \tag{7.37}$$

where F has expectation zero. Here α and β are unknown, α being a location parameter and β indicating the rate at which the center of F_i increases with i. If F is normal, the standard test of the hypothesis $\beta = 0$ is a t-test. The Pitman efficiency of D relative to this t-test is obtained in the normal case by Stuart (1954, 1956) and for general F by Aiyar (1969), who shows it to be equal to $[e_{W,t}(F)]^{1/3}$. Here the quantity $e_{W,t}(F)$ is the asymptotic efficiency of the Wilcoxon test relative to the t-test and is given by (A.240) of the Appendix. In the normal case, the value of this efficiency is $(3/\pi)^{1/3} = .98$, and by (A.241) it is greater than or equal to .95 for all F. These results are generalized to alternatives (7.37) with general F by Guillier (1972). Aiyar and Guillier also investigate the corresponding efficiency against the so-called autoregressive alternatives according to which the Z's are no longer independent but each depends on the preceding observation through the equation

$$Z_i = \gamma Z_{i-1} + U_i \tag{7.38}$$

where the U's are identically and independently distributed.

5B. Estimating the Regression Coefficient β

A close competitor to D or D' in testing randomness against an upward or downward trend is the statistic

$$B = \sum_{i<j} U_{ij} \tag{7.39}$$

where U_{ij} is given by (7.6). The test based on B and its properties are discussed by Mann (1945) and Kendall (1970). The latter also provides a table of the null distribution of B. The Pitman efficiency of B relative to D against the alternatives (7.37) is shown to be 1 by Stuart (1954) and Aiyar (1969), and against more general alternatives by Guillier (1972).

An advantage of B is its easy invertibility in the model (7.37), which leads to simple confidence statements for β. To perform this inversion, denote the statistic (7.39) by $B(Z_1, \ldots, Z_N)$. Putting $M = \binom{N}{2}$, let

$$D_{(1)} < \cdots < D_{(M)}$$

denote the ordered slopes

$$\frac{Z_j - Z_i}{j-i}, \qquad (i < j)$$

Then in analogy with Chap. 2, Theorem 5, it follows that

$$D_{(k)} \leqslant b \quad \text{if and only if} \quad B(Z_1 - \alpha - b, \ldots, Z_N - \alpha - Nb) \leqslant M - k \tag{7.40}$$

To see this, note that the first inequality of (7.40) holds if and only if at least k of the ratios

$$\frac{Z_j - Z_i}{j-i} - b = \frac{(Z_j - \alpha - jb) - (Z_i - \alpha - ib)}{j-i}$$

are less than or equal to zero, and hence if at most $M-k$ of these differences are greater than zero. Since $j-i > 0$, this will be the case if for at most $M-k$ pairs $i < j$,

$$Z_i - \alpha - ib < Z_j - \alpha - jb$$

and thus if the right-hand side of (7.40) holds.

One can now proceed as in Chap. 2, Sec. 6. In particular, it follows from (7.40) that

$$D_{(k)} \leqslant \beta < D_{(k+1)} \quad \text{if and only if} \quad B(Z_1 - \alpha - \beta, \ldots, Z_N - \alpha - N\beta) = M - k \tag{7.41}$$

where $D_{(0)} = -\infty$ and $D_{(M+1)} = \infty$. Since the distribution of $Z_i - \alpha - i\beta$ is independent of α and β, the relation (7.41) shows that for any values of α and β,

$$P_{\alpha,\beta}(D_{(k)} \leqslant \beta < D_{(k+1)}) = P_{0,0}(B = M-k) \qquad \text{for all } k = 0, 1, \ldots, M \tag{7.42}$$

where the subscripts indicate the values of α and β for which the probability is being computed. The statistics $D_{(1)}, \ldots, D_{(M)}$ therefore divide the real line into $M+1$ random intervals into which β falls with known probabilities, independent of α, β, and F, and given exactly by the null distribution of B. Confidence intervals for β based on (7.42) were proposed by Theil (1950), who also suggested the median of the slopes $(Z_j - Z_i)/(j-i)$ as a point estimate of β. Some extensions of Theil's result and references to other estimation methods for β are given by Sen (1968). It is interesting to note, as is done by Bhattacharyya (1968), that the same point estimate of β is obtained by applying the method of Chap. 2, Sec. 6, to the test statistic D' instead of to B. The estimate is also derived from a completely different point of view by Beran (1971).

When one suspects a linear trend and hence a model of the form (7.37), it may be of interest to test the appropriateness of this model. Two simple tests for this purpose are proposed by Olshen (1967).

Important problems also arise in the comparison of several regression coefficients when one is dealing with more than one series of observations for each of which one assumes a model of the form (7.37) with common distribution function F. This situation is considered, for example, by Sen (1969), Hollander (1970), Adichie (1974), and Potthoff (1974).

For further work on inference concerning regression parameters, including multiple regression and other approaches to robust estimation, see Adichie (1967), Bickel (1971), Jureckova (1971), and Koul (1969).

5C. Tests of Randomness Based on Runs

The alternatives of an upward or downward trend considered in Sec. 2 are not the only alternatives to randomness that may be of interest. Instead, a trend might be cyclical (for example, seasonal, or following some other pattern), or successive observations may be dependent, as is the case in model (7.38). The problem of testing for randomness against these or other less clearly specified alternatives arises, for example, in quality control if one wishes to know whether the quality of successively produced items behaves like a sequence of identically, independently distributed random variables, in the study of economic time series, or when considering a sequence of physiological or psychological measurements taken on the same individual over a period of time.

Although the alternatives to randomness are often not clearly defined, a common feature of a large class of alternatives is a tendency toward clustering so that high (or low) values tend to occur together. This can be exploited by considering various kinds of runs of like elements exhibited by the series of observations and rejecting the hypothesis of randomness when the number of runs is too small, or when too many long runs occur.

Consider first the important special case in which the response is dichotomous, so that each observation represents either a success or a failure. If these two outcomes are denoted by 1 and 0, respectively, the N observations form a sequence of ones and zeros. The hypothesis of randomness states that this sequence constitutes a sequence of N binomial trials with unknown success probability, say, p. A natural test statistic to test this hypothesis against the alternatives of clustering is the total number R of runs of ones and zeros. For example, in the series

$$1 \quad 0 \quad 0 \quad 0 \quad 1 \quad 1 \quad 0 \quad 0 \quad 1 \quad 1$$

there are three runs of ones (of lengths 1, 2, and 2) and two runs of zeros (of lengths 3 and 2), so that $R = 5$.

The null distribution of R depends on p; however, the conditional null distribution, given that the number of zeros is m and the number of ones is n, does not since the conditional probability of any particular arrangement of the m zeros and n ones is $1\Big/\binom{N}{n}$. The conditional distribution of R has in fact the simple expression

$$P(R = 2k) = \frac{2\binom{m-1}{k-1}\binom{n-1}{k-1}}{\binom{N}{n}} \tag{7.43}$$

and

$$P(R = 2k+1) = \frac{\binom{m-1}{k-1}\binom{n-1}{k}+\binom{m-1}{k}\binom{n-1}{k-1}}{\binom{N}{n}} \tag{7.44}$$

As m and n tend to infinity with m/n tending to γ, it is shown by Wald and Wolfowitz (1940) that the distribution of

$$\frac{R-2m/(1+\gamma)}{\sqrt{4\gamma m/(1+\gamma)^3}}$$

tends to the Standard Normal distribution. (Wald and Wolfowitz discuss R in the context of the two-sample problem. For this problem, the R-test is now known to be very inefficient.) Tables of the null distribution of R are given by Swed and Eisenhart (1943). As pointed out by David (1947) and Lehmann (1959, pp. 155–156), the R-test has certain optimum properties against the alternatives that the observations come from a stationary Markov chain. For further work on the application of run tests to Markov chains, see Goodman (1958) and a survey paper by Billingsley (1961).

Other run statistics that have been considered for testing randomness in a sequence of zeros and ones are the number of runs of length greater than l and the length of the longest run. For a discussion of some interesting early applications of runs, see von Mises (1931, Sec. 4.2). A systematic account of the theory of runs of two kinds of elements and its extension to the case of more than two is given by Mood (1940). Additional material can be found in the book by David and Barton (1962).

Let us now return to the general problem of a series $Z_1, \ldots, Z_N$ for which the hypothesis of randomness (in the population model, the hypothesis that the Z's are identically and independently distributed) is to be tested against the alternative of clustering of like values. A possible test is obtained by replacing each observation by a zero if it falls below and by a one if it falls above the median of the Z's and by taking as test statistic the number R of runs in the resulting series of ones and zeros. The statistic R clearly depends only on the ranks of the observations. Both when $N = 2n+1$ and when $N = 2n$, the null distribution of R is given by (7.43) with $m = n$ and N replaced by $2n$.

A class of run statistics different from those based on runs above and below the median is obtained by considering the signs of the successive differences $Z_2 - Z_1$, $Z_3 - Z_2$, ..., $Z_N - Z_{N-1}$. This again constitutes a sequence of two kinds of elements (a plus sign if $Z_i - Z_{i-1} > 0$, a minus sign if $Z_i - Z_{i-1} < 0$) but the associated runs up (i.e., runs of plus signs) and runs down (i.e., of minus signs) have a quite different null distribution. A simple test statistic is the total number of runs up and down, which is essentially the number of turning points or of peaks and troughs in the series, considered by Wallis and Moore (1941). A table of the null distribution is given by Edgington (1961). The distribution of the number of runs of given length and the joint distribution of numbers of runs of several different lengths has been studied by, among others, Kermack and McKendrick (1937a, b), Levene and Wolfowitz (1944), Wolfowitz (1944), and Olmstead (1946). The power of the associated tests and the problem of choosing a test that is appropriate against a specified class of alternatives is discussed by Levene (1952).

5D. Other Tests of Independence

The statistic $B = \sum_{i<j} U_{ij}$ discussed in Sec. 5B above for testing randomness against an upward or downward trend can also be adapted to the problem of testing independence against positive or negative association of two variables. If the N pairs of observations $(X_1, Y_1), \ldots, (X_N, Y_N)$ are arranged in increasing order of the X's, B counts the number of pairs (i,j) with $i < j$ for which $Y_i < Y_j$. If the X's are not first ordered, it follows that B is the number of pairs (i,j) with $X_i < X_j$ for which $Y_i < Y_j$. Thus, in general, B is the number of pairs (i,j) for which the differences $X_j - X_i$ and $Y_j - Y_i$ have the same sign, or equivalently, the number of pairs for which

$$(X_j - X_i)(Y_j - Y_i) > 0 \tag{7.45}$$

Such pairs are said to be *concordant*. The probability p of the event (7.45), which is independent of i and j, is a measure of the strength of the association of X and Y. Because it is conventional to have such measures vary between -1 and $+1$, one may instead consider as a measure of association the quantity

$$\tau = p-(1-p) \tag{7.46}$$

known as *Kendall's tau*, which goes from -1 to $+1$ as p increases from 0 to 1. A detailed discussion of the history of tau, and of the relative advantages of this and other measures of association, is given by Kruskal (1958). Properties of tau and the test based on B are studied by Kendall (1970), where there is also a table of the null distribution of B and a discussion of the normal approximation to this distribution. [For more extensive tables, see Kaarsemaker and van Wijngaarden (1953) and Best (1973).] The case of ties in one of the variables is treated by Sillitto (1947) and that of ties in both variables by Burr (1960).

Other tests of independence are the corner test of Olmstead and Tukey (1947), the quadrant test of Blomquist (1950), and the Normal Scores test studied by Fieller and Pearson (1961) and Bhuchongkul (1964). A test against omnibus alternatives is considered by Hoeffding (1948b) and by Blum, Kiefer, and Rosenblatt (1961).

5E. Power and Efficiency of Tests of Independence

The study of the power and efficiency of tests of independence is complicated by the difficulty of defining natural classes of alternatives to the hypothesis of independence. Some qualitative results requiring only a concept of positive and negative association are given by Lehmann (1966) and by Yanagimoto and Okamoto (1969). In the normal case, natural alternatives to independence are provided by the bivariate normal distributions with nonzero correlation. Some power values in this case are given by Bhattacharyya, Johnson, and Neave (1970) [see also Kraemer (1974)]. The Pitman efficiency of the test based on Spearman's coefficient r_s, as well as that based on the statistic B, relative to the test based on the correlation coefficient

$$R = \frac{\sum(X_i-\bar{X})(Y_i-\bar{Y})}{\sqrt{\sum(X_i-\bar{X})^2\sum(Y_i-\bar{Y})^2}}$$

is $(3/\pi)^2 = .912$.

Other classes of alternatives are considered by Konijn (1956), Farlie (1960), Bhuchongkul (1964), Plackett (1965), and Gokhale (1966). The last of these authors defines his alternatives as

$$X = U+\Delta Z_1 \quad , \quad Y = V+\Delta Z_2 \tag{7.47}$$

where U, V, and $Z=(Z_1,Z_2)$ are independently distributed with distributions F, G, and M possessing densities f, g, and m, respectively. Under certain additional regularity conditions he shows that the Pitman efficiency of r_s to R is given by

$$e_{r_s,R}(F,G) = [12\,\sigma\tau\textstyle\int f^2(x)\,dx\int g^2(y)\,dy]^2 \tag{7.48}$$

where σ^2 and τ^2 are the variances of U and V and where it is assumed that the covariance of Z_1 and Z_2 is different from zero. It is interesting to note that the efficiency is independent of M. It follows from the results on the Pitman efficiency $e_{W,t}(F)$ of the Wilcoxon to the t-test that

$$e_{r_s,R}(F,G) \geqslant (.864)^2 = .746$$

for all pairs (F,G) satisfying the required regularity conditions. Interesting extensions of Gokhale's results are obtained by Behnen (1971). Efficiencies against Plackett's alternatives are found by Mardia (1969). For the Normal Scores test mentioned in Sec. 5D, it is shown by Bhuchongkul (1964) and Gokhale (1966) that its Pitman efficiency relative to R is greater than or equal to 1 against the alternatives (7.47). The Bahadur efficiency of the tests of Spearman and Kendall and of the Normal Scores test is discussed by Woodworth (1970).

5F. Contingency Tables

The asymptotic power of the L-test of independence in the population models, where either $n_1, \ldots, n_s$ subjects are drawn at random from s subpopulations or N subjects are drawn from the population at large, is discussed by Meng and Chapman (1966). Various measures of the strength of an association in an $s \times t$ table are studied in a series of papers by Goodman and Kruskal (1954, 1959, 1963, 1972).

The analysis of two-way (and higher-dimensional) contingency tables is a large and fascinating subject, which, however, is only marginally related to the subject of the present book. An excellent, detailed exposition of the subject is provided in a forthcoming book by Bishop, Fienberg, and Holland (1975); see also the book by Lancaster (1969) and the survey papers of Cochran (1954), Mosteller (1968), and Mosteller and Tukey (1968). Some recent references are the papers by Bhapkar and Koch (1968), Bishop (1969, 1971), Fienberg (1970, 1972), Fienberg and Gilbert (1970), Goodman (1968, 1970a, b, 1971), Grizzle, Starmer, and Koch (1969), Haberman (1973), Light and Margolin (1971), Mantel (1970), Plackett (1969), and Williams and Grizzle (1972).

6. PROBLEMS

(A)

Section 1

1. What is the probability that the T's are in the "right order," i.e., that $T_1 = 1$, $T_2 = 2$, and so on, if (i) $N = 4$; (ii) $N = 5$?

2. If $N = 4$, find (i) $P_H(T_1 = 3)$; (ii) $P_H(T_1 = 1, T_2 = 4)$.

3. For arbitrary N find (i) $P_H(T_1 = 1)$; (ii) $P_H(T_1 = 1, T_2 = 2)$.

4. If $N = 3$, find $P_H(\sum iT_i = 13)$.

Section 2

5. By enumeration of cases obtain the null distribution of D when (i) $N = 3$; (ii) $N = 4$.

6. Suppose that in Example 3, the measure of interest is not the total number of illegitimate births but the number of such births per 1,000 total live births. These ratios are as follows:[1]

	1940	1945	1947	1950	1955	1957
Alabama	17.0	16.2	12.5	12.9	13.2	14.2
Delaware	29.3	25.1	23.5	24.0	18.6	18.2
District of Columbia	23.8	34.9	32.6	31.7	43.3	58.8
Kansas	11.5	18.3	13.2	12.6	13.9	14.5
Lousiana	17.4	14.7	17.7	12.4	13.3	13.1

(i) Test whether these ratios have tended to increase in time.

(ii) Test whether the ratios tend to form a consistent pattern from state to state by applying Friedman's statistic Q with the states as blocks and the years as "treatments." Discuss the appropriateness of the test statistic and its null distribution.

7. Given the following hypothetical IQ scores of a child at the indicated ages (in years), test the hypothesis of randomness against the alternative of an upward trend.

Age	3	$3\frac{1}{2}$	4	$4\frac{1}{2}$	5	$5\frac{1}{2}$	6	7	8	9	10	11	12
IQ	130	132	136	133	135	138	141	140	137	139	150	153	160

8. The following are the (smoothed) IQ scores of a child at the indicated ages (in years).[2] Test the hypothesis of randomness against the alternative of an upward trend.

Age	3	$3\frac{1}{2}$	4	$4\frac{1}{2}$	5	$5\frac{1}{2}$	6	7	8	9	10	11	12
IQ	105	110	109	114	114	115	114	118	123	128	128	127	126

9. Let $N = 5$ and let the values of the five midranks be 1, 2.5, 2.5, 4, 5. (i) Find the number of distinct arrangements of these midranks; (ii) find $P_H(T_1^* = T_2^* = 2.5)$; (iii) find $P_H(T_1^* = 2.5)$.

10. Solve the three parts of the preceding problem if the five midranks are 1, 2.5, 2.5, 4.5, 4.5.

11. Let $N = 7$ and let the values of the seven midranks be 1, 2, 4, 4, 4, 6, 7. Find (i) the number of distinct arrangements of the midranks; (ii) $P_H(T_1^* = 1, T_2^* = 2)$; (iii) $P_H(T_1^* = T_2^* = 4)$.

12. Obtain the null distribution of D^* when (i) $N = 5, d_1 = 3, d_2 = 2$; (ii) $N = 5, d_1 = 2, d_2 = 1, d_3 = 2$; (iii) $N = 6, d_1 = d_2 = d_3 = 2$; (iv) $N = 6, d_1 = 4, d_2 = d_3 = 1$.

Section 3

13. Compute the null distribution of D^{**} for the situation described in the text where $N = 5$ and the two sets of midranks are 1, 2.5, 2.5, 4, 5 and 1, 3, 3, 3, 5. [*Hint:* Fix the order of the first set and consider all possible arrangements of the second set.]

14. Compute the null distribution of D^{**} if $N = 5$ and if the midranks of the two sets are (i) 1, 3, 3, 3, 5 and 1, 2.5, 2.5, 4.5, 4.5; (ii) 1.5, 1.5, 3.5, 3.5, 5 and 1, 2.5, 2.5, 4.5, 4.5; (iii) 1, 2, 4, 4, 4 and 1, 2.5, 2.5, 4, 5.

[1] Data from same source as in Example 3.

[2] Read (approximately) from the graph of Case 10, one of 140 cases shown by Sontag, Baker, and Nelson, "Mental Growth and Personality Development: A Longitudinal Study," *Monograph. Soc. Res. Child Develop.* **23** serial no. 68, no. 2 (1958).

15. Twenty mediation cases involving small businesses were ranked according to the success of the mediation effort and the amount of hostility shown in the early part of the mediation session, with the following results:[1]

Hostility	1	2	3	4	5	6	7	8	9	10	11	12
Success	12	7	10	11	4	1	8	9	3	6	2	5

(Here 1 corresponds in the first row to the greatest amount of hostility and in the second row to the greatest degree of success.) Test the hypothesis of independence against the alternative of a negative association of the two sets of ranks.

16. Fourteen factory supervisors were ranked on their potential for promotion by a psychologist (on the basis of an information blank and various psychological tests) and by an executive who had observed the work of the supervisors, with the following results:[2]

Supervisor	A	B	C	D	E	F	G	H	I	J	K	L	M	N
Executive	7	2	1	5	4	3	6	9	11	14	10	12	8	13
Psychologist	9	3	10	5	4	8	1	6	7	12	14	11	2	13

Test the hypothesis of independence against the alternatives of a positive association of the two sets of ranks.

17. The following are the test scores (the totals of a number of psychological tests) of 13 dizygous (i.e., nonidentical) male twins.[3] Test the hypothesis of independence against the alternatives of positive dependence:

1	2	3	4	5	6	7	8	9	10	11	12	13
277	169	157	139	108	213	232	229	114	232	161	149	128
256	118	137	144	146	221	184	188	97	231	114	187	230

18. For a set of 16 male babies aged forty-eight weeks, both the height and head circumference measurements were taken, with the following results:[4]

Height	77.3	73.0	73.9	71.7	79.6	75.4	77.6	72.0	76.4
Head circumference	47.5	46.9	45.9	46.3	47.5	47.4	47.1	47.3	48.2
Height	75.6	74.9	70.5	71.6	73.3	70.9	75.0		
Head circumference	46.5	46.4	48.2	48.2	45.0	46.1	47.4		

Test the hypothesis of independence against the alternatives of positive dependence.

19. The following scores were obtained by 21 monozygous (identical) male twins on a psychological test (the Progressive Matrices test):[5]

45	40	48	42	45	44	40	37	27	45	51	44	44	29	27	48	33	39	47	54	32
37	41	50	46	31	39	37	42	27	48	49	52	27	36	29	44	44	43	34	48	33

[1] Data from Landsberger, "A Study of Mediation through an Analysis of the Background of Disputes and the Activities of the Mediator," unpublished Ph.D. thesis, Cornell University, Ithaca, N.Y., 1954, as quoted in McCarthy, *Introduction to Statistical Reasoning*, McGraw-Hill Book Company, New York, 1957.

[2] Part of the data reported by Edwards, "Statistical Methods for the Behavioral Sciences," from Dulsky and Krout, "Predicting Promotion Potential on the Basis of Psychological Tests," *Personnel Psych.* **3**:345–351 (1950).

[3] Part of the data of Clark, Vandenberg, and Proctor, "On the Relationship of Scores on Certain Psychological Tests with a Number of Anthropometric Characters and Birth Order in Twins." *Human Biol.* **33**:163–180 (1961).

[4] Part of the data of Thompson, "Data on the Growth of Children during the First Year of Life," *Human Biol.* **23**:75–92 (1951).

[5] From the same source as Prob. 17.

Test the hypothesis of independence against the alternatives of positive dependence.

Section 4

20. Display all possible 3×3 tables with $N = 5, n_1 = n_2 = 1, n_3 = 3; d_1 = 1, d_2 = 2, d_3 = 2$, and use (7.29) to find the probability of each.

21. Solve the preceding problem if $N = 6; n_1 = n_2 = 1, n_3 = 4; d_1 = 1, d_2 = 2, d_3 = 3$.

22. Find the null distribution of K^* under the assumptions of (i) Prob. 20; (ii) Prob. 21.

23. Check the value $K^* = 5.5009$ given in Example 7.

24. Check the value $D^{**} = 521{,}459.5$ given in Example 7 (continued).

25. Find the null distribution of D^{**} under the assumptions of (i) Prob. 20; (ii) Prob. 21.

26. Verify the values of $E_H(D^{**})$ and $\text{Var}_H(D^{**})$ given in Example 7 (continued).

27. Verify the value $L = 6.358$ given in Example 7 (second continuation).

28. In the 3×4 table of Example 7, suppose the first row is 6, 8, 20, 36 instead of 8, 8, 19, 35. With this change, compute the significance probability of (i) K^*; (ii) D^{**}; (iii) L.

29. Find the null distribution of L under the assumptions of (i) Prob. 20; (ii) Prob. 21.

30. The data of Chap. 5, Prob. 21 combine the two categories "reached on second call" and reached on third or later call" of the original data into a single category. The original division into the two subcategories is shown in the following table:

Second call	122	115	67	34
Third or later call	95	85	75	44
Total	217	200	142	78

Compute the significance probability of the resulting 3×4 table using (i) K^*; (ii) L.

31. The following table shows the distribution of a Swedish sample of 25,263 couples who had been married for at most 5 years according to income and number of children:[1]

	Income				
Children	0–1,000	1,000–2,000	2,000–3,000	3,000+	Total
0	2,161	3,577	2,184	1,636	9,558
1	2,755	5,081	2,222	1,052	11,110
2	936	1,753	640	306	3,635
3 or more	264	517	127	52	960
Total	6,116	10,928	5,173	3,046	25,263

Test the hypothesis of independence (i) using the L-test; (ii) using the D^{**}-test against the alternatives that the couples with higher incomes tend to have fewer children.

32. In a public opinion poll, the following replies were obtained from a sample of 1,464 persons on the question (among others): "Do you think tax money should, or should not, be spent on nursery schools for children less than four and a half years old?"[2]

[1] Taken from H. Cramér, *Mathematical Methods of Statistics*, Princeton University Press, Princeton, N.J., 1946.

[2] From Rope, "Opinion Conflict and School Support," as cited by Walker and Lev, *Statistical Inference*, Henry Holt, New York, 1953. (Actually, the sample in this study was not simple random but highly stratified.)

Age	Reply Favorable	No Opinion	Unfavorable	Total
20–34	153	35	377	565
35–54	182	50	417	649
Over 54	65	25	160	250
Total	400	110	954	1,464

Test the hypothesis that the reply is independent of the age of the respondent (i) using the L-test; (ii) using the D^{**}-test against the alternatives that with increasing age there is a tendency to be less favorably disposed to this use of tax money.

33. The following table classifies a number of students by the conditions under which their homework was carried out (A_1 = best, ..., A_5 = worst) and the teacher's assessment of the quality of the work (B_1 = best, ..., B_3 = worst):[1]

	A_1	A_2	A_3	A_4	A_5	Total
B_1	141	67	114	79	39	440
B_2	131	66	143	72	35	447
B_3	36	14	38	28	16	132
Total	308	147	295	179	90	1,019

(i) Compute the significance probability using D^{**}.
(ii) Compare the result obtained in (i) with the significance probabilities of K^* and L.

(B)

Section 2

34. (i) Show that the null distribution of $\sum iT_i$ is symmetric about $N(N+1)^2/4$.
(ii) Show that the null distribution of D is symmetric about $(N^3-N)/6$.
[*Hint* for (i): If $T_i' = N+1-T_i$, then $\sum iT_i$ and $\sum iT_i'$ have the same null distribution.]

35. Find the smallest and largest value that the statistics D and $\sum iT_i$ can take on. [*Hint:* Use the result of the preceding problem.]

36. Show that the relationship between D and D' defined by (7.7), (7.8), and (7.9) extends to D^* and $D^{*\prime} = \sum(j-i)U_{ij}^*$ where $U_{ij}^* = 1, \frac{1}{2}$, or 0 as T_j^* is greater than, equal to, or less than T_i^*. [*Hint:* As in the earlier case, one finds that T_j^*-1 is the sum of the U_{ij}^* for $i \neq j$, and that $U_{ji}^* = 1-U_{ij}^*$. The argument then proceeds as before.]

Section 3

37. Suppose the ranks $T_1, \ldots, T_N$ should have been $1, \ldots, N$ in this order. However, the first observation is misread and as a result the ranks are

$$T_1 = N,\ T_2 = 1,\ T_3 = 2,\ \ldots,\ T_N = N-1$$

For large N, and a typical value of the significance level α, determine whether the hypothesis of randomness will be accepted or rejected against the alternative of an upward trend.

[1] Data of Chapman quoted by Yates (1948). "The Analysis of Contingency Tables with Groupings Based on Quantitative Characters," *Biometrika* **35**:176–181.

38. Suppose that $N = 2n$ and that $T_1 = 1, T_3 = 3, \ldots, T_{2n-1} = 2n-1$, and the n-tuple $T_2, \ldots, T_{2n}$ is a random permutation of the integers $2, \ldots, 2n$, the probability of each such permutation being $1/n!$. Find the limiting distribution of D as n tends to infinity.

Section 4

39. Verify the expression (7.33) for D^{**}.

40. When $s = 2$, show that (i) the statistic D^{**} is equivalent to the one-sided Wilcoxon statistic; (ii) the statistic K^* is equivalent to the two-sided Wilcoxon statistic; (iii) the statistic L is equivalent to the statistic $\bar{K}^*$ given by (5.18). [*Hint* for (i): Use formula (7.33).]

41. When $t = 2$, show that the tests based on K^* and L are equivalent.

7. REFERENCES

Adichie, J. N. (1967): "Asymptotic Efficiency of a Class of Nonparametric Tests for Regression Scores," *Ann. Math. Statist.* **38:**884–893.

——— (1974): "Rank Score Comparison of Several Regression Parameters," *Ann. Statist.* **2:**396–402.

Aiyar, Radhakrishnan J. (1969): "On Some Tests for Trend and Autocorrelation," unpublished Ph.D. thesis, University of California at Berkeley.

Armitage, P. (1955): "Tests for Linear Trends in Proportions and Frequencies," *Biometrics* **11:**375–386. [Discusses alternative tests for trend in $2 \times t$ tables.]

Barnard, G. A. (1947): "Significance Tests for 2×2 Tables," *Biometrika* **34:**123–138.

Behnen, Konrad (1971): "Asymptotic Optimality and ARE of Certain Rank Order Tests under Contiguity," *Ann. Math. Statist.* **42:**325–329.

Beran, R. J. (1971): "On Distribution-free Statistical Inference with Upper and Lower Probabilities," *Ann. Math. Statist.* **42:**157–168.

Best, D. J. (1973): "Extended Tables for Kendall's Tau," *Biometrika* **60:**429–430.

Bhapkar, V. P. and Koch, Gary C. (1968): "On the Hypothesis of No Interaction, in Contingency Tables," *Biometrics* **24:**567–594.

Bhattacharyya, G. K. (1968): "Robust Estimates of Linear Trend in Multivariate Time Series," *Ann. Inst. Statist. Math.* **20:**299–310.

———, Johnson, R. A., and Neave, H. R. (1970): "Percentage Points of Non-parametric Tests for Independence," *J. Am. Statist. Assoc.* **65:**976–983.

Bhuchongkul, S. (1964): "A Class of Nonparametric Tests for Independence in Bivariate Populations," *Ann. Math. Statist.* **35:**138–149.

Bickel, P. J. (1971): "Analogues of Linear Combinations of Order Statistics in the Linear Model," in *Statistical Decision Theory and Related Topics*, Academic Press, New York.

Billingsley, Patrick (1961): "Statistical Methods in Markov Chains," *Ann. Math. Statist.* **32:**12–40.

Bishop, Yvonne M. M. (1969): "Full Contingency Tables, Logits and Split Contingency Tables," *Biometrics* **25:**383–400.

——— (1971): "Effects of Collapsing Multidimensional Contingency Tables," *Biometrics* **27**:545–562.

———, Fienberg, S., and Holland, P. (1975): *Discrete Multivariate Analysis*, M.I.T. Press, Cambridge, Mass.

Blomquist, Nils (1950): "On a Measure of Dependence between Two Random Variables," *Ann. Math. Statist.* **21**:593–600.

Blum, J. R., Kiefer, J., and Rosenblatt, M. (1961): "Distribution-free Tests of Independence Based on the Sample Distribution Functions," *Ann. Math. Statist.* **32**:485–498.

Burr, E. J. (1960): "The Distribution of Kendall's Score *S* for a Pair of Tied Rankings," *Biometrika* **47**:151–171.

Chapman, D. G. and Nam, Jun-Mo (1968): "Asymptotic Power of Chi Square Tests for Linear Trends in Proportions," *Biometrics* **24**:315–327. [Discusses the power and required sample size of the tests considered by Armitage (1955).]

Cochran, William G. (1954): "Some Methods for Strengthening the Common χ^2 Test," *Biometrics* **10**:417–451.

Cramér, H. (1946): *Mathematical Methods of Statistics*, Princeton Univ. Press, Princeton, N.J.

Daniels, H. E. (1950): "Rank Correlation and Population Models," *J. Roy. Statist. Soc.* **B12**:171–181.

David, F. N. (1947): "A Power Function for Tests of Randomness in a Sequence of Alternatives," *Biometrika* **34**:335–339.

——— and Barton, D. E. (1962): *Combinatorial Chance*, Charles Griffin, London.

Edgington, Eugene S. (1961): "Probability Table for Number of Runs of Signs of First Differences in Ordered Series," *J. Am. Statist. Assoc.* **56**:156–159.

Farlie, D. J. G. (1960): "The Performance of Some Correlation Coefficients for a General Bivariate Distribution," *Biometrika* **47**:307–323.

Fieller, E. C. and Pearson, E. S. (1961): "Tests for Rank Correlation Coefficients II," *Biometrika* **48**:29–40.

Fienberg, Stephen E. (1970): "An Iterative Procedure for Estimation in Contingency Tables," *Ann. Math. Statist.* **41**:907–917.

——— (1972): "The Analysis of Incomplete Multiway Contingency Tables," *Biometrics* **28**:177–202.

——— and Gilbert, J. P. (1970): "Geometry of a Two by Two Contingency Table," *J. Am. Statist. Assoc.* **65**:694–701.

Gokhale, D. V. (1966): "Some Problems in Independence and Dependence," unpublished Ph.D. thesis, University of California at Berkeley.

Goodman, Leo A. (1958): "Simplified Run Tests and Likelihood Ratio Tests for Markoff Chains," *Biometrika* **45**:181–197.

——— (1968): "The Analysis of Cross-classified Data: Independence, Quasi-independence, and Interactions in Contingency Tables with or without Missing Entries," *J. Am. Statist. Assoc.* **63**:1091–1131.

——— (1970a): "The Multivariate Analysis of Qualitative Data: Interactions among Multiple Classifications," *J. Am. Statist. Assoc.* **65**:226–256.

——— (1970b): "The Analysis of Multidimensional Contingency Tables: Stepwise Procedures and Direct Estimation Methods for Building Models for Multiple Classifications," *Technometrics* **12**:33–61.

——— (1971): "Partitioning of Chi-square, Analysis of Marginal Contingency Tables, and Estimation of Expected Frequencies in Multidimensional Contingency Tables," *J. Am. Statist. Assoc.* **66**:339–344.

——— (1973): "The Analysis of Multidimensional Contingency Tables When Some Variables are Posterior to Others: A Modified Path Analysis Approach," *Biometrika* **60**:179–192.

——— and Kruskal, William H. (1954, 1959, 1963, 1972): "Measures of Association for Cross Classifications," *J. Am. Statist. Assoc.* **49**:732–764; **54**:123–163; **58**:310–364; **67**:415–421.

Grizzle, J. E., Starmer, C. F., and Koch, G. G. (1969): "Analysis of Categorical Data by Linear Models," *Biometrics* **25**:489–504.

Guillier, Claude L. (1972): "Asymptotic Relative Efficiencies of Rank Tests for Trend Alternatives," unpublished Ph.D. thesis, University of California at Berkeley.

Haberman, S. J. (1973): "The Analysis of Residuals in Cross-classified Tables," *Biometrics* **29**:205–220.

Haldane, J. B. S. and Smith, Cedric A. B. (1948): "A Simple Exact Test for Birth-order Effect," *Ann. Eugenics* **14**:117–124.

Hoeffding, Wassily (1948a): "A Class of Statistics with Asymptotically Normal Distribution," *Ann. Math. Statist.* **19**:293–325. [Proves the asymptotic normality of Spearman's statistic r_s when the hypothesis of independence is not assumed to be true.]

——— (1948b): "A Non-parametric Test of Independence," *Ann. Math. Statist.* **19**:546–557.

Hollander, Myles (1970): "A Distribution-free Test for Parallelism," *J. Am. Statist. Assoc.* **65**:387–394.

Hotelling, Harold and Pabst, Margaret (1936): "Rank Correlation and Tests of Significance Involving No Assumption of Normality," *Ann. Math. Statist.* **7**:29–43. [Proves the asymptotic normality of r_s under the hypothesis of independence.]

Jureckova, Jane (1971): "Nonparametric Estimate of Regression Coefficients," *Ann. Math. Statist.* **42**:1328–1338.

Kaarsemaker, L. and van Wijngaarden, A. (1953): "Tables for Use in Rank Correlation," *Statistica Neerlandika* **7**:41–54.

Kendall, Maurice G. (1938): "A New Measure of Rank Correlation," *Biometrika* **30**:81–93. [Points to the distinction between tests of association and tests of trend, which is further elaborated by Moran (1950) and Daniels (1950).]

——— (1948, 1955, 1962, 1970): *Rank Correlation Methods*, Griffin, London.

Kermack, W. O. and McKendrick, A. G. (1937a): "Tests for Randomness in a Series of Numerical Observations," *Proc. Roy. Soc. Edinburgh* **57**:228–240.

——— and ——— (1937b): "Some Distributions Associated with a Randomly Arranged Set of Numbers," *Proc. Roy. Soc. Edinburgh* **57**:332–376.

Konijn, H. S. (1956): "On the Power of Certain Tests for Independence in Bivariate Populations," *Ann. Math. Statist.* **27**:300–323. "Errata," ibid. **29**:935 (1958).

Koul, Hira Lal (1969): "Asymptotic Behavior of Wilcoxon-type Confidence Regions in Multiple Linear Regression," *Ann. Math. Statist.* **40**:1950–1979.

Kraemer, Helena Chmura (1974): "The Non-null Distribution of the Spearman Rank Correlation Coefficient," *J. Am. Statist. Assoc.* **69:**114–117.

Kruskal, William H. (1958): "Ordinal Measures of Association," *J. Am. Statist. Assoc.* **53:**814–861.

Lancaster, H. O. (1969): *The Chi-squared Distribution*, John Wiley & Sons, New York.

Lehmann, E. L. (1953): "The Power of Rank Tests," *Ann. Math. Statist.* **24:**23–43.

——— (1959): *Testing Statistical Hypotheses*, John Wiley & Sons, New York.

——— (1966): "Some Concepts of Dependence," *Ann. Math. Statist.* **37:**1137–1153.

Levene, Howard (1952): "On the Power Function of Tests of Randomness Based on Runs Up and Down," *Ann. Math. Statist.* **23:**34–56.

——— and Wolfowitz, J. (1944): "The Covariance Matrix of Runs Up and Down," *Ann. Math. Statist.* **15:**58–69.

Light, R. J. and Margolin, B. H. (1971): "Analysis of Variance for Categorical Data," *J. Am. Statist. Assoc.* **66:**534–544.

Mann, Henry B. (1945): "Nonparametric Tests against Trend," *Econometrika* **13:**245–259.

Mantel, N. (1970): "Incomplete Contingency Tables," *Biometrics* **26:**291–304.

Mardia, K. V. (1969): "The Performance of Some Tests of Independence for Contingency Tables," *J. Am. Statist. Assoc.* **61:**967–975.

McNemar, I. (1947): "Note on the Sampling Errors of the Differences between Correlated Proportions of Percentages," *Psychometrika* **12:**153–157.

Meng, R. C. and Chapman, D. G. (1966): "The Power of Chi-square Tests for Contingency Tables," *J. Am. Statist. Assoc.* **61:**967–975.

von Mises, R. (1931): *Wahrscheinlichkeitsrechnung und ihre Anwendung in der Statistik und Theoretischen Physik*, F. Deuticke, Leipzig and Vienna.

Mood, Alexander M. (1940): "The Distribution Theory of Runs," *Ann. Math. Statist.* **11:**367–392.

Moran, P. A. P. (1950): "Recent Developments in Ranking Theory," *J. Roy. Statist. Soc.* **B12:**153–162.

Mosteller, F. (1952): "Some Statistical Problems in Measuring the Subjective Response to Drugs," *Biometrics* **8:**220–226.

——— (1968): "Association and Estimation in Contingency Tables," *J. Am. Statist. Assoc.* **63:**1–28.

——— and Tukey, John W. (1968): "Data Analysis, Including Statistics," in *Handbook of Social Psychology*, rev. ed. (Lindzay and Anderson, eds.), Addison-Wesley, Reading, Mass.

Nam, Jun-Mo (1971): "On Two Tests for Comparing Matched Proportions," *Biometrics* **24:**945–959.

Olmstead, P. S. (1946): "Distribution of Sample Arrangements for Runs Up and Down," *Ann. Math. Statist.* **17:**24–33.

——— and Tukey, John W. (1947): "A Corner Test for Association," *Ann. Math. Statist.* **18:**495–513.

Olshen, Richard A. (1967): "Sign and Wilcoxon Tests for Linearity," *Ann. Math. Statist.* **38:**1759–1769.

Pearson, E. S. (1947): "The Choice of Statistical Tests Illustrated on the Interpretation of Data Classed in a 2 × 2 Table," *Biometrika* **34:**139–167.

Plackett, R. L. (1965): "A Class of Bivariate Distributions," *J. Am. Statist. Assoc.* **60**:516–522.

——— (1969): "Multidimensional Contingency Tables," *Bull. Int. Statist. Inst.* **43**(1):133–142.

Potthoff, Richard F. (1974): "A Non-parametric Test of Whether Two Simple Regression Lines Are Parallel," *Ann. Statist.* **2**:295–310.

Rao, C. R. (1965): *Linear Statistical Inference and Its Applications*, John Wiley & Sons, New York.

Roy, S. N. and Mitra, S. K. (1956): "An Introduction to Some Nonparametric Generalizations of Analysis of Variance and Multivariate Analysis," *Biometrika* **43**:361–376. [In generalization of the discussion of 2×2 tables by Barnard (1947) and Pearson (1947), considers the different models for an $s \times t$ contingency table.]

Sen, Pranab Kumar (1968): "Estimates of the Regression Coefficient Based on Kendall's Tau," *J. Am. Statist. Assoc.* **63**:1379–1389.

——— (1969): "On a Class of Rank-order Tests for the Parallelism of Several Regression Lines," *Ann. Math. Statist.* **40**:1668–1683.

Sillitto, G. P. (1947): "The Distribution of Kendall's Coefficient of Rank Correlation in Rankings Containing Ties," *Biometrika* **34**:36–40.

Spearman, C. (1904): "The Proof and Measurement of Association between Two Things," *Am. J. Psychol.* **15**:72–101.

Stuart, Alan (1954): "Asymptotic Relative Efficiencies of Distribution-free Tests of Randomness against Normal Alternatives," *J. Am. Statist. Assoc.* **49**:147–157.

——— (1956): "The Efficiencies of Tests of Randomness against Normal Regression," *J. Am. Statist. Assoc.* **51**:285–287.

Swed, Frieda and Eisenhart, C. (1943): "Tables for Testing Randomness of Grouping in a Sequence of Alternatives," *Ann. Math. Statist.* **14**:66–87.

Theil, H. (1950): "A Rank-invariant Method of Linear and Polynomial Regression Analysis," *Koninkl. Ned. Akad. Wetenschap. Proc.* **53**:386–392, 521–525, 1397–1412.

Wald, A. and Wolfowitz, J. (1940): "On a Test Whether Two Samples Are from the Same Population," *Ann. Math. Statist.* **11**:147–162.

Wallis, W. Allen and Moore, Geoffrey H. (1941): "A Significance Test for Time Series," *J. Am. Statist. Assoc.* **36**:401–409. [See also Moore and Wallis (1943), "Time Series Significance Tests Based on Signs of Differences," *J. Am. Statist. Assoc.* **38**:153–164.]

Williams, O. Dale and Grizzle, James E. (1972): "Analysis of Contingency Tables," *Biometrics* **28**:177–202.

Wolfowitz, J. (1944): "Asymptotic Distribution of Runs Up and Down," *Ann. Math. Statist.* **15**:163–172.

Woodworth, George G. (1970): "Large Deviations and Bahadur Efficiency of Linear Rank Statistics," *Ann. Math. Statist.* **41**:251–283.

Yanagimoto, Takemi and Okamoto, Masashi (1969): "Partial Orderings of Permutations and Monotonicity of a Rank Correlation Statistic," *Ann. Inst. Statist. Math.* **21**:489–506.

APPENDIX

1. EXPECTATION AND VARIANCE FORMULAS

The normal approximations and limit theorems that play such an important role throughout the text and which constitute the main topic of this appendix require for their application the expectations and variances of the statistics under consideration. In the present section we shall briefly recall some of the main facts concerning these measures of location and dispersion and then apply them to derive the expectation and variance formulas that are stated in the text without proof. Since the distributions in question are all discrete we shall restrict attention to this case.

Definition 1. Let K be a random variable taking on values $k_1, \ldots, k_a$. Then the *expectation* $E(K)$ of K is

$$E(K) = k_1 P(K = k_1) + \cdots + k_a P(K = k_a) \tag{A.1}$$

Typically, the expectation of K is a reasonable measure of the center of its distribution. Expectation has the obvious properties that for any constant c

$$E(K+c) = E(K)+c \qquad \text{and} \qquad E(cK) = cE(K) \tag{A.2}$$

Further, for any random variables $K_1, \ldots, K_s$ (not necessarily independent), we have the important addition law of expectation

$$E(K_1+\cdots+K_s) = E(K_1)+\cdots+E(K_s) \tag{A.3}$$

As an illustration of this law, consider s trials, each of which may be a success or failure, and which are not necessarily independent. Suppose that the probability of success on the ith trial is p_i, and let S be the total number of successes in the s trials. Then S is a random variable, and we shall now show that its expectation is

$$E(S) = p_1+\cdots+p_s \tag{A.4}$$

To see (A.4), let I_j be 1 or 0 as the jth trial is a success or failure. Then

$$S = I_1+\cdots+I_s$$

since the right-hand side counts the number of ones among the I's and hence the number of successes. From (A.1) it is seen that $E(I_j) = P(I_j = 1) = p_j$, and (A.4) now follows from (A.3).

As a last property of expectation, we mention that any two *independent* random variables K and L satisfy

$$E(KL) = E(K)\,E(L) \tag{A.5}$$

Definition 2. If μ denotes the expectation of K, the variance $\text{Var}(K)$ of K is

$$\text{Var}(K) = E(K-\mu)^2 \tag{A.6}$$

On squaring the right-hand side and simplifying, one finds the alternative form

$$\text{Var}(K) = E(K^2)-\mu^2 \tag{A.7}$$

The variance of a random variable K is often a reasonable measure of the dispersion of its distribution, i.e., of the degree to which the distribution is spread out. In particular, it follows from (A.6) that $\text{Var}(K) = 0$ if and only if K is a constant, or more precisely, if there exists a value μ such that K is equal to μ with probability 1.

EXAMPLE 1. Suppose that a population consists of N numbers $v_1, \ldots, v_N$. Let one of these numbers be selected at random and denoted by V. If the v's are distinct, it follows from (A.1) that

$$E(V) = \frac{v_1+\cdots+v_N}{N} = \bar{v} \tag{A.8}$$

where $\bar{v}$ denotes the average of the v's. If the v's are not distinct, suppose that n_1 are equal to $a_1, \ldots, n_c$ equal to a_c. Then (A.1) gives

$$E(V) = \frac{n_1 a_1+\cdots+n_c a_c}{N} = \bar{v}$$

so that (A.8) continues to hold.

To obtain the variance of V, suppose once more that the v's are distinct. Then by (A.6)

$$\text{Var}(V) = \tau^2 \tag{A.9}$$

where

$$\tau^2 = \frac{1}{N}\sum_{i=1}^{N}(v_i - \bar{v})^2 \tag{A.10}$$

or equivalently, by (A.7),

$$\tau^2 = \frac{1}{N}\sum_{i=1}^{N} v_i^2 - \bar{v}^2 \tag{A.11}$$

The argument used for the expectation of V shows that (A.9) to (A.11) remain valid when the v's are not necessarily distinct.

Let us now consider two important special cases.

(i) Suppose that the population values $v_1, \ldots, v_N$ are the integers $1, \ldots, N$. Then it follows from (1.12) that

$$\bar{v} = \frac{N+1}{2} \tag{A.12}$$

and hence from (1.68) that τ^2 is

$$\frac{(N+1)(2N+1)}{6} - \left(\frac{N+1}{2}\right)^2$$

which simplifies to

$$\tau^2 = \frac{N^2-1}{12} \tag{A.13}$$

(ii) Let $d_1, \ldots, d_e$ be integers adding up to N, and suppose that d_1 of the v's are equal to the average of the integers $1, \ldots, d_1$, that d_2 of the v's are equal to the average of $d_1+1, \ldots, d_1+d_2$, and so on. The v's (which are the midranks discussed in Chap. 1, Sec. 4) are then given by

$$\begin{aligned} v_1 = \cdots = v_{d_1} &= \tfrac{1}{2}(d_1+1) \\ v_{d_1+1} = \cdots = v_{d_1+d_2} &= d_1 + \tfrac{1}{2}(d_2+1) \\ v_{d_1+d_2+1} = \cdots = v_{d_1+d_2+d_3} &= d_1 + d_2 + \tfrac{1}{2}(d_3+1) \end{aligned} \tag{A.14}$$

and so forth. To determine $\bar{v}$, note that the sum of the first group of v's is the same as that of the integers $1, \ldots, d_1$, that the sum of the second group is the same as that of the integers $d_1+1, \ldots, d_1+d_2$, and so on. Thus, the sum, and hence the average, of the v's is the same as in case (i), so that again

$$\bar{v} = \frac{N+1}{2} \tag{A.15}$$

To obtain τ^2, consider the identity

$$\sum_{i=1}^{k} a_i^2 = \sum_{i=1}^{k}(a_i - \bar{a})^2 + k\bar{a}^2 \tag{A.16}$$

which holds for any real numbers $a_1, \ldots, a_k$. Apply this first to $a_1 = 1, \ldots, a_{d_1} = d_1$. Then $\bar{a} = v_1 = \cdots = v_{d_1}$, and by (A.13) we have $\sum(a_i - \bar{a})^2 = d_1(d_1{}^2-1)/12$, so that

$$1^2 + \cdots + d_1{}^2 = \sum_{j=1}^{d_1} v_j{}^2 + \frac{d_1(d_1{}^2-1)}{12}$$

Applying (A.16) next to $a_1 = d_1+1, \ldots, a_{d_2} = d_1+d_2$, we similarly find

$$(d_1+1)^2+\cdots+(d_1+d_2)^2 = \sum_{j=d_1+1}^{d_1+d_2} v_j^2 + \frac{d_2(d_2^2-1)}{12}$$

Continuing in this manner, and adding the resulting equations, we get

$$1^2+\cdots+N^2 = \sum_{j=1}^{N} v_j^2 + \sum_i \frac{d_i(d_i^2-1)}{12}$$

and hence by (1.68)

$$\text{(A.17)} \qquad \sum_{j=1}^{N} v_j^2 = \frac{N(N+1)(2N+1)}{6} - \frac{\sum_i d_i(d_i^2-1)}{12}$$

Subtracting $N\bar{v}^2$ and dividing by N, we finally obtain

$$\text{(A.18)} \qquad \tau^2 = \frac{N^2-1}{12} - \frac{\sum d_i(d_i^2-1)}{12N}$$

The properties for variance analogous to (A.2) are

$$\text{(A.19)} \qquad \mathrm{Var}(K+c) = \mathrm{Var}(K) \quad \text{and} \quad \mathrm{Var}(cK) = c^2\,\mathrm{Var}(K)$$

However, the addition law corresponding to (A.3) holds only under additional assumptions. The most important case is that of pairwise independent variables

$$\text{(A.20)} \qquad \mathrm{Var}(K_1+\cdots+K_a) = \mathrm{Var}(K_1)+\cdots+\mathrm{Var}(K_a)$$

if K_i, K_j are independent for all pairs i, j. As a special case of (A.20) one finds with the help of (A.19) that if X and Y are independent, then

$$\text{(A.21)} \qquad \mathrm{Var}(Y-X) = \mathrm{Var}(Y)+\mathrm{Var}(X)$$

EXAMPLE 2. Suppose that a population consists of N numbers $u_1, \ldots, u_N$. A sample is drawn by the following method of *binomial sampling*: each of the u's is included in the sample with probability p, the decisions on the N numbers being independent. Let the u-values included in the sample be denoted by $A_1, \ldots, A_B$. The number B of included values is a random variable, which has the binomial distribution corresponding to N trials and success probability p, given by (A.86) below.

The sum of the u-values that are included in the sample can be written as

$$\text{(A.22)} \qquad S = A_1+\cdots+A_B = u_1I_1+\cdots+u_NI_N$$

where I_j is 1 or 0 as u_j is included or not. Since by (A.1)

$$\text{(A.23)} \qquad E(I_j) = P(I_j = 1) = p$$

it follows from (A.3) that

$$\text{(A.24)} \qquad E(S) = p(u_1+\cdots+u_N)$$

Similarly, by (A.7)

$$\text{(A.25)} \qquad \mathrm{Var}(I_j) = p-p^2 = pq \qquad (\text{where } q = 1-p)$$

and hence by (A.19) and (A.20)

(A.26) $$\operatorname{Var}(S) = (u_1{}^2 + \cdots + u_N{}^2)pq$$

Important special cases of (A.24) and (A.26) are obtained by giving particular values to the u's.

(i) If all the u's are equal to 1, the sum S is simply the number B of items included in the sample, which has the binomial distribution (A.86) below. It follows from (A.24) and (A.26) that the expectation and variance of this binomial distribution are, respectively,

(A.27) $$E(B) = Np \qquad \text{and} \qquad \operatorname{Var}(B) = Npq$$

(ii) If the u's are the integers $1, \ldots, N$ and if $p = \frac{1}{2}$, the sum S has the null distribution of the signed-rank Wilcoxon statistic V_s. It follows from (1.12) and (A.24) that

(A.28) $$E(V_s) = \frac{N(N+1)}{4}$$

and from (1.68) and (A.26) that

(A.29) $$\operatorname{Var}(V_s) = \frac{N(N+1)(2N+1)}{24}$$

(iii) The next example has the aim of finding the expectation and variance of the signed-midrank statistic V_s^* of Chap. 3. Let $d_0, d_1, \ldots, d_e$ be integers adding up to N, let $N' = N - d_0$, and suppose that of the N' values $u_1, \ldots, u_{N'}, d_1$ are equal to $d_0 + \frac{1}{2}(d_1 + 1)$, d_2 are equal to $d_0 + d_1 + \frac{1}{2}(d_2 + 1)$, and so on. Then V_s^* has the distribution of the statistic S given by (A.22) with N' in place of N and with $p = \frac{1}{2}$. To obtain $\sum u_i$ and $\sum u_i{}^2$, note that the set of u's augmented by d_0 values equal to $\frac{1}{2}(d_0 + 1)$ essentially coincides with the set of N v's defined by (A.14). There is only a slight change of notation: the integers adding up to N, which were previously denoted by $d_1, \ldots, d_e$, are now denoted by $d_0, \ldots, d_e$. It follows from (A.15) that

$$\sum_{i=1}^{N'} u_i + \tfrac{1}{2} d_0(d_0 + 1) = \sum_{j=1}^{N} v_j = \frac{N(N+1)}{2}$$

and hence from (A.24) that

(A.30) $$E(V_s^*) = \tfrac{1}{4}[N(N+1) - d_0(d_0 + 1)]$$

Similarly, by (A.17)

$$\sum_{i=1}^{N'} u_i{}^2 + \tfrac{1}{4} d_0(d_0 + 1)^2 = \sum_{j=1}^{N} v_j{}^2 = \frac{N(N+1)(2N+1)}{6} - \frac{\sum_{i=0}^{e} d_i(d_i{}^2 - 1)}{12}$$

Since

$$\frac{d_0(d_0{}^2 - 1)}{12} + \frac{d_0(d_0 + 1)^2}{4} = \frac{d_0(d_0 + 1)(2d_0 + 1)}{6}$$

formula (A.26) gives

(A.31) $$\operatorname{Var}(V_s^*) = \frac{N(N+1)(2N+1) - d_0(d_0 + 1)(2d_0 + 1)}{24} - \frac{\sum_{i=1}^{e} d_i(d_i{}^2 - 1)}{48}$$

Many statistics of interest can be represented as sums of dependent variables for which the variance law (A.20) is not valid. To obtain a generalization of (A.20), consider two random variables K and L which may be dependent, and denote their expectations by $E(K) = \mu$ and $E(L) = \nu$.

Definition 3. The covariance $\text{Cov}(K,L)$ is

(A.32) $$\text{Cov}(K,L) = E[(K-\mu)(L-\nu)]$$

On multiplying the two factors on the right-hand side, one finds in analogy with (A.7) the alternative form

(A.33) $$\text{Cov}(K,L) = E(KL) - \mu\nu$$

Corresponding to (A.19), covariance satisfies for any constants c and d

(A.34) $$\text{Cov}(K+c, L+d) = \text{Cov}(K,L)$$

and

(A.35) $$\text{Cov}(cK,dL) = cd\,\text{Cov}(K,L)$$

Finally, from the definition of covariance and (A.3) one obtains the following addition law for covariance (Prob. 2):

(A.36) $$\text{Cov}\left(\sum_{i=1}^{r} K_i, \sum_{j=1}^{s} L_j\right) = \sum_{i=1}^{r}\sum_{j=1}^{s} \text{Cov}(K_i,L_j)$$

Using the fact that $\text{Cov}(K,K) = \text{Var}(K)$ and putting $r = s$ and $K_i = L_i$, it is seen that (A.36) specializes to the general addition law for variance, which states that for any random variables $K_1, \ldots, K_s$

(A.37) $$\text{Var}\left(\sum K_i\right) = \sum \text{Var}(K_i) + \sum_{i \neq j}\sum \text{Cov}(K_i,K_j)$$

In particular, it is seen that the earlier law (A.20) holds when $\text{Cov}(K_i,K_j) = 0$ for each pair i and j (the variables K_i and K_j are then said to be *uncorrelated*). By (A.5), this condition is satisfied whenever K_i and K_j are independent, which establishes (A.20) as a special case of (A.37).

EXAMPLE 3. Under the assumptions of Example 1, suppose that n of the v's are selected by the method of simple random sampling, i.e., so that all $\binom{N}{n}$ possible choices are equally likely. Let $V_1, \ldots, V_n$ denote the n selected v-values and let

(A.38) $$T = V_1 + \cdots + V_n$$

Since all of the V's have the same distribution, it follows from (A.8) that their common expectation is $\bar{v}$ and hence that

(A.39) $$E(T) = n\bar{v}$$

To determine the variance of T, note that by (A.9) the common variance of the V's is τ^2, given by (A.10) or (A.11). Similarly, $\text{Cov}(V_i,V_j)$ is independent of i and j since the pair (V_i,V_j)

takes on the pairs of values (v_k, v_l), $k \neq l$, each with probability $1/N(N-1)$ for every i and j. If this common covariance is λ, it follows from (A.37) that

(A.40) $$\operatorname{Var}(T) = n\tau^2 + n(n-1)\lambda$$

To obtain the value of λ, put $n = N$ in (A.40). Then T becomes the constant $v_1 + \cdots + v_N$, and hence

$$\operatorname{Var}(T) = N\tau^2 + N(N-1)\lambda = 0$$

This shows that

(A.41) $$\lambda = \operatorname{Cov}(V_i, V_j) = -\frac{\tau^2}{N-1}$$

Substitution of this value in (A.40) gives, after some simplification,

(A.42) $$\operatorname{Var}(T) = \frac{n(N-n)}{N-1}\tau^2$$

Let us now again consider some special cases.

(i) If r of the v's are equal to 1 and the remaining $N-r$ are equal to 0, then T is just the number D of ones included in the sample, which has the hypergeometric distribution given by (A.88) below. By (A.39) the expectation of D is

(A.43) $$E(D) = \frac{nr}{N}$$

From (A.11) it is seen that

$$\tau^2 = \frac{r}{N} - \left(\frac{r}{N}\right)^2 = \frac{r}{N}\left(1 - \frac{r}{N}\right)$$

and hence from (A.42) that

(A.44) $$\operatorname{Var}(D) = \frac{n(N-n)}{N-1}\frac{r}{N}\left(1 - \frac{r}{N}\right)$$

(ii) Suppose next that the v's are the integers $1, \ldots, N$. The statistic T then becomes the Wilcoxon rank-sum statistic W_s defined in Chap. 1, Sec. 2. By (A.39) and (A.12), the expectation of W_s is

(A.45) $$E(W_s) = \frac{n(N+1)}{2}$$

Similarly, (A.42) and (A.13) show the variance of W_s to be

(A.46) $$\operatorname{Var}(W_s) = \frac{n(N-n)(N+1)}{12}$$

(iii) If the v's have the values (A.14), the statistic T becomes the Wilcoxon midrank statistic W_s^* of Chap. 1, Sec. 4. By (A.39) and (A.15) the expectation of W_s^* is still $n(N+1)/2$, while (A.42) and (A.18) show the variance of W_s^* to be given by (1.35).

The next example is a generalization of Example 3 and provides the basis for determining the expectation and variance of the statistics D, D^*, and D^{**} of Chap. 7.

EXAMPLE 4. Let $a(1), \ldots, a(N)$ and $c_1, \ldots, c_N$ be two sets of constants, with averages $\bar{a}$ and $\bar{c}$. Suppose that $(T_1, \ldots, T_N)$ is a random permutation of the integers $1, \ldots, N$, that is, $(T_1, \ldots, T_N)$ is one of the $N!$ possible permutations, with probability $1/N!$ for each. Consider the statistic

$$T = \sum c_i a(T_i) \tag{A.47}$$

which reduces to (A.38) if $a(1) = \cdots = a(n) = 1$, $a(n+1) = \cdots = a(N) = 0$ and if $c_i = v_i$ for all i. It follows from (A.8) that $a(T_i)$ has expectation $\bar{a}$, and hence from (A.2) and (A.3) that $E(T) = \bar{a}\sum c_i$, so that

$$E(T) = N\bar{a}\bar{c} \tag{A.48}$$

To obtain the variance of T, note that by (A.9) and (A.10)

$$\operatorname{Var}[a(T_i)] = \tau^2 = \frac{\sum [a(i) - \bar{a}]^2}{N}$$

and by (A.41)

$$\operatorname{Cov}[a(T_i), a(T_j)] = -\frac{\tau^2}{N-1}$$

Hence, by (A.19), (A.35), and (A.37),

$$\operatorname{Var}(T) = \sum c_i^2 \tau^2 - \sum_{i \neq j}\sum c_i c_j \tau^2/(N-1)$$

Since

$$\sum_{i \neq j}\sum c_i c_j = \Big(\sum_i c_i\Big)^2 - \sum c_i^2 = N^2\bar{c}^2 - \sum c_i^2$$

it follows after some simplification that

$$\operatorname{Var}(T) = \frac{\sum (c_i - \bar{c})^2 \sum [a(i) - \bar{a}]^2}{N-1} \tag{A.49}$$

Consider now two special cases.

(i) Let $a(i) = c_i = i$, so that

$$T = \sum iT_i \tag{A.50}$$

It follows from (A.12) that $\bar{a} = \bar{c} = \frac{1}{2}(N+1)$, and hence from (A.48) that

$$E(T) = \frac{N(N+1)^2}{4} \tag{A.51}$$

Similarly, it is seen from (A.10) and (A.13) that

$$\sum [a(i) - \bar{a}]^2 = \sum (c_i - \bar{c})^2 = \frac{N(N^2-1)}{12}$$

and hence that

$$\operatorname{Var}(T) = \frac{N^2(N+1)^2(N-1)}{144} \tag{A.52}$$

(ii) Let $c_i = i$ as before, but suppose that the a's are the numbers $v_1, \ldots, v_N$ defined in (A.14). Then T is the statistic $\sum iT_i^*$ of Chap. 7, Sec. 2. From (A.48), (A.12), and (A.15) it follows that

$$E(T) = \frac{N(N+1)^2}{4} \tag{A.53}$$

Similarly, from (A.49), (A.13), and (A.18), it is seen after some simplification that

$$\text{(A.54)} \qquad \operatorname{Var}(T) = \frac{N^2(N+1)}{144}\left[(N^2-1) - \frac{1}{N}\sum(d_i^3 - d_i)\right]$$

The purpose of the remaining two examples is to provide the expectation and variance of the Wilcoxon statistics W_{XY} and V_s under alternatives to the null hypothesis.

EXAMPLE 5. Let $X_1, \ldots, X_m$ and $Y_1, \ldots, Y_n$ be independent, the X's identically distributed with distribution F and the Y's with distribution G. Let φ be a function of two real variables, and let

$$\text{(A.55)} \qquad W = \sum_{i=1}^{m} \sum_{j=1}^{n} \varphi(X_i, Y_j)$$

If the common expectation of the variables

$$U_{ij} = \varphi(X_i, Y_j)$$

is denoted by θ, then by (A.3)

$$\text{(A.56)} \qquad E(W) = mn\theta$$

The variance of W is, by (A.37),

$$\text{(A.57)} \qquad \operatorname{Var}(W) = \sum_{i=1}^{m} \sum_{j=1}^{n} \operatorname{Var}(U_{ij}) + \sum\sum\sum\sum \operatorname{Cov}(U_{ij}, U_{kl})$$

where the fourfold sum is over all quadruples (i,j,k,l) for which either $i \neq k$, or $j \neq l$, or both. Since the variance terms are all equal, their contribution is $mn \operatorname{Var}(U_{ij})$. The covariance terms are of three types. First, if $i \neq k$ and $j \neq l$, it follows from the independence of the variables X_i, Y_j, X_k, Y_l that U_{ij} and U_{kl} are independent, so that their covariance is zero. Second, there are terms with $i \neq k$ but $j = l$, and these are all equal. To determine their number, note that there are m possible choices for i and, once i has been chosen, $m-1$ choices for k. For each of the $m(m-1)$ possible values of i and k, there are n choices for j. The total number of such covariance terms is therefore $(m-1)mn$. Third, there are the terms with $i = k$ and $j \neq l$. These are again all equal and their number is $(n-1)mn$. It thus follows that

$$\text{(A.58)} \qquad \operatorname{Var}(W) = mn \operatorname{Var}(U_{ij}) + mn(m-1)\operatorname{Cov}(U_{ij}, U_{kj}) + nm(n-1)\operatorname{Cov}(U_{ij}, U_{il})$$

An important special case is obtained by assuming that the distributions F and G are continuous and that

$$\text{(A.59)} \qquad \varphi(x,y) = \begin{cases} 1 & \text{if } x < y \\ 0 & \text{otherwise} \end{cases}$$

The variable W then becomes the rank-sum Wilcoxon statistic in its Mann-Whitney form, that is, the statistic W_{XY} defined by (1.24). The expectation θ of the U_{ij} then reduces to

$$\text{(A.60)} \qquad p_1 = P(X_i < Y_j)$$

so that

$$\text{(A.61)} \qquad E(W_{XY}) = mnp_1$$

The variance of W_{XY} requires the variance of U_{ij} and the covariances on the right-hand side of (A.58). The variance of U_{ij}, by (A.25), is given by

(A.62) $$\text{Var}(U_{ij}) = p_1(1-p_1)$$

To find the covariance of U_{ij} and U_{il}, note that

$$E(U_{ij}.U_{il}) = P(U_{ij} = U_{il} = 1)$$

is equal to

(A.63) $$p_2 = P(X_i < Y_j \text{ and } X_i < Y_l)$$

Thus, by (A.33),

(A.64) $$\text{Cov}(U_{ij}, U_{il}) = p_2 - p_1{}^2$$

In the same way it is seen that

(A.65) $$\text{Cov}(U_{ij}, U_{kj}) = p_3 - p_1{}^2$$

where

(A.66) $$p_3 = P(X_i < Y_j \text{ and } X_k < Y_j)$$

Substitution of (A.62), (A.64), and (A.65) into (A.58) gives for the variance of W_{XY} the formula stated as (2.21).

EXAMPLE 6. Let $Z_1, \ldots, Z_N$ be identically, independently distributed, and let

$$V' = \sum_{i<j}\sum \varphi(Z_i, Z_j)$$

where φ is a function satisfying

(A.67) $$\varphi(a,b) = \varphi(b,a) \qquad \text{for all } a \text{ and } b$$

If the common expectation of the variables

$$V_{ij} = \varphi(Z_i, Z_j)$$

is denoted by ξ, then by (A.3)

(A.68) $$E(V') = \binom{N}{2}\xi$$

The variance of V', by (A.37), is

(A.69) $$\text{Var}(V') = \sum_{i<j}\sum \text{Var}(V_{ij}) + \sum_{i<j}\sum\sum_{k<l}\sum \text{Cov}(V_{ij}, V_{kl})$$

Since the variance terms are all equal, their contribution is $\binom{N}{2}\text{Var}(V_{ij})$. The covariance terms are of two types. First, there are those in which all four subscripts i, j, k, l are distinct. The number of such terms is $\binom{N}{2}\binom{N-2}{2}$, and they are all equal to zero since V_{ij} depends only on Z_i and Z_j, and V_{kl} only on Z_k and Z_l, so that V_{ij} and V_{kl} are independent. Second, there are those terms in which only three of the four subscripts are distinct. Since the total number of terms on the right-hand side of (A.69) is $\binom{N}{2}^2$, the number of such terms is

$$\binom{N}{2}^2 - \binom{N}{2}\binom{N-2}{2} - \binom{N}{2} = N(N-1)(N-2)$$

Furthermore, because of (A.67), these terms are all equal. Thus

(A.70) $$\operatorname{Var}(V') = \tfrac{1}{2}N(N-1)\operatorname{Var}(V_{ij}) + N(N-1)(N-2)\operatorname{Cov}(V_{ij}, V_{il})$$

Actually, the statistic of principal interest in the present example is not V' but

(A.71) $$V = \sum_{i \leqslant j}\sum \varphi(Z_i, Z_j) = V' + S$$

where

(A.72) $$S = \sum_{i=1}^{N} \varphi(Z_i, Z_i)$$

If the common expectation of $\varphi(Z_i, Z_i) = V_{ii}$ is denoted by η, it follows from (A.3) and (A.68) that

(A.73) $$E(V) = \binom{N}{2}\xi + N\eta$$

The variance of V is

(A.74) $$\operatorname{Var}(V) = \operatorname{Var}(V') + 2\operatorname{Cov}(V', S) + \operatorname{Var}(S)$$

Here the variance of V' is given by (A.70) while the variance of S is, by (A.20),

(A.75) $$\operatorname{Var}(S) = N\operatorname{Var}(V_{ii})$$

The covariance term of (A.74) is by (A.36)

$$\operatorname{Cov}(V', S) = \operatorname{Cov}\left(\sum_{i<j}\sum V_{ij}, \sum_{k} V_{kk}\right) = \sum_{i<j}\sum\sum_{k}\operatorname{Cov}(V_{ij}, V_{kk})$$

If k is different from both i and j, the variables V_{ij} and V_{kk} are independent and their covariance is zero. The remaining terms are all equal. Since there are $\frac{1}{2}N(N-1)$ pairs (i,j) with $i < j$, this is also the number of terms $\operatorname{Cov}(V_{ij}, V_{ii})$ and the number of terms $\operatorname{Cov}(V_{ij}, V_{jj})$, and this shows that

(A.76) $$\operatorname{Var}(V) = \operatorname{Var}(V') + N\operatorname{Var}(V_{ii}) + N(N-1)\operatorname{Cov}(V_{ij}, V_{ii})$$

An important special case is obtained by assuming that the distribution of the Z's is continuous and that

(A.77) $$\varphi(z_i, z_j) = \begin{cases} 1 & \text{if } z_i + z_j > 0 \\ 0 & \text{otherwise} \end{cases}$$

The variable V then becomes the signed-rank Wilcoxon statistic V_s of Chap. 3, Sec. 2. In this case, the expectations of V_{ij} and V_{ii} reduce to

(A.78) $$\xi = p_1' = P(Z_i + Z_j > 0) \qquad \text{and} \qquad \eta = p = P(Z_i > 0)$$

so that

$$E(V_s) = \binom{N}{2} p'_1 + Np \tag{A.79}$$

The variance of V_s requires expressions for $\mathrm{Var}(V_{ij})$, $\mathrm{Cov}(V_{ij}, V_{il})$, $\mathrm{Var}(V_{ii})$, and $\mathrm{Cov}(V_{ij}, V_{ii})$. Of these, the two variances are, by (A.25),

$$\mathrm{Var}(V_{ij}) = p'_1(1-p'_1) \qquad \text{and} \qquad \mathrm{Var}(V_{ii}) = p(1-p) \tag{A.80}$$

The covariance of V_{ij} and V_{il} by the argument leading to (A.64) is

$$\mathrm{Cov}(V_{ij}, V_{il}) = p'_2 - p'^2_1 \tag{A.81}$$

where

$$p'_2 = P(Z_i + Z_j > 0 \text{ and } Z_i + Z_l > 0) \tag{A.82}$$

Similarly,

$$\mathrm{Cov}(V_{ij}, V_{ii}) = p''_2 - pp'_1 \tag{A.83}$$

where

$$p''_2 = P(Z_i + Z_j > 0 \text{ and } Z_i > 0) \tag{A.84}$$

To complete the derivation of $\mathrm{Var}(V_s)$, we shall now show that p''_2 can be expressed in terms of the probabilities p and p'_1. By definition, p''_2 is the probability of the region in the (Z_i, Z_j)-plane above the line $Z_i + Z_j = 0$ and to the right of the line $Z_i = 0$. It is thus equal to $P(B) + P(C)$, where B and C are the regions indicated in Fig. A.1. The regions A and C are defined by the inequalities $A: Z_i + Z_j > 0$ and $Z_i < 0$; $C: Z_i + Z_j > 0$ and $Z_j < 0$, and since Z_i and Z_j are independent and have the same distribution, $P(C) = P(A)$. From the equations

$$P(B) = p^2 \qquad \text{and} \qquad P(A) + P(B) + P(C) = p'_1$$

it therefore follows that $P(C) = \frac{1}{2}(p'_1 - p^2)$, so that

$$p''_2 = \tfrac{1}{2}(p'_1 + p^2) \tag{A.85}$$

Substitution of the above values into (A.70) and (A.76) gives formula (4.27).

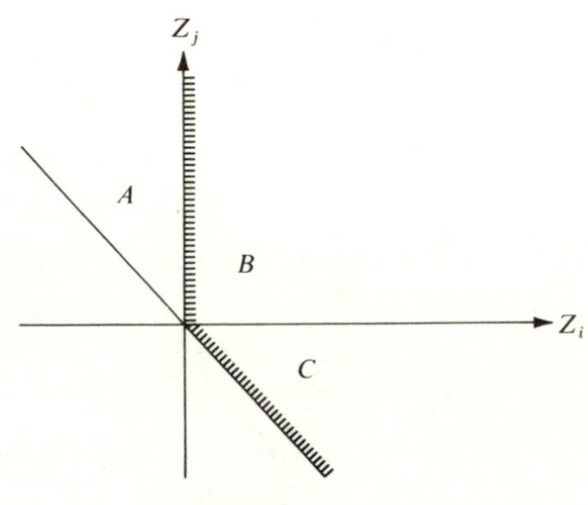

FIGURE A.1
Events related to p, p'_1, p''_2.

2. SOME STANDARD DISTRIBUTIONS

In the present section we briefly describe some standard distributions, both as a reference for their use in the text and as background material for the asymptotic results of the following sections.

2A. The Binomial Distribution

Consider n independent trials, each having two possible outcomes that (purely as a convention) we shall call success and failure. If the probability of success is the same for all n trials, say,

$$P(\text{success}) = p$$

the trials are said to constitute a sequence of *binomial trials.* Let X denote the number of successes in such a sequence of n binomial trials. The distribution of X, the *binomial distribution* $b(p,n)$, is then given by

$$P(X = x) = \binom{n}{x} p^x q^{n-x} \qquad x = 0, 1, \ldots, n \tag{A.86}$$

where $q = 1-p$ is the probability of failure. By (A.27) the expectation and variance of a random variable X with distribution (A.86) are

$$E(X) = np \qquad \text{and} \qquad \operatorname{Var}(X) = npq \tag{A.87}$$

2B. The Hypergeometric Distribution

Suppose a population consists of N items, each of which does or does not have a certain characteristic, say, good or bad. A sample of size n is chosen from the population at random, that is, in such a way that all $\binom{N}{n}$ possible samples are equally likely. Let the number of good items in the population be d and the number of good items in the sample be X. Then X is a random variable whose distribution, the *hypergeometric distribution* $H(N,d,n)$, is given by

$$P(X = x) = \binom{d}{x}\binom{N-d}{n-x} \Big/ \binom{N}{n} \qquad x = 0, 1, \ldots, d \tag{A.88}$$

[Actually, not all values of x between 0 and d may be possible, in which case the corresponding probabilities (A.88) reduce to zero. See Prob. 6.] Formula (A.88) follows from the fact that the number of ways of selecting x items for the sample from the d available good items is $\binom{d}{x}$ while $\binom{N-d}{n-x}$ is the number of ways of selecting the remaining $n-x$ items of the sample from the $N-d$ available bad items. The expectation and variance of a random variable X with distribution (A.88) are by (A.43) and (A.44)

$$E(X) = n\frac{d}{N} \qquad \text{and} \qquad \operatorname{Var}(X) = \frac{N-n}{N-1} n \frac{d}{N}\left(1 - \frac{d}{N}\right) \tag{A.89}$$

To obtain a random sample of size n one often resorts to the following process of *ordered random sampling without replacement.* A first item is drawn at random (i.e., so that each of the

N items has probability $1/N$ of being selected). Then a second item is obtained by drawing at random an item from the remaining $N-1$; and so on until the sample of n is completed. It is not difficult to see that the resulting sample, when its order is neglected, is in fact a random sample.

Consider now the different scheme of ordered sampling *with replacement*, in which at each stage the selected item is replaced before the next item is drawn. Thus at each stage an item is selected from the full population of N items, with each of the N items having the same probability, $1/N$, of being drawn. Let us consider each drawing as a trial, which is a success if the selected item is good. The n drawings then constitute n binomial trials with success probability $p = d/N$ and the number X of good items in the sample has the binomial distribution $b(d/N, n)$. As is seen from (A.87), the expectation of X continues to be given by (A.89); but the variance is missing the "correction factor" $(N-n)/(N-1)$ required in (A.89) to compensate for the decrease in variance resulting from the dependence of the trials when the sampling is without replacement.

It is plausible that for large N and small n/N there is not much difference between sampling with and without replacement, and that the binomial distribution $b(d/N, n)$ should then provide a good approximation to the hypergeometric distribution (A.88). For a limit theorem supporting this approximation, see Prob. 7.

The distributions (A.86) and (A.88) are discrete, and we now turn to the consideration of some continuous distributions. A continuous distribution is characterized by the fact that its cumulative distribution function

$$F(x) = P(X \leqslant x) \tag{A.90}$$

is a continuous function of x, or equivalently (Prob. 8), by the fact that

$$P(X = x) = 0 \qquad \text{for all } x \tag{A.91}$$

An important property of continuous distribution is described by the following theorem, the proof of which is sketched in Probs. 9 and 10.

Theorem 1. Let $X_1, \ldots, X_n$ be n independent random variables with continuous distributions $F_1, \ldots, F_n$. Then the probability is 1 that no two of the X's will be equal.

The continuous distributions with which we shall be concerned have the additional property of possessing a density; that is, there exists a nonnegative function p such that

$$F(x) = \int_{-\infty}^{x} p(t)\,dt$$

Thus $F(x)$ is the area (shown in Fig. A.2) to the left of the point x under the graph of p.

2C. The Normal Distribution

A central role in what follows is played by the normal distribution. Its importance for large-sample theory stems from the fact that (under assumptions which will be discussed in Sec. 4) it is the limit distribution of a sum of independent random variables as the number of terms tends to infinity. Related to this is the observation that the distribution of measurements taken under

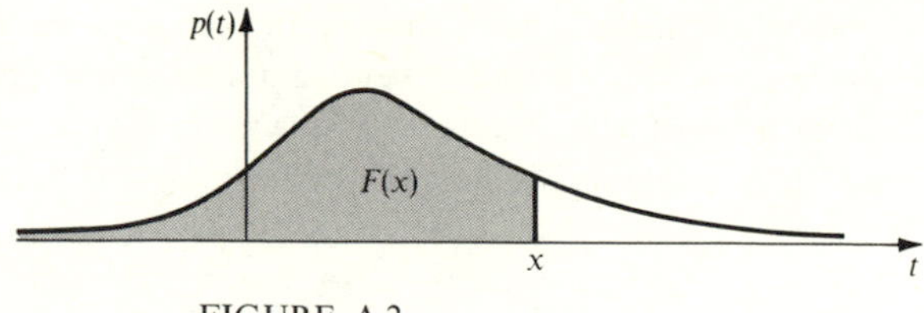

FIGURE A.2
$F(x)$ as area under the density curve.

sufficiently well-controlled conditions is often approximately normal. (It used to be believed that the normal distribution is practically universal as a distribution of measurements, but it has been realized more recently that this tends to be true only for the central part of the distribution, with the tail of the distribution often being distorted by mistakes or other exceptional circumstances—so-called gross errors.)

The density of a normal distribution is

(A.92) $$p(x) = \frac{1}{\sqrt{2\pi}\,\sigma} e^{-(x-\mu)^2/2\sigma^2}$$

If X is a random variable with this density, it is said to be normally distributed or, more precisely, to have the normal distribution $N(\mu,\sigma^2)$. The expectation and variance of such a random variable are

$$E(X) = \mu \qquad \text{and} \qquad \mathrm{Var}(X) = \sigma^2$$

A random variable with expectation μ and variance $\sigma^2 = 0$ assigns probability 1 to μ. The associated distribution can be viewed as the degenerate case of a normal distribution with $\sigma = 0$.

It is easy to see that if X is normally distributed, so is any linear function $Y = aX + b$. If the distribution of X is $N(\mu,\sigma^2)$, that of Y is $N(a\mu + b, a^2\sigma^2)$. In particular, the random variable $X^* = (X - \mu)/\sigma$ then has the *standard normal distribution* with density

(A.93) $$\varphi(x) = \frac{1}{\sqrt{2\pi}} e^{-x^2/2}$$

obtained from (A.92) by setting $\mu = 0$ and $\sigma = 1$.

Another important property of the family (A.92) of normal distributions is the fact that the sum of two independent normally distributed random variables is again normal. Combining this with the property mentioned in the preceding paragraph, we see that if X and Y are independently distributed as $N(\mu,\sigma^2)$ and $N(\nu,\tau^2)$, respectively, the distribution of $aX + bY$ is

(A.94) $$N(a\mu + b\nu, a^2\sigma^2 + b^2\tau^2)$$

The normal density (A.92) is symmetric about μ and unimodal. It therefore takes on its maximum value at μ and decreases as x tends away from μ on either side. The feature that often makes the distribution unrealistic is the rapidity with which the density tends to zero as x tends to infinity. In the standardized form (A.93), for example, the density is inversely proportional to $e^{x^2/2}$, which tends to infinity much more rapidly even than e^x. Let us now consider briefly some other densities that have the same general shape (symmetric, unimodal) as the normal density but which decrease more slowly. In each case, we shall give only one member of the family of densities, which is centered at 0 and in which the scale is chosen so as to give the density a simple form. If X is a random variable with this density, the more general associated location-scale family is obtained as the family of densities of $aX+b$ for arbitrary b and positive a.

2D. The Cauchy, Logistic, and Double-exponential Distributions

The density of the *Cauchy distribution* is

$$p(x) = \frac{1}{\pi(1+x^2)} \tag{A.95}$$

which tends to zero at the very slow rate of $1/x^2$. An indication of the heaviness of the tails of this distribution is the fact that not only the variance but even the expected value of $|X|$ is infinite, so that a random variable with a Cauchy distribution does not have an expectation. A curious consequence of this tail behavior concerns the average $\bar{X}$ of n independent, identically distributed Cauchy random variables $X_1, \ldots, X_n$. Typically, the average of n independent observations from the same distribution has a distribution that is much more concentrated than that of a single observation X_1. However, in the Cauchy case, the distribution of the average $\bar{X}$ is the same as that of a single observation, regardless of the number n of observations being averaged (Prob. 12). It is often of interest to test the performance of a statistical procedure in the Cauchy case, not because one believes in the realism of the Cauchy distribution, but to see how well the procedure stands up under such extreme tail behavior.

A distribution with tail behavior intermediate between that of the normal and Cauchy distribution is the *logistic distribution*, whose density is

$$p(x) = \frac{e^{-x}}{(1+e^{-x})^2} \tag{A.96}$$

This is again symmetric about zero and unimodal (Prob. 13), and tends to zero at the rate of e^{-x} as x tends to infinity, more slowly than $e^{-x^2/2}$, but much faster than $1/x^2$. The same tail behavior is shown by the *double-exponential distribution* with density

$$p(x) = \tfrac{1}{2} e^{-|x|} \tag{A.97}$$

which, however, has a quite different shape in the center where it has a spike.

We have discussed the tail behavior of distributions in terms of the rate at which the density $p(x)$ tends to zero as x tends to $\pm\infty$. Although this is reasonable for the particular distributions considered here, the general concept of "heavy tails" must be formulated somewhat differently. Interesting definitions of heavy tails are given by Van Zwet (1964) and Doksum (1969).

For each of the densities (A.95), (A.96), and (A.97), the function

$$\frac{1}{a} p\left(\frac{x-b}{a}\right) \tag{A.98}$$

defines a probability density provided $a > 0$. These more general densities and their associated distributions are again referred to as Cauchy, logistic, and double-exponential, respectively.

2E. The Rectangular (Uniform) and Exponential Distributions

The totality of densities (A.98), for fixed p but varying a and b, constitutes a location-scale family, the parameter a being a measure of scale and b a measure of location. Two further important location-scale families are provided by the uniform and exponential distributions defined, respectively, by the densities

(A.99)
$$p(x) = \begin{cases} \frac{1}{a} & \text{if } b-\frac{1}{2}a < x < b+\frac{1}{2}a \\ 0 & \text{otherwise} \end{cases}$$

and

(A.100)
$$p(x) = \begin{cases} \frac{1}{a}e^{-(x-b)/a} & \text{if } b < x \\ 0 & \text{otherwise} \end{cases}$$

The first of these is the probability density of a point selected at random from the interval $(b-\frac{1}{2}a, b+\frac{1}{2}a)$; the second plays an important role in reliability theory (see Chap. 2, Sec. 7K).

2F. The χ^2-distribution

There are a number of important distributions related to the normal distribution such as the t-, F-, and χ^2-distributions. We shall here consider only the last of these.

Let $X_1, \ldots, X_f$ be f independent random variables each with distribution $N(0,1)$. Then the distribution of the sum of squares

(A.101)
$$Y = X_1^2 + \cdots + X_f^2$$

is called the χ^2-distribution with f degrees of freedom. It has the probability density

(A.102)
$$p(y) = Cy^{\frac{1}{2}f-1}e^{-\frac{1}{2}y} \qquad 0 < y < \infty$$

where the constant C is determined so that the area under the curve from 0 to ∞ is equal to 1. The expectation and variance of Y are given (Prob. 14) by

$$E(Y) = f \quad \text{and} \quad \text{Var}(Y) = 2f$$

Like the normal distribution, the χ^2-distribution plays an important role as a limit distribution. In this connection, it is useful to know that not only the sum of squares of independent standard normal variables has a χ^2-distribution, but also certain sums of squares of dependent normal variables, although the number of degrees of freedom may then be smaller than the number of terms in the sum of squares. The following is an example of this phenomenon, which is proved in many standard statistical texts.

If $X_1, \ldots, X_n$ are independent, each with distribution $N(\mu,\sigma^2)$, then

(A.103)
$$\frac{1}{\sigma^2}\sum_{i=1}^{n}(X_i-\bar{X})^2$$

is distributed as χ^2 with $n-1$ degrees of freedom.

The following slight extension of this result is an easy consequence. Let $X_{i1}, \ldots, X_{in_i}$ $(i = 1, \ldots, s)$ be independent, each with distribution $N(\mu, \sigma^2)$, and let

$$X_{i.} = \sum_{j=1}^{n_i} \frac{X_{ij}}{n_i} \qquad \text{and} \qquad X_{..} = \frac{\sum\sum X_{ij}}{\sum n_i}$$

then

$$\frac{\sum n_i (X_{i.} - X_{..})^2}{\sigma^2}$$

is distributed as χ^2 with $s-1$ degrees of freedom.

If Y is defined by (A.101), the distribution of $\sqrt{Y}$ is the χ-distribution with f degrees of freedom. In particular, the χ-distribution with one degree of freedom is the distribution of the absolute value $|Z|$ when Z is distributed according to $N(0,1)$. The expectation of such a random variable is (Prob. 15)

$$E(|Z|) = \sqrt{\frac{2}{\pi}}$$

2G. Order Statistics

Let $X_1, \ldots, X_n$ be independent, identically distributed according to a distribution F with density f, and let $X_{(1)} < \cdots < X_{(n)}$ denote the n ordered observations, the so-called *order statistics*. Then $X_{(s)}$ has a probability density which is given by

$$n\binom{n-1}{s-1}[F(x)]^{s-1}[1-F(x)]^{n-s}f(x) \tag{A.104}$$

This formula is suggested by noting that the sth smallest must be a particular one of the X's whose density is $f(x)$; that $[F(x)]^{s-1}$ is the probability that there are $s-1$ smaller observations and $[1-F(x)]^{n-s}$ the probability that there are $n-s$ larger ones; finally, that n is the number of ways of selecting an observation to be the sth smallest and $\binom{n-1}{s-1}$ is the number of ways of dividing the remaining $n-1$ observations into those that are smaller and those that are larger than $X_{(s)}$. [For a formal proof of (A.104) see Prob. 16.]

An important special case occurs when the distribution of $X_1, \ldots, X_n$ is the uniform distribution on (0,1). Then $F(x) = x$ and $f(x) = 1$ for $0 < x < 1$, and (A.104) reduces to

$$n\binom{n-1}{s-1}x^{s-1}(1-x)^{n-s}, \quad 0 < x < 1$$

This is the density of the beta distribution $B(s, n-s+1)$. The expectation and variance of a random variable Y with this density are (Prob. 17)

$$E(Y) = \frac{s}{n+1} \qquad \text{and} \qquad \operatorname{Var}(Y) = \frac{s(n-s+1)}{(n+1)^2(n+2)} \tag{A.105}$$

3. CONVERGENCE IN PROBABILITY AND IN LAW

Consider a sequence of random variables $T_1, \ldots, T_n$. It sometimes happens that as n increases the T's become less variable and that for sufficiently large n they become essentially constant. For example, if T_n is the frequency of success in n binomial trials with success probability p, it is nearly certain that T_n will be very close to p (and hence essentially constant) for sufficiently large n. More generally, if T_n is an estimate of a parameter θ based on n observations, one hopes that T_n will be nearly certain to be close to θ for large n.

Definition 4. A sequence $\{T_n\}$ of random variables is said to *converge in probability* to a constant θ if for every positive number a

(A.106) $$P(|T_n - \theta| < a) \to 1 \qquad \text{as } n \to \infty$$

This definition states that for any positive a, no matter how small, we can be nearly certain that T_n differs from θ by less than a if n is sufficiently large. Condition (A.106) is clearly equivalent to

(A.107) $$P(|T_n - \theta| \geqslant a) \to 0 \qquad \text{as } n \to \infty$$

A convenient tool for proving convergence in probability is provided by the following inequality.

Chebyshev Inequality. For any random variable Z and any positive number a, we have

(A.108) $$E(Z^2) \geqslant a^2 P(|Z| \geqslant a)$$

Proof. We shall give the proof only for the case of discrete random variables. Except for notation, exactly the same proof applies in the general case. Let Z take on values z_1, z_2, etc., with probabilities p_1, p_2, etc., so that

$$E(Z^2) = \sum p_i z_i^2$$

Consider the contribution to this sum of those terms for which $z_i^2 \geqslant a^2$ and denote the resulting sum by d; that is,

$$d = \sum p_i z_i^2 \quad \text{summed over all } i \text{ for which } z_i^2 \geqslant a^2$$

Then $E(Z^2) \geqslant d$. If in the sum for d we replace z_i^2 by a^2, the result will be less than or equal to d, so that

$$d \geqslant a^2 \sum p_i \quad \text{summed over all } i \text{ for which } z_i^2 \geqslant a^2$$

The coefficient of a^2 in this last inequality is just $P(Z^2 \geqslant a^2)$, so that

$$E(Z^2) \geqslant d \geqslant a^2 P(Z^2 \geqslant a^2)$$

and this completes the proof since $P(Z^2 \geqslant a^2) = P(|Z| \geqslant a)$.

Through substitution of $T_n - \theta$ for Z in (A.108), the inequality can be written as

(A.109) $$P(|T_n - \theta| \geqslant a) \leqslant \frac{1}{a^2} E(T_n - \theta)^2$$

Comparison of (A.109) with (A.107) proves the important fact,

(A.110) $$E(T_n-\theta)^2 \to 0 \quad \text{implies that } T_n \to \theta \text{ in probability}$$

An interesting special case of (A.110) obtains when the random variables T_n have a common expectation μ. Replacing θ by μ in (A.110) one then finds that

(A.111) $$E(T_n) = \mu \qquad \text{and} \qquad \text{Var}(T_n) \to 0 \qquad \text{imply that } T_n \to \mu \text{ in probability}$$

EXAMPLE 7. Let $X_1, \ldots, X_n$ be independent with common mean μ and finite variance σ^2, and let $\overline{X}_n = (X_1 + \cdots + X_n)/n$ denote their average. Then $E(\overline{X}_n) = \mu$, and $\text{Var}(\overline{X}_n) = \sigma^2/n$ tends to zero as n tends to infinity. Thus, by (A.111)

(A.112) $$\overline{X}_n \to \mu \quad \text{in probability}$$

a fact known as the *weak law of large numbers*. [If the X's are identically distributed with finite mean μ, (A.112) continues to hold even when σ^2 is infinite. However, the proof of this theorem of Khinchine is more difficult.]

EXAMPLE 8. Let $X_1, \ldots, X_n$ be independent, identically distributed with finite variance σ^2. Then

$$S^2 = \frac{\sum (X_i - \overline{X})^2}{(n-1)}$$

tends in probability to σ^2 (Prob. 20).

EXAMPLE 9. Let $X_1, \ldots, X_m$ and $Y_1, \ldots, Y_n$ be independent, with distributions F and G, respectively, and let $p_1 = P(X < Y)$. If W_{XY} is the random variable defined by Eq. (1.24), then W_{XY}/mn tends to p_1 in probability (Prob. 21) as m and n tend to infinity.

Convergence of T_n to θ in probability means that the differences $T_n - \theta$ tend to be very small for large n, but it says nothing about how small. Are they, for example, of order $1/n$, $1/n^2$, or even $1/e^n$, or are they of larger order such as $1/\sqrt{n}$, $1/\sqrt[4]{n}$, or $1/\log n$?

Let us consider once more Example 7, for which inequality (A.109) becomes

(A.113) $$P(|\overline{X}_n - \mu| \geqslant a) \leqslant \frac{\sigma^2}{a^2 n}$$

In an effort to throw some light on the question of the order of the differences $\overline{X}_n - \mu$, let us see what happens if we put for example $a = c/\sqrt[4]{n}$. Then (A.113) becomes

$$P\left(|\overline{X}_n - \mu| \geqslant \frac{c}{\sqrt[4]{n}}\right) \leqslant \frac{\sigma^2}{c^2\sqrt{n}}$$

The right-hand side tends to zero as n tends to infinity, and this shows that for any c and any given n which is sufficiently large, it is practically certain that $|\overline{X}_n - \mu|$ will be smaller than $c/\sqrt[4]{n}$, that is, of order less than $1/\sqrt[4]{n}$. Exactly the same argument shows that if b_n is any sequence of constants tending to infinity more slowly than $\sqrt{n}$ (that is, in such a way that $b_n/\sqrt{n} \to 0$ as $n \to \infty$), then for any $c > 0$

(A.114) $$P\left(|\bar{X}_n - \mu| \geqslant \frac{c}{b_n}\right) \to 0 \qquad \text{as } n \to \infty$$

The question of the order of the differences $\bar{X}_n - \mu$ is closely related to another problem. To get a better idea of the detailed behavior of the random variables $\bar{X}_n - \mu$, one may try to magnify them through multiplication by suitable constants b_n tending to infinity. The appropriate magnification would be such that the variables $b_n(\bar{X}_n - \mu)$ tend neither to zero nor to infinity but take on intermediate values, thus opening up the possibility that the distribution of $b_n(\bar{X}_n - \mu)$ might tend to a nondegenerate limit distribution in a sense made precise by the following definition.

Definition 5. Let $\{T_n\}$ be a sequence of random variables and F a continuous distribution. Then T_n is said to have the limit distribution F (or to be asymptotically distributed according to F), if for all t

(A.115) $$P(T_n \leqslant t) \to F(t) \qquad \text{as } n \to \infty$$

If T is a random variable with distribution F, relation (A.115) is expressed alternatively by saying that T_n tends to T *in law*. [Note: If F is not continuous, condition (A.115) is required to hold only at all continuity points of F.]

Let us now return to Example 7 and the magnified differences $b_n(\bar{X}_n - \mu)$. It follows from (A.114) that multiplication by any sequence b_n tending to infinity more slowly than $\sqrt{n}$ is not enough since $b_n(\bar{X}_n - \mu)$ then still tends to zero in probability. The following theorem shows not only that the correct magnification is $b_n = \sqrt{n}$, but also determines the resulting limit distribution. It is one of the simplest versions of the so-called central limit theorem; for a proof see, for example, Cramér (1946), Sec. 17.4, or Feller (1966), Sec. VIII. 4, Theorem 1.

Theorem 2. If $X_1, \ldots, X_n$ are independent, identically distributed with expectation $E(X_i) = \mu$ and finite variance $\text{Var}(X_i) = \sigma^2$, then $\sqrt{n}(\bar{X}_n - \mu)$ is asymptotically normally distributed with distribution $N(0, \sigma^2)$.

This result can be given a slightly different form by introducing the notion of standardization. A random variable Y is said to be expressed in *standard units* or to be *standardized* if its expectation is subtracted and the difference is divided by the standard deviation of Y. The result is the standardized variable

(A.116) $$Y^* = \frac{Y - E(Y)}{\sqrt{\text{Var}(Y)}}$$

In particular, the variable $\bar{X}_n$ expressed in standard units is

(A.117) $$\bar{X}_n^* = \frac{\sqrt{n}(\bar{X}_n - \mu)}{\sigma}$$

and Theorem 2 states that the limit distribution of $\bar{X}_n^*$ is the standard normal distribution $N(0,1)$.

If a sequence $\{T_n\}$ is such that the distribution of the standardized variables $T_n^* = [T_n - E(T_n)]/\sqrt{\mathrm{Var}(T_n)}$ tends to $N(0,1)$, we shall say that the variables T_n are *asymptotically normal.*

EXAMPLE 10. Let X be the number of successes in n binomial trials with success probability p, so that X has the binomial distribution $b(p,n)$. To express X as a sum of independent, identically distributed random variables, let $I_j = 1$ or 0 as the jth trial is a success or failure. Then

$$X = I_1 + \cdots + I_n$$

the I's are independent, and clearly have the same distribution with mean $E(I_j) = p$ and variance $\mathrm{Var}(I_j) = pq$. The standardized variable $\bar{I}_n^*$ defined by (A.117) with the I's in place of the X's becomes

$$\frac{\sqrt{n}(X/n - p)}{\sqrt{pq}} \tag{A.118}$$

and Theorem 2 shows that (A.118) is asymptotically normally distributed with zero mean and unit variance.

In applications of asymptotic theory, it often happens that the value of t on the left-hand side of (A.115) depends on n. This possibility is covered by the following theorem, which is due to Polya. [A hint for an easy proof is given by Feller (1966), Chap. 8, Prob. 5.]

Theorem 3. Let $\{T_n\}$ be a sequence of random variables with a continuous limit distribution F. Then for any value $-\infty \leqslant t \leqslant \infty$ and any sequence $t_n \to t$

$$P(T_n \leqslant t_n) \to F(t) \qquad \text{as } n \to \infty \tag{A.119}$$

As an application, let us see what happens in Theorem 2 if we use a factor different from $\sqrt{n}$. Consider first a sequence b_n of constants such that $b_n/\sqrt{n} \to 1$. Then

$$P[b_n(\bar{X}_n - \mu) \leqslant t] = P[\sqrt{n}(\bar{X}_n - \mu) \leqslant t\sqrt{n}/b_n] = P[\sqrt{n}(\bar{X}_n - \mu) \leqslant t_n]$$

Here $t_n = t(1 + r_n)$, where $r_n = (\sqrt{n}/b_n) - 1 \to 0$ as $n \to \infty$. It thus follows from Theorem 3 that $b_n(\bar{X}_n - \mu)$ has the same limit distribution as $\sqrt{n}(\bar{X}_n - \mu)$. Other cases can be handled quite analogously (Prob. 24).

To establish the asymptotic distribution of complicated statistics from the known limit distribution of a simpler statistic, the following theorem is often helpful. [For a proof see Cramér (1946), Sec. 20.6.]

Theorem 4. If T_n tends in law to T, and if B_n and D_n are random variables converging in probability to b and d, respectively, then $B_n T_n + D_n$ tends in law to $bT + d$. (*Note:* It is important to realize that this theorem does not require the variables B_n and D_n to be independent of T_n.)

EXAMPLE 11. Under the assumptions of Theorem 2 with $\mu = 0$, consider the large-sample behavior of $T_n = \sqrt{n}\bar{X}_n/S$, where S is defined in Example 8. Let us rewrite T_n as

$$T_n = \frac{\sqrt{n}\bar{X}_n/\sigma}{S/\sigma}$$

The numerator has the limit distribution $N(0,1)$ by Theorem 2; the denominator tends to 1 in probability by Example 8. It therefore follows from Theorem 4 that T_n has the limit distribution $N(0,1)$. This shows that T_n is asymptotically "distribution-free," i.e., that its limit distribution is independent of the distribution of the variables X_i under the sole assumption that σ^2 is finite.

A simple and useful consequence of Theorem 4 is

Corollary 1. If T_n tends in law to T and if

(A.120) $$E(T_n - S_n)^2 \to 0 \qquad \text{as } n \to \infty$$

then S_n also tends in law to T.

Proof. Let $S_n - T_n = R_n$. Then R_n tends in probability to zero by (A.110), and the result follows from Theorem 4.

The following important modification of Corollary 1 is due to Hajek.

Corollary 2. If $[T_n - E(T_n)]/\sqrt{\text{Var}(T_n)}$ has the limit distribution L and if

(A.121) $$\frac{E(T_n - S_n)^2}{\text{Var}(T_n)} \to 0 \qquad \text{as } n \to \infty$$

then $[S_n - E(S_n)]/\sqrt{\text{Var}(S_n)}$ also has the limit distribution L.

Proof. Let us prove first that $[S_n - E(S_n)]/\sqrt{\text{Var}(T_n)}$ has the limit distribution L. This follows from Corollary 1 since

(A.122) $$E\left[\frac{T_n - E(T_n)}{\sqrt{\text{Var}(T_n)}} - \frac{S_n - E(S_n)}{\sqrt{\text{Var}(T_n)}}\right]^2 \leqslant \frac{E(T_n - S_n)^2}{\text{Var}(T_n)} \to 0$$

The result will therefore follow from Theorem 4 if we can show that

(A.123) $$\frac{\text{Var}(T_n)}{\text{Var}(S_n)} \to 1 \qquad \text{as } n \to \infty$$

Now it follows from (A.122) that (A.121) remains valid if T_n and S_n are replaced by $T_n - E(T_n)$ and $S_n - E(S_n)$, respectively. We can therefore without loss of generality assume that $E(T_n) = E(S_n) = 0$. Condition (A.121) then becomes

(A.124) $$1 - \frac{2E(S_n T_n)}{E(T_n^2)} + \frac{E(S_n^2)}{E(T_n^2)} \to 0 \qquad \text{as } n \to \infty$$

Since $E(S_n T_n) \leqslant \sqrt{E(S_n^2)E(T_n^2)}$, the left-hand side of (A.124) is greater than or equal to

$$1 - 2\sqrt{\frac{E(S_n^2)}{E(T_n^2)}} + \frac{E(S_n^2)}{E(T_n^2)} = \left[1 - \sqrt{\frac{E(S_n^2)}{E(T_n^2)}}\right]^2$$

which therefore also tends to zero. This completes the proof of (A.123).

Theorem 2 concerns the limit behavior of $\bar{X}_n$ when $X_1, \ldots, X_n$ are both independent and identically distributed. We shall conclude this section with a central limit theorem (due to Lindeberg) for sums of independent variables that are not necessarily identically distributed. [For a proof, see, for example, Feller (1966), Sec. VIII. 4, Theorem 3.] Some more special problems involving dependent variables will be considered in the next two sections.

Suppose for the sake of simplicity that the variables $X_1, X_2, \ldots$ are discrete, with X_1 taking on values $a_1, a_2, \ldots$; X_2 taking on values $b_1, b_2, \ldots$; etc. If $E(X_i) = \mu_i$ and $\operatorname{Var}(X_i) = \sigma_i^2$, then, for example,

$$\sigma_1^2 = \sum_j P(X_1 = a_j)(a_j - \mu_1)^2$$

Consider the contribution to this sum of only those terms for which

$$(a_j - \mu_1)^2 > t(\sigma_1^2 + \cdots + \sigma_n^2) \tag{A.125}$$

and denote this contribution to the variance by

(A.126) $\tau_{n1}^2(t) = \sum P(X_1 = a_j)(a_j - \mu_1)^2$ summed over all values of j for which (A.125) holds.

The corresponding contribution to σ_2^2 we denote by $\tau_{n2}^2(t)$, so that

$$\tau_{n2}^2(t) = \sum P(X_2 = b_j)(b_j - \mu_2)^2 \text{ summed over all } j \text{ for which } (b_j - \mu_2)^2 > t(\sigma_1^2 + \cdots + \sigma_n^2)$$

$\tau_{n3}^2(t), \ldots, \tau_{nn}^2(t)$ are defined analogously.

Theorem 5. Let $X_1, \ldots, X_n$ be independent, with means $E(X_i) = \mu_i$ and variances $\operatorname{Var}(X_i) = \sigma_i^2$, and let S_n^* be the standardized sum of the X's

$$S_n^* = \frac{\sum_{i=1}^{n} (X_i - \mu_i)}{\sqrt{\sum_{i=1}^{n} \sigma_i^2}} \tag{A.127}$$

Then the limit distribution of S_n^* is the standard normal distribution $N(0,1)$ provided for any $t > 0$

$$\frac{\tau_{n1}^2(t) + \cdots + \tau_{nn}^2(t)}{\sigma_1^2 + \cdots + \sigma_n^2} \to 0 \quad \text{as } n \to \infty \tag{A.128}$$

Condition (A.128) may seem rather forbidding to a reader who has not seen it before. However, its application often turns out to be much easier than one might expect, as is illustrated by the following two examples.

EXAMPLE 12. Let the X's be identically distributed with finite variance σ^2. Then $\tau_{n1}^2 = \cdots = \tau_{nn}^2$ and $\sigma_1^2 = \cdots = \sigma_n^2 = \sigma^2$, so that condition (A.128) reduces to

$$\frac{\tau_{n1}^2(t)}{\sigma^2} \to 0 \quad \text{as } n \to \infty \tag{A.129}$$

In the definition of $\tau_{n1}{}^2(t)$, the sum (A.126) now extends over all values j for which $(a_j-\mu)^2 > nt\sigma^2$. Thus

$$\frac{\tau_{n1}{}^2(t)}{\sigma^2} = \sum \frac{(a_j-\mu)^2}{\sigma^2} P(X_1 = a_j) \quad \text{summed over all } j \text{ for which } (a_j-\mu)^2/\sigma^2 > tn.$$

The series on the right-hand side, when summed over all j, is a convergent series whose sum is 1. The contribution from terms for which $(a_j-\mu)^2/\sigma^2$ exceeds tn therefore tends to zero as n tends to infinity. This completes the verification of (A.128) in the present case, and shows that Theorem 2 is a special case of Theorem 5.

EXAMPLE 13. As another application of Theorem 5, let us prove asymptotic normality of the sum

$$S_n = \sum_{i=1}^{n} iK_i$$

where the K's are independent and K_i takes on the values 1 and 0 with probabilities p and q, respectively. Putting $X_i = iK_i$, we have $\mu_i = ip$, $\sigma_i{}^2 = i^2pq$, and hence

$$\sigma_1{}^2 + \cdots + \sigma_n{}^2 = pq(1^2 + \cdots + n^2) = \tfrac{1}{6}n(n+1)(2n+1)pq$$

Since X_i takes on only two values, the sum $\tau_{ni}{}^2(t)$ consists of two, one, or no terms. It has no terms, and thus is equal to zero, if both $(i-\mu_i)^2 = i^2q^2$ and $(0-\mu_i)^2 = i^2p^2$ are less than or equal to $tn(n+1)(2n+1)pq/6$. Now i^2q^2 and i^2p^2 are less than n^2 for all $i = 1, \ldots, n$, and n^2 is less than $tpqn(n+1)(2n+1)/6$ for sufficiently large n, say, for $n > n_0$ (which of course depends on t). Thus, for $n > n_0$ it follows that $\tau_{n1}{}^2(t) = \cdots = \tau_{nn}{}^2(t) = 0$, and this completes the verification of (A.128) and hence the proof of the asymptotic normality of S_n.

For $p = \frac{1}{2}$, this example proves the asymptotic normality of the one-sample Wilcoxon statistic V_S discussed in Chap. 3, Sec. 2, to which S_n reduces in this case (with N in place of n). In other applications, the value of p, rather than being fixed, may depend on n. This situation is not covered by Theorem 5 as stated, since the random variables K_i and hence X_i (and their distributions) then depend on n.

To state the required extension of Theorem 5, consider a random variable X_{11} (with distribution F_{11}); random variables X_{21} and X_{22} (independent, with distributions F_{21} and F_{22}); random variables X_{31}, X_{32}, and X_{33} (independent, with distributions F_{31}, F_{32}, and F_{33}); and so on. Denote the mean and variance of X_{ni} by μ_{ni} and $\sigma_{ni}{}^2$, respectively, and let

$$S_n^* = \frac{\sum_{i=1}^{n} (X_{ni} - \mu_{ni})}{\sqrt{\sum_{i=1}^{n} \sigma_{ni}{}^2}}$$

The quantity $\tau_{n1}{}^2(t)$ is defined as in (A.125) and (A.126) but with a_i replaced by a_{ni} (the values taken on by X_{n1}), μ_1 by μ_{n1}, and $\sigma_i{}^2$ by $\sigma_{ni}{}^2$. The definition of $\tau_{ni}{}^2(t)$ for $i > 1$ is modified analogously. With these modifications, we have

Theorem 5′. If for any $t > 0$

$$\frac{\tau_{n1}{}^2(t) + \cdots + \tau_{nn}{}^2(t)}{\sigma_{n1}{}^2 + \cdots + \sigma_{nn}{}^2} \to 0 \qquad \text{as } n \to \infty \tag{A.130}$$

the limit distribution of S_n^* is the standard normal distribution $N(0,1)$.

This theorem establishes the asymptotic normality of S_n in Example 13 even when p is allowed to depend on n, provided $p_n q_n$ does not tend to zero too fast. No changes are required in the proof.

EXAMPLE 14. Let $X_{ni} = v_{ni} K_i$ where the v's are constants and K_i is defined as in Example 13. Then the standardized sum of the X's

$$S_n^* = \frac{\sum (X_{ni} - p v_{ni})}{\sqrt{pq \sum v_{ni}{}^2}}$$

has the limit distribution $N(0,1)$ provided the following condition is satisfied: $\left(\max\limits_i v_{ni}{}^2\right) \Big/ \left(\sum\limits_i v_{ni}{}^2\right) \to 0$ as $n \to \infty$ (Prob. 27).

For $p = \frac{1}{2}$, this example can be used to prove the asymptotic normality of the one-sample Wilcoxon statistic V_s^* in the presence of ties (Prob. 28).

4. SAMPLING FROM A FINITE POPULATION

In the present section we shall obtain the null distribution of the one- and two-sample Wilcoxon statistics and of the trend and rank correlation statistics of Chap. 7, Secs. 2 and 3, both with and without ties. The method used is due to Hajek (1961). However, we shall not develop Hajek's results in their full generality but only as far as is necessary for these particular applications.

Let a population consist of a number of elements, or members, to each of which is attached a numerical value (such as age, blood pressure, or bank balance). Actually, we shall be interested not in a single such population but a sequence of populations of increasing size. Let these populations be $\Pi_1, \Pi_2, \ldots$, with Π_1 consisting of a single element to which is attached the value v_{11}; Π_2 of two elements with values $v_{21}, v_{22}; \ldots; \Pi_N$ of N elements with values $v_{N1}, \ldots, v_{NN}$.

Suppose that from each population a random sample is drawn and that $n = n(N)$ is the size of the sample from the Nth population. Let the n values included in the sample be denoted by $A_{N1}, \ldots, A_{Nn}$, the sum of these n sample values by

$$S_N = A_{N1} + \cdots + A_{Nn} \tag{A.131}$$

and the average of the N values attached to the elements of Π_N by

$$v_{N.} = \frac{v_{N1} + \cdots + v_{NN}}{N} \tag{A.132}$$

Since the terms of S_N are dependent, Theorem 5 is not directly applicable. Nevertheless, S_N is asymptotically normal under conditions made precise by the following theorem, which is a special case of a theorem of Hajek (1961).

Theorem 6. A sufficient condition for the standardized variables $S_N^* = [S_N - E(S_N)]/\sqrt{\operatorname{Var}(S_N)}$ to be asymptotically normally distributed according to $N(0,1)$ is that

$$n \text{ and } m = N-n \text{ both tend to infinity} \tag{A.133}$$

and that

$$\frac{\max(v_{Ni}-v_{N.})^2}{\sum(v_{Nj}-v_{N.})^2}\max\left(\frac{N-n}{n},\frac{n}{N-n}\right)\to 0 \qquad \text{as } N\to\infty \tag{A.134}$$

Proof. To prove this theorem, we shall construct another sequence, which is the sum of independent random variables, which is asymptotically equivalent to S_N in the sense of Corollary 2, and to which we can apply Theorem 5. To this end, let $U_1,\ldots,U_N$ be independent random variables, each uniformly distributed on the interval (0,1). In terms of these variables, the process of drawing a random sample of size n from Π_N can be defined by including v_{Ni} in the sample if and only if U_i is one of the n smallest U's; that is, if

$$R_i \leqslant n \tag{A.135}$$

where R_i denotes the rank of U_i. Since each set of n U's is equally likely to constitute the set of the n smallest U's, each of the $\binom{N}{n}$ possible samples is then equally likely. If we put

$$J_i = \begin{cases} 1 & \text{if } R_i \leqslant n \\ 0 & \text{otherwise} \end{cases} \tag{A.136}$$

the sum S_N of the v-values in the sample can be written as

$$S_N = \sum v_{Ni}J_i \tag{A.137}$$

Since the J's are dependent, let us also consider the independent variables

$$K_i = \begin{cases} 1 & \text{if } U_i \leqslant \dfrac{n}{N} \\ 0 & \text{otherwise} \end{cases} \tag{A.138}$$

and the sum

$$T_N = \sum(v_{Ni}-v_{N.})K_i + nv_{N.} \tag{A.139}$$

Lemma 1. Let S_N and T_N be defined by (A.137) and (A.139), respectively, and suppose that condition (A.133) holds. Then

$$\frac{E(T_N-S_N)^2}{\operatorname{Var}(T_N)}\to 0 \qquad \text{as } N\to\infty \tag{A.140}$$

Before proving this lemma, we shall show how it provides a basis for proving Theorem 6, and give some illustrations of this theorem.

Proof of Theorem 6. By Corollary 2, it is enough to show that the standardized variables

T_N^* have the limit distribution $N(0,1)$. Since the variables K_i are independent, we can apply Theorem 5′ to the variables

(A.141) $$X_{Ni} = (v_{Ni} - v_{N.})K_i$$

The expectation and variance of X_{Ni} are

$$\mu_{Ni} = (v_{Ni} - v_{N.})\frac{n}{N} \qquad \text{and} \qquad \sigma_{Ni}^2 = (v_{Ni} - v_{N.})^2 \frac{n}{N}\left(1 - \frac{n}{N}\right)$$

The variable $X_{Ni} - \mu_{Ni}$ takes on only two values:

$$(v_{Ni} - v_{N.})\left(1 - \frac{n}{N}\right) \qquad \text{with probability } \frac{n}{N}$$

and

$$-(v_{Ni} - v_{N.})\frac{n}{N} \qquad \text{with probability } 1 - \frac{n}{N}$$

Thus $\tau_{Ni}^2(t) = 0$ provided both

$$(v_{Ni} - v_{N.})^2\left(1 - \frac{n}{N}\right)^2 \qquad \text{and} \qquad (v_{Ni} - v_{N.})^2\left(\frac{n}{N}\right)^2$$

are less than or equal to $t\frac{n}{N}\left(1 - \frac{n}{N}\right)\sum(v_{Ni} - v_{N.})^2$, that is, provided the left-hand side of (A.134) is less than or equal to t. If condition (A.134) holds, it follows that for any given $t > 0$ the sum $\tau_{N1}^2(t) + \cdots + \tau_{NN}^2(t)$ will be zero for N sufficiently large, and this completes the proof.

Corollary 3. The conclusion of Theorem 6 holds if (A.133) holds and if

(A.142) $$\frac{\max(v_{Ni} - v_{N.})^2}{\sum(v_{Nj} - v_{N.})^2/N} \text{ is bounded as } N \to \infty$$

that is, if there exists a constant $M < \infty$ such that the ratio displayed in (A.142) is less than or equal to M for all sufficiently large N.

Proof. Suppose without loss of generality that $n \geqslant N - n$, so that

$$\max\left(\frac{N-n}{n}, \frac{n}{N-n}\right) = \frac{n}{N-n}$$

Since the left-hand side of (A.134) is then obtained from the quantity in (A.142) through multiplication by $n/N(N-n)$, it is enough to show that this factor tends to zero. However, n/N is less than or equal to 1 and $N-n$ tends to infinity, and this completes the proof.

In applications of Corollary 3, the v's are given constants and it is then only necessary to check condition (A.142).

EXAMPLE 15. Suppose that d_N of the v-values in Π_N are equal to 1 and the remaining $N - d_N$

equal to zero. The sum S_N is then just the hypergeometric random variable X of Sec. 2B. Since

$$\max(v_{Ni}-v_{N.})^2 = \max\left[\left(1-\frac{d_N}{N}\right)^2, \left(\frac{d_N}{N}\right)^2\right]$$

and

$$\sum(v_{Nj}-v_{N.})^2 = N\frac{d_N}{N}\left(1-\frac{d_N}{N}\right)$$

the ratio in (A.142) reduces to

$$\frac{\max[(1-d_N/N)^2,(d_N/N)^2]}{(d_N/N)(1-d_N/N)}$$

This is clearly bounded as $N \to \infty$ provided d_N/N is bounded away from both 0 and 1; that is, if there exist constants λ_0 and λ_1 such that

(A.143) $$0 < \lambda_0 < \frac{d_N}{N} < \lambda_1 < 1 \qquad \text{for all sufficiently large } N$$

We have therefore proved that the standardized hypergeometric random variable has the limit distribution $N(0,1)$ provided (A.133) holds and (A.143) holds for some λ_0 and λ_1.

EXAMPLE 16. Let the values of Π_N be the integers $1, \ldots, N$ so that

$$v_{N1} = 1, \ldots, v_{NN} = N \qquad \text{for all } N$$

It was shown in (A.13) that $\sum(v_{Ni}-v_{N})^2/N$ is equal to $\tau^2 = (N^2-1)/12$. Furthermore $v_{N.} = \frac{1}{2}(N+1)$ and

$$\max|v_{Ni}-v_{N.}| = |N-\tfrac{1}{2}(N+1)| = \tfrac{1}{2}|N-1|$$

The ratio in (A.142) thus reduces to $\frac{1}{4}(N-1)^2/\frac{1}{12}(N^2-1)$. This is clearly bounded as N tends to infinity, and we have therefore proved the asymptotic normality of S_N provided (A.133) holds. This establishes asymptotic normality for the null distribution of the Wilcoxon statistic W_s defined in Chap. 1, Sec. 2.†

EXAMPLE 17. This example is motivated by the desire to prove asymptotic normality for the null distribution of the Wilcoxon statistic in the presence of ties. Let $d_1, \ldots, d_e$ be integers whose sum is N, and let the N v-values $v_{N1}, \ldots, v_{NN}$ be the midranks given by (A.14) with $v_{Ni} = v_i$. The numbers $e, d_1, \ldots, d_e$ will typically depend on N but for the sake of simplicity we delete the additional subscript N.

To apply Corollary 3, note that $v_{N.} = \frac{1}{2}(N+1)$ by (A.15), so that

$$\max(v_{Ni}-v_{N.})^2 \leqslant \tfrac{1}{4}(N-1)^2$$

as in the preceding example. On the other hand, it is seen from (A.10) and (A.18) that the denominator in (A.142) is

$$\frac{1}{N}\sum(v_{Ni}-v_{N.})^2 = \frac{N^2-1}{12} - \frac{\sum(d_i^3-d_i)}{12N} = \frac{N^2}{12} - \frac{\sum d_i^3}{12N}$$

† For an interesting alternative proof by a quite different method, see Haigh (1971).

For $e = 1$, $d_1 = N$, this is zero (as it must since S_N is then a constant) and condition (A.142) does not hold. Let us therefore impose condition (1.36) that $\max(d_i/N)$ is bounded away from 1 as N tends to infinity. This postulates the existence of a positive number $\varepsilon < 1$ such that $d_i/N \leqslant 1-\varepsilon$ for all i, or

(A.144) $$d_i \leqslant (1-\varepsilon)N \qquad \text{for all } i$$

Since the d's add up to N, it follows that

$$\sum d_i^3 \leqslant (1-\varepsilon)^2 N^2 \sum d_i = (1-\varepsilon)^2 N^3 \leqslant (1-\varepsilon) N^3$$

Thus,

$$\frac{1}{N}\sum (v_{Ni}-v_{N.})^2 \geqslant \frac{N^2}{12}[1-(1-\varepsilon)] = \frac{N^2}{12}\varepsilon$$

and the ratio (A.142) is less than or equal to

$$\frac{N^2/4}{\varepsilon N^2/12} = \frac{3}{\varepsilon}$$

This shows that the v-values of the present example satisfy (A.142), and hence proves the asymptotic normality of the Wilcoxon statistic W_s^* defined in Chap. 1, Sec. 4, provided (A.144) holds.

To complete the proof of Theorem 6, and thereby validate Examples 15 to 17, we must now prove Lemma 1.

Proof of Lemma 1. It is convenient to introduce the functions

(A.145) $$a_N(u) = \begin{cases} 1 & \text{if } u \leqslant \dfrac{n}{N} \\ 0 & \text{otherwise} \end{cases}$$

Then

$$J_i = a_N\left(\frac{R_i}{N}\right) \qquad \text{and} \qquad K_i = a_N(U_i) = a_N(U_{(R_i)})$$

where $U_{(1)} < \cdots < U_{(N)}$ denote the ordered U's. The last equation follows from the fact that R_i is the rank of U_i. We therefore have

(A.146) $$T_N - S_N = \sum (v_{Ni}-v_{N.})\left[a_N(U_{(R_i)}) - a_N\left(\frac{R_i}{N}\right)\right]$$

To compute $E(T_N - S_N)^2$ we shall compute first the conditional expectation given the order statistics $U_{(1)}, \ldots, U_{(N)}$, that is, the N values of the U's but not which U_i belongs to which value. Since $E[(K_i - J_i)|U_{(1)}, \ldots, U_{(N)}]$ is independent of i and since $\sum(v_{Ni}-v_{N.}) = 0$, the conditional expectation of $T_N - S_N$ given $U_{(1)}, \ldots, U_{(N)}$ is zero and hence

$$E[(T_N - S_N)^2|U_{(1)}, \ldots, U_{(N)}] = \mathrm{Var}[(T_N - S_N)|U_{(1)}, \ldots, U_{(N)}]$$

Now given $U_{(1)}, \ldots, U_{(N)}$, all $N!$ possible assignments of $U_1, \ldots, U_N$ to these values are equally likely. This shows that, given $U_{(1)}, \ldots, U_{(N)}$, the vector of ranks $(R_1, \ldots, R_N)$ is one of the $N!$

permutations of $(1, \ldots, N)$, each permutation having probability $1/N!$. It follows from (A.49) with $c_i = v_{Ni} - v_{N.}$ and $a(R_i) = a_N(U_{(R_i)}) - a_N(R_i/N) = K_i - J_i$ that the conditional variance of $T_N - S_N$ satisfies

$$\text{(A.147)} \qquad \operatorname{Var}\left[(T_N - S_N)|U_{(1)}, \ldots, U_{(N)}\right] \leqslant \frac{1}{N-1}\sum(v_{Ni} - v_{N.})^2 \sum(K_i - J_i)^2$$

Now it is easy to verify (Prob. 32) that

$$\sum(K_i - J_i)^2 = |D - n|$$

where D is the number of U's less than or equal to n/N. Taking the expectation of the conditional expectation of $(T_N - S_N)^2$, we therefore find that

$$E(T_N - S_N)^2 \leqslant \frac{1}{N-1}\sum(v_{Ni} - v_{N.})^2 E(|D - n|)$$

Now for any random variable Y with zero expectation

$$(E|Y|)^2 \leqslant E(Y^2) = \operatorname{Var}(Y)$$

Since D has the binomial distribution $b(n/N, N)$, application of this inequality to $Y = D - n$ gives

$$E(|D - n|) \leqslant \sqrt{N\frac{n}{N}\left(1 - \frac{n}{N}\right)}$$

and hence

$$E(T_N - S_N)^2 \leqslant \frac{1}{N-1}\sum(v_{Ni} - v_{N.})^2 \sqrt{N\frac{n}{N}\left(1 - \frac{n}{N}\right)}$$

On the other hand

$$\operatorname{Var}(T_N) = \sum(v_{Ni} - v_{N.})^2 \frac{n}{N}\left(1 - \frac{n}{N}\right)$$

so that

$$\frac{E(T_N - S_N)^2}{\operatorname{Var}(T_N)} \leqslant \frac{N}{N-1}\sqrt{\frac{N}{n(N-n)}}$$

Suppose without loss of generality that $n \geqslant N - n$, so that $n \geqslant \frac{1}{2}N$. Then

$$\sqrt{\frac{N}{n(N-n)}} \leqslant \sqrt{\frac{2}{N-n}} \to 0$$

since $N - n$ tends to infinity, and this completes the proof of Lemma 1.

The proof actually carries us somewhat further than Lemma 1. For let $a_N(u)$ be functions defined on $(0,1)$ but not necessarily given by (A.145). Let

$$\text{(A.148)} \qquad S_N = \sum_{i=1}^{N} v_{Ni} a_N\left(\frac{R_i}{N}\right)$$

and

$$\text{(A.149)} \qquad T_N = \sum_{i=1}^{N} (v_{Ni} - v_{N.}) a_N(U_i) + v_{N.} \sum_{i=1}^{N} a_N\left(\frac{i}{N}\right)$$

so that $T_N - S_N$ is given by (A.146). Then the argument leading from (A.146) to (A.147) shows that

$$\text{(A.150)} \qquad \operatorname{Var}(T_N - S_N) \leqslant \frac{1}{N-1} \sum (v_{Ni} - v_{N.})^2 E\left\{\sum \left[a_N(U_i) - a_N\left(\frac{R_i}{N}\right)\right]^2\right\}$$

$$= \frac{N}{N-1} \sum (v_{Ni} - v_{N.})^2 E\left[a_N(U_1) - a_N\left(\frac{R_1}{N}\right)\right]^2$$

while

$$\text{(A.151)} \qquad \operatorname{Var}(T_N) = \sum (v_{Ni} - v_{N.})^2 \operatorname{Var}[a_N(U_1)]$$

We have therefore proved the following extension of Lemma 1.

Lemma 2. The statistics S_N and T_N defined by (A.148) and (A.149), respectively, satisfy condition (A.140) provided

$$\text{(A.152)} \qquad \frac{E[a_N(U_1) - a_N(R_1/N)]^2}{\operatorname{Var}[a_N(U_1)]} \to 0 \qquad \text{as } N \to \infty$$

A simple consequence of this lemma is the following extension of Theorem 6.

Theorem 7. Let the constants v_{Ni} satisfy condition (A.142), and let a_N be a sequence of functions defined on (0,1) satisfying (A.152) and

$$\text{(A.153)} \qquad \frac{\max_u \{a_N(u) - E[a_N(U)]\}^2}{N \operatorname{Var}[a_N(U)]} \to 0 \qquad \text{as } N \to \infty$$

where U is uniformly distributed over (0,1). Then if $(R_1, \ldots, R_N)$ is a random permutation of $(1, \ldots, N)$ and S_N is defined by (A.148), the distribution of the standardized variable $[S_N - E(S_N)]/\sqrt{\operatorname{Var}(S_N)}$ tends to the standard normal distribution $N(0,1)$ as N tends to infinity.

Proof. If T_N is defined by (A.149), it follows from Lemma 2 that (A.140) is satisfied. Consider now the variables

$$X_{Ni} = (v_{Ni} - v_{N.}) a_N(U_i)$$

and apply Theorem 5′. As in the proof of Theorem 6, it is seen that $\tau_{Ni}^2(t)$ is zero provided

$$\frac{\max (v_{Ni} - v_{N.})^2}{\frac{1}{N}\sum (v_{Nj} - v_{N.})^2} \, \frac{\max \{a_N(u) - E[a_N(U)]\}^2}{N \operatorname{Var}[a_N(U)]} \leqslant t$$

By (A.142) and (A.153), the left-hand side of this inequality tends to zero as N tends to infinity. The assumptions of Theorem 5′ are therefore satisfied, and this completes the proof.

For the applications below, the following corollary is slightly more convenient.

Corollary 4. The conclusion of Theorem 7 holds provided the constants v_{Ni} satisfy (A.142) and the functions a_N satisfy the following three conditions:

(i) The functions a_N are uniformly bounded; i.e., there exist constants $-\infty < m < M < \infty$

such that

$$m \leqslant a_N(u) \leqslant M \qquad \text{for all } u \text{ and all } N$$

(ii) the variances $\operatorname{Var}[a_N(U)]$ are bounded away from zero as N tends to infinity;
(iii) the expectations $E[a_N(U_1)-a_N(R_1/N)]^2$ tend to zero as N tends to infinity.

Proof. Assumptions (A.152) and (A.153) are consequences, respectively, of (ii) and (iii), and of (i) and (ii).

EXAMPLE 18. The purpose of this example is to prove asymptotic normality for the trend statistic D of Chap. 7, Sec. 2. Let

(A.154) $$a_N(u) = \frac{j}{N} \qquad \text{if } \frac{j-1}{N} < u \leqslant \frac{j}{N} \qquad (j = 1, \ldots, N)$$

Then $a_N(R_i/N) = R_i/N$ and

(A.155) $$S_N = \frac{\sum v_{Ni} R_i}{N}$$

With

(A.156) $$v_{Ni} = i \qquad \text{for } i = 1, \ldots, N$$

S_N is seen to be equivalent to the statistic D of Chap. 7, Sec. 2 (where T_i is used to denote the present R_i).

To prove asymptotic normality of S_N, let us now check the three conditions of Corollary 4. The first condition is obviously satisfied with $m = 0$ and $M = 1$. For the second condition, note that the random variable $Na_N(U_1)$ takes on the values $1, \ldots, N$ with probability $1/N$ each. It was seen in Example 1 that the variance of such a random variable is $(N^2-1)/12$, so that

(A.157) $$\operatorname{Var}[a_N(U_1)] = \frac{N^2-1}{12N^2}$$

This proves condition (ii). It remains to show that the expectation in condition (iii) tends to zero. This follows easily from the inequality

$$\frac{1}{3}\left[a_N(U_1)-a_N\left(\frac{R_1}{N}\right)\right]^2 \leqslant [a_N(U_1)-U_1]^2 + \left(U_1-\frac{R_1}{N}\right)^2 + \left[\frac{R_1}{N}-a_N\left(\frac{R_1}{N}\right)\right]^2$$

From (A.154) it is clear that $[a_N(u)-u]^2 \leqslant 1/N^2$ for all u and that the third term on the right-hand side is zero. The result will therefore follow if we can show that

(A.158) $$E\left(U_1-\frac{R_1}{N}\right)^2 \to 0 \qquad \text{as } N \to \infty$$

Let us compute this expectation by taking first the conditional expectation given R_1. The condition $R_1 = i$ means that $U_1 = U_{(i)}$ and hence that the conditional expectation is

$$E\left(U_{(i)}-\frac{i}{N}\right)^2$$

As was discussed in Sec. 2G, the ith-order statistic $U_{(i)}$ of $U_1, \ldots, U_N$ has a beta distribution with expectation $i/(N+1)$ and variance less than $1/N$. Furthermore,

$$E\left(U_{(i)}-\frac{i}{N}\right)^2 = E\left(U_{(i)}-\frac{i}{N+1}\right)^2+\left(\frac{i}{N}-\frac{i}{N+1}\right)^2$$

and this completes the proof of (iii).

The above argument shows that S_N defined by (A.155) is asymptotically normal whenever the coefficients v_{Ni} satisfy (A.142). In particular, it was shown in Example 16 that (A.142) holds when $v_{Ni} = i$ for all $i = 1, \ldots, N$. Thus, the statistic $\sum iR_i$ is asymptotically normal, and so therefore is the statistic D of Chap. 7, Sec. 2.

EXAMPLE 19. As a last result, we shall now discuss the asymptotic normality of the statistics D^* and D^{**} of Chap. 7, Secs. 2 and 3, which are modifications of D in the presence of ties. If R_i^* denotes the ith midrank, we have

$$D^* = \sum(R_i^*-i)^2$$

Since $(R_1^*, \ldots, R_N^*)$ is a random permutation of the N midranks, which in their natural order are given by (A.14), it is clear that D^* has the same distribution as the statistic

$$\sum(R_i-v_{Ni})^2$$

where $(R_1, \ldots, R_N)$ is a random permutation of the integers $(1, \ldots, N)$ and the v's are given by (A.14). In the preceding example, it was seen that $\sum v_{Ni}R_i$ is asymptotically normal whenever the v_{Ni} satisfy condition (A.142). That the v's given by (A.14) satisfy this condition was shown in Example 17 under assumption (A.144). Under this assumption, the distribution of $\sum v_{Ni}R_i$ and hence of D^* is therefore asymptotically normal.

Consider next the statistic

$$D^{**} = \sum(S_i^*-R_i^*)^2$$

where $(R_1^*, \ldots, R_N^*)$ is a random permutation of the constants v_{Ni} defined by (A.14) and where $(S_1^*, \ldots, S_N^*)$ is a random permutation of constants $(v'_{N1}, \ldots, v'_{NN})$ defined by replacing e and $d_1, \ldots, d_e$ in (A.14) by e' and $d'_1, \ldots, d'_{e'}$. The constants v'_{Ni} will be assumed to satisfy (A.144). Concerning the constants v_{Ni}, we shall assume that the integer e in (A.14) is independent of N. (Typically e depends on N although this is not indicated by the notation.) This assumption is satisfied in the applications to contingency tables considered in Chap. 7. Under the additional assumption that $d_1, \ldots, d_e$ satisfy (A.144), we shall now prove the asymptotic normality of D^{**} using Corollary 4.

Instead of D^{**}, it is clearly enough to consider the statistic $\sum(R_i^*-v'_{Ni})^2$, which has the same distribution, or equivalently the statistic

$$S'_N = \frac{\sum v'_{Ni}R_i^*}{N}$$

Let $U_1, \ldots, U_N$ be independent, uniformly distributed on (0,1) as before, let R_i denote the rank of

U_i, and let the functions a_N be defined by

$$Na_N(u) = \begin{cases} \frac{1}{2}(d_1+1) & \text{if } 0 < u \leqslant \frac{d_1}{N} \\ d_1+\frac{1}{2}(d_2+1) & \text{if } \frac{d_1}{N} < u \leqslant \frac{d_1+d_2}{N} \\ \cdots & \cdots \end{cases}$$

Then the N-tuple $[a_N(R_1/N), \ldots, a_N(R_N/N)]$ has the same distribution as $(R_1^*/N, \ldots, R_N^*/N)$, and hence S'_N has the same distribution as

$$S_N = \sum v'_{Ni} a_N\left(\frac{R_i}{N}\right)$$

As in Example 18, the first condition of Corollary 4 is satisfied with $m = 0$, $M = 1$. For the second condition note that the distribution of $Na_N(U)$ is that of a v-value selected at random from $v_{N1}, \ldots, v_{NN}$, so that

$$N^2 \operatorname{Var}[a_N(U)] = \frac{\sum(v_{Ni}-v_{N.})^2}{N}$$

Under condition (A.144), it was shown in Example 17 that the right-hand side is greater than or equal to $N^2\varepsilon/12$, and this proves the validity of condition (ii).

To prove that the functions a_N satisfy condition (iii), define e auxiliary functions $a_N^{(1)}, \ldots, a_N^{(e)}$ as follows:

$$a_N^{(1)}(u) = \begin{cases} 1 & \text{if } 0 < u \leqslant \frac{d_1}{N} \\ 0 & \text{otherwise} \end{cases}$$

$$a_N^{(2)}(u) = \begin{cases} 1 & \text{if } \frac{d_1}{N} < u \leqslant \frac{d_1+d_2}{N} \\ 0 & \text{otherwise} \end{cases}$$

and so on

Then $a_N(u)$ can be represented as the sum

$$a_N(u) = \frac{1}{2N}(d_1+1)a_N^{(1)}(u) + \frac{1}{N}[d_1+\tfrac{1}{2}(d_2+1)]a_N^{(2)}(u) + \cdots + \frac{1}{N}[d_1+\cdots+d_{e-1}+\tfrac{1}{2}(d_e+1)]a_N^{(e)}(u)$$

From the inequality

$$(x_1+\cdots+x_e)^2 \leqslant e(x_1^2+\cdots+x_e^2)$$

which holds for any numbers $x_1, \ldots, x_e$ [since τ^2 defined by (A.11) is $\geqslant 0$], and from the fact that the coefficients of $a_N^{(1)}(u), \ldots, a_N^{(e)}(u)$ in the representation of $a_N(u)$ are all less than or equal to 1, it follows that

$$\left[a_N(U_1) - a_N\left(\frac{R_1}{N}\right)\right]^2 \leqslant e \sum_{i=1}^{e} \left[a_N^{(i)}(U_1) - a_N^{(i)}\left(\frac{R_1}{N}\right)\right]^2$$

However, the proof of Lemma 1 shows that

$$E\left[a_N^{(i)}(U_1) - a_N^{(i)}\left(\frac{R_1}{N}\right)\right]^2 \to 0 \qquad \text{as } N \to \infty$$

This establishes the validity of condition (iii), and hence the asymptotic normality of D^{**} under the stated conditions.

5. *U*-STATISTICS

In the preceding section the null distributions of a number of rank statistics were shown to be asymptotically normal. We shall now investigate the limit behavior of the corresponding distributions when the null hypothesis does not necessarily hold. The assumptions for the two sets of results will, however, be somewhat different. In the preceding section, we essentially assumed a randomization model and covered both cases with and without ties. Now we shall be concerned with population models (since we defined alternatives to the hypothesis only within such models), and the results will be of interest only in the continuous case, which is not compatible with the presence of ties. The present approach is due to Hoeffding (1948), and is applicable to a large class of statistics, the so-called U-statistics. Theorems 9 and 10 below deal with two important special cases of this class.

Let $X_1, \ldots, X_m$ and $Y_1, \ldots, Y_n$ be independent, the X's identically distributed with distribution F and the Y's with distribution G. Let φ be a function of two real variables, and let

$$U = \frac{1}{mn} \sum_{i=1}^{m} \sum_{j=1}^{n} \varphi(X_i, Y_j) \tag{A.159}$$

Then we shall prove that U is asymptotically normal as m and n tend to infinity. The method of proof, like that in the preceding section, consists in replacing the variable U by a sum of independent random variables, which is asymptotically equivalent to U and to which the central limit theorem can then be applied. It is natural for this purpose to try a sum of the form

$$S = \sum a_i(X_i) + \sum b_j(Y_j) \tag{A.160}$$

but how should one choose the functions a_i and b_j? The following "projection method," introduced in a different context by Hajek (1961), produces the a_i and b_j most likely to succeed in the sense of Corollary 1 to Theorem 4, that is, which will minimize $E(T-S)^2$ for a given statistic T. A discussion of this method requires the basic properties of conditional expectation, which can be found, for example, in Bickel (1971).

Let $Z_1, \ldots, Z_N$ be independently distributed, Z_i with distribution F_i, and let $T = T(Z_1, \ldots, Z_N)$ be any statistic satisfying

$$E(T) = 0 \tag{A.161}$$

Consider the problem of finding the sum

$$S = \sum k_i(Z_i) \tag{A.162}$$

for which $E(T-S)^2$ is as small as possible; the minimizing S may be considered the "projection" of T onto the linear space formed by the functions S. Let

(A.163) $$r_i(z_i) = E(T|Z_i = z_i)$$

be the conditional expectation of T given $Z_i = z_i$, and let

(A.164) $$T^* = \sum r_i(Z_i)$$

That T^* is the desired minimizing function is an immediate consequence of the following identity, which holds for all statistics T and S satisfying (A.161) and (A.162) for which the required expectations exist:

(A.165) $$E(T-S)^2 = E(T-T^*)^2 + E(T^*-S)^2$$

To prove this identity, write

$$E(T-S)^2 = E[(T-T^*)+(T^*-S)]^2$$

Squaring the right-hand side proves (A.165) if it can be shown that

(A.166) $$E[(T-T^*)(T^*-S)] = 0$$

Since the left-hand side of (A.166) is the sum of the expectations of

(A.167) $$[r_i(Z_i)-k_i(Z_i)](T-T^*)$$

it is enough to show that the expectation of (A.167) is zero for all i. We shall prove this by showing that the conditional expectation of (A.167) given Z_i is zero. In the conditional expectation of this product, the first factor can be taken out of the expectation sign since it depends only on Z_i, so that it is finally only necessary to show that the conditional expectation of $T-T^*$ given Z_i is zero. Now

$$E[(T - T^*)|Z_i] = E\{[T - r_i(Z_i) - \sum_{j \neq i} r_j(Z_j)]|Z_i\}$$

From the definition of $r_i(Z_i)$, it is seen that the conditional expectation of $T-r_i(Z_i)$ given Z_i is zero. On the other hand, since Z_i and Z_j are independent, the conditional expectation of $r_j(Z_j)$ given Z_i is equal to the unconditional expectation of $r_j(Z_j)$, which by the definition of r_j is equal to $E(T)$ and hence equal to zero. This completes the proof of (A.166) and therefore of (A.165).

A useful special case of (A.165) is obtained by putting $S = 0$, which gives after rearrangement

(A.168) $$E(T-T^*)^2 = E(T^2)-E(T^{*2}) = \text{Var}(T)-\text{Var}(T^*)$$

Theorem 8. If T satisfies (A.161) and S is of the form (A.162), then a necessary and sufficient condition for S to minimize $E(T-S)^2$ is that for all i

(A.169) $$k_i(Z_i) = r_i(Z_i) \qquad \text{with probability 1}$$

Proof. The sufficiency part of the theorem, which is the part required for applications, is obvious from (A.164) since the right-hand side of (A.165) is minimized for $S = T^*$. For the necessity part, see Prob. 34.

Let us now apply Theorem 8 to the U-statistic (A.159). Put

(A.170) $$E(U) = E[\varphi(X,Y)] = \theta$$

and

(A.171) $$\psi(x,y) = \varphi(x,y) - \theta$$

To compute the projection of

(A.172) $$mn[U - E(U)] = \sum\sum \psi(X_\alpha, Y_\beta)$$

onto the space of functions (A.160), consider first the conditional expectation of (A.172) given $X_i = x_i$. Since $E\psi(X, Y) = 0$, it follows that

$$E[\psi(X_\alpha, Y_\beta)|X_i = x_i] = \begin{cases} E[\psi(x_i, Y_\beta)] & \text{if } \alpha = i \\ 0 & \text{if } \alpha \neq i \end{cases}$$

where in $E\psi(x, Y)$ the value of x is fixed and the expectation is taken with respect to Y. Similarly,

$$E[\psi(X_\alpha, Y_\beta)|Y_j = y_j] = \begin{cases} E[\psi(X_\alpha, y_j)] & \text{if } \beta = j \\ 0 & \text{if } \beta \neq j \end{cases}$$

Putting

$$\psi_{10}(x) = E\psi(x, Y) \qquad \text{and} \qquad \psi_{01}(y) = E\psi(X, y)$$

we therefore see that the expectation of (A.172) given $X_i = x_i$ is $n\psi_{10}(x_i)$ and the expectation of (A.172) given $Y_j = y_j$ is $m\psi_{01}(y_j)$. The desired projection of (A.172) is therefore

(A.173) $$n \sum_{i=1}^{m} \psi_{10}(X_i) + m \sum_{j=1}^{n} \psi_{01}(Y_j)$$

We are now in a position to state a limit theorem for U. It involves the variances

(A.174) $$\sigma_{10}{}^2 = \text{Var}[\psi_{10}(X_i)] \qquad \text{and} \qquad \sigma_{01}{}^2 = \text{Var}[\psi_{01}(Y_j)]$$

Theorem 9. Let U be defined by (A.159), and assume without loss of generality that

(A.175) $$m \leqslant n \qquad \text{and} \qquad m/n \to \lambda \qquad \text{as } m \text{ and } n \to \infty$$

where λ may be zero. Then

(A.176) $$T = \sqrt{m}(U - \theta)$$

is asymptotically normal with mean zero and variance

(A.177) $$\sigma^2 = \sigma_{10}{}^2 + \lambda\sigma_{01}{}^2$$

Proof. (i) By (A.173) the projection T^* of T is

(A.178) $$T^* = \sqrt{m}\left[\frac{1}{m}\sum_{i=1}^{m} \psi_{10}(X_i) + \frac{1}{n}\sum_{j=1}^{n} \psi_{01}(Y_j)\right]$$

and we shall now show that the expected squared difference (A.168) tends to zero as m and n tend to infinity. The variance of T^* is by (A.178)

(A.179) $$\text{Var}(T^*) = \sigma_{10}{}^2 + \frac{m}{n}\sigma_{01}{}^2 \to \sigma^2$$

On the other hand, the variance of T is given by (A.58) multiplied by $m/m^2n^2 = 1/mn^2$. Thus

(A.180) $$\text{Var}(T) \to \text{Cov}(U_{ij}, U_{il}) + \lambda\,\text{Cov}(U_{ij}, U_{kj})$$

where $U_{ij} = \varphi(X_i, Y_j)$. To determine

$$\mathrm{Cov}(U_{ij}, U_{il}) = E[\psi(X_i, Y_j)\psi(X_i, Y_l)]$$

compute first the conditional expectation of the right-hand side given $X_i = x_i$, which is

$$E[\psi(x_i, Y_j)\psi(x_i, Y_l)] = \psi_{10}{}^2(x_i)$$

It follows that

$$\mathrm{Cov}(U_{ij}, U_{il}) = \sigma_{10}{}^2 \tag{A.181}$$

and similarly that

$$\mathrm{Cov}(U_{ij}, U_{kj}) = \sigma_{01}{}^2 \tag{A.182}$$

so that

$$\mathrm{Var}(T) \to \sigma^2 \tag{A.183}$$

and the difference (A.168) tends to zero, as was to be shown.

(ii) To complete the proof, it is only necessary to find the limiting distribution of T^*. Now by Theorem 2, the distributions of $\Sigma\psi_{10}(X_i)/\sqrt{m}$ and $\Sigma\psi_{01}(Y_j)/\sqrt{n}$ tend to normal distributions with mean zero and variances $\sigma_{10}{}^2$ and $\sigma_{01}{}^2$, respectively. Then by (A.94) and Theorem 16 below, it follows that T^* tends to the normal distribution with mean zero and variance σ^2, and this completes the proof.

Corollary 5. Under the assumptions of Theorem 9, the distribution of

$$U^* = \frac{U - E(U)}{\sqrt{\mathrm{Var}(U)}}$$

tends to the standard normal distribution provided one of the following two conditions holds:

$$\lambda = 0 \quad \text{and} \quad \sigma_{10}{}^2 > 0 \tag{A.184}$$

or

$$\lambda > 0 \quad \text{and at least one of } \sigma_{10}{}^2 \text{ and } \sigma_{01}{}^2 \text{ is positive} \tag{A.185}$$

Proof. It follows from (A.183) that the distribution of U^* tends to the standard normal distribution provided the limiting variance σ^2 is positive. Condition (A.184) ensures that $\sigma^2 > 0$ if m tends to infinity more slowly than n so that $\lambda = 0$; condition (A.185) ensures that $\sigma^2 > 0$ if m and n tend to infinity at the same rate so that $\lambda > 0$.

EXAMPLE 20. Let $\varphi(x, y)$ be given by (A.59) so that mnU is the Wilcoxon rank-sum statistic in its Mann-Whitney form as given by (1.24). To prove asymptotic normality of U or W_{XY}, it is, by Corollary 5, only necessary to check condition (A.184) or (A.185). From (A.64) and (A.181) it is seen that

$$\sigma_{10}{}^2 = p_2 - p_1{}^2 \tag{A.186}$$

and from (A.65) and (A.182) that

$$\sigma_{01}{}^2 = p_3 - p_1{}^2 \tag{A.187}$$

where p_1, p_2, and p_3 are defined by (A.60), (A.63), and (A.66), respectively.

Let us now consider the meaning of the condition

(A.188) $$\sigma_{01}^2 = 0$$

under the additional assumption that F and G are continuous. It follows from (A.60) and (A.66) that p_1 and p_3 can be written as

$$p_1 = E[F(Y)] \quad \text{and} \quad p_3 = E[F^2(Y)]$$

so that

(A.189) $$\sigma_{01}^2 = \text{Var}[F(Y)]$$

This is zero if and only if the random variable $F(Y)$ is a constant with probability 1, say, if

(A.190) $$P[F(Y) = c] = 1$$

For any continuous cumulative distribution function F, the set of y-values for which $F(y) = c$ is a closed interval, say, $a \leqslant y \leqslant b$. Under condition (A.190) the probability is zero that Y takes on values outside of $[a,b]$ [where $F(y) \neq c$] so that (A.190) implies

$$P(a \leqslant Y \leqslant b) = 1$$

On the other hand, since $F(a) = F(b)$, we have $P(a \leqslant X \leqslant b) = 0$. Thus, condition (A.190) implies the existence of constants $a \leqslant b$ such that

(A.191) $$P(a \leqslant Y \leqslant b) = 1 \quad \text{and} \quad P(X < a) + P(X > b) = 1$$

Since the distribution of Y is continuous, it follows that $a < b$. Conversely, (A.191) implies that (A.190) holds, with $c = P(X < a)$. Hence $\sigma_{01}^2 = 0$ if and only if there exist constants $a < b$ such that (A.191) holds.

Analogously, the condition $\sigma_{10}^2 = 0$ is equivalent to the existence of constants $a' < b'$ such that

(A.192) $$P(a' \leqslant X \leqslant b') = 1 \quad \text{and} \quad P(Y < a') + P(Y > b') = 1$$

Finally, the two conditions (A.191) and (A.192) together mean that either $P(X < Y) = 1$ or $P(X > Y) = 1$, i.e., at least one of σ_{01}^2 and σ_{10}^2 is positive provided

(A.193) $$0 < P(X < Y) < 1$$

The next result is quite similar in nature to Theorem 9, but concerns only a single sample. Let $X_1, \ldots, X_N$ be independent and identically distributed and let

(A.194) $$U = \frac{1}{\binom{N}{2}} \sum_{i<j} \varphi(X_i, X_j)$$

where the function φ is symmetric in its two arguments, that is, it satisfies

$$\varphi(x_i, x_j) = \varphi(x_j, x_i)$$

for all x_i, x_j. Then we shall prove that U is asymptotically normal as $N \to \infty$.

We begin again by obtaining the projection of U through Theorem 8. Put

$$E(U) = E\varphi(X_i, X_j) = \theta$$

and

$$\psi(x_i,x_j) = \varphi(x_i,x_j) - \theta$$

To compute the projection of

(A.195) $$\binom{N}{2}[U - E(U)] = \sum_{\alpha<\beta} \psi(X_\alpha, X_\beta)$$

onto the space of functions $S = \sum k_i(X_i)$, note that the conditional expectation of $\psi(X_\alpha,X_\beta)$ given $X_i = x_i$ is

$$E[\psi(X_\alpha,X_\beta)|X_i = x_i] = \begin{cases} E\psi(x_i,X_\beta) & \text{if } \alpha = i \\ E\psi(X_\alpha,x_i) & \text{if } \beta = i \\ 0 & \text{otherwise} \end{cases}$$

Putting

$$\psi_1(x) = E\psi(X_\alpha,x) = E\psi(x,X_\beta)$$

we see that the expectation of (A.195) given $X_i = x_i$ is

$$(N-1)\psi_1(x_i)$$

and the desired projection is therefore

(A.196) $$(N-1)\sum_{i=1}^{N} \psi_1(X_i)$$

We can now state a limit theorem for U; it involves the variance

(A.197) $$\text{Var}\,[\psi_1(X)] = \sigma_1^2$$

Theorem 10. Let U be defined by (A.194).

(*a*) Then

(A.198) $$T = \sqrt{N}(U-\theta)$$

is asymptotically normal with mean zero and variance $4\sigma_1^2$.

(*b*) If $\sigma_1^2 > 0$, the standardized variable

$$U^* = \frac{U - E(U)}{\sqrt{\text{Var}(U)}}$$

is asymptotically normal with mean zero and variance one.

Proof. (i) By (A.196), the projection T^* of T is

(A.199) $$T^* = \frac{2}{\sqrt{N}}\sum_{i=1}^{N} \psi_1(X_i)$$

and we shall now show that the expected squared difference (A.168) tends to zero as $N \to \infty$. The variance of T^* is by (A.199)

(A.200) $$\text{Var}\,(T^*) = 4\sigma_1^2$$

On the other hand, the variance of T is given by (A.70) multiplied by $N\Big/\binom{N}{2}^2 = 4/N(N-1)^2$.

Thus

$$\text{(A.201)} \qquad \operatorname{Var}(T) \to 4 \operatorname{Cov}(V_{ij}, V_{il})$$

where $V_{ij} = \varphi(X_i, X_j)$. To determine

$$\operatorname{Cov}(V_{ij}, V_{il}) = E[\psi(X_i, X_j), \psi(X_i, X_l)]$$

condition first on X_i and then take the expected value of this conditional expectation. In this way, it is seen as in (A.181) that

$$\text{(A.202)} \qquad \operatorname{Cov}(V_{ij}, V_{il}) = \sigma_1^{\,2}$$

so that

$$\text{(A.203)} \qquad \operatorname{Var}(T) \to 4\sigma_1^{\,2}$$

and the difference (A.168) tends to zero as was to be shown.

(ii) To complete the proof of (*a*), it is only necessary to note that by Theorem 2, the distribution of T^* tends to the normal distribution with zero mean and variance $4\sigma_1^{\,2}$. The result now follows from Corollary 1. Finally, statement (*b*) is an immediate consequence of (*a*) and (A.203).

EXAMPLE 21. Let

$$\varphi(x_i, x_j) = \begin{cases} 1 & \text{if } x_i + x_j > 0 \\ 0 & \text{otherwise} \end{cases}$$

so that $\binom{N}{2}U$ is the statistic V' of Example 6, with the random variables $Z_1, \ldots, Z_N$ now denoted by $X_1, \ldots, X_N$. To prove asymptotic normality of U or V', it is by Theorem 10(*b*) only necessary to check that $\sigma_1^{\,2} > 0$. From (A.81) and (A.202) it is seen that

$$\text{(A.204)} \qquad \sigma_1^{\,2} = p_2' - p_1'^2$$

where p_1' and p_2' are defined by (A.78) and (A.82). We shall discuss the meaning of the condition

$$\text{(A.205)} \qquad \sigma_1^{\,2} = 0$$

only for the case that the common distribution L of $X_1, \ldots, X_N$ is continuous. Writing p_1' and p_2' as

$$p_1' = P(X' > -X) \qquad \text{and} \qquad p_2' = P(X' > -X \quad \text{and} \quad X'' > -X)$$

we see that p_1' and p_2' are equal to the probabilities p_1 and p_2 of Example 20 if the distribution L of X is identified with G and the distribution L^* of $-X$ with F. It follows from (A.192) that $p_2' - p_1'^2 = 0$ if and only if there exist constants $a' < b'$ such that

$$P(a' \leqslant -X \leqslant b') = 1 \qquad \text{and} \qquad P(X < a') + P(X > b') = 1$$

It is easy to see (Prob. 39) that the first of these conditions implies the second if either a' is positive or b' negative, but that the two conditions are incompatible if $a' < 0 < b'$. Thus $p_2' - p_1'^2 = 0$ is equivalent to the condition that $P(X < 0) = 0$ or 1, and we have therefore proved asymptotic normality of V' under the assumptions that L is continuous and that

$$\text{(A.206)} \qquad 0 < P(X < 0) < 1$$

The statistic of principal interest in Chaps. 3 and 4 is not V' but

$$V = V' + S_N$$

where S_N is the number of positive X's. The asymptotic normality of the standardized variable V^* will be established—in view of Corollary 2 and the asymptotic normality of V'^*—if we can show that

$$\frac{E(V-V')^2}{\operatorname{Var}(V')} = \frac{E(S_N{}^2)}{\operatorname{Var}(V')} \to 0$$

From the fact that S_N has the binomial distribution $b(p,N)$ where $p = P(X_i > 0)$, it follows that

$$E(S_N{}^2) = Npq + N^2p^2$$

On the other hand,

$$\operatorname{Var}(V') = \binom{N}{2}^2 \operatorname{Var}(U) = \frac{N(N-1)^2}{4} N \operatorname{Var}(U)$$

Since $N \operatorname{Var}(U) \to 4\sigma_1{}^2$ by (A.203), it follows that the distribution of V^* tends to the standard normal distribution provided (A.206) holds.

Theorem 10 easily generalizes to statistics of the form

$$U = \frac{1}{\binom{N}{m}} \sum \varphi(X_{i_1}, \ldots, X_{i_m}) \tag{A.207}$$

where the function φ is assumed to be symmetric in its m arguments, and where the summation extends over all $\binom{N}{m}$ m-tuples $1 \leqslant i_1 < i_2 < \cdots < i_m \leqslant N$.

The argument is completely analogous to that of Theorem 10, and is sketched in Prob. 41. The function ψ is again defined as φ minus its expectation, ψ_1 is now given by

$$\psi_1(x_1) = E[\psi(x_1, X_2, \ldots, X_m)] \tag{A.208}$$

and $\sigma_1{}^2$ is defined by (A.197) as before. Then the distribution of the standardized variable $U^* = [U - E(U)]/\sqrt{\operatorname{Var}(U)}$ tends to the standard normal distribution as $N \to \infty$ provided $\sigma_1{}^2 > 0$.

EXAMPLE 22. The random variables $X_1, \ldots, X_N$ in Theorem 10 and in the extension just stated need not be univariate (as they were in Example 21), and we shall now apply the extended theorem with $m = 3$ to a situation in which $X_1 = (Y_1, Z_1), \ldots, X_N = (Y_N, Z_N)$ are a sample from a continuous bivariate distribution F. (For some general definitions relating to bivariate distributions, see Sec. 7.) If $R_1, \ldots, R_N$ and $S_1, \ldots, S_N$ denote the ranks of the Y's and Z's, respectively, let r_s be the rank correlation coefficient defined in Chap. 7, Sec. 3 by

$$r_s = \frac{12}{N^3 - N} \sum [R_i - \tfrac{1}{2}(N+1)][S_i - \tfrac{1}{2}(N+1)] \tag{A.209}$$

To relate this to a U-statistic, it is convenient to introduce the functions

$$c(u) = \begin{cases} 0 & \text{if } u < 0 \\ \frac{1}{2} & \text{if } u = 0 \\ 1 & \text{if } u > 0 \end{cases}$$

and

$$s(u) = \begin{cases} -1 & \text{if } u < 0 \\ 0 & \text{if } u = 0 \\ 1 & \text{if } u > 0 \end{cases}$$

which are connected by the equation

$$c(u) = \tfrac{1}{2}s(u) + \tfrac{1}{2}$$

Consider now the variable Y_i. Since its rank is R_i, there are $R_i - 1$ among the Y's that are smaller than Y_i, one that is equal to Y_i, and $N - R_i$ that are larger than Y_i. Hence

$$R_i = \sum_{j=1}^{N} c(Y_i - Y_j) + \tfrac{1}{2}$$

or equivalently

$$R_i = \tfrac{1}{2} \sum_{j=1}^{N} s(Y_i - Y_j) + \tfrac{1}{2}(N+1) \tag{A.210}$$

The rank correlation coefficient can therefore be written as

$$r_s = \frac{3}{N^3 - N} \sum_{i=1}^{N} \sum_{j=1}^{N} \sum_{k=1}^{N} s(Y_i - Y_j)\, s(Z_i - Z_k) \tag{A.211}$$

In order to prove asymptotic normality of r_s, let us introduce the closely related statistic

$$U = \frac{1}{\binom{N}{3}} \sum s(Y_i - Y_j)\, s(Z_i - Z_k) \tag{A.212}$$

where the summation extends not over all N^3 triples of subscripts as before, but only over the $N(N-1)(N-2)$ triples of distinct subscripts. The statistic (A.212) is a U-statistic of the form (A.207) with $m = 3$; the function φ is given by

$$\begin{aligned} \varphi(x_i, x_j, x_k) = {} & s(y_i - y_j)\, s(z_i - z_k) + s(y_i - y_k)\, s(z_i - z_j) + s(y_j - y_i)\, s(z_j - z_k) \\ & + s(y_j - y_k)\, s(z_j - z_i) + s(y_k - y_i)\, s(z_k - z_j) + s(y_k - y_j)\, s(z_k - z_i) \end{aligned}$$

which has the required symmetry. It follows from the extension of Theorem 10 that U is asymptotically normal provided ${\sigma_1}^2 > 0$. This condition is rather complicated, and we shall give no further analysis of it here.

Let us now return to the statistic r_s. From (A.211) and (A.212) it is seen that, after some simplification,

$$2r_s = \frac{N-2}{N+1} U + \frac{6}{N^3 - N} D \tag{A.213}$$

and hence that

(A.214) $$2[r_s - E(r_s)] - [U - E(U)] = \frac{6}{N^3 - N}[D - E(D)] - \frac{3}{N+1}[U - E(U)]$$

where

$$D = \sum s(Y_i - Y_j)\, s(Z_i - Z_k)$$

summed over all

$$N^3 - N(N-1)(N-2) = 3N^2 - 2N$$

triples $1 \leqslant i, j, k \leqslant N$ that are not distinct. From the asymptotic normality of U and Corollary 2 it will then follow that the distribution of the standardized variable r_s^* tends to the standard normal distribution, if we can show that the expected square of the right-hand side of (A.214) divided by $\mathrm{Var}(U)$ tends to zero as N tends to infinity. This is, however, an easy consequence of the facts that (i) $(a-b)^2 \leqslant 2(a^2+b^2)$ for any a,b; (ii) $N\,\mathrm{Var}(U)$ is bounded away from 0 by assumption; (iii) $N\,\mathrm{Var}(D)/(N^3-N)^2 \to 0$ as $N \to \infty$. Only the third of these statements requires proof. Since $|s(u)| \leqslant 1$ for all u, $D - E(D)$ is the sum of $3N^2 - 2N < 3N^2$ terms each of which is $\leqslant 2$ in absolute value, and the third statement follows.

This completes the proof of the asymptotic normality of r_s provided ${\sigma_1}^2 > 0$.

6. PITMAN EFFICIENCY

The asymptotic normality of U-statistics, established in the preceding section, forms the basis for the normal approximation to the power of the one- and two-sample Wilcoxon tests against a given alternative. For certain applications. this approximation has the disadvantage that the power against a fixed alternative of interest typically tends to 1 as the sample sizes tend to infinity. This fact suggests the possibility of obtaining another useful approximation by fixing not the alternative but the value of the power being approximated. The resulting approximation tends to be simpler than the earlier one but less accurate. It is being considered here primarily as a basis for determining Pitman efficiencies.

Without restricting attention to U-statistics, consider a sequence of statistics T_k, $k = 1, 2, \ldots$, based on N_k observations with N_k tending to infinity as $k \to \infty$. The distribution of T_k is assumed to depend on a real-valued parameter θ, and we shall be concerned with the limiting behavior of the random variables

(A.215) $$\frac{T_k - \mu(\theta)}{\sigma_k(\theta)}$$

where $\mu(\theta)$ and $\sigma_k(\theta)$ are suitable normalizing constants. In applications, they will often but not always be the expectation and standard deviation of T_k. Note that μ is assumed to be independent of k, as will be the case for example for the expectation of the U-statistics (A.159) or (A.194).

Suppose that the hypothesis H: $\theta = \theta_0$ is being tested against the alternatives $\theta > \theta_0$ and

(1) the function μ is differentiable at θ_0 with derivative $\mu'(\theta_0) \neq 0$; and

(2) $\sigma_k(\theta_0)$ is of order $1/\sqrt{N_k}$

An important quantity in what follows is the limit

(A.216)
$$c = \lim_{k \to \infty} \frac{\mu'(\theta_0)}{\sqrt{N_k}\sigma_k(\theta_0)}$$

Let the hypothesis be rejected when

(A.217)
$$\frac{T_k - \mu(\theta_0)}{\sigma_k(\theta_0)} \geqslant C_k$$

If the distribution of the left-hand side tends to the standard normal distribution when $\theta = \theta_0$, then $C_k \to u_k$ where u_k is the upper α-point of the standard normal distribution.

As the sample size increases, the alternative θ at which the power of the test achieves a fixed value Π will change and typically will tend to the hypothetical value θ_0 as $N_k \to \infty$. We shall therefore consider sequences θ_k of alternatives for which $\theta_k \to \theta_0$ and for which we shall assume the following:

(3) the distribution of $[T_k - \mu(\theta_k)]/\sigma_k(\theta_k)$ tends to the standard normal distribution
(4) $\sigma_k(\theta_k)/\sigma_k(\theta_0) \to 1$

Consider now the power Π_k of the test (A.217) against θ_k. This can be written as

$$\Pi_k = P\left\{\frac{T_k - \mu(\theta_k)}{\sigma_k(\theta_k)} \geqslant \left[C_k - \frac{\mu(\theta_k) - \mu(\theta_0)}{\sigma_k(\theta_0)}\right]\frac{\sigma_k(\theta_0)}{\sigma_k(\theta_k)}\right\}$$

We are interested in the circumstances under which Π_k tends to a fixed value Π, where $\alpha < \Pi < 1$. Let l be the unique solution of the equation

(A.218)
$$\Phi(u_\alpha - l) = 1 - \Pi$$

Then it follows from assumptions 3 and 4 that $\Pi_k \to \Pi$ if and only if

(A.219)
$$\frac{\mu(\theta_k) - \mu(\theta_0)}{\sigma_k(\theta_0)} \to l$$

In view of assumptions 1 and 2, condition (A.219) implies that $\theta_k - \theta_0$ is of order $1/\sqrt{N_k}$, and this motivates considering alternatives of the form

(A.220)
$$\theta_k = \theta_0 + \frac{\delta}{\sqrt{N_k}} + o\left(\frac{1}{\sqrt{N_k}}\right) \qquad (\delta > 0)$$

that is, satisfying

(A.221)
$$\sqrt{N_k}(\theta_k - \theta_0) \to \delta$$

For such alternatives we have the following result.

Theorem 11. If assumptions 1 and 2 are satisfied, and if assumptions 3 and 4 hold for any alternatives θ_k satisfying (A.220), then the limit of the power Π_k of the test (A.217) against the alternatives (A.220) is given by

(A.222)
$$\Pi_k \to \Phi[c\delta - u_\alpha]$$

where c is given by (A.216). Conversely, under the same assumptions (A.222) implies (A.220).

Proof. Let θ_k satisfy (A.220) and hence (A.221). Then by (A.216) the left-hand side of

(A.219), on multiplication and division by $\sqrt{N_k}(\theta_k - \theta_0)$, is seen to tend to $l = c\delta$, and hence (A.222) follows from (A.218). The converse is seen by following the argument in the inverse order.

Theorem 11 shows that the limiting power of (A.217) against the alternatives (A.220) depends on the test only through c and is an increasing function of c when $\delta > 0$. The quantity c defined by (A.216) is called the *efficacy* of the test sequence (A.217).

EXAMPLE 23. As a first application consider the two-sample translation problem in which $X_1, \ldots, X_m$ and $Y_1, \ldots, Y_n$ are independently distributed according to distributions

$$P(X_i \leqslant x) = F(x) \qquad \text{and} \qquad P(Y_j \leqslant y) = F(y - \Delta) \tag{A.223}$$

where we shall assume that F has a density f. We shall be concerned with a sequence of sample sizes m_k and n_k satisfying

$$\frac{m_k}{N_k} \to \gamma_1 \quad , \quad \frac{n_k}{N_k} \to \gamma_2 \qquad (N_k = m_k + n_k,\ \gamma_1 + \gamma_2 = 1) \tag{A.224}$$

Let the statistic T_k of Theorem 11 be the Wilcoxon statistic U_k with $m = m_k$ and $n = n_k$, let $\mu(\Delta)$ and $\sigma_k{}^2(\Delta)$ be the expectation and variance of U_k and let $\theta = \Delta$ and $\theta_0 = 0$. To check condition 1 and at the same time compute $\mu'(0)$ note from Example 20 that

$$\mu(\Delta) = P(X < Y) = \int_{-\infty}^{\infty} F(y) f(y - \Delta)\, dy = \int F(y + \Delta) f(y)\, dy \tag{A.225}$$

Thus

$$\frac{\mu(\Delta) - \mu(0)}{\Delta} = \int_{-\infty}^{\infty} \frac{F(y + \Delta) - F(y)}{\Delta} f(y)\, dy$$

If the limit of the left-hand side exists as $\Delta \to 0$ and can be obtained by passing to the limit under the integral sign, it follows that $\mu'(0)$ exists and is given by

$$\mu'(0) = \int_{-\infty}^{\infty} f^2(y)\, dy \tag{A.226}$$

Formula (A.226) does in fact hold under mild assumptions on the density f. It was proved by Olshen (1967) and by Mehra and Sarangi (1967) under the sole condition that

$$\int_{-\infty}^{\infty} f^2(y)\, dy < \infty \tag{A.227}$$

This condition holds in particular whenever f is bounded since $0 \leqslant f(x) \leqslant M$ implies that

$$\int_{-\infty}^{\infty} f^2(x)\, dx \leqslant M \int_{-\infty}^{\infty} f(x)\, dx = M$$

A somewhat simpler proof of (A.226) under additional regularity assumptions on F is given by Hodges and Lehmann (1961).

Condition 2 of Theorem 11 follows from Example 20 and assumption (A.224), and condition 3 also follows from Example 20. Finally, condition 4 is an easy consequence of the variance formula (2.21) and Theorem 16 below. To complete the evaluation of the efficacy (A.216) of the Wilcoxon rank-sum test, it is now only necessary to find $\lim \sqrt{N_k}\sigma_k(0)$, which by (1.30) is $1/\sqrt{12\gamma_1\gamma_2}$. The desired efficacy is therefore

(A.228)
$$c = \sqrt{12\gamma_1\gamma_2}\int_{-\infty}^{\infty} f^2(y)\,dy$$

The limit relation (A.222) with c given by (A.228) provides a simple approximation to the power of the Wilcoxon test. Given a particular value of N and of the alternative Δ against which the power is to be computed, one identifies the latter with the value $\delta/\sqrt{N}$ and hence obtains $\delta = \sqrt{N}\Delta$. Substituting this value of δ, and $\gamma_1 = m/N$, $\gamma_2 = n/N$ in (A.228) leads to the approximation

(A.229)
$$\Pi \approx \Phi\left[\Delta\sqrt{\frac{12mn}{N}}\int f^2(y)\,dy - u_\alpha\right]$$

This is essentially the same as the approximation (2.29) (see Prob. 47).

EXAMPLE 24. As a second application of Theorem 11 let us now obtain the efficacy and the analogue of (A.229) for the signed-rank Wilcoxon test in the one-sample location problem. Let $Z_1, \ldots, Z_N$ be independently distributed according to a distribution

(A.230)
$$P(Z_i \leqslant z) = E(z-\Delta)$$

where E is assumed to be symmetric about zero and to have a density e. Let T_k of Theorem 11 be the statistic $U = U_k$ of Example 21 based on $N = N_k$ observations, let $\mu(\Delta)$ be its expectation and $\sigma_k{}^2(\Delta)$ its variance, and let Π_k be the power of the associated test (A.217) for testing $H: \Delta = 0$ against $\Delta > 0$ with C_k tending to u_α.

To check condition 1 and compute $\mu'(0)$, note from Example 21 that by an argument similar to that leading to (A.225) we have

(A.231)
$$\mu(\Delta) = P(Z' > -Z) = \int_{-\infty}^{\infty} E(z+2\Delta)e(z)\,dz$$

and hence

$$\frac{\mu(\Delta)-\mu(0)}{\Delta} = 2\int_{-\infty}^{\infty} \frac{E(z+2\Delta)-E(z)}{2\Delta}e(z)\,dz$$

If the limit of the left-hand side exists as $\Delta \to 0$ and can be obtained by passing to the limit under the integral sign, it follows that $\mu'(0)$ exists and is given by

(A.232)
$$\mu'(0) = 2\int_{-\infty}^{\infty} e^2(z)\,dz$$

As was discussed in the preceding example, a sufficient condition for (A.232) is that (A.227) holds with f replaced by e. Condition 2 follows from Example 21, and conditions 3 and 4 are proved as in the preceding example.

To complete the evaluation of the efficacy (A.216) it is only necessary to find $\lim \sqrt{N_k}\sigma_k(0)$, which from Example 21 can be seen to be $\sqrt{1/3}$. Thus the desired efficacy is

$$c = \sqrt{12}\int_{-\infty}^{\infty} e^2(z)\,dz \tag{A.233}$$

The statistic U considered above is not the signed-rank Wilcoxon statistic V_s, but rather the statistic V' of Example 6 so that

$$\frac{V_s - \mu(\Delta)}{\sigma_k(\Delta)} = \frac{V' - \mu(\Delta)}{\sigma_k(\Delta)} + \frac{S_N}{\sigma_k(\Delta)}$$

Since it was shown in Example 21 that $E(S_N{}^2)/\sigma_k{}^2(\Delta) \to 0$, it follows that the validity of conditions 1 to 4 and the value (A.233) of c remain unchanged if V' is replaced by V_s.

As in the preceding example, the limit relation (A.222) with c given by (A.233) provides a simple approximation to the power Π of the Wilcoxon test. For given values of N and Δ put $\delta = \sqrt{N}\Delta$ as before to find

$$\Pi \approx \Phi\left[\Delta\sqrt{12N}\int_{-\infty}^{\infty} e^2(z)\,dz - u_\alpha\right] \tag{A.234}$$

Theorem 11 provides not only a simple approximation to the power of certain tests but also the basis for obtaining the asymptotic relative efficiency of two test sequences with respect to each other. Let $T = \{T_k\}$ and $T' = \{T_k'\}$ be two sequences of tests for the same problem based on sample sizes N_k and N_k', respectively. The limiting ratio of the sample sizes N_k' and N_k required to achieve the same limiting power Π, against the same sequence of alternatives when the significance levels of the two test sequences also have the same limit, is the Pitman efficiency of T relative to T',

$$e_{T,T'} = \lim \frac{N_k'}{N_k} \tag{A.235}$$

If, for example, $e_{T,T'} = \frac{1}{2}$, this means that the sequence T' requires approximately half as many observations as the sequence T to achieve the same asymptotic results.

Theorem 12. Let T and T' be two sequences of tests for testing $H\colon \theta = \theta_0$, based on sample sizes N_k and N_k', respectively, with significance levels α_k and α_k' both tending to α, and with powers Π_k and Π_k' tending to a common limit $\alpha < \Pi < 1$ against a sequence of alternatives θ_k. If conditions 1 to 4 of Theorem 11 hold for both sequences, and if the efficacies of the two sequences are c and c', respectively, the Pitman efficiency of T relative to T' exists and is equal to

$$e_{T,T'} = \left(\frac{c}{c'}\right)^2 \tag{A.236}$$

Proof. Since Π_k tends to a limit Π between α and 1, it follows from Theorem 11 that θ_k satisfies (A.220). Further, since Π_k' tends to the same limit Π, Theorem 11 shows that $c\delta = c'\delta'$. On the other hand, by (A.221)

$$\sqrt{\frac{N_k'}{N_k}} = \frac{\sqrt{N_k'}(\theta_k - \theta_0)}{\sqrt{N_k}(\theta_k - \theta_0)} \to \frac{\delta'}{\delta}$$

and this completes the proof.

EXAMPLE 25. As an application of this theorem, let us obtain the Pitman efficiency of the Wilcoxon rank-sum test to the two-sample t-test. For this purpose, it is necessary to check conditions 1 to 4 for the t-test, and determine its efficacy. Let $X_1, \ldots, X_m$ and $Y_1, \ldots, Y_n$ be distributed as in Example 23 and let the t-test reject when

$$\frac{\bar{Y}-\bar{X}}{S\sqrt{1/m_k+1/n_k}} \geqslant C'_k \tag{A.237}$$

where

$$S^2 = \frac{\sum(X_i-\bar{X})^2+\sum(Y_j-\bar{Y})^2}{N_k-2} \tag{A.238}$$

and where $C'_k \to u_\alpha$. Suppose that F has finite variance σ^2 and let us apply Theorem 11 to the statistic

$$T = \frac{\bar{Y}-\bar{X}}{S}$$

with $\mu(\Delta) = \Delta/\sigma$ and $\sigma_k(\Delta) = \sqrt{1/m_k+1/n_k}$ (independent of Δ). (Note that these are not the expectation and standard deviation of T.) Then conditions 1, 2, and 4 are trivially satisfied. To verify condition 3, we must show that

$$\frac{\bar{Y}-\bar{X}}{S\sqrt{1/m_k+1/n_k}} - \frac{\Delta_k}{\sigma\sqrt{1/m_k+1/n_k}}$$

tends in law to the standard normal distribution, where $\sqrt{N_k}\Delta_k$ tends to some finite positive limit δ by (A.220). Since

$$\frac{\bar{Y}-\bar{X}-\Delta_k}{S\sqrt{1/m_k+1/n_k}}$$

tends in law to $N(0,1)$, it is only necessary to show that the difference

$$\frac{\Delta_k}{\sqrt{1/m_k+1/n_k}}\left(\frac{1}{S}-\frac{1}{\sigma}\right)$$

between these two random variables tends to zero in probability. The first factor tends to $\delta/\sqrt{1/\gamma_1+1/\gamma_2}$ by (A.224) and $1/S-1/\sigma$ tends to zero in probability. This completes the proof and shows the efficacy of the t-test to be

$$c = \frac{1}{\sigma\sqrt{1/\gamma_1+1/\gamma_2}} = \frac{\sqrt{\gamma_1\gamma_2}}{\sigma} \tag{A.239}$$

Combining (A.239) with the efficacy (A.228) of the Wilcoxon test, we see from Theorem 12 that the Pitman efficiency of the Wilcoxon rank-sum test to Student's t-test in the two-sample location problem exists and is given by

$$e_{W,t}(F) = 12\sigma^2\left[\int f^2(x)\,dx\right]^2 \tag{A.240}$$

It is a remarkable fact that this efficiency does not depend on the values of α and Π.

As an example, consider the case that f is the standard normal density. Then $\sigma^2 = 1$ and

$$\int f^2(x)\,dx = \frac{1}{2\pi}\int e^{-x^2}\,dx$$

Writing

$$e^{-x^2} = e^{-\frac{1}{2}x^2/\frac{1}{2}}$$

it is seen from the fact that the integral of the normal density with variance $\frac{1}{2}$ is 1 that

$$\int f^2(x)\,dx = \frac{1}{2\pi}\sqrt{2\pi\tfrac{1}{2}} = \frac{1}{2\sqrt{\pi}}$$

The Pitman efficiency (A.240) in this case is therefore

$$e_{W,t}(\Phi) = \frac{12}{4\pi} = \frac{3}{\pi} \approx .955$$

Since $e_{W,t}(F)$ is independent of the choice of location and scale (Prob. 48), this is also the value of the efficiency for any other normal distribution. The value of (A.240) is easily obtained also, for example, for the double-exponential, logistic, uniform, and exponential distributions (see Prob. 49), and in each of these cases is found to be greater than or equal to 1. (Of course, a high value of the efficiency $e_{W,t}(F)$ does not mean that the Wilcoxon test is a good test of the hypothesis $\Delta = 0$ when F is the true distribution but only that it is good relative to the t-test.) Quite generally, it can be shown that

$$e_{W,t}(F) \geqslant .864 \tag{A.241}$$

for all F with finite variance. (Without the assumption of finite variance the comparison of the two tests becomes meaningless since the asymptotic significance level of the t-test is then in doubt.) We shall now prove (A.241) under the additional restriction that F has a density f satisfying (A.227).†

Since $e_{W,t}(F)$ is independent of location and scale, suppose without loss of generality that F has expectation 0 and variance 1. Subject to these conditions, let us minimize $e_{W,t}(F)$, or equivalently,

$$\int f^2(x)\,dx \tag{A.242}$$

Since f is a probability density, the problem becomes that of minimizing (A.242) subject to the restrictions

$$\int f(x)\,dx = 1 \quad , \quad \int xf(x)\,dx = 0 \quad , \quad \int x^2 f(x)\,dx = 1 \tag{A.243}$$

as well as

$$f(x) \geqslant 0 \qquad \text{for all } x \tag{A.244}$$

Using the method of undetermined multipliers (Prob. 53), we shall instead minimize

$$\int [f^2(x) + 2ax^2 f(x) + 2bf(x)]\,dx \tag{A.245}$$

subject to (A.244), and then determine a and b so as to satisfy (A.243). Here (A.245) should really also include a term $2cxf(x)$ corresponding to the middle condition of (A.243). However, it will turn out that the density f minimizing (A.245) will automatically satisfy this middle condition, so that the associated multiplier c can be taken to be zero.

† If this assumption is not satisfied, it can be shown that $e_{W,t}(F)$ is infinite and hence certainly satisfies (A.241). See Hodges and Lehmann (1956), "The Efficiency of Some Nonparametric Competitors of the t-test," *Ann. Math. Statist.* **27**:324–355.

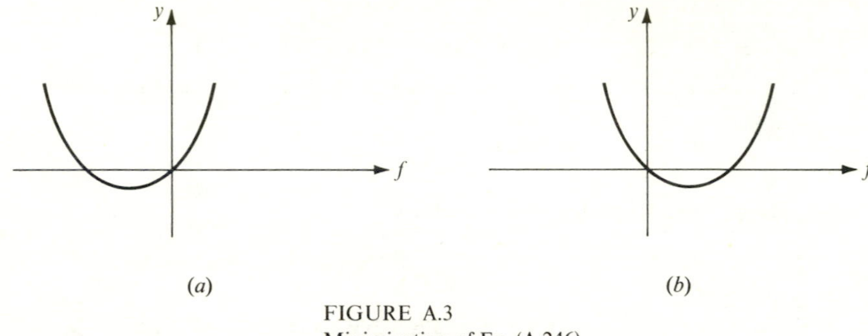

FIGURE A.3
Minimization of Eq. (A.246).

The integral (A.245) is minimized, subject to (A.244), by minimizing the integrand for each x. Writing f for $f(x)$, the problem therefore reduces to minimizing, for each x,

(A.246) $$y = f^2 + 2ax^2f + 2bf$$

subject to $f \geqslant 0$. The quantity y defined by (A.246) is a quadratic function of f, which has its minimum value at $f = -(ax^2+b)$. Depending on the value of x, one must now distinguish two cases. These correspond to the possibilities of $-(ax^2+b)$ being positive or negative, as shown in Fig. A.3.

In case (b), the minimizing f is positive and, since it satisfies the restriction $f \geqslant 0$, is therefore the desired solution. On the other hand, in case (a) the minimizing f is negative and hence inadmissible. In this case, it is seen from the figure that y attains its minimum value, subject to $f \geqslant 0$, at $f = 0$. We have therefore shown that y is minimized subject to $f \geqslant 0$ by

(A.247) $$f(x) = \begin{cases} -(ax^2+b) & \text{if } ax^2+b<0 \\ 0 & \text{if } ax^2+b>0 \end{cases}$$

It now remains to check that a and b can be found so that the function f defined by (A.247) satisfies conditions (A.243). It is easy to see (Prob. 54) that (A.247) cannot satisfy these conditions unless $a > 0$ and $b < 0$, in which case (A.247) can be rewritten as

(A.248) $$f(x) = \begin{cases} A(B^2-x^2) & \text{if } |x|<B \\ 0 & \text{otherwise} \end{cases}$$

A straightforward calculation (Prob. 55) shows that this function satisfies (A.243) for $B = \sqrt{5}$ and $A = 3/20\sqrt{5}$ and that with these values (A.242) is equal to $3/5\sqrt{5}$, and hence (A.240) is equal to $108/125 = .864$, as was to be proved.

EXAMPLE 26. As another application of Theorem 12 let us determine the Pitman efficiency of the Wilcoxon signed-rank test to the one-sample t-test. This requires checking conditions 1 to 4 for the t-test, and determining its efficacy. Let $Z_1, \ldots, Z_N$ be distributed as in Example 24, and let the t-test reject when

(A.249) $$\frac{\sqrt{N_k}\bar{Z}}{S} \geqslant c'_k$$

where $S^2 = \sum (Z_i - \bar{Z})^2/(N_k - 1)$ and where $c'_k \to u_\alpha$. Then Theorem 11 applies exactly as in the preceding example to the statistic $T_k = \bar{Z}/S$, with $\mu(\Delta) = \Delta/\tau$ and $\sigma_k(\Delta) = 1/\sqrt{N_k}$, where τ^2 is the variance, assumed to be finite, of the Z's. The checking of conditions 1 to 4 now proceeds as in the preceding example and leads to the efficacy, for the one-sample t-test

$$c = \frac{1}{\tau} \tag{A.250}$$

Combining this with the efficacy (A.233) of the signed-rank Wilcoxon test, we see from Theorem 12 that the Pitman efficiency of the Wilcoxon signed-rank test to the one-sample t-test is

$$e_{V,t}(E) = 12\tau^2\left[\int_{-\infty}^{\infty} e^2(z)\,dz\right]^2 \tag{A.251}$$

Since this is just the efficiency (A.240) with E and τ^2 in place of F and σ^2, the earlier results concerning this efficiency continue to hold. In particular, its value is $3/\pi \approx .955$ if E is normal, and the efficiency is greater than or equal to .864 for all E with finite variance.

EXAMPLE 27. As a last application of Theorem 12, let us obtain for the one-sample location problem of Examples 24 and 26 the Pitman efficiency of the sign test relative to the t-test and Wilcoxon test. The sign test based on N_k observations rejects when

$$\frac{S_k - \frac{1}{2}N_k}{\frac{1}{2}\sqrt{N_k}} \geqslant c_k \tag{A.252}$$

where S_k is the number of Z's that are greater than zero. Let us first check conditions 1 to 4 of Theorem 11 with $T_k = S_k/N_k$ and with $\mu(\Delta)$ and $\sigma_k{}^2(\Delta)$ given by the expectation and variance of T_k. To check condition 1 and determine $\mu'(0)$, note that

$$\mu(\Delta) = p = P(Z > 0) = 1 - E(-\Delta) = E(\Delta)$$

so that $\mu'(0) = e(0)$. On the other hand, $\sigma_k{}^2(\Delta) = pq/N_k$ and, since $p = \frac{1}{2}$ when $\Delta = 0$, it follows that $\lim \sqrt{N_k}\sigma_k(0) = \frac{1}{2}$. Conditions 3 and 4 are obviously satisfied, and the efficacy of the sign test is therefore

$$c = 2e(0) \tag{A.253}$$

Combining this with the efficacy (A.250) of the t-test, we find the Pitman efficiency of the sign to the t-test to be

$$e_{s,t}(E) = 4\tau^2 e^2(0) \tag{A.254}$$

In the particular case that E is the standard normal distribution, $\tau^2 = 1$ and $e(0) = 1/\sqrt{2\pi}$, so that

$$e_{s,t}(\Phi) = \frac{2}{\pi} \approx .637$$

Since the efficiency (A.254) is again independent of scale (Prob. 56), this same value applies to any normal distribution with zero mean. The implications of this result are discussed in Chap. 4, Sec. 3.

A similar comparison of (A.253) with the efficacy (A.233) of the Wilcoxon test shows that the Pitman efficiency of the sign to the Wilcoxon test is given by

(A.255) $$e_{s,V}(E) = \frac{e^2(0)}{3[\int e^2(z)\,dz]^2}$$

7. SOME MULTIVARIATE DISTRIBUTIONS

The remainder of the appendix will be concerned with extensions of the results of Secs. 2 to 4 to the joint distribution of several variables. We begin by considering multivariate generalizations of some of the distributions discussed in Sec. 2.

7A. The Multinomial Distribution

Consider n independent trials, each having r possible outcomes, say, $O_1, \ldots, O_r$ (for example, n throws with a die, each throw resulting in one of the six outcomes "one point," . . ., "six points"). If the probabilities of a trial resulting in outcomes $O_1, \ldots, O_r$ are the same for all of the trials, say,

$$P(O_i) = p_i \qquad (p_1 + \cdots + p_r = 1)$$

the trials are said to constitute a sequence of *multinomial trials.* Let X_i denote the number of trials in such a sequence that result in outcome O_i. Then the joint distribution of $X_1, \ldots, X_r$, the *multinomial distribution* $M(p_1, \ldots, p_r; n)$, is given by

(A.256) $$P(X_1 = x_1, \ldots, X_r = x_r) = \binom{n}{x_1, \ldots, x_r} p_1^{x_1} \ldots p_r^{x_r}$$

where $(x_1, \ldots, x_r)$ is any set of r nonnegative integers adding up to n and where the coefficient on the right-hand side is the *multinomial coefficient*

(A.257) $$\binom{n}{x_1, \ldots, x_r} = \frac{n!}{x_1! \cdots x_r!}$$

This coefficient is just the numbers of ways of dividing n items into r groups of sizes $x_1, \ldots, x_r$.

To obtain the marginal distribution of X_i, consider a trial to be a success if O_i occurs and a failure if any one of the remaining $r-1$ outcomes occurs. Then X_i is seen to be the number of successes in n binomial trials with success probability p_i and hence has the binomial distribution $b(p_i, n)$. In particular, the expectation and variance of X_i are therefore given by (A.87) with p_i in place of p.

The covariance of X_i and X_j can be obtained by putting

$$X_i = I_1 + \cdots + I_n \qquad \text{and} \qquad X_j = J_1 + \cdots + J_n$$

where

$$I_k = \begin{cases} 1 & \text{if the } k\text{th trial results in outcome } O_i \\ 0 & \text{otherwise} \end{cases}$$

and

$$J_k = \begin{cases} 1 & \text{if the } k\text{th trial results in outcome } O_j \\ 0 & \text{otherwise} \end{cases}$$

Then

$$E(I_k J_l) = \begin{cases} p_i p_j & \text{if } k \neq l \\ 0 & \text{if } k = l \end{cases}$$

and hence

$$\begin{aligned} \operatorname{Cov}(X_i, X_j) &= \sum\sum [E(I_k J_l) - p_i p_j] \\ &= n(0 - p_i p_j) + n(n-1) \times 0 \end{aligned}$$

so that

(A.258) $$\operatorname{Cov}(X_i, X_j) = -np_i p_j \qquad (i \neq j)$$

7B. The Multiple Hypergeometric Distribution

As in the case of the hypergeometric distribution, suppose that a finite population consists of N items, but assume now that each item belongs to one of c exclusive and exhaustive categories instead of belonging to just two such categories. (Students in a class, for example, might be classified as being in their first, second, third, or fourth year, or as being graduate students.) Suppose that d_i of the N items belong to the ith category ($d_1 + \cdots + d_c = N$), and that a random sample of size m is drawn from the population. If $X_1, \ldots, X_c$ denote the numbers of items in the sample belonging to categories $1, \ldots, c$ (so that $X_1 + \cdots + X_c = m$), the joint distribution of the X's is the *multiple hypergeometric distribution*

(A.259) $$P(X_1 = x_1, \ldots, X_c = x_c) = \binom{d_1}{x_1} \cdots \binom{d_c}{x_c} \bigg/ \binom{N}{m}$$

To see this, one need only note that $\binom{d_i}{x_i}$ is the number of ways of selecting x_i items from the d_i items of category i. An alternative expression for this distribution is

(A.260) $$P(X_1 = x_1, \ldots, X_c = x_c) = \frac{\binom{m}{x_1, \ldots, x_c}\binom{n}{y_1, \ldots, y_c}}{\binom{N}{d_1, \ldots, d_c}}$$

where $n = N - m$ and $y_i = d_i - x_i$. This is seen by writing the right-hand side of (A.259) as

$$\frac{d_1! \ldots d_c! m! (N-m)!}{x_1!(d_1 - x_1)! \ldots x_c!(d_c - x_c)! N!}$$

and comparing this expression with (A.260), using (A.257).

To see another way in which this distribution arises, suppose that each of the N items belongs to one of two exclusive and exhaustive categories A and B, with m belonging to A and n to B. Suppose successive samples of sizes $d_1, \ldots, d_c$ ($\sum d_i = N$) are drawn, and let X_i denote the number of items in the ith sample belonging to category A. Then the distribution of the X's is again given by (A.260). This is seen by noting that the denominator of (A.260) gives the total number of possible assignments of the N items to the c samples, while the numerator provides the number of such assignments for which $X_1 = x_1, \ldots, X_c = x_c$.

The marginal distribution of X_i is clearly the hypergeometric distribution $H(N, d_i, m)$, and it follows from (A.89) that $E(X_i) = md_i/N$ and

$$\text{(A.261)} \qquad \operatorname{Var}(X_i) = \frac{N-m}{N-1} m \frac{d_i}{N}\left(1 - \frac{d_i}{N}\right)$$

The covariance of X_i and X_j can be obtained by putting

$$X_i = I_1 + \cdots + I_m \qquad \text{and} \qquad X_j = J_1 + \cdots + J_m$$

where

$$I_k = \begin{cases} 1 & \text{if the } k\text{th item drawn belongs to category } i \\ 0 & \text{otherwise} \end{cases}$$

and J_k is defined analogously with respect to the jth category. Then

$$E(I_k J_l) = \begin{cases} \dfrac{d_i d_j}{N(N-1)} & \text{if } k \neq l \\ 0 & \text{if } k = l \end{cases}$$

and hence

$$\operatorname{Cov}(X_i, X_j) = (m^2 - m)\frac{d_i d_j}{N(N-1)} - m^2 \frac{d_i d_j}{N^2}$$

so that

$$\text{(A.262)} \qquad \operatorname{Cov}(X_i, X_j) = -\frac{N-m}{N-1} m \frac{d_i d_j}{N^2} \qquad (i \neq j)$$

It is interesting to note that the variance and covariances are just $(N-m)/(N-1)$ times what they would be if the sampling were with replacement, since in this latter case the joint distribution of the X's is the multinomial distribution $M(d_1/N, \ldots, d_c/N; m)$.

For large N and small m/N, the multinomial distribution $M(d_1/N, \ldots, d_c/N; m)$ will typically provide a reasonable approximation to the distribution (A.259) since there is then little difference between sampling with and without replacement. For a limit theorem supporting this approximation, see Prob. 67.

Drawing a sample of size m (and thereby leaving a remainder of $n = N - m$) can be viewed as dividing the population at random into two groups of sizes m and n. Let us now consider the more general problem of dividing the population at random into r groups of sizes $n_1, \ldots, n_r$ $(\sum n_i = N)$. Let $X_{ij} (i = 1, \ldots, r; j = 1, \ldots, c)$ denote the number of items in the ith group belonging to category j, so that $d_j = \sum_{i=1}^{r} X_{ij}$ and the X_{ij} can be displayed in the following tableau with r rows and c columns:

$$\begin{array}{cccc|c} X_{11} & X_{12} & \cdots & X_{1c} & n_1 \\ X_{21} & X_{22} & \cdots & X_{2c} & n_2 \\ \vdots & \vdots & \vdots & \vdots & \vdots \\ X_{r1} & X_{r2} & \cdots & X_{rc} & n_r \\ \hline d_1 & d_2 & \cdots & d_c & N \end{array}$$

where the n's and the d's are the row and column totals. The joint distribution of the X_{ij}, that is, the probability that $X_{ij} = x_{ij}$ for all i and j, is the *generalized multiple hypergeometric distribution*, given by

(A.263)
$$\frac{\binom{d_1}{x_{11},\ldots,x_{r1}}\binom{d_2}{x_{12},\ldots,x_{r2}}\cdots\binom{d_c}{x_{1c},\ldots,x_{rc}}}{\binom{N}{n_1,\ldots,n_r}}$$

This result is easy to see. The denominator is the total number of ways of dividing the population into r groups of sizes $n_1,\ldots,n_r$, and the numerator is the number of ways of dividing the d_1 items in the first category into groups of sizes $x_{11},\ldots,x_{r1}$; the d_2 items in the second category into groups of sizes $x_{12},\ldots,x_{r2}$; and so on.

An alternative expression for (A.263) is

(A.264)
$$\frac{\binom{n_1}{x_{11},\ldots,x_{1c}}\binom{n_2}{x_{21},\ldots,x_{2c}}\cdots\binom{n_r}{x_{r1},\ldots,x_{rc}}}{\binom{N}{d_1,\ldots,d_c}}$$

This is seen by using (A.257) to express the multinomial coefficients in (A.263) in terms of factorials so that (A.263) becomes

(A.265)
$$\frac{n_1!\cdots n_r!\,d_1!\cdots d_c!}{N!\prod_{i,j} x_{ij}!}$$

and then recombining the factors to produce (A.264).

The marginal distribution of $(X_{i1},\ldots,X_{ic})$ is obtained by considering the n_i items in the ith group as a random sample of size n_i. This shows the joint distribution of $(X_{i1},\ldots,X_{ic})$ to be given by (A.260) with $n_i = m$ and $N-n_i = n$; with X_j replaced by X_{ij} and Y_j by $(X_{1j}+\cdots+X_{rj})-X_{ij}$. It follows in particular from the results found in connection with (A.260) that the marginal distribution of X_{ij} is the hypergeometric distribution $H(N,d_j,n_i)$, so that $E(X_{ij}) = n_i d_j/N$ and

(A.266)
$$\operatorname{Var}(X_{ij}) = \frac{N-n_i}{N-1} n_i \frac{d_j}{N}\left(1-\frac{d_j}{N}\right)$$

while the covariance of X_{ij} and X_{ik} is

(A.267)
$$\operatorname{Cov}(X_{ij},X_{ik}) = -\frac{N-n_i}{N-1} n_i \frac{d_j d_k}{N^2}$$

Because of the symmetry between rows and columns exhibited by (A.265), it follows further that

(A.268)
$$\operatorname{Cov}(X_{ik},X_{jk}) = -\frac{N-d_k}{N-1} d_k \frac{n_i n_j}{N^2}$$

Finally, the covariance of two X's that are neither in the same row or column can be seen to be equal to

(A.269)
$$\operatorname{Cov}(X_{ij},X_{kl}) = \frac{n_i n_k d_j d_l}{(N-1)N^2}$$

by an argument completely analogous to that leading to (A.262).

The distribution (A.263) arises not only as the distribution of the X_{ij} under the assumptions stated above, but also as a conditional distribution in certain multinomial models, as follows.

Theorem 13. (i) Let $(X_{i1}, \ldots, X_{ic})$, $i = 1, \ldots, r$ be independent, distributed according to the multinomial distribution $M(p_1, \ldots, p_c; n_i)$, so that the sample size may vary with i but the probabilities $p_1, \ldots, p_c$ are independent of i. Then the conditional distribution of the X_{ij} given $\sum_{i=1}^{r} X_{ij} = d_j$ for $j = 1, \ldots, c$ is the generalized multiple hypergeometric distribution (A.263).

(ii) Let the rc variables X_{ij} $(i = 1, \ldots, r; j = 1, \ldots, c)$ have the multinomial distribution $M(p_{11}, \ldots, p_{rc}; N)$ with the probabilities p_{ij} being of the form

$$p_{ij} = p_i' p_j'' \qquad \left(\sum p_i' = \sum p_j'' = 1\right)$$

Then the conditional distribution of the X_{ij} given that $\sum_{j=1}^{c} X_{ij} = n_i$ for $i = 1, \ldots, r$ and $\sum_{i=1}^{r} X_{ij} = d_j$ for $j = 1, \ldots, c$ is the generalized multiple hypergeometric distribution (A.263).

Proof. (i) Since the distribution of $\left(\sum_{i=1}^{r} X_{i1}, \ldots, \sum_{i=1}^{r} X_{ic}\right)$ is the multinomial distribution $M(p_1, \ldots, p_c; N)$, it follows that

$$P\left(X_{ij} = x_{ij} \quad \text{for all} \quad i,j \,\middle|\, \sum_{i=1}^{r} X_{ik} = d_k \quad \text{for all } k\right) = \frac{\left[\prod_{i=1}^{r} \binom{n_i}{x_{i1}, \ldots, x_{ic}}\right] p_1^{d_1} \ldots p_c^{d_c}}{\binom{N}{d_1, \ldots, d_c} p_1^{d_1} \ldots p_c^{d_c}}$$

This is equal to (A.264) and hence to (A.263).

(ii) The joint distribution of the X_{ij} is given by

$$\binom{N}{x_{11}, \ldots, x_{rc}} \prod_{i,j} (p_i' p_j'')^{x_{ij}} = \binom{N}{x_{11}, \ldots, x_{rc}} \prod_{i=1}^{r} (p_i')^{n_i} \prod_{j=1}^{c} (p_j'')^{d_j} \tag{A.270}$$

To obtain the required conditional distribution, let us first impose only the conditions $\sum_{j=1}^{c} X_{ij} = n_i$ for $i = 1, \ldots, r$. The marginal distribution of $(\sum X_{1j}, \ldots, \sum X_{rj})$ is the multinomial distribution corresponding to N trials and probabilities

$$p_{11} + \cdots + p_{1c} = p_1', \quad \ldots, \quad p_{r1} + \cdots + p_{rc} = p_r'$$

for the r possible outcomes. Thus

$$P(\sum X_{1j} = n_1, \ldots, \sum X_{rj} = n_r) = \binom{N}{n_1, \ldots, n_r} (p_1')^{n_1} \ldots (p_r')^{n_r} \tag{A.271}$$

The conditional distribution of the X_{ij} given $\sum_j X_{ij} = n_i$ for $i = 1, \ldots, r$ is the ratio of (A.270) to (A.271), which after some simplification is seen to be equal to

$$\prod_{i=1}^{r} \binom{n_i}{x_{i1}, \ldots, x_{ic}} \prod_{j=1}^{c} (p_j'')^{d_j}$$

Using the fact that $d_j = \sum_{i=1}^{r} x_{ij}$, we see that the r vectors $(X_{i1}, \ldots, X_{ic})$ are independent, distributed according to the multinomial distributions $M(p_1'', \ldots, p_c''; n_i)$, and the result now follows from part (i) of the lemma.

7C. The Multivariate Normal Distribution

A full understanding of this distribution requires a knowledge of linear algebra. We shall sketch here (without assuming such knowledge and hence without proofs) some of the main results as a background to the multivariate limit theorems of the next section.

The simplest example of a p-variate normal distribution is the joint distribution of p independent normal random variables. The most general p-variate distribution is the joint distribution of p linear functions of any number of independent normal random variables, say, the joint distribution of

(A.272) $$X_i = \sum_{k=1}^{r} a_{ik} Y_k \qquad (i = 1, \ldots, p)$$

where the Y_k are independent with normal distributions $N(\mu_k, \sigma_k^2)$. Two immediate consequences of this definition are:

(i) If $(X_1, \ldots, X_p)$ have a p-variate normal distribution, then the marginal distribution of any subset $(X_{i_1}, \ldots, X_{i_q})$, $q < p$, is q-variate normal.

(ii) If $(X_1, \ldots, X_p)$ have a p-variate normal distribution, and if

$$V_j = \sum_{i=1}^{p} b_{ji} X_i \qquad (j = 1, \ldots, s)$$

where the b's are any constants, then $(V_1, \ldots, V_s)$ have an s-variate normal distribution. This follows from the fact that the X's are linear functions of the Y's and hence the V's can be expressed as linear functions of the Y's.

Let us denote the expectations, variances, and covariances of $(X_1, \ldots, X_p)$ in (A.272) by

(A.273) $$\xi_i = E(X_i) \quad , \quad \tau_i^2 = \text{Var}(X_i) \quad , \quad \lambda_{ij} = \text{Cov}(X_i, X_j)$$

and refer to the set of numbers τ_i^2 and λ_{ij} as the *covariance structure* of the distribution. These quantities are related to the corresponding quantities of the Y's through the equations

(A.274) $$\xi_i = \sum_{k=1}^{r} a_{ik}\mu_k \quad , \quad \tau_i^2 = \sum_{k=1}^{r} a_{ik}^2\sigma_k^2 \quad , \quad \lambda_{ij} = \sum_{k=1}^{r} a_{ik}a_{jk}\sigma_k^2$$

Starting out with quite different values of the μ_k, σ_k^2, and the constants a_{ik}, one may through Eqs. (A.274) arrive at the same values for the means, variances, and covariances of the X_i. It follows from Theorem 14 below that then not only the means and covariance structure of the X's are the same but that they in fact have the same distributions.

Theorem 14. Given any set of p numbers $\xi_1, \ldots, \xi_p$, and any set of $p + \binom{p}{2}$ numbers τ_i^2 and $\lambda_{ij} (i < j)$ that are possible variances and covariances of a set of p random variables, there exists one and only one p-variate normal distribution with these means and this covariance structure.

Thus, a multivariate normal distribution is completely specified by its means and its covariance structure. In particular, the only p-variate normal distributions with $p = 1$ are the univariate normal distributions $N(\xi_i, \tau_i^2)$. To introduce another aspect of multivariate normal distributions, let us consider the following example.

EXAMPLE 28. Suppose that $Y_1, \ldots, Y_p$ are independent, each distributed according to $N(0,1)$, and let $X_i = Y_i - \bar{Y}$, where $\bar{Y} = \sum Y_j/p$. The variables $X_1, \ldots, X_p$ satisfy the linear equation $\sum X_i = 0$ and thus can take on values only in the $(p-1)$-dimensional linear subspace of p-space defined by this equation. The means of the X's are zero, and from (A.274) it is easily checked (Prob. 61) that

$$\tau_i^2 = \frac{p-1}{p} \qquad \text{and} \qquad \lambda_{ij} = -\frac{1}{p}$$

The feature of special interest in this example is that the distribution assigns probability 1 to a lower-dimensional linear subspace, or equivalently that the random variables $X_1, \ldots, X_p$ are connected by one or more linear equations. A multivariate normal distribution with this property is called *singular*. A p-variate normal distribution not of this type has positive probability density at all points of p-space, of the form

$$p(x_1, \ldots, x_p) = C \exp\left[-\frac{1}{2} \sum_{i=1}^{p} \sum_{j=1}^{p} d_{ij}(x_i - \xi_i)(x_j - \xi_j) \right] \tag{A.275}$$

Here the ξ's are the expectations

$$E(X_i) = \xi_i$$

and the d's are the unique solutions of the p^2 equations

$$\sum_{j=1}^{p} \lambda_{ij} d_{jk} = \begin{cases} 1 & \text{if } i = k \\ 0 & \text{if } i \neq k \end{cases}$$

where for convenience the variances τ_i^2 have been denoted by λ_{ii} and where $\lambda_{ji} = \lambda_{ij}$. That these equations have a unique solution is a consequence of the fact that the distribution of the X's is nonsingular.

8. CONVERGENCE OF RANDOM VECTORS

In the present section we shall generalize some of the results of Secs. 3 and 4 to the joint distribution of several random variables.

Definition 6. A sequence $T_n = (T_{1n}, \ldots, T_{pn})$ of random p-vectors *converges in probability* to a constant vector $(\theta_1, \ldots, \theta_p)$ if for each $i = 1, \ldots, p$ the random variables T_{in} converge in probability to θ_i as $n \to \infty$.

Since this definition reduces the p-dimensional problem to the one-dimensional problem considered in Sec. 3, it is not necessary to generalize the results given by (A.107), (A.110), and (A.111) of that section, and we can go directly to the extension of Definition 5.

Definition 7. Let $T_n = (T_{1n}, \ldots, T_{pn})$, $n = 1, 2, \ldots$, be a sequence of random p vectors and F a continuous distribution in p-space. Then T_n is said to have the limit distribution F if for all $(t_1, \ldots, t_p)$

(A.276) $$P(T_{1n} \leqslant t_1, \ldots, T_{pn} \leqslant t_p) \to F(t_1, \ldots, t_p) \qquad \text{as } n \to \infty$$

If T is a random vector with distribution F, relation (A.276) is expressed alternatively by saying that T_n tends to T *in law.* As before, if F is not continuous, (A.276) is required to hold only at all continuity points $(t_1, \ldots, t_p)$ of F.

In generalization of Sec. 3, Theorem 2, we have the following *multivariate central limit theorem.* [For a proof, see Cramér (1946), Sec. 24.7 or Feller (1966), Sec. VIII. 4, Theorem 2.]

Theorem 15. If the vectors $X_n = (X_{1n}, \ldots, X_{pn})$ are independent, identically distributed with expectation $(\mu_1, \ldots, \mu_p)$ and covariance structure given by

(A.277) $$\operatorname{Var}(X_{in}) = \sigma_i^2 \qquad \operatorname{Cov}(X_{in}, X_{jn}) = \lambda_{ij} \qquad (i, j = 1, \ldots, p)$$

and if $X_{i.} = \sum_{\nu=1}^{n} X_{i\nu}/n$, then the sequence of vectors

$$[\sqrt{n}(X_{1.} - \mu_1), \ldots, \sqrt{n}(X_{p.} - \mu_p)]$$

has a p-variate normal limit distribution with mean zero and covariance structure (A.277).

The results up to this point were just the natural extensions of the corresponding one-dimensional results of Sec. 3. We now turn to a question which it was not necessary to consider in the one-dimensional case. Suppose that T_n tends in law to T, where T_n and T are vectors in p-space; then if $\mathscr{S}$ is a set in p-space, it will be important to know whether the probability that T_n falls in $\mathscr{S}$ tends to the corresponding probability that T falls in $\mathscr{S}$; that is, whether

(A.278) $$P(T_n \in \mathscr{S}) \to P(T \in \mathscr{S})$$

In the case $p = 1$, the sets $\mathscr{S}$ of principal interest are intervals, for which the answer is trivially in the affirmative if F is continuous. When p is greater than 1, the following theorem covers all cases with which we shall be concerned. [A proof of the theorem can be found in Billingsley (1968), Theorem 2.1.]

Theorem 16. If T_n and T are random vectors in p-space such that $T_n \to T$ in law, and if $\mathscr{S}$ is a set in p-space for which $P(T_n \in \mathscr{S})$ and $P(T \in \mathscr{S})$ are defined, then (A.278) holds provided the boundary of $\mathscr{S}$ has probability zero under the distribution of T. (A boundary point of $\mathscr{S}$ is a point which either belongs to or is a limit point of $\mathscr{S}$ and also of its complement. The boundary of $\mathscr{S}$ is the set of all its boundary points.)

EXAMPLE 29. Let $R_n = (R_{1n}, \ldots, R_{sn})$, $n = 1, 2, \ldots$, be a sequence of independent random vectors, with a common distribution that specifies that the s components of a vector are a permutation of the integers $(1, \ldots, s)$, each of the $s!$ permutations having probability $1/s!$. This is the model for ranks in randomized blocks considered in Chap. 6, Sec. 1. It follows from (A.12) that

$$\mu_i = E(R_{in}) = \tfrac{1}{2}(s+1) \tag{A.279}$$

and from (A.13) and (A.41) that

$$\sigma_i^2 = \operatorname{Var}(R_{in}) = \tfrac{1}{12}(s^2-1) \qquad \text{and} \qquad \lambda_{ij} = \operatorname{Cov}(R_{in}, R_{jn}) = -\tfrac{1}{12}(s+1) \tag{A.280}$$

If $R_{i.} = \sum_{v=1}^{n} R_{iv}/n$, then by Theorem 15, the sequence of random vectors

$$\sqrt{n}[R_{1.} - \tfrac{1}{2}(s+1)], \ldots, \sqrt{n}[R_{s.} - \tfrac{1}{2}(s+1)]$$

has as its limit distribution the s-variate normal distribution with zero means and covariance structure given by (A.280).

Consider now the sequence of random variables

$$Q_n = \frac{12n}{s(s+1)} \sum_{i=1}^{s} [R_{i.} - \tfrac{1}{2}(s+1)]^2 \tag{A.281}$$

the test statistics for the Friedman test of Chap. 6, Sec. 2. To determine the limit distribution of Q_n, let

$$R'_{in} = \sqrt{\frac{12n}{s(s+1)}}[R_{in} - \tfrac{1}{2}(s+1)]$$

and let $R'_{i.} = \sum_{v=1}^{n} R'_{iv}/n$, so that

$$Q_n = \sum_{i=1}^{s} R'^2_{i.}$$

It is seen from (A.280) that the components of the random vector $(R'_{1.}, \ldots, R'_{s.})$ have zero means and covariance structure

$$\operatorname{Var}(R'_{i.}) = \frac{s-1}{s} \qquad \text{and} \qquad \operatorname{Cov}(R'_{i.}, R'_{j.}) = -\frac{1}{s} \tag{A.282}$$

Let us now compare this with the following quite different situation. Let the random variables $Y_i (i = 1, \ldots, s)$ be independent, each with the standard normal distribution $N(0,1)$, and let

$$Z_i = Y_i - \bar{Y}$$

where $\bar{Y} = \sum_{j=1}^{s} Y_j/s$. Then it was pointed out in Example 28 that $E(Z_i) = 0$ and that

$$\operatorname{Var}(Z_i) = \frac{s-1}{s} \qquad \text{and} \qquad \operatorname{Cov}(Z_i, Z_j) = -\frac{1}{s}$$

Since the Z's have a joint normal distribution, comparison with (A.282) shows that the joint limit distribution of $(R'_{1.}, \ldots, R'_{s.})$ coincides with the distribution of $(Z_1, \ldots, Z_s)$.

Suppose now that $\mathscr{S}$ is the sphere in s-space centered at the origin and with radius $\sqrt{C}$, and let $P(\mathscr{S})$ denote the probability assigned to this sphere by the limit distribution of $(R'_{1.}, \ldots, R'_{s.})$. Then

$$P(\mathscr{S}) = P[\sum Z_i^2 \leqslant C] = P[\sum (Y_i - \bar{Y})^2 \leqslant C]$$

Now it follows from Sec. 2F that $\sum (Y_i - \bar{Y})^2$ has a χ^2-distribution with $s-1$ degrees of freedom, so that

$$P(\mathscr{S}) = P(\chi_{s-1}^2 \leqslant C)$$

Since the boundary of $\mathscr{S}$ has the probability

$$P[\sum (Y_i - \bar{Y})^2 = C] = P(\chi_{s-1}^2 = C) = 0$$

Theorem 16 shows that $P(\chi_{s-1}^2 \leqslant C)$ is also the limit of the probability $P(Q_n \leqslant C)$. This proves the desired result: that the limit distribution of Q_n is the χ^2-distribution with $s-1$ degrees of freedom.

We next state generalizations of Theorem 4 of Sec. 3 and of its corollaries.

Theorem 17. If $T_n = (T_{1n}, \ldots, T_{pn})$ tends in law to $T = (T_1, \ldots, T_p)$ and if B_{in} and D_{in} are random variables tending in probability to constants b_i and d_i, respectively, for $i = 1, \ldots, p$, then the sequence

$$(B_{1n} T_{1n} + D_{1n}, \ldots, B_{pn} T_{pn} + D_{pn})$$

tends in law to $(b_1 T_1 + d_1, \ldots, b_p T_p + d_p)$.

The following are two immediate corollaries.

Corollary 6. Let $T_n = (T_{1n}, \ldots, T_{pn})$ and $S_n = (S_{1n}, \ldots, S_{pn})$. If T_n tends in law to T, and if

$$E(T_{in} - S_{in})^2 \to 0 \qquad \text{as } n \to \infty, \text{ for all } i = 1, \ldots, p \tag{A.283}$$

then S_n tends in law to T.

Corollary 7. If the distribution of

$$\left(\frac{T_{1n} - E(T_{1n})}{\sqrt{\operatorname{Var}(T_{1n})}}, \ldots, \frac{T_{pn} - E(T_{pn})}{\sqrt{\operatorname{Var}(T_{pn})}}\right)$$

tends to a limit distribution L, and if

$$\frac{E(T_{in} - S_{in})^2}{\operatorname{Var}(T_{in})} \to 0 \qquad \text{for all } i = 1, \ldots, p \tag{A.284}$$

then the distribution of

$$\left(\frac{S_{1n} - E(S_{1n})}{\sqrt{\operatorname{Var}(S_{1n})}}, \ldots, \frac{S_{pn} - E(S_{pn})}{\sqrt{\operatorname{Var}(S_{pn})}}\right)$$

also tends to L.

The proofs of these corollaries are exactly analogous to those of the corresponding univariate results.

We finally require an extension of the Lindeberg central limit theorem (Theorem 5 of Sec. 3).

Theorem 18. Let $X_1 = (X_{11}, \ldots, X_{p1})$, $X_2 = (X_{12}, \ldots, X_{p2})$, ..., be a sequence of independent random p-vectors, not necessarily identically distributed. Let their means be $\mu_1 = (\mu_{11}, \ldots, \mu_{p1})$, $\mu_2 = (\mu_{12}, \ldots, \mu_{p2})$, ..., and their covariance structure be given by

$$\operatorname{Cov}(X_{i1}, X_{j1}) = \lambda_{ij}^{(1)} \quad , \quad \operatorname{Cov}(X_{i2}, X_{j2}) = \lambda_{ij}^{(2)} \quad , \quad \cdots$$

and suppose that as $N \to \infty$ the following two conditions are satisfied:

(i)
$$\frac{\sum_{\nu=1}^{N} \lambda_{ij}^{(\nu)}}{N} \to \lambda_{ij}$$

(ii) The Lindeberg condition (A.128), with the obvious changes in notation, holds for each of the p sequences of random variables $(X_{i1}, X_{i2}, X_{i3}, \ldots)$ $i = 1, \ldots p$.

Then the joint distribution of

$$\sqrt{N}(X_{1.} - \mu_{1.}), \ldots, \sqrt{N}(X_{p.} - \mu_{p.})$$

tends to the p-variate normal distribution with means zero and covariances λ_{ij}. (Here $X_{i.} = \sum_{\nu=1}^{N} X_{i\nu}/N$ and $\mu_{i.} = \sum_{\nu=1}^{N} \mu_{i\nu}/N$, as usual.)

Corollary 8. Let the correlation coefficient of $S_{iN} = X_{i1} + \cdots + X_{iN}$ and $S_{jN} = X_{j1} + \cdots + X_{jN}$ be

$$\rho_{ij}^{(N)} = \frac{\sum_{\nu=1}^{N} \lambda_{ij}^{(\nu)}}{\sqrt{\sum_{\nu=1}^{N} \lambda_{ii}^{(\nu)} \sum_{\nu=1}^{N} \lambda_{jj}^{(\nu)}}} \tag{A.285}$$

and suppose that the limit

$$\rho_{ij} = \lim_{N \to \infty} \rho_{ij}^{(N)} \tag{A.286}$$

exists. Then if

$$S_{iN}^* = \frac{S_{iN} - E(S_{iN})}{\sqrt{\operatorname{Var}(S_{iN})}}$$

the joint distribution of $(S_{1N}^*, \ldots, S_{pN}^*)$ tends to the p-variate normal distribution with covariance structure ρ_{ij}, provided condition (ii) holds.

Proof. Let

$$Y_{i\nu} = \frac{X_{i\nu} - \mu_{i\nu}}{\sqrt{\frac{1}{N} \sum_{\nu=1}^{N} \lambda_{ii}^{(\nu)}}}$$

Then the covariance of $(Y_{i\nu}, Y_{j\nu})$ is

$$\rho_{ij}^{*(\nu)} = \frac{\lambda_{ij}^{(\nu)}}{\sqrt{\frac{1}{N}\sum_{\nu=1}^{N} \lambda_{ii}^{(\nu)} \frac{1}{N}\sum_{\nu=1}^{N} \lambda_{jj}^{(\nu)}}}$$

Thus condition (i) of Theorem 18 applies to the variables Y_{ij}. Furthermore, since $S_{iN}^* = \sqrt{N}\, Y_{i.}$, the result follows.[1]

EXAMPLE 30. Let us generalize Example 29 to allow for the presence of ties. The random vector $(R_{1n}, \ldots, R_{sn})$ is then replaced by $(R_{1n}^*, \ldots, R_{sn}^*)$, which is a random permutation not of the integers $(1, \ldots, s)$ but of the midranks in the nth block, say $(a_{1n}, \ldots, a_{sn})$, as discussed in Chap. 6, Sec. 2. From the formulas given in Examples 1 and 4 we find that

$$\mu_{in} = E(R_{in}^*) = a_{.n} = \frac{\sum_{j=1}^{s} a_{jn}}{s}$$

and that

$$\lambda_{ii}^{(n)} = \operatorname{Var}(R_{in}^*) = \sigma_n^2 = \frac{\sum_{j=1}^{s} (a_{jn} - a_{.n})^2}{s}$$

and

$$\lambda_{ij}^{(n)} = \operatorname{Cov}(R_{in}^*, R_{jn}^*) = -\frac{\sigma_n^2}{s-1} \qquad \text{for } i \neq j$$

Consider now the joint distribution of

$$\frac{S_{1n} - E(S_{1n})}{\sqrt{\operatorname{Var}(S_{1n})}}, \ldots, \frac{S_{sn} - E(S_{sn})}{\sqrt{\operatorname{Var}(S_{sn})}} \tag{A.287}$$

where $S_{in} = \sum_{\nu=1}^{n} R_{i\nu}^*$. To prove joint asymptotic normality of these variables, we shall apply Corollary 8. The covariances $\rho_{ij}^{(n)}$ defined by (A.285) are seen to be equal to $\rho_{ij}^{(n)} = -1/(s-1)$. Since this is independent of n, condition (A.286) is automatically satisfied. To check condition (ii) of Theorem 18, note that $1 \leqslant a_{in} \leqslant s$ and hence $1 \leqslant R_{in}^* \leqslant s$ for all i and n. Since the variables R_{in}^* are thus uniformly bounded, it follows from Prob. 26 that (ii) holds provided the sum

$$\sigma_1^2 + \cdots + \sigma_n^2 \to \infty \qquad \text{as } n \to \infty \tag{A.288}$$

Now the a_{in} can only take on a finite number of distinct values since each is an average of from 1 to s of the integers $1, \ldots, s$. It follows that the variances $\sigma_1^2, \sigma_2^2, \ldots,$ can only take on a finite number of values. The smallest of these is zero; let the next smallest be b. Then

$$\sigma_1^2 + \cdots + \sigma_n^2 \geqslant b\,[\text{no. of nonzero terms among } \sigma_1^2, \ldots, \sigma_n^2]$$

[1] This theorem, although presumably known to various workers in the field, does not seem to be in the literature. A proof can be found in Aiyar (1969), "On Some Tests for Trend and Autocorrelation," unpublished thesis, University of California at Berkeley.

Hence (A.288) holds provided the number of nonzero variances among $\sigma_1{}^2, \ldots, \sigma_n{}^2$ tends to infinity as $n \to \infty$. Since $\sigma_\nu{}^2 = 0$ if and only if $a_{1\nu} = \cdots = a_{s\nu}$, we have proved the following result: the joint distribution of the variables (A.287) tends to the s-variate normal distribution with zero means, unit variances, and covariances $\rho_{ij} = -1/(s-1)$, provided the number of blocks ν in which the observations (and hence the midranks $a_{i\nu}$) take on at least two distinct values tends to infinity as $n \to \infty$.

To define the statistic of interest, we require some additional notation. Suppose the midranks $a_{1\nu}, \ldots, a_{s\nu}$ in the νth block take on e_ν distinct values; that $d_{1\nu}$ of the a's are equal to the smallest of these values, $d_{2\nu}$ to the next smallest, and so on. In terms of the d's, it was shown in formula (A.18) that

$$\sigma_\nu{}^2 = \frac{s^2-1}{12} - \frac{\sum_{j=1}^{e_\nu} (d_{j\nu}^3 - d_{j\nu})}{12s} \tag{A.289}$$

The modification of the statistic Q_n of Example 29, which was considered in Chap. 6, Sec. 2, is

$$Q_n^* = \frac{12n}{s(s+1)} \frac{\sum_{i=1}^{s} [R_{i.}^* - \frac{1}{2}(s+1)]^2}{[1 - \sum\sum (d_{j\nu}^3 - d_{j\nu})/ns(s^2-1)]} \tag{A.290}$$

To determine the limit distribution of $Q_{n.}^*$ let

$$R_{i\nu}^{*\prime} = \sqrt{\frac{12n}{s(s+1)}} \frac{R_{i\nu}^* - \frac{1}{2}(s+1)}{\sqrt{1 - \sum\sum (d_{j\nu}^3 - d_{j\nu})/ns(s^2-1)}}$$

and let $R_{i.}^{*\prime} = \sum R_{i\nu}^{*\prime}/n$ so that

$$Q_n^* = \sum (R_{i.}^{*\prime})^2$$

It is seen from the covariance structure of the vectors $(R_{1\nu}^*, \ldots, R_{s\nu}^*)$, the independence of these vectors, and formula (A.289) that the components of the random vector $(R_{1.}^{*\prime}, \ldots, R_{s.}^{*\prime})$ have zero means and covariance structure (A.282) with $R_{i.}^{*\prime}$ in place of $R_{i.}'$. Their joint asymptotic normality follows from that of the variables (A.287) from which they only differ by a scale factor. The discussion of the second part of Example 29 now shows that the limit distribution of Q_n^*, like that of Q_n, is χ^2 with $s-1$ degrees of freedom.

As was the case with Theorem 5, Theorem 18 permits a slight generalization which is important for applications. It is obtained by considering not a single sequence of p-vectors $X_1, X_2, \ldots$, with distributions $F_1, F_2, \ldots$, but a double sequence $X_1{}^{(1)} = (X_{11}{}^{(1)}, \ldots, X_{p1}{}^{(1)})$ with distribution F_{11}; $X_1{}^{(2)} = (X_{11}{}^{(2)}, \ldots, X_{p1}{}^{(2)})$, $X_2{}^{(2)} = (X_{12}{}^{(2)}, \ldots, X_{p2}{}^{(2)})$ with distributions F_{21} and F_{22}; $X_1{}^{(3)}, X_2{}^{(3)}, X_3{}^{(3)}$ with distributions F_{31}, F_{32}, and F_{33}; and so on. The covariances now require a double superscript. Let us denote them by

$$\operatorname{Cov}(X_{i1}{}^{(\nu)}, X_{j1}{}^{(\nu)}) = \lambda_{ij}{}^{(1\nu)} \quad , \quad \operatorname{Cov}(X_{i2}{}^{(\nu)}, X_{j2}{}^{(\nu)}) = \lambda_{ij}{}^{(2\nu)} \quad \cdots$$

In analogy with Theorem 5′, we then have the following result.

Theorem 18′. Suppose that as $N \to \infty$ the following two conditions are satisfied:

(i) The averages $\sum_{k=1}^{N} \lambda_{ij}{}^{(kN)}/N$ tend to limits, say, λ_{ij}.

(ii) The Lindeberg condition (A.130), with the obvious changes in notation, holds for each of the p double sequences of random variables

$$X_{i1}^{(1)} \quad , \quad (X_{i1}^{(2)}, X_{i2}^{(2)}) \quad , \quad (X_{i1}^{(3)}, X_{i2}^{(3)}, X_{i3}^{(3)}) \quad , \quad \cdots \quad , \quad i = 1, \ldots, p$$

Then if

$$X_{i.}^{(N)} = [X_{i1}^{(N)} + \cdots + X_{iN}^{(N)}]/N$$

the joint distribution of

$$\sqrt{N}(X_{1.}^{(N)} - \mu_{1.}^{(N)}), \ldots, \sqrt{N}(X_{p.}^{(N)} - \mu_{p.}^{(N)})$$

tends to the p-variate normal distribution with means zero and covariances λ_{ij}. An analogous generalization holds for Corollary 8, and we shall refer to it as Corollary 8′.

In Sec. 4, we considered the asymptotic behavior of the sample sum in a sample from a finite population. We are now in a position to generalize these results to the simultaneous consideration of several samples. As before, let Π_N be a population consisting of the N elements, $v_{N1}, \ldots, v_{NN}$. Suppose that a random sample of size $n_1 = n_1(N)$ is drawn, then from the remainder a second sample of size $n_2 = n_2(N)$ is drawn, and so on, until finally only n_s elements are left in the population to constitute the sth sample $(n_1 + \cdots + n_s = N)$. We shall assume that s is fixed and that as $N \to \infty$

(A.291) $$n_i \to \infty \qquad \text{for all } i$$

and the proportions n_i/N will each tend to a limit, say,

(A.292) $$n_i/N \to \gamma_i < 1$$

where some of the γ's may be equal to zero.

Let S_{1N} denote the sum of the n_1 v-values constituting the first sample, S_{2N} the sum of the n_2 v-values constituting the second sample, and so on. The correlation coefficient of S_{iN} and S_{jN} can be shown to be equal to (Prob. 69)

(A.293) $$-\sqrt{\frac{n_i n_j}{(N-n_i)(N-n_j)}}$$

and hence will tend to

(A.294) $$\lim\left[-\sqrt{\frac{n_i n_j}{(N-n_i)(N-n_j)}}\right] = -\sqrt{\frac{\gamma_i \gamma_j}{(1-\gamma_i)(1-\gamma_j)}} = \rho_{ij}$$

provided none of the γ's are equal to 1.

Theorem 19. A sufficient condition for the joint distribution of the standardized variables

$$\frac{S_{1N} - E(S_{1N})}{\sqrt{\text{Var}(S_{1N})}} \quad , \ldots , \quad \frac{S_{sN} - E(S_{sN})}{\sqrt{\text{Var}(S_{sN})}}$$

to tend to the s-variate normal distribution with zero means, unit variances and covariances ρ_{ij} is that

(A.295) $$\frac{\max_\nu (v_{N\nu} - v_{N.})^2}{\sum_{\nu=1}^{N} (v_{N\nu} - v_{N.})^2} \max_k \left(\frac{N-n_k}{n_k}\right) \to 0 \qquad \text{as } N \to \infty$$

Proof. The proof is quite analogous to that of Theorem 6. Again, let $U_1, \ldots, U_N$ be independent random variables, each uniformly distributed on the unit interval. In terms of these variables, the process of drawing from Π_N successive samples of sizes $n_1, \ldots, n_s$ can be defined by including $v_{N\nu}$ in the first sample if the rank R_ν of U_ν is less than or equal to n_1, in the second sample, if $n_1 < R_\nu \leqslant n_1 + n_2$, and so on. This procedure generates random samples of the required sizes. For each $\nu = 1, \ldots, N$, let

$$\begin{array}{ll} J_{1\nu} = 1 & \text{if } R_\nu \leqslant n_1 \\ J_{2\nu} = 1 & \text{if } n_1 < R_\nu \leqslant n_1 + n_2 \\ \cdots\vdots\cdots & \vdots\cdots\cdots\cdots\vdots\cdots\cdots\cdots \\ J_{s\nu} = 1 & \text{if } n_1 + \cdots + n_{s-1} < R_\nu \leqslant N \end{array}$$

and the J's equal to zero otherwise.

Then

$$S_{1N} = \sum_{\nu=1}^{N} J_{1\nu} v_{N\nu}, \quad S_{2N} = \sum_{\nu=1}^{N} J_{2\nu} v_{N\nu}, \quad \ldots, \quad S_{sN} = \sum_{\nu=1}^{N} J_{s\nu} v_{N\nu} \tag{A.296}$$

Let us similarly put

$$\begin{array}{ll} K_{1\nu} = 1 & \text{if } U_\nu \leqslant \dfrac{n_1}{N} \\ K_{2\nu} = 1 & \text{if } \dfrac{n_1}{N} < U_\nu \leqslant \dfrac{n_1 + n_2}{N} \\ \cdots\cdots & \cdots \quad \cdots\cdots\cdots\cdots \\ K_{s\nu} = 1 & \text{if } \dfrac{n_1 + \cdots + n_{s-1}}{N} < U_\nu \leqslant 1 \end{array}$$

and the K's equal to zero otherwise, and let

$$T_{1N} = \sum_{\nu=1}^{N} K_{1\nu}(v_{N\nu} - v_{N.}) + n_1 v_{N.}, \ldots, T_{sN} = \sum_{\nu=1}^{N} K_{s\nu}(v_{N\nu} - v_{N.}) + n_s v_{N.} \tag{A.297}$$

To prove the theorem let us apply Corollary 8′ (mentioned following Theorem 18′) to the variables

$$X_{i\nu}^{(N)} = (v_{N\nu} - v_{N.}) K_{i\nu}$$

The expectations and variances of the X's are by the corresponding one-sample results

$$\mu_{i\nu}^{(N)} = E(X_{i\nu}^{(N)}) = (v_{N\nu} - v_{N.}) \frac{n_i}{N}$$

and

$$\lambda_{ii}^{(\nu N)} = \operatorname{Var}(X_{i\nu}^{(N)}) = (v_{N\nu} - v_{N.})^2 \frac{n_i}{N}\left(1 - \frac{n_i}{N}\right)$$

Furthermore, if $i \neq j$,

$$\operatorname{Cov}(K_{i\nu}, K_{j\nu}) = E(K_{i\nu} K_{j\nu}) - \frac{n_i n_j}{N^2} = -\frac{n_i n_j}{N^2}$$

and hence

(A.298) $$\lambda_{ij}^{(vN)} = -\frac{(v_{Nv} - v_{N.})^2 n_i n_j}{N^2}$$

It follows that

(A.299) $$\rho_{ij}^{(N)} = -\frac{n_i n_j/N^2}{\sqrt{\frac{n_i}{N}\left(1-\frac{n_i}{N}\right)\frac{n_j}{N}\left(1-\frac{n_j}{N}\right)}} = -\sqrt{\frac{n_i n_j}{(N-n_i)(N-n_j)}} \to \rho_{ij}$$

The proof of Theorem 6 shows that the Lindeberg condition holds for the double sequence $(X_{i1}^{(N)}, \ldots, X_{iN}^{(N)})$, $N = 1, \ldots,$ as $N \to \infty$ provided (A.134) holds with n_i in place of n. Now $n_i \leqslant N - n_j$ when $j \neq i$, and hence

$$\frac{n_i}{N-n_i} \leqslant \frac{N-n_j}{n_j}$$

Thus, (A.295) implies that (A.134) holds, with n_i in place of n, for all i. This shows that the joint distribution of

$$\left(\frac{T_{1N} - E(T_{1N})}{\sqrt{\mathrm{Var}(T_{1N})}}, \ldots, \frac{T_{sN} - E(T_{sN})}{\sqrt{\mathrm{Var}(T_{sN})}}\right)$$

tends to the s-variate normal distribution with zero means, unit variances, and correlation coefficients ρ_{ij}.

Lemma 1 of Sec. 4 shows that

$$\frac{E(T_{iN} - S_{iN})^2}{\mathrm{Var}(T_{iN})} \to 0 \qquad \text{as } N \to \infty \text{ for all } i$$

and the result now follows from Corollary 7.

Corollary 9. Condition (A.295) and hence the conclusion of Theorem 19 hold if the constants v_{Nv} satisfy (A.142) and if

(A.300) $$n_i \to \infty \qquad \text{as } N \to \infty \text{ for all } i = 1, \ldots, s$$

Proof. It is clearly only necessary to show that $\max_k[(N-n_k)/Nn_k] \to 0$. Since $(N-n_k)/N \leqslant 1$ for all k, this is an obvious consequence of (A.300).

Condition (A.142) was checked in Examples 15 to 17 for three different sets of v-values. It now follows from Corollary 9 that for these three sets the result of Theorem 19 holds.

Corollary 10. Suppose that as $N \to \infty$, the sample sizes $n_i \to \infty$ and each of the ratios n_i/N tends to a limit $\gamma_i < 1$. Then the distribution of the statistic

(A.301) $$Q_N = \frac{N-1}{\sum_{v=1}^{N} (v_{Nv} - v_{N.})^2} \sum_{i=1}^{s} \frac{1}{n_i}(S_{iN} - n_i v_{N.})^2$$

tends to the χ^2-distribution with $s-1$ degrees of freedom provided (A.295) holds, and hence in particular under the assumptions of Corollary 9.

Proof. Let

$$S'_{iN} = \sqrt{\frac{n_i(N-1)}{\sum (v_{N\nu} - v_{N.})^2}} \left(\frac{S_{iN}}{n_i} - v_{N.} \right) \tag{A.302}$$

so that

$$Q_N = \sum_{i=1}^{s} S'^2_{iN}$$

The expectations of the variables S'_{iN} are zero and their covariance structure is given by (Prob. 70)

$$\operatorname{Var}(S'_{iN}) = \frac{N-n_i}{N}, \quad \operatorname{Cov}(S'_{iN}, S'_{jN}) = -\frac{\sqrt{n_i n_j}}{N} \tag{A.303}$$

Let us compare this with the following quite different situation. Let the random variables $Y_{i\nu}$ ($\nu = 1, \ldots, n_i; i = 1, \ldots, s$) be independent, each with standard normal distribution $N(0,1)$, and let

$$Z_i = \sqrt{n_i}(Y_{i.} - Y_{..})$$

where

$$Y_{i.} = \sum_{\nu=1}^{n_i} \frac{Y_{i\nu}}{n_i} \quad \text{and} \quad Y_{..} = \frac{\sum\sum Y_{i\nu}}{\sum n_i}$$

Then an easy calculation shows [Prob. 70(ii)] that $E(Z_i) = 0$ and that the covariance structure of the Z's is also given by (A.303), with Z_i in place of S'_{iN}. Since the Z's have a joint normal distribution, it follows that the joint limit distribution of $(S'_{1N}, \ldots, S'_{sN})$ coincides with the distribution of $(Z_1, \ldots, Z_s)$. The proof is now completed in exact analogy with Example 29.

EXAMPLE 31. For $v_{N\nu} = \nu$ $(\nu = 1, \ldots, N)$, the statistic (A.301) reduces to the Kruskal-Wallis statistic K defined by Eq. (5.6) (Prob. 71). Since the $v_{N\nu}$ then satisfy (A.142) by Example 16, it follows from Corollary 10 that the limit distribution of K is χ^2 with $s-1$ degrees of freedom as the group sizes n_i tend to infinity, provided condition (A.292) holds.

If instead the constants $v_{N\nu}$ are defined as in Example 17, the statistic (A.301) reduces to the Kruskal-Wallis statistic K^* defined by Eq. (5.11) (Prob. 72). It follows from Corollary 10 and Example 17 that the limit distribution of K^* is χ^2 with $s-1$ degrees of freedom, provided condition (1.36) is added to the assumptions of the preceding paragraph.

9. PROBLEMS

Section 1

1. Show that (A.20) is not necessarily true for dependent variables by putting $a = 2$ and $K_2 = K_1$.

2. Prove (A.36).

3. (i) Use (A.51) and (A.52) of Sec. 1, and relation (7.10) to prove formulas (7.11) and (7.12).
(ii) Use (A.17), (A.53), and (A.54) of Sec. 1, and relation (7.18) to prove formulas (7.16) and (7.17).

4. Use the method of the preceding example to prove formulas (7.25) and (7.26).

5. Let $X_{i\alpha}$ $(\alpha = 1, \ldots, n_i; i = 1, \ldots, s)$ be independently distributed according to the same continuous distribution F. Let W_{ij} be the number of pairs (α, β) for which $X_{i\alpha} < X_{j\beta}$, and let $W = \sum W_{ij}$, summed over all pairs $i < j$. Show that the variance of W is given by (5.61). [*Hint:* The variance of W_{ij} is given by (1.30). By using the method of Example 5 one finds that

$$\mathrm{Cov}\,(W_{ij}, W_{ik}) = -\mathrm{Cov}\,(W_{ij}, W_{jk}) = \frac{n_i n_j n_k}{12}$$

Section 2

6. If X is the number of good items in a sample of size n from a population of size N, which contains d good items, show that the possible values x of X satisfy

$$\max\,(0, d+n-N) \leqslant x \leqslant \min\,(d, n)$$

7. Consider a sequence of hypergeometric distributions $H(N, d_N, n)$ with n fixed and d_N satisfying $d_N/N \to p\;(0 < p < 1)$. Show that the probabilities (A.88) then tend to the binomial probabilities (A.86).

8. (i) For any random variable X and any sequence of positive numbers ε_n tending to zero, show that as n tends to infinity

$$P(X \leqslant x+\varepsilon_n) \to P(X \leqslant x) \qquad \text{and} \qquad P(X \leqslant x-\varepsilon_n) \to P(X < x)$$

(ii) Use the results of part (i) to prove that a cumulative distribution function F is continuous for all x if and only if (A.91) holds.

9. If X and Y are independent random variables with continuous distributions F and G, show that $P(X = Y) = 0$. {*Hint:* For any positive Δ

$$P(X = Y) \leqslant P(|X-Y| < \Delta) \leqslant E[F(Y+\Delta)-F(Y-\Delta)]$$

Since F is continuous, monotone, and bounded, it is uniformly continuous, i.e., given any $\varepsilon > 0$, there exists $\Delta > 0$ such that $F(y+\Delta)-F(y-\Delta) < \varepsilon$ for all y.}

10. Use the preceding problem to prove Theorem 1. [*Hint:* The probability that at least one pair of observations will be equal is $\leqslant \sum P(X_i = Y_j)$, summed over all pairs (i,j).]

11. If X and Y are independent random variables with distributions F and G, then $P(X = Y) = 0$ if and only if F and G have no common point of discontinuity. [*Hint:* "Only if" is easy. To see "if," let X and Y have discontinuities at $a_1, a_2, \ldots,$ and $b_1, b_2, \ldots,$ respectively. The result follows by considering separately the cases that the common value of X and Y is one of the a's or b's or that it is neither of these, and by applying Prob. 9 to the second of these possibilities.]

12. (i) Let X_1, X_2 be independent, each distributed according to the Cauchy distribution (A.95). Find the distribution of $a_1X_1 + a_2X_2$.

(ii) Use (i) to prove (by induction) that if $X_1, \ldots, X_n$ are independent, each distributed according to (A.95), then $\bar{X}$ is also distributed according to (A.95).

13. Show that the density (A.96) is symmetric about zero and unimodal.

14. Let Y be defined by (A.101). Use the fact that $E(X_i) = E(X_i^3) = 0$, $E(X_i^2) = 1$, and $E(X_i^4) = 3$ to find $E(Y)$ and $\mathrm{Var}(Y)$.

15. If Z is distributed according to the standard normal distribution $N(0,1)$, show that $E(|Z|) = \sqrt{2/\pi}$.

16. Let $X_1, \ldots, X_n$ be independent. each distributed with cumulative distribution function F. which is assumed to have a density f. Let $X_{(1)} < \cdots < X_{(n)}$ denote the ordered X's.

(i) If P_Δ denotes the probability that $X_1, \ldots, X_{i-1}$ are the $i-1$ smallest, X_i is the ith smallest, and $X_{i+1}, \ldots, X_n$ are the $n-i$ largest of the X's and that $x_0 < X_i < x_0 + \Delta$, show that

$$P(x_0 < X_{(i)} < x_0 + \Delta) = n\binom{n-1}{i-1} P_\Delta$$

(ii) Obtain a lower bound for P_Δ by noting that P_Δ is greater than or equal to the probability that $X_1, \ldots, X_{i-1}$ are less than x_0, $X_{i+1}, \ldots, X_n$ are greater than $x_0 + \Delta$, and $x_0 < X_i < x_0 + \Delta$.

(iii) Obtain an upper bound for P_Δ by noting that P_Δ is less than or equal to the probability that $X_1, \ldots, X_{i-1}$ are $\leqslant x_0 + \Delta$, $X_{i+1}, \ldots, X_n$ are $\geqslant x_0$, and $x_0 < X_i < x_0 + \Delta$.

(iv) Prove formula (A.104) by dividing the upper and lower bounds of the preceding parts by Δ and letting Δ tend to zero.[1]

17. Prove formula (A.105).

Section 3

18. If A_n, B_n tend in probability to constants a, b, respectively, show that (i) $A_n + B_n$ tends in probability to $a+b$; (ii) $A_n B_n$ tends in probability to ab; (iii) $1/A_n$ tends in probability to $1/a$ if $a \neq 0$.

19. (i) Give a suitable definition of T_n tending in probability to infinity.

(ii) With this definition, show that Prob. 18(iii) remains valid even for $a = 0$.

20. Under the assumptions of Example 8, prove that S^2 tends to σ^2 in probability. {*Hint:* Write $S^2 = [1/(n-1)] \sum X_i^2 - [n/(n-1)] \bar{X}^2$, and apply Khinchine's theorem (stated in Example 7) to show that the first term tends in probability to $E(X_i^2)$ and the second one to $[E(X_i)]^2$. [Alternatively, the result follows from the weak law of large numbers provided $E(X_i^4)$ is finite.]}

21. Prove the result stated in Example 9 by showing that $\mathrm{Var}(W_{XY}/mn)$ tends to zero. [*Hint:* See Eq. (1.30).]

22. Show that if T_n tends in probability to d, then it converges in law to a random variable which takes on the value d with probability 1.

23. (i) If the random variables T_n are distributed according to the normal distributions $N(\mu_n, \sigma_n^2)$ and if $\mu_n \to \mu$ and $\sigma_n^2 \to \sigma^2 > 0$, show that T_n has the limit distribution $N(\mu, \sigma^2)$.

[1] This proof of formula (A.104) was communicated to me by Louis Jaeckel.

(ii) Determine whether this statement remains correct when $\sigma = 0$.

24. Under the assumptions of Example 7, show that $b_n(\overline{X}_n - \mu)$ tends in probability to zero if $b_n/\sqrt{n}$ tends to zero.

25. Show that (A.121) implies that the correlation coefficient of S_n and T_n tends to 1.

26. Let $X_1, X_2, \ldots,$ be a sequence of independent random variables which is uniformly bounded, say, satisfying $|X_i| \leqslant A$ for all i. Show that if $\sigma_1^2 + \cdots + \sigma_n^2 \to \infty$ as $n \to \infty$, condition (A.128) holds for all $t > 0$.

27. Prove the statement made in Example 14.

28. (i) Let $d_0, \ldots, d_e$ be nonnegative integers adding to n, and suppose that of the n numbers $v_{n1}, \ldots, v_{nn}$, d_0 are equal to zero, and the remaining $N' = n - d_0$ are the numbers $u_1, \ldots, u_{N'}$ of Example 2(iii), with N replaced by n.

Show that the condition of Example 14 is satisfied provided

(A.304) $$n - d_0 \quad \text{tends to infinity as } n \to \infty$$

(ii) Show that the Wilcoxon statistic V_s^* defined in Chap. 3, Sec. 2, has a normal limit distribution provided (A.304) holds.

[*Hint* for (i): The sum of squares of the v's is equal to the sum of squares of the u's evaluated in Example 2(iii), with n in place of N. To complete the proof use the fact that $\sum_{i=1}^{e} d_i^3 \leqslant (\sum d_i)^3 = (n - d_0)^3$.]

29. If $\hat{W}_s$ is the statistic defined by (3.29), prove that the distribution of $[\hat{W}_s - E(\hat{W}_s)]/\sqrt{\operatorname{Var}(\hat{W}_s)}$ tends to the standard normal distribution as s tends to infinity, when the block sizes N_i are bounded and when each block contains, after alignment, at least one positive and one negative observation.

{*Hint:* The proof is based on the fact that $\hat{W}_s$ is the sum of the s independent variables $\hat{W}_s^{(i)}$, and is analogous to the argument of Example 14. Let k_{ij}, $j = 1, \ldots, N_i$ denote the aligned ranks of the ith block, let $N = \sum N_i$, and suppose that $N_i \leqslant A$ for all i. To show that condition (A.130) of Theorem 5′ is satisfied, it is enough to prove that there exist positive constants a and b such that

(i) $$(k_{ij} - k_{i.})^2 \leqslant as^2 \qquad \text{for all } i \text{ and } j$$

(ii) $$\sum_{i=1}^{s} E[\hat{W}_s^{(i)} - E(\hat{W}_s^{(i)})]^2 \geqslant bs^3$$

The first of these is obvious since $N \leqslant sA$. To see (ii), note that the variance (3.31) is minimized when $n_i = 1$, so that it is enough to prove $\sum \sigma_i^2 \geqslant bs^3$ where $\sigma_i^2 = \sum_{j=1}^{N_i} (k_{ij} - k_{i.})^2/N_i$. Now of $k_{i1}, \ldots, k_{iN_i}$ at least one lies on either side of k_0, where k_0 is the rank of the aligned observation closest to zero. (For one value of i, one of these two ranks may be k_0 itself.) Hence

$$\sigma_i^2 \geqslant \frac{\min_j (k_{ij} - k_0)^2}{N_i} \geqslant \frac{\min_j (k_{ij} - k_0)^2}{A}$$

Since the differences $|k_{ij} - k_0|$ are integers, and since the k_{ij} are distinct, it follows that

$$\sum \sigma_i^2 \geqslant \frac{0^2 + 1^2 + \cdots + \left[\dfrac{s}{2}\right]^2}{A}$$

where $[s/2]$ denotes the greatest integer less than or equal to $s/2$, and this proves (ii).}

30. To show that the result of the preceding problem is not necessarily true without some condition on the method of alignment, consider the case in which all blocks are of size two and the aligned ranks are $(2i,2i+1)$ for the ith block when $i = 1,\ldots,s-1$ and $(1,2s)$ for the sth block. Find the limiting distribution of $[\hat{W}_s - E(\hat{W}_s)]/\sqrt{\mathrm{Var}(\hat{W}_s)}$ and note that it is not normal. {*Hint:* When divided by $\sqrt{\mathrm{Var}(\hat{W}_s)}$, the contribution of the first $s-1$ terms of

$$\hat{W}_s - E(\hat{W}_s) = \sum_{i=1}^{s} [\hat{W}_s^{(i)} - E(\hat{W}_s^{(i)})]$$

tends to zero in probability.}

Section 4

31. Show that condition (A.134) holds if

$$\frac{\max_i (v_{Ni} - v_{N.})^2}{\sum (v_{Nj} - v_{N.})^2} \to 0 \qquad \text{as } N \to \infty$$

and if n/N is bounded away from 0 and 1 as $N \to \infty$.

32. If J_i, K_i are defined by (A.136) and (A.138), respectively, show that $\sum (K_i - J_i)^2 = |D - n|$, where D is the number of U's that are less than or equal to n/N. [*Hint:* Suppose that $D \leqslant n$ and let J^* and K^* denote the values of J and K associated with $U^{(i)}$. Then $J^* = K^* = 1$ if $i \leqslant D$; $J^* = 1$, $K^* = 0$ for $i = D+1,\ldots,n$; $J^* = K^* = 0$ for $i > n$. The case $D > n$ can be analyzed analogously.]

33. Let $X_{Ni} = (c_{Ni} - c_{N.})Y_{Ni}$ where the c's are constants and the Y's are independent and identically distributed with common variance σ_N^2. Show that condition (A.130) holds provided

$$\frac{\max_i (c_{Ni} - c_{N.})^2}{\sigma_N^2 \sum (c_{Nj} - c_{N.})^2} \to 0 \qquad \text{as } N \to \infty \tag{A.305}$$

[*Hint:* See the proof of Example 12.]

Section 5

34. Prove the necessity part of Theorem 8. [*Hint:* Note that $E(T-S)^2 \geqslant \mathrm{Var}(T-S)$ and apply the remark that follows Eq. (A.7).]

35. If X and Y are independently distributed with density $p(x)$ and $p(y-\Delta)$, respectively, prove the following formulas for the probability p_1 defined by (A.60):

(i) If p is the exponential density (A.100) with $a = 1$,

$$p_1 = 1 - \tfrac{1}{2}e^{-\Delta} \qquad \text{if } \Delta > 0 \tag{A.306}$$

(ii) If p is the uniform density (A.99) with $a = 1$,

$$p_1 = \tfrac{1}{2} + \Delta - \tfrac{1}{2}\Delta^2 \qquad \text{if } 0 < \Delta < 1 \tag{A.307}$$

(iii) If p is the double-exponential density (A.97)

$$p_1 = 1 - \tfrac{1}{2}e^{-\Delta}(1 + \tfrac{1}{2}\Delta) \qquad \text{if } \Delta > 0 \tag{A.308}$$

(iv) If p is the logistic density (A.96)

(A.309)
$$p_1 = \frac{1-(1+\Delta)e^{-\Delta}}{(1-e^{-\Delta})^2}$$

36. If X, X' and Y, Y' are independently distributed with densities $p(x)$ and $p(y-\Delta)$, respectively, prove the following formulas for $p_2 = p_3$, given by (A.63) or (A.66):

(i) Rectangular: $p_2 = p_3 = \frac{1}{3}(1+3\Delta-\Delta^3)$ if $0 < \Delta < 1$, $a = 1$

(ii) Double-exponential: $p_2 = p_3 = 1-(\frac{7}{12}+\frac{1}{2}\Delta)e^{-\Delta}-\frac{1}{12}e^{-2\Delta}$

(iii) Logistic: $p_2 = p_3 = \dfrac{1-2\Delta e^{-\Delta}-e^{-2\Delta}}{(1-e^{-\Delta})^3}$

(iv) If p is the exponential density (A.100) with $a = 1$, show that
$$p_2 = 1-\tfrac{2}{3}e^{-\Delta} \quad \text{and} \quad p_3 = 1-e^{-\Delta}+\tfrac{1}{3}e^{-2\Delta} \quad \text{if } \Delta > 0$$

37. In the shift model (2.7) with continuous F, prove that: (i) $p_1 > 1/2$ when $\Delta > 0$; (ii) $p_1 < 1/2$ when $\Delta < 0$. {*Hint:* Show that

(A.310)
$$p_1 - \tfrac{1}{2} = \tfrac{1}{2}P(|X_2 - X_1| < \Delta)$$

where X_1 and X_2 are independent variables with distribution F, and use the fact that the probability on the right-hand side is greater than or equal to

$$\sum_{i=-\infty}^{\infty} P[i\Delta < X_1, X_2 < (i+1)\Delta].\}$$

38. Generalize the result of part (i) of the preceding problem by showing that $p_1 > 1/2$ for any continuous distributions satisfying (2.6) of Chap. 2. {*Hint:* Since

$$p_1 - \tfrac{1}{2} = E[F(Y) - G(Y)]$$

it is only necessary to show that $F(y) > G(y)$ on a set of y-values that has positive probability under G.}

39. Show that $P(a \leqslant -X \leqslant b) = 1$ implies $P(X < a) + P(X > b) = 1$ if either $a > 0$ or $b < 0$ but that the two conditions are incompatible if $a < 0 < b$.

40. Show that the distribution of the statistic W_{XY}^* defined by (1.37), and hence of W_s^*, is asymptotically normal in the population model of Chap. 2, Sec. 1. [*Hint:* Express W_{XY}^* as a U-statistic (A.159) by letting $\varphi(x, y)$ be 1, 1/2, or 0 as $y > x$, $y = x$, or $y < x$.]

41. If U is given by (A.207) and $E(U) = \theta$, show that

(i) The projection of $T_N = \sqrt{N}(U-\theta)$ onto the space of functions $S = \sum K_i(X_i)$ is

$$T_N = \frac{m\sum\psi_1(X_i)}{\sqrt{N}}$$

where ψ_1 is given by (A.208)

(ii) The expected squared difference (A.168) tends to zero as $N \to \infty$

(iii) The distribution of $[U - E(U)]/\sqrt{\text{Var}(U)}$ tends to the standard normal distribution provided $\sigma_1{}^2 > 0$ where $\sigma_1{}^2 = \text{Var}[\psi_1(X_i)]$

42. The following is a generalization of Theorems 9 and 10 and of the preceding problem. We begin by establishing some notation. Let $X_1, \ldots, X_{n_1}; Y_1, \ldots, Y_{n_2}; \ldots;$ be independently distributed according to distributions $F, G, \ldots$. Let

(A.311) $$U = \frac{1}{\binom{n_1}{m_1}\binom{n_2}{m_2}\cdots} \sum \cdots \sum \varphi(x_{i_1}, \ldots, x_{i_{m_1}}; y_{j_1}, \ldots, y_{j_{m_2}}; \ldots)$$

where φ is a function of $m_1 + m_2 + \cdots + m_s$ variables, which is symmetric in its first m_1 arguments, symmetric in its second m_2 arguments, etc. The summation extends over all $\binom{n_1}{m_1}\binom{n_2}{m_2}\cdots\binom{n_s}{m_s}$ choices of subscripts $1 \leqslant i_1 < \cdots < i_{m_1} \leqslant n_1; 1 \leqslant j_1 < \cdots < j_{m_2} \leqslant n_2$; etc. Let $E(U) = \theta$ and put

$$\psi(x_1, \ldots, x_{m_1}; y_1, \ldots, y_{m_2}; \ldots) = \varphi(x_1, \ldots, x_{m_1}; y_1, \ldots, y_{m_2}; \ldots) - \theta$$

Let

$$\psi_{10\ldots0}(x_1) = E[\psi(x_1, X_2, \ldots, X_{m_1}; Y_1, \ldots, Y_{m_2}; \ldots)]$$
$$\psi_{010\ldots0}(y_1) = E[\psi(X_1, X_2, \ldots, X_{m_1}; y_1, Y_2, \ldots, Y_{m_2}; \ldots)]$$

and so on,

and let

$$\sigma^2_{10\ldots0} = \operatorname{Var}[\psi_{10\ldots0}(X_1)], \quad \sigma^2_{010\ldots0} = \operatorname{Var}[\psi_{010\ldots0}(Y_1)]$$

and so on.

(i) If $N = n_1 + \cdots + n_s$ show that the projection of $T = \sqrt{N}(U - \theta)$ onto the space of functions $\sum a_i(X_i) + \sum b_j(Y_j) + \cdots$ is

$$T^* = \sqrt{N}\left[\frac{m_1}{n_1}\sum \psi_{10\ldots0}(X_i) + \frac{m_2}{n_2}\sum \psi_{010\ldots0}(Y_j) + \cdots\right]$$

(ii) Show that $E(T - T^*)^2 \to 0$ as $n_1, n_2, \ldots, n_s \to \infty$.

(iii) If $\lambda_i = \lim(n_i/N)$ exists and is not equal to zero for any i, show that T has a normal limit distribution with zero mean and variance

$$\sigma^2 = \frac{{m_1}^2}{\lambda_1}\sigma^2_{10\ldots0} + \frac{{m_2}^2}{\lambda_2}\sigma^2_{010\ldots0} + \cdots$$

43. In part (iii) of the preceding problem, the restriction that the λ's are different from zero can be dropped if the following changes are made. Let $n = \min(n_1, \ldots, n_s)$. In the definition of the λ's, of T, and of T^* replace N by n.

44. Use the results of Prob. 42 or Prob. 43 to prove asymptotic normality of the statistic W defined by Eq. (5.58). [*Hint:* Let

$$\varphi(x_{11}; x_{21}; \ldots; x_{s1}) = \frac{1}{n_1 n_2 \cdots n_s} \sum_{i<j}\sum n_i n_j I_{x_{i1}, x_{j1}}$$

where

$$I_{u,v} = \begin{cases} 1 & \text{if } u < v \\ 0 & \text{otherwise} \end{cases}$$

and note that

$$W = \sum_{\alpha=1}^{n_1} \sum_{\beta=1}^{n_2} \cdots \sum_{\delta=1}^{n_s} \varphi(X_{1\alpha}; X_{2\beta}; \ldots; X_{s\delta})]$$

Section 6

45. Give an example of a distribution F not satisfying (A.227).

46. Let X and Y be independently distributed according to a distribution F with density f and let the distribution of $Y-X$ be F^* with density f^*. Then if condition (3) of Sec. 6 holds,

(A.312) $$f^*(0) = \int f^2(x)\,dx$$

[*Hint:* By definition, $f^*(0)$ is the limit of $[F^*(\Delta)-F^*(0)]/\Delta$ as $\Delta \to 0$.]

47. Show that formula (A.229) is equivalent to formula (2.29) provided (A.312) holds.

48. Show that the efficiency (A.240) is unchanged under location and scale changes of F, that is, when $F(x)$ is replaced by $F[(x-b)/a]$, $a > 0$.

49. Evaluate the efficiency (A.240) when the distribution F is (i) double-exponential; (ii) logistic; (iii) uniform; (iv) exponential.

50. Consider the gross error model

$$F(x) = (1-\varepsilon)\Phi(x) + \varepsilon\Phi(x/\tau)$$

with $\varepsilon < \frac{1}{2}$. This represents the distribution of observations from a standard normal distribution that with probability ε are gross errors coming from a normal distribution with mean zero and variance τ^2. Show that the efficiency (A.240) tends to infinity as τ tends to infinity.

51. Let F be the Cauchy distribution with density f given by (A.95) and let F_A denote this distribution truncated at $\pm A$, so that its density is of the form $f_A(x) = Cf(x)$ if $|x| \leqslant A$ and $f_A(x) = 0$ if $|x| > A$. Show that the efficiency (A.240) with f_A in place of f tends to infinity as $A \to \infty$.

52. Generalize the preceding problem to other distributions with infinite variance.

53. Consider the problem of minimizing $g_0(F)$ subject to the r side conditions $g_1(F) = a_1, \ldots, g_r(F) = a_r$. This problem can sometimes be reduced to a minimization problem without side conditions by the following method of *undetermined multipliers*:

Let F_0 minimize the quantity

$$g_0(F) + c_1 g_1(F) + \cdots + c_r g_r(F) \qquad \text{where } c_1, \ldots, c_r \text{ are constant}$$

Show that if F_0 also satisfies the side conditions, it is a solution of the original problem.

54. Show that (A.247) cannot be a probability density unless $a > 0$ and $b < 0$.

55. (i) Show that (A.248) satisfies (A.243) if $B = \sqrt{5}$ and $A = 3/20\sqrt{5}$.

(ii) With these values of A, B show that (A.240) has the value .864.

56. Show that the efficiencies (A.254) and (A.255) are independent of scale, that is, are unchanged if $E(z)$ is replaced by $E(z/a)$, $a > 0$.

57. Evaluate the efficiency (A.254) when E is (i) double-exponential; (ii) logistic; (iii) uniform.

58. Evaluate the efficiency (A.255) when E is (i) Cauchy; (ii) double-exponential; (iii) logistic; (iv) uniform.

59. Suppose that E coincides with the distribution F of Prob. 50. Find the efficiencies (A.254) and (A.255) and their limits as τ tends to infinity.

60. If E is unimodal, or more generally satisfies the condition

$$e(z) \leqslant e(0) \qquad \text{for all } z$$

show that the efficiency (A.254) is greater than or equal to 1/3. [*Hint:* Without loss of generality let $e(0) = 1$. Use the method of undetermined multipliers to show that the minimizing distribution is the uniform distribution on $(-\frac{1}{2}, \frac{1}{2})$.]

Section 7

61. Let $Y_1, \ldots, Y_p$ be independent, each distributed according to the standard normal distribution $N(0,1)$, and let $X_i = Y_i - \bar{Y}$. Determine: (i) Var (X_i); (ii) Cov (X_i, X_j).

62. Under the assumptions of Prob. 16, let $i < j$ and let P_{Δ_1,Δ_2} denote the probability that $X_1, \ldots, X_{i-1}$ are the $i-1$ smallest, X_i is the ith smallest, $X_{i+1}, \ldots, X_{j-1}$ are the $(i+1)$st to $(j-1)$st smallest, X_j is the jth smallest of the X's, and $X_{j+1}, \ldots, X_n$ are the $n-j$ largest of the X's, and that

$$x_i < X_i < x_i + \Delta_1 \qquad \text{and} \qquad x_j < X_j < x_j + \Delta_2$$

(i) Show that

$$P(x_i < X_{(i)} < x_i + \Delta_1, x_j < X_{(j)} < x_j + \Delta_2) = \frac{n!}{(i-1)!(j-i)!(n-j)!} P_{\Delta_1,\Delta_2}$$

(ii) Generalize the lower and upper bounds of Prob. 16 and use these bounds to find the joint probability density of $X_{(i)}$ and $X_{(j)}$.

63. Adapt the argument of Prob. 16 to show (under the assumptions of that problem) that the joint probability density of $X_{(1)}, \ldots, X_{(n)}$ is $n! f(x_1) \cdots f(x_n)$ when $x_1 < x_2 < \cdots < x_n$ and is zero otherwise.

64. Let $X_1, \ldots, X_n$ be independent, each distributed according to the exponential distribution with density (A.100) with $a = 1$, $b = 0$, and let $Y_1 = X_{(1)}$, $Y_2 = X_{(2)} - X_{(1)}, \ldots, Y_n = X_{(n)} - X_{(n-1)}$. Use the result of the preceding problem to show that the joint density of the Y's is

$$n! \exp\{-[ny_1 + (n-1)y_2 + (n-2)y_3 + \cdots + y_n]\} \qquad 0 < y_1, y_2, \ldots, y_n$$

and hence that the variables

$$Z_1 = nY_1 \quad , \quad Z_2 = (n-1)Y_2 \quad , \quad Z_3 = (n-2)Y_3 \quad , \quad \cdots$$

are independent, with the probability density of each being given by (A.100) with $a = 1$, $b = 0$.

65. (i) If the probability density of X is given by (A.100) with $a = 1$, $b = 0$, show that $E(X) = 1$.

(ii) With the notation of the preceding problem show that $E(Y_1) = 1/n$, $E(Y_2) = 1/(n-1)$, etc.

(iii) If $X_1, \ldots, X_n$ are distributed as in Prob. 64, use part (ii) of the present problem to show that

$$E[X_{(s)}] = \frac{1}{n} + \frac{1}{n-1} + \cdots + \frac{1}{n-s+1}$$

Section 8

66. Let $(X_1, \ldots, X_r)$ be distributed according to the multinomial distribution (A.256). Show that

$$\frac{X_1 - p_1}{\sqrt{n}}, \ldots, \frac{X_r - p_r}{\sqrt{n}}$$

have a multivariate normal limit distribution with zero mean, and determine its covariance structure. [*Hint:* Represent each X_i as a sum of independent 0,1-variables.]

67. Consider a sequence of random vectors $(X_1^{(N)}, \ldots, X_c^{(N)})$ $N = 1, 2, \ldots$, each distributed according to a multiple hypergeometric distribution (A. 259), with m fixed and with $d_i = d_i^{(N)}$ satisfying $d_i^{(N)}/N \to p_i$, $(0 < p_i < 1)$. Show that the probabilities (A.259) then tend to the multinomial probabilities (A.256) with m in place of n.

68. Let X_n, U_n be independent, distributed according to $N(0,1)$ and $N(0,1/n)$, respectively, and let $Y_n = X_n + U_n$. If $T_n = (X_n, Y_n)$ and $\mathscr{S}$ is the set of points (x,y) for which $x = y$, show that the conclusion (A.278) of Theorem 16 does not hold.

69. Prove formula (A.293) for the correlation coefficient of S_{iN} and S_{jN}.

70. (i) If the variables S_{iN} are defined as in the preceding problem, and if S'_{iN} is given by (A.302), check the variance and covariance formulas (A.303).

(ii) Determine the expectations, variances and covariances of the variables Z_i defined in the proof of Corollary 10.

71. Show that for $v_{N\nu} = \nu$, the statistic (A.301) reduces to the Kruskal-Wallis statistic K.

72. If the constants $v_{N\nu}$ are defined as in Example 17, show that (A.301) reduces to the statistic K^* defined by Eq. (5.11).

73. Under the assumptions of Prob. 42, let $U^{(1)}, \ldots, U^{(r)}$ be r U-statistics defined in terms of functions $\varphi^{(1)}, \ldots, \varphi^{(r)}$. Here the values of $m_1, m_2, \ldots$, may change from one of these statistics to the next, but $n_1, \ldots, n_s$ are the same for all. Let $\theta^{(i)} = E[U^{(i)}]$ and $T_N^{(i)} = \sqrt{N}[U^{(i)} - \theta^{(i)}]$. Then the joint distribution of $(T_N^{(1)}, \ldots, T_N^{(r)})$ has an r-variate normal limit distribution, and the variances and covariances of the limit distribution are the limits of the variances and covariances of the T's. [*Hint:* For each $T_N^{(i)}$, its projection is found exactly as in the proofs of Theorems 9 and 10. Asymptotic normality of $T_N^{(1)}, \ldots, T_N^{(r)}$ is then proved by applying Theorem 15 to the vectors of projections.]

74. Let $S_{1N}, \ldots, S_{sN}$ be the variables of Theorem 19 with $v_{N\nu} = \nu$. Use the preceding theorem to give an alternative proof of the joint asymptotic normality of these variables. [*Hint:* The S_{iN} are rank-sums and can be expressed in terms of U-statistics by utilizing the relationship between W_s and W_{XY} given as (1.13).]

75. Show that the null distribution of the statistic $\hat{Q}$ defined by Eq. (6.15) is χ^2 with $s-1$ degrees of freedom, provided the block sizes are bounded and that each block, after alignment, contains at least one positive and one negative observation. [*Hint:* Apply Corollary 8 with $\hat{R}_{ij}$ in place of X_{ij} and use the fact that $\text{Cov}(\hat{R}_{ik}, \hat{R}_{jk}) = -\text{Var}(\hat{R}_{ik})/(s-1)$ by (A.41). To check condition (ii) of Theorem 18, refer to Prob. 29. To complete the proof, apply the argument given in Prob. 29.]

10. REFERENCES

Bickel, P. J. (1971): *Mathematical Statistics*, pt. I, prel. ed., Holden-Day, San Francisco.

Billingsley, P. (1968): *Weak Convergence of Probability Measures*, John Wiley & Sons, New York.

Cramér, Harald (1946): *Mathematical Methods of Statistics*, Princeton University Press, Princeton, N.J.

Doksum, Kjell (1969): "Star-shaped Transformations and the Power of Rank Tests," *Ann. Math. Statist.* **40**:1167–1176.

Feller, William (1966): *An Introduction to Probability Theory and Its Applications*, John Wiley & Sons, New York, vol. II.

Haigh, J. (1971): "A Neat Way to Prove Asymptotic Normality," *Biometrika* **58**:677–678.

Hajek, Jaroslav (1961): "Some Extensions of the Wald-Wolfowitz-Noether Theorem," *Ann. Math. Statist.* **32**:506–523.

Hodges, J. L., Jr., and Lehmann, E. L. (1961): "Comparison of the Normal Scores and Wilcoxon Tests," in *Proceedings of the Fourth Berkeley Symposium on Mathematical Statistics and Probability*, University of California Press, Berkeley, Calif., vol. 1, pp. 307–317.

Hoeffding, Wassily (1948): "A Class of Statistics with Asymptotically Normal Distribution," *Ann. Math. Statist.* **19**:293–325.

Mehra, K. L. and Sarangi, J. (1967): "Asymptotic Efficiency of Certain Rank Tests for Comparative Experiments," *Ann. Math. Statist.* **38**:90–107.

Olshen, Richard A. (1967): "Sign and Wilcoxon Tests for Linearity," *Ann. Math. Statist.* **38**:1759–1769.

Van Zwet, W. R. (1964): "Convex Transformations of Random Variables," *Math. Centrium*, Amsterdam.

TABLE A. Number of combinations $\binom{N}{n}$ of N things taken n at a time

N \ n	2	3	4	5	6	7	8	9	10	11	12
2	1										
3	3	1									
4	6	4	1								
5	10	10	5	1							
6	15	20	15	6	1						
7	21	35	35	21	7	1					
8	28	56	70	56	28	8	1				
9	36	84	126	126	84	36	9	1			
10	45	120	210	252	210	120	45	10	1		
11	55	165	330	462	462	330	165	55	11	1	
12	66	220	495	792	924	792	495	220	66	12	1
13	78	286	715	1,287	1,716	1,716	1,287	715	286	78	13
14	91	364	1,001	2,002	3,003	3,432	3,003	2,002	1,001	364	91
15	105	455	1,365	3,003	5,005	6,435	6,435	5,005	3,003	1,365	455
16	120	560	1,820	4,368	8,008	11,440	12,870	11,440	8,008	4,368	1,820
17	136	680	2,380	6,188	12,376	19,448	24,310	24,310	19,448	12,376	6,188
18	153	816	3,060	8,568	18,564	31,824	43,758	48,620	43,758	31,824	18,564
19	171	969	3,876	11,628	27,132	50,388	75,582	92,378	92,378	75,582	50,388
20	190	1,140	4,845	15,504	38,760	77,520	125,970	167,960	184,756	167,960	125,970
21	210	1,330	5,985	20,349	54,264	116,280	203,490	293,930	352,716	352,716	293,930
22	231	1,540	7,315	26,334	74,613	170,544	319,770	497,420	646,646	705,432	646,646
23	253	1,771	8,855	33,649	100,947	245,157	490,314	817,190	1,144,066	1,352,078	1,352,078
24	276	2,024	10,626	42,504	134,596	346,104	735,471	1,307,504	1,961,256	2,496,144	2,704,156
25	300	2,300	12,650	53,130	177,100	480,700	1,081,575	2,042,975	3,268,760	4,457,400	5,200,300

TABLE B. Wilcoxon rank-sum distribution: $P(W_{XY} \leqslant a)$

k_1	a	$k_2 = 3$	$k_2 = 4$	$k_2 = 5$	$k_2 = 6$	$k_2 = 7$	$k_2 = 8$	$k_2 = 9$	$k_2 = 10$	$k_2 = 11$	$k_2 = 12$
3	0	.0500	.0286	.0179	.0119	.0083	.0061	.0045	.0035	.0027	.0022
	1	.1000	.0571	.0357	.0238	.0167	.0121	.0091	.0070	.0055	.0044
	2	.2000	.1143	.0714	.0476	.0333	.0242	.0182	.0140	.0110	.0088
	3	.3500	.2000	.1250	.0833	.0583	.0424	.0318	.0245	.0192	.0154
	4	.5000	.3143	.1964	.1310	.0917	.0667	.0500	.0385	.0302	.0242
	5	.6500	.4286	.2857	.1905	.1333	.0970	.0727	.0559	.0440	.0352
	6	.8000	.5714	.3929	.2738	.1917	.1394	.1045	.0804	.0632	.0505
	7	.9000	.6857	.5000	.3571	.2583	.1879	.1409	.1084	.0852	.0681
	8	.9500	.8000	.6071	.4524	.3333	.2485	.1864	.1434	.1126	.0901
	9	1.0000	.8857	.7143	.5476	.4167	.3152	.2409	.1853	.1456	.1165
	10		.9429	.8036	.6429	.5000	.3879	.3000	.2343	.1841	.1473
	11		.9714	.8750	.7262	.5833	.4606	.3636	.2867	.2280	.1824
	12		1.0000	.9286	.8095	.6667	.5394	.4318	.3462	.2775	.2242
	13			.9643	.8690	.7417	.6121	.5000	.4056	.3297	.2681
	14			.9821	.9167	.8083	.6848	.5682	.4685	.3846	.3165
	15			1.0000	.9524	.8667	.7515	.6364	.5315	.4423	.3670
	16				.9762	.9083	.8121	.7000	.5944	.5000	.4198
	17				.9881	.9417	.8606	.7591	.6538	.5577	.4725
	18				1.0000	.9667	.9030	.8136	.7133	.6154	.5275
4	0		.0143	.0079	.0048	.0030	.0020	.0014	.0010	.0007	.0005
	1		.0286	.0159	.0095	.0061	.0040	.0028	.0020	.0015	.0011
	2		.0571	.0317	.0190	.0121	.0081	.0056	.0040	.0029	.0022
	3		.1000	.0556	.0333	.0212	.0141	.0098	.0070	.0051	.0038
	4		.1714	.0952	.0571	.0364	.0242	.0168	.0120	.0088	.0066
	5		.2429	.1429	.0857	.0545	.0364	.0252	.0180	.0132	.0099
	6		.3429	.2063	.1286	.0818	.0545	.0378	.0270	.0198	.0148
	7		.4429	.2778	.1762	.1152	.0768	.0531	.0380	.0278	.0209
	8		.5571	.3651	.2381	.1576	.1071	.0741	.0529	.0388	.0291
	9		.6571	.4524	.3048	.2061	.1414	.0993	.0709	.0520	.0390
	10		.7571	.5476	.3810	.2636	.1838	.1301	.0939	.0689	.0516
	11		.8286	.6349	.4571	.3242	.2303	.1650	.1199	.0886	.0665
	12		.9000	.7222	.5429	.3939	.2848	.2070	.1518	.1128	.0852
	13		.9429	.7937	.6190	.4636	.3414	.2517	.1868	.1399	.1060
	14		.9714	.8571	.6952	.5364	.4040	.3021	.2268	.1714	.1308
	15		.9857	.9048	.7619	.6061	.4667	.3552	.2697	.2059	.1582
	16		1.0000	.9444	.8238	.6758	.5333	.4126	.3177	.2447	.1896
	17			.9683	.8714	.7364	.5960	.4699	.3666	.2857	.2231
	18			.9841	.9143	.7939	.6586	.5301	.4196	.3304	.2604
	19			.9921	.9429	.8424	.7152	.5874	.4725	.3766	.2995
	20			1.0000	.9667	.8848	.7697	.6448	.5275	.4256	.3418
	21				.9810	.9182	.8162	.6979	.5804	.4747	.3852
	22				.9905	.9455	.8586	.7483	.6334	.5253	.4308
	23				.9952	.9636	.8929	.7930	.6823	.5744	.4764
	24				1.0000	.9788	.9232	.8350	.7303	.6234	.5236

TABLE B. **Wilcoxon rank-sum distribution:** $P(W_{XY} \leqslant a)$ *(Continued)*

k_1	a	$k_2 = 5$	$k_2 = 6$	$k_2 = 7$	$k_2 = 8$	$k_2 = 9$	$k_2 = 10$
5	0	.0040	.0022	.0013	.0008	.0005	.0003
	1	.0079	.0043	.0025	.0016	.0010	.0007
	2	.0159	.0087	.0051	.0031	.0020	.0013
	3	.0278	.0152	.0088	.0054	.0035	.0023
	4	.0476	.0260	.0152	.0093	.0060	.0040
	5	.0754	.0411	.0240	.0148	.0095	.0063
	6	.1111	.0628	.0366	.0225	.0145	.0097
	7	.1548	.0887	.0530	.0326	.0210	.0140
	8	.2103	.1234	.0745	.0466	.0300	.0200
	9	.2738	.1645	.1010	.0637	.0415	.0276
	10	.3452	.2143	.1338	.0855	.0559	.0376
	11	.4206	.2684	.1717	.1111	.0734	.0496
	12	.5000	.3312	.2159	.1422	.0949	.0646
	13	.5794	.3961	.2652	.1772	.1199	.0823
	14	.6548	.4654	.3194	.2176	.1489	.1032
	15	.7262	.5346	.3775	.2618	.1818	.1272
	16	.7897	.6039	.4381	.3108	.2188	.1548
	17	.8452	.6688	.5000	.3621	.2592	.1855
	18	.8889	.7316	.5619	.4165	.3032	.2198
	19	.9246	.7857	.6225	.4716	.3497	.2567
	20	.9524	.8355	.6806	.5284	.3986	.2970
	21	.9722	.8766	.7348	.5835	.4491	.3393
	22	.9841	.9113	.7841	.6379	.5000	.3839
	23	.9921	.9372	.8283	.6892	.5509	.4296
	24	.9960	.9589	.8662	.7382	.6014	.4765
	25	1.0000	.9740	.8990	.7824	.6503	.5235
6	0		.0011	.0006	.0003	.0002	.0001
	1		.0022	.0012	.0007	.0004	.0002
	2		.0043	.0023	.0013	.0008	.0005
	3		.0076	.0041	.0023	.0014	.0009
	4		.0130	.0070	.0040	.0024	.0015
	5		.0206	.0111	.0063	.0038	.0024
	6		.0325	.0175	.0100	.0060	.0037
	7		.0465	.0256	.0147	.0088	.0055
	8		.0660	.0367	.0213	.0128	.0080
	9		.0898	.0507	.0296	.0180	.0112
	10		.1201	.0688	.0406	.0248	.0156
	11		.1548	.0903	.0539	.0332	.0210
	12		.1970	.1171	.0709	.0440	.0280
	13		.2424	.1474	.0906	.0567	.0363
	14		.2944	.1830	.1142	.0723	.0467
	15		.3496	.2226	.1412	.0905	.0589
	16		.4091	.2669	.1725	.1119	.0736
	17		.4686	.3141	.2068	.1361	.0903
	18		.5314	.3654	.2454	.1638	.1099
	19		.5909	.4178	.2864	.1942	.1317
	20		.6504	.4726	.3310	.2280	.1566
	21		.7056	.5274	.3773	.2643	.1838
	22		.7576	.5822	.4259	.3035	.2139
	23		.8030	.6346	.4749	.3445	.2461
	24		.8452	.6859	.5251	.3878	.2811
	25		.8799	.7331	.5741	.4320	.3177
	26		.9102	.7774	.6227	.4773	.3564
	27		.9340	.8170	.6690	.5227	.3962
	28		.9535	.8526	.7136	.5680	.4374
	29		.9675	.8829	.7546	.6122	.4789
	30		.9794	.9097	.7932	.6555	.5211

k_1	a	$k_2 = 7$	$k_2 = 8$	$k_2 = 9$	$k_2 = 10$
7	0	.0003	.0002	.0001	.0001
	1	.0006	.0003	.0002	.0001
	2	.0012	.0006	.0003	.0002
	3	.0020	.0011	.0006	.0004
	4	.0035	.0019	.0010	.0006
	5	.0055	.0030	.0017	.0010
	6	.0087	.0047	.0026	.0015
	7	.0131	.0070	.0039	.0023
	8	.0189	.0103	.0058	.0034
	9	.0265	.0145	.0082	.0048
	10	.0364	.0200	.0115	.0068
	11	.0487	.0270	.0156	.0093
	12	.0641	.0361	.0209	.0125
	13	.0825	.0469	.0274	.0165
	14	.1043	.0603	.0356	.0215
	15	.1297	.0760	.0454	.0277
	16	.1588	.0946	.0571	.0351
	17	.1914	.1159	.0708	.0439
	18	.2279	.1405	.0869	.0544
	19	.2675	.1678	.1052	.0665
	20	.3100	.1984	.1261	.0806
	21	.3552	.2317	.1496	.0966
	22	.4024	.2679	.1755	.1148
	23	.4508	.3063	.2039	.1349
	24	.5000	.3472	.2349	.1574
	25	.5492	.3894	.2680	.1819
	26	.5976	.4333	.3032	.2087
	27	.6448	.4775	.3403	.2374
	28	.6900	.5225	.3788	.2681
	29	.7325	.5667	.4185	.3004
	30	.7721	.6106	.4591	.3345
	31	.8086	.6528	.5000	.3698
	32	.8412	.6937	.5409	.4063
	33	.8703	.7321	.5815	.4434
	34	.8957	.7683	.6212	.4811
	35	.9175	.8016	.6597	.5189

TABLE B. Wilcoxon rank-sum distribution: $P(W_{XY} \leqslant a)$ (*Continued*)

k_1	a	$k_2 = 8$	$k_2 = 9$	$k_2 = 10$	k_1	a	$k_2 = 9$	$k_2 = 10$	k_1	a	$k_2 = 10$
8	0	.0001	.0000	.0000	9	0	.0000	.0000	10	0	.0000
	1	.0002	.0001	.0000		1	.0000	.0000		1	.0000
	2	.0003	.0002	.0001		2	.0001	.0000		2	.0000
	3	.0005	.0003	.0002		3	.0001	.0001		3	.0000
	4	.0009	.0005	.0003		4	.0002	.0001		4	.0001
	5	.0015	.0008	.0004		5	.0004	.0002		5	.0001
	6	.0023	.0012	.0007		6	.0006	.0003		6	.0002
	7	.0035	.0019	.0010		7	.0009	.0005		7	.0002
	8	.0052	.0028	.0015		8	.0014	.0007		8	.0004
	9	.0074	.0039	.0022		9	.0020	.0011		9	.0005
	10	.0103	.0056	.0031		10	.0028	.0015		10	.0008
	11	.0141	.0076	.0043		11	.0039	.0021		11	.0010
	12	.0190	.0103	.0058		12	.0053	.0028		12	.0014
	13	.0249	.0137	.0078		13	.0071	.0038		13	.0019
	14	.0325	.0180	.0103		14	.0094	.0051		14	.0026
	15	.0415	.0232	.0133		15	.0122	.0066		15	.0034
	16	.0524	.0296	.0171		16	.0157	.0086		16	.0045
	17	.0652	.0372	.0217		17	.0200	.0110		17	.0057
	18	.0803	.0464	.0273		18	.0252	.0140		18	.0073
	19	.0974	.0570	.0338		19	.0313	.0175		19	.0093
	20	.1172	.0694	.0416		20	.0385	.0217		20	.0116
	21	.1393	.0836	.0506		21	.0470	.0267		21	.0144
	22	.1641	.0998	.0610		22	.0567	.0326		22	.0177
	23	.1911	.1179	.0729		23	.0680	.0394		23	.0216
	24	.2209	.1383	.0864		24	.0807	.0474		24	.0262
	25	.2527	.1606	.1015		25	.0951	.0564		25	.0315
	26	.2869	.1852	.1185		26	.1112	.0667		26	.0376
	27	.3227	.2117	.1371		27	.1290	.0782		27	.0446
	28	.3605	.2404	.1577		28	.1487	.0912		28	.0526
	29	.3992	.2707	.1800		29	.1701	.1055		29	.0615
	30	.4392	.3029	.2041		30	.1933	.1214		30	.0716
	31	.4796	.3365	.2299		31	.2181	.1388		31	.0827
	32	.5204	.3715	.2574		32	.2447	.1577		32	.0952
	33	.5608	.4074	.2863		33	.2729	.1781		33	.1088
	34	.6008	.4442	.3167		34	.3024	.2001		34	.1237
	35	.6395	.4813	.3482		35	.3332	.2235		35	.1399
	36	.6773	.5187	.3809		36	.3652	.2483		36	.1575
	37	.7131	.5558	.4143		37	.3981	.2745		37	.1763
	38	.7473	.5926	.4484		38	.4317	.3019		38	.1965
	39	.7791	.6285	.4827		39	.4657	.3304		39	.2179
	40	.8089	.6635	.5173		40	.5000	.3598		40	.2406
						41	.5343	.3901		41	.2644
						42	.5683	.4211		42	.2894
						43	.6019	.4524		43	.3153
						44	.6348	.4841		44	.3421
						45	.6668	.5159		45	.3697
										46	.3980
										47	.4267
										48	.4559
										49	.4853
										50	.5147

TABLE C. Area $\Phi(z)$ under the normal curve to the left of z

z	.00	.01	.02	.03	.04	.05	.06	.07	.08	.09
.0	.5000	.5040	.5080	.5120	.5160	.5199	.5239	.5279	.5319	.5359
.1	.5398	.5438	.5478	.5517	.5557	.5596	.5636	.5675	.5714	.5753
.2	.5793	.5832	.5871	.5910	.5948	.5987	.6026	.6064	.6103	.6141
.3	.6179	.6217	.6255	.6293	.6331	.6368	.6406	.6443	.6480	.6517
.4	.6554	.6591	.6628	.6664	.6700	.6736	.6772	.6808	.6844	.6879
.5	.6915	.6950	.6985	.7019	.7054	.7088	.7123	.7157	.7190	.7224
.6	.7257	.7291	.7324	.7357	.7389	.7422	.7454	.7486	.7517	.7549
.7	.7580	.7611	.7642	.7673	.7704	.7734	.7764	.7794	.7823	.7852
.8	.7881	.7910	.7939	.7967	.7995	.8023	.8051	.8078	.8106	.8133
.9	.8159	.8186	.8212	.8238	.8264	.8289	.8315	.8340	.8365	.8389
1.0	.8413	.8438	.8461	.8485	.8508	.8531	.8554	.8577	.8599	.8621
1.1	.8643	.8665	.8686	.8708	.8729	.8749	.8770	.8790	.8810	.8830
1.2	.8849	.8869	.8888	.8907	.8925	.8944	.8962	.8980	.8997	.9015
1.3	.9032	.9049	.9066	.9082	.9099	.9115	.9131	.9147	.9162	.9177
1.4	.9192	.9207	.9222	.9236	.9251	.9265	.9279	.9292	.9306	.9319
1.5	.9332	.9345	.9357	.9370	.9382	.9394	.9406	.9418	.9429	.9441
1.6	.9452	.9463	.9474	.9484	.9495	.9505	.9515	.9525	.9535	.9545
1.7	.9554	.9564	.9573	.9582	.9591	.9599	.9608	.9616	.9625	.9633
1.8	.9641	.9649	.9656	.9664	.9671	.9678	.9686	.9693	.9699	.9706
1.9	.9713	.9719	.9726	.9732	.9738	.9744	.9750	.9756	.9761	.9767
2.0	.9772	.9778	.9783	.9788	.9793	.9798	.9803	.9808	.9812	.9817
2.1	.9821	.9826	.9830	.9834	.9838	.9842	.9846	.9850	.9854	.9857
2.2	.9861	.9864	.9868	.9871	.9875	.9878	.9881	.9884	.9887	.9890
2.3	.9893	.9896	.9898	.9901	.9904	.9906	.9909	.9911	.9913	.9916
2.4	.9918	.9920	.9922	.9925	.9927	.9929	.9931	.9932	.9934	.9936
2.5	.9938	.9940	.9941	.9943	.9945	.9946	.9948	.9949	.9951	.9952
2.6	.9953	.9955	.9956	.9957	.9959	.9960	.9961	.9962	.9963	.9964
2.7	.9965	.9966	.9967	.9968	.9969	.9970	.9971	.9972	.9973	.9974
2.8	.9974	.9975	.9976	.9977	.9977	.9978	.9979	.9979	.9980	.9981
2.9	.9981	.9982	.9982	.9983	.9984	.9984	.9985	.9985	.9986	.9986
3.0	.9987	.9987	.9987	.9988	.9988	.9989	.9989	.9989	.9990	.9990
3.1	.9990	.9991	.9991	.9991	.9992	.9992	.9992	.9992	.9993	.9993
3.2	.9993	.9993	.9994	.9994	.9994	.9994	.9994	.9995	.9995	.9995
3.3	.9995	.9995	.9995	.9996	.9996	.9996	.9996	.9996	.9996	.9997
3.4	.9997	.9997	.9997	.9997	.9997	.9997	.9997	.9997	.9997	.9998

Auxiliary table of z in terms of $\Phi(z)$

$\Phi(z)$	z	$\Phi(z)$	z	$\Phi(z)$	z
.50	0	.91	1.341	.995	2.576
.55	.126	.92	1.405	.999	3.090
.60	.253	.93	1.476	.9995	3.291
.65	.385	.94	1.555	.9999	3.719
.70	.524	.95	1.645	.99995	3.891
.75	.674	.96	1.751	.99999	4.265
.80	.842	.97	1.881	.999995	4.417
.85	1.036	.98	2.054	.999999	4.753
.90	1.282	.99	2.326	.9999999	5.199

TABLE D. Square roots

n	$\sqrt{n}$	$\sqrt{10n}$	n	$\sqrt{n}$	$\sqrt{10n}$	n	$\sqrt{n}$	$\sqrt{10n}$
1	1.0000	3.162	34	5.8310	18.439	67	8.1854	25.884
2	1.4142	4.472	35	5.9161	18.708	68	8.2462	26.077
3	1.7321	5.477	36	6.0000	18.974	69	8.3066	26.268
4	2.0000	6.325	37	6.0828	19.235	70	8.3666	26.458
5	2.2361	7.071	38	6.1644	19.494	71	8.4261	26.646
6	2.4495	7.746	39	6.2450	19.748	72	8.4853	26.833
7	2.6458	8.367	40	6.3246	20.000	73	8.5440	27.019
8	2.8284	9.944	41	6.4031	20.248	74	8.6023	27.203
9	3.0000	9.487	42	6.4807	20.494	75	8.6603	27.386
10	3.1623	10.000	43	6.5574	20.736	76	8.7178	27.568
11	3.3166	10.488	44	6.6332	20.976	77	8.7750	27.749
12	3.4641	10.954	45	6.7082	21.213	78	8.8318	27.928
13	3.6056	11.402	46	6.7823	21.448	79	8.8882	28.107
14	3.7417	11.832	47	6.8557	21.679	80	8.9443	28.284
15	3.8730	12.247	48	6.9282	21.909	81	9.0000	28.460
16	4.0000	12.649	49	7.0000	22.136	82	9.0554	28.636
17	4.1231	13.038	50	7.0711	22.361	83	9.1104	28.810
18	4.2426	13.416	51	7.1414	22.583	84	9.1652	28.983
19	4.3589	13.784	52	7.2111	22.804	85	9.2195	29.155
20	4.4721	14.142	53	7.2801	23.022	86	9.2736	29.326
21	4.5826	14.491	54	7.3485	23.238	87	9.3274	29.496
22	4.6904	14.832	55	7.4162	23.452	88	9.3808	29.665
23	4.7958	15.166	56	7.4833	23.664	89	9.4340	29.833
24	4.8990	15.492	57	7.5498	23.875	90	9.4868	30.000
25	5.0000	15.811	58	7.6158	24.083	91	9.5394	30.166
26	5.0990	16.125	59	7.6811	24.290	92	9.5917	30.332
27	5.1962	16.432	60	7.7460	24.495	93	9.6437	30.496
28	5.2915	16.733	61	7.8102	24.698	94	9.6954	30.659
29	5.3852	17.029	62	7.8740	24.900	95	9.7468	30.822
30	5.4772	17.321	63	7.9373	25.100	96	9.7980	30.984
31	5.5678	17.607	64	8.0000	25.298	97	9.8489	31.145
32	5.6569	17.889	65	8.0623	25.495	98	9.8995	31.305
33	5.7446	18.166	66	8.1240	25.690	99	9.9499	31.464

TABLE E. Smirnov exact upper-tail probabilities: $P(D_{n,n} \geqslant a/n)$

a \ n	1	2	3	4	5	6
1	1	1	1	1	1	1
2		0.3333	0.6000	0.7714	0.8730	0.9307
3			0.1000	0.2286	0.3571	0.4740
4				0.0286	0.0794	0.1429
5					0.0079	0.0260
6						0.0022

a \ n	7	8	9	10	11	12
1	1	1	1	1	1	1
2	0.9627	0.9801	0.9895	0.9945	0.9971	0.9985
3	0.5752	0.6601	0.7301	0.7869	0.8326	0.8690
4	0.2121	0.2827	0.3517	0.4175	0.4792	0.5361
5	0.0530	0.0870	0.1259	0.1678	0.2115	0.2558
6	0.0082	0.0186	0.0336	0.0524	0.0747	0.0995
7	0.0006	0.0025	0.0063	0.0123	0.0207	0.0314
8		0.0002	0.0007	0.0021	0.0044	0.0079
9			0.0000	0.0002	0.0007	0.0015
10				0.0000	0.0001	0.0002
11					0.0000	0.0000

a \ n	13	14	15	16	17	18
1	1	1	1	1	1	1
2	0.9992	0.9996	0.9998	0.9999	0.9999	1.0000
3	0.8978	0.9205	0.9383	0.9523	0.9631	0.9715
4	0.5882	0.6355	0.6781	0.7164	0.7506	0.7810
5	0.2999	0.3433	0.3855	0.4263	0.4654	0.5026
6	0.1265	0.1549	0.1844	0.2145	0.2450	0.2754
7	0.0443	0.0590	0.0755	0.0933	0.1124	0.1324
8	0.0126	0.0188	0.0262	0.0350	0.0450	0.0560
9	0.0029	0.0049	0.0077	0.0112	0.0156	0.0207
10	0.0005	0.0010	0.0018	0.0030	0.0046	0.0067
11	0.0001	0.0002	0.0004	0.0007	0.0012	0.0018
12	0.0000	0.0000	0.0001	0.0001	0.0002	0.0004
13			0.0000	0.0000	0.0000	0.0001
14						0.0000

TABLE E. Smirnov exact upper-tail probabilities: $P(D_{n,n} \geqslant a/n)$ (*Continued*)

a \ n	19	20	21	22	23	24
1	1	1	1	1	1	1
2	1.0000	1.0000	1.0000	1.0000	1.0000	1.0000
3	0.9781	0.9831	0.9870	0.9901	0.9924	0.9942
4	0.8081	0.8320	0.8531	0.8717	0.8880	0.9024
5	0.5379	0.5713	0.6028	0.6324	0.6601	0.6860
6	0.3057	0.3356	0.3650	0.3937	0.4218	0.4490
7	0.1532	0.1745	0.1963	0.2184	0.2406	0.2628
8	0.0681	0.0811	0.0948	0.1093	0.1243	0.1398
9	0.0267	0.0335	0.0411	0.0493	0.0583	0.0678
10	0.0092	0.0123	0.0159	0.0200	0.0247	0.0299
11	0.0028	0.0040	0.0055	0.0073	0.0095	0.0120
12	0.0007	0.0011	0.0017	0.0024	0.0032	0.0043
13	0.0002	0.0003	0.0004	0.0007	0.0010	0.0014
14	0.0000	0.0001	0.0001	0.0002	0.0003	0.0004
15		0.0000	0.0000	0.0000	0.0001	0.0001
16					0.0000	0.0000

a \ n	25	26	27	28	29	30
1	1	1	1	1	1	1
2	1.0000	1.0000	1.0000	1.0000	1.0000	1.0000
3	0.9955	0.9966	0.9974	0.9980	0.9985	0.9988
4	0.9150	0.9260	0.9357	0.9441	0.9514	0.9578
5	0.7102	0.7327	0.7537	0.7732	0.7912	0.8080
6	0.4755	0.5010	0.5256	0.5494	0.5722	0.5941
7	0.2850	0.3071	0.3290	0.3506	0.3720	0.3929
8	0.1558	0.1720	0.1886	0.2053	0.2221	0.2391
9	0.0779	0.0885	0.0996	0.1110	0.1229	0.1350
10	0.0356	0.0418	0.0484	0.0555	0.0630	0.0709
11	0.0148	0.0181	0.0217	0.0256	0.0299	0.0346
12	0.0056	0.0071	0.0089	0.0109	0.0131	0.0156
13	0.0019	0.0026	0.0033	0.0043	0.0053	0.0065
14	0.0006	0.0008	0.0011	0.0015	0.0020	0.0025
15	0.0002	0.0002	0.0004	0.0005	0.0007	0.0009
16	0.0000	0.0001	0.0001	0.0001	0.0002	0.0003
17		0.0000	0.0000	0.0000	0.0001	0.0001
18				0.0000	0.0000	0.0000

TABLE F. Smirnov limiting distribution: $K(z) = \lim P[\sqrt{mn/(m+n)}D_{m,n} \geqslant z]$

z	$K(z)$	z	$K(z)$	z	$K(z)$	z	$K(z)$
0.32	1.0000	0.72	0.6777	1.12	0.1626	1.54	0.0174
0.33	0.9999	0.73	0.6609	1.13	0.1555	1.56	0.0154
0.34	0.9998	0.74	0.6440	1.14	0.1486	1.58	0.0136
0.35	0.9997	0.75	0.6272	1.15	0.1420	1.60	0.0120
0.36	0.9995	0.76	0.6104	1.16	0.1356	1.62	0.0105
0.37	0.9992	0.77	0.5936	1.17	0.1294	1.64	0.0092
0.38	0.9987	0.78	0.5770	1.18	0.1235	1.66	0.0081
0.39	0.9981	0.79	0.5605	1.19	0.1177	1.68	0.0071
0.40	0.9972	0.80	0.5441	1.20	0.1122	1.70	0.0062
0.41	0.9960	0.81	0.5280	1.21	0.1070	1.72	0.0054
0.42	0.9945	0.82	0.5120	1.22	0.1019	1.74	0.0047
0.43	0.9926	0.83	0.4962	1.23	0.0970	1.76	0.0041
0.44	0.9903	0.84	0.4806	1.24	0.0924	1.78	0.0035
0.45	0.9874	0.85	0.4653	1.25	0.0879	1.80	0.0031
0.46	0.9840	0.86	0.4503	1.26	0.0836	1.82	0.0027
0.47	0.9800	0.87	0.4355	1.27	0.0794	1.84	0.0023
0.48	0.9753	0.88	0.4209	1.28	0.0755	1.86	0.0020
0.49	0.9700	0.89	0.4067	1.29	0.0717	1.88	0.0017
0.50	0.9639	0.90	0.3927	1.30	0.0681	1.90	0.0015
0.51	0.9572	0.91	0.3791	1.31	0.0646	1.92	0.0013
0.52	0.9497	0.92	0.3657	1.32	0.0613	1.94	0.0011
0.53	0.9415	0.93	0.3527	1.33	0.0582	1.96	0.0009
0.54	0.9325	0.94	0.3399	1.34	0.0551	1.98	0.0008
0.55	0.9228	0.95	0.3275	1.35	0.0522	2.00	0.0007
0.56	0.9124	0.96	0.3154	1.36	0.0495	2.02	0.0006
0.57	0.9013	0.97	0.3036	1.37	0.0469	2.04	0.0005
0.58	0.8896	0.98	0.2921	1.38	0.0444	2.06	0.0004
0.59	0.8772	0.99	0.2809	1.39	0.0420	2.08	0.0003
0.60	0.8643	1.00	0.2700	1.40	0.0397	2.10	0.0003
0.61	0.8508	1.01	0.2594	1.41	0.0375	2.12	0.0002
0.62	0.8368	1.02	0.2492	1.42	0.0354	2.14	0.0002
0.63	0.8222	1.03	0.2392	1.43	0.0335	2.16	0.0002
0.64	0.8073	1.04	0.2296	1.44	0.0316	2.18	0.0001
0.65	0.7920	1.05	0.2202	1.45	0.0298	2.20	0.0001
0.66	0.7764	1.06	0.2111	1.46	0.0282	2.22	0.0001
0.67	0.7604	1.07	0.2024	1.47	0.0266	2.24	0.0001
0.68	0.7442	1.08	0.1939	1.48	0.0250	2.26	0.0001
0.69	0.7278	1.09	0.1857	1.49	0.0236	2.28	0.0001
0.70	0.7112	1.10	0.1777	1.50	0.0222	2.30	0.0001
0.71	0.6945	1.11	0.1701	1.52	0.0197	2.32	0.0000

TABLE G. Distribution of sign-test statistic: $P(S_N \leqslant a)$

a \ N	2	3	4	5	6	7	8	9
0	.2500	.1250	.0625	.0313	.0156	.0078	.0039	.0020
1	.7500	.5000	.3125	.1875	.1094	.0625	.0352	.0195
2	1.0000	.8750	.6875	.5000	.3438	.2266	.1445	.0898
3		1.0000	.9375	.8125	.6562	.5000	.3633	.2539
4			1.0000	.9687	.8906	.7734	.6367	.5000

a \ N	10	11	12	13	14	15	16	17
0	.0010	.0005	.0002	.0001	.0001	.0000	.0000	.0000
1	.0107	.0059	.0032	.0017	.0009	.0005	.0003	.0001
2	.0547	.0327	.0193	.0112	.0065	.0037	.0021	.0012
3	.1719	.1133	.0730	.0461	.0287	.0176	.0106	.0064
4	.3770	.2744	.1938	.1334	.0898	.0592	.0384	.0245
5	.6230	.5000	.3872	.2905	.2120	.1509	.1051	.0717
6	.8281	.7256	.6128	.5000	.3953	.3036	.2272	.1662
7	.9453	.8867	.8062	.7095	.6047	.5000	.4018	.3145
8	.9893	.9673	.9270	.8666	.7880	.6964	.5982	.5000

a \ N	18	19	20	21	22	23	24	25
0	.0000	.0000	.0000	.0000	.0000	.0000	.0000	.0000
1	.0001	.0000	.0000	.0000	.0000	.0000	.0000	.0000
2	.0007	.0004	.0002	.0001	.0001	.0000	.0000	.0000
3	.0038	.0022	.0013	.0007	.0004	.0002	.0001	.0001
4	.0154	.0096	.0059	.0036	.0022	.0013	.0008	.0005
5	.0481	.0318	.0207	.0133	.0085	.0053	.0033	.0020
6	.1189	.0835	.0577	.0392	.0262	.0173	.0113	.0073
7	.2403	.1796	.1316	.0946	.0669	.0466	.0320	.0216
8	.4073	.3238	.2517	.1917	.1431	.1050	.0758	.0539
9	.5927	.5000	.4119	.3318	.2617	.2024	.1537	.1148
10	.7597	.6762	.5881	.5000	.4159	.3388	.2706	.2122
11	.8811	.8204	.7483	.6682	.5841	.5000	.4194	.3450
12	.9519	.9165	.8684	.8083	.7383	.6612	.5806	.5000

TABLE G. Distribution of sign-test statistic: $P(S_N \leqslant a)$ (*Continued*)

a \ N	26	27	28	29	30	31	32	33
0	.0000	.0000	.0000	.0000	.0000	.0000	.0000	.0000
1	.0000	.0000	.0000	.0000	.0000	.0000	.0000	.0000
2	.0000	.0000	.0000	.0000	.0000	.0000	.0000	.0000
3	.0000	.0000	.0000	.0000	.0000	.0000	.0000	.0000
4	.0003	.0002	.0001	.0001	.0000	.0000	.0000	.0000
5	.0012	.0008	.0005	.0003	.0002	.0001	.0001	.0000
6	.0047	.0030	.0019	.0012	.0007	.0004	.0003	.0002
7	.0145	.0096	.0063	.0041	.0026	.0017	.0011	.0007
8	.0378	.0261	.0178	.0121	.0081	.0053	.0035	.0023
9	.0843	.0610	.0436	.0307	.0214	.0147	.0100	.0068
10	.1635	.1239	.0925	.0680	.0494	.0354	.0251	.0175
11	.2786	.2210	.1725	.1325	.1002	.0748	.0551	.0401
12	.4225	.3506	.2858	.2291	.1808	.1405	.1077	.0814
13	.5775	.5000	.4253	.3555	.2923	.2366	.1885	.1481
14	.7214	.6494	.5747	.5000	.4278	.3601	.2983	.2434
15	.8365	.7790	.7142	.6445	.5722	.5000	.4300	.3642
16	.9157	.8761	.8275	.7709	.7077	.6399	.5700	.5000

a \ N	34	35	36	37	38	39	40
0	.0000	.0000	.0000	.0000	.0000	.0000	.0000
1	.0000	.0000	.0000	.0000	.0000	.0000	.0000
2	.0000	.0000	.0000	.0000	.0000	.0000	.0000
3	.0000	.0000	.0000	.0000	.0000	.0000	.0000
4	.0000	.0000	.0000	.0000	.0000	.0000	.0000
5	.0000	.0000	.0000	.0000	.0000	.0000	.0000
6	.0001	.0001	.0000	.0000	.0000	.0000	.0000
7	.0004	.0003	.0002	.0001	.0001	.0000	.0000
8	.0015	.0009	.0006	.0004	.0002	.0001	.0001
9	.0045	.0030	.0020	.0013	.0008	.0005	.0003
10	.0122	.0083	.0057	.0038	.0025	.0017	.0011
11	.0288	.0205	.0144	.0100	.0069	.0047	.0032
12	.0607	.0448	.0326	.0235	.0168	.0119	.0083
13	.1147	.0877	.0662	.0494	.0365	.0266	.0192
14	.1958	.1553	.1215	.0939	.0717	.0541	.0403
15	.3038	.2498	.2025	.1620	.1279	.0998	.0769
16	.4321	.3679	.3089	.2557	.2088	.1684	.1341
17	.5679	.5000	.4340	.3714	.3136	.2612	.2148
18	.6962	.6321	.5660	.5000	.4357	.3746	.3179
19	.8042	.7502	.6911	.6286	.5643	.5000	.4373
20	.8853	.8447	.7975	.7443	.6864	.6254	.5627

TABLE H. Wilcoxon signed-rank distribution: $P(V \leqslant v)$

v \ N	1	2	3	4	5	6	7
0	.5000	.2500	.1250	.0625	.0313	.0156	.0078
1	1.0000	.5000	.2500	.1250	.0625	.0313	.0156
2		.7500	.3750	.1875	.0938	.0469	.0234
3		1.0000	.6250	.3125	.1563	.0781	.0391
4			.7500	.4375	.2188	.1094	.0547
5			.8750	.5625	.3125	.1563	.0781
6			1.0000	.6875	.4063	.2188	.1094
7				.8125	.5000	.2813	.1484
8				.8750	.5937	.3438	.1875
9				.9375	.6875	.4219	.2344
10				1.0000	.7812	.5000	.2891
11					.8437	.5781	.3438
12					.9062	.6562	.4063
13					.9375	.7187	.4688
14					.9687	.7812	.5312

v \ N	8	9	10	11	12	13	14
0	.0039	.0020	.0010	.0005	.0002	.0001	.0001
1	.0078	.0039	.0020	.0010	.0005	.0002	.0001
2	.0117	.0059	.0029	.0015	.0007	.0004	.0002
3	.0195	.0098	.0049	.0024	.0012	.0006	.0003
4	.0273	.0137	.0068	.0034	.0017	.0009	.0004
5	.0391	.0195	.0098	.0049	.0024	.0012	.0006
6	.0547	.0273	.0137	.0068	.0034	.0017	.0009
7	.0742	.0371	.0186	.0093	.0046	.0023	.0012
8	.0977	.0488	.0244	.0122	.0061	.0031	.0015
9	.1250	.0645	.0322	.0161	.0081	.0040	.0020
10	.1563	.0820	.0420	.0210	.0105	.0052	.0026
11	.1914	.1016	.0527	.0269	.0134	.0067	.0034
12	.2305	.1250	.0654	.0337	.0171	.0085	.0043
13	.2734	.1504	.0801	.0415	.0212	.0107	.0054
14	.3203	.1797	.0967	.0508	.0261	.0133	.0067
15	.3711	.2129	.1162	.0615	.0320	.0164	.0083
16	.4219	.2480	.1377	.0737	.0386	.0199	.0101
17	.4727	.2852	.1611	.0874	.0461	.0239	.0123
18	.5273	.3262	.1875	.1030	.0549	.0287	.0148
19	.5781	.3672	.2158	.1201	.0647	.0341	.0176
20	.6289	.4102	.2461	.1392	.0757	.0402	.0209
21	.6797	.4551	.2783	.1602	.0881	.0471	.0247
22	.7266	.5000	.3125	.1826	.1018	.0549	.0290
23	.7695	.5449	.3477	.2065	.1167	.0636	.0338
24	.8086	.5898	.3848	.2324	.1331	.0732	.0392
25	.8437	.6328	.4229	.2598	.1506	.0839	.0453
26	.8750	.6738	.4609	.2886	.1697	.0955	.0520
27	.9023	.7148	.5000	.3188	.1902	.1082	.0594
28	.9258	.7520	.5391	.3501	.2119	.1219	.0676
29	.9453	.7871	.5771	.3823	.2349	.1367	.0765
30	.9609	.8203	.6152	.4155	.2593	.1527	.0863
31	.9727	.8496	.6523	.4492	.2847	.1698	.0969
32	.9805	.8750	.6875	.4829	.3110	.1879	.1083

TABLE H. Wilcoxon signed-rank distribution: $P(V \leqslant v)$ (*Continued*)

v \ N	8	9	10	11	12	13	14
33	.9883	.8984	.7217	.5171	.3386	.2072	.1206
34	.9922	.9180	.7539	.5508	.3667	.2274	.1338
35	.9961	.9355	.7842	.5845	.3955	.2487	.1479
36	1.0000	.9512	.8125	.6177	.4250	.2709	.1629
37		.9629	.8389	.6499	.4548	.2939	.1788
38		.9727	.8623	.6812	.4849	.3177	.1955
39		.9805	.8838	.7114	.5151	.3424	.2131
40		.9863	.9033	.7402	.5452	.3677	.2316
41		.9902	.9199	.7676	.5750	.3934	.2508
42		.9941	.9346	.7935	.6045	.4197	.2708
43		.9961	.9473	.8174	.6333	.4463	.2915
44		.9980	.9580	.8398	.6614	.4730	.3129
45		1.0000	.9678	.8608	.6890	.5000	.3349
46			.9756	.8799	.7153	.5270	.3574
47			.9814	.8970	.7407	.5537	.3804
48			.9863	.9126	.7651	.5803	.4039
49			.9902	.9263	.7881	.6066	.4276
50			.9932	.9385	.8098	.6323	.4516
51			.9951	.9492	.8303	.6576	.4758
52			.9971	.9585	.8494	.6823	.5000

v \ N	15	16	17	18	19	20
0	.0000	.0000	.0000	.0000	.0000	.0000
1	.0001	.0000	.0000	.0000	.0000	.0000
2	.0001	.0000	.0000	.0000	.0000	.0000
3	.0002	.0001	.0000	.0000	.0000	.0000
4	.0002	.0001	.0001	.0000	.0000	.0000
5	.0003	.0002	.0001	.0000	.0000	.0000
6	.0004	.0002	.0001	.0001	.0000	.0000
7	.0006	.0003	.0001	.0001	.0000	.0000
8	.0008	.0004	.0002	.0001	.0000	.0000
9	.0010	.0005	.0003	.0001	.0001	.0000
10	.0013	.0007	.0003	.0002	.0001	.0000
11	.0017	.0008	.0004	.0002	.0001	.0001
12	.0021	.0011	.0005	.0003	.0001	.0001
13	.0027	.0013	.0007	.0003	.0002	.0001
14	.0034	.0017	.0008	.0004	.0002	.0001
15	.0042	.0021	.0010	.0005	.0003	.0001
16	.0051	.0026	.0013	.0006	.0003	.0002
17	.0062	.0031	.0016	.0008	.0004	.0002
18	.0075	.0038	.0019	.0010	.0005	.0002
19	.0090	.0046	.0023	.0012	.0006	.0003
20	.0108	.0055	.0028	.0014	.0007	.0004
21	.0128	.0065	.0033	.0017	.0008	.0004
22	.0151	.0078	.0040	.0020	.0010	.0005
23	.0177	.0091	.0047	.0024	.0012	.0006
24	.0206	.0107	.0055	.0028	.0014	.0007
25	.0240	.0125	.0064	.0033	.0017	.0008
26	.0277	.0145	.0075	.0038	.0020	.0010
27	.0319	.0168	.0087	.0045	.0023	.0012

TABLE H. Wilcoxon signed-rank distribution: $P(V \leq v)$ *(Continued)*

v \ N	15	16	17	18	19	20
28	.0365	.0193	.0101	.0052	.0027	.0014
29	.0416	.0222	.0116	.0060	.0031	.0016
30	.0473	.0253	.0133	.0069	.0036	.0018
31	.0535	.0288	.0153	.0080	.0041	.0021
32	.0603	.0327	.0174	.0091	.0047	.0024
33	.0677	.0370	.0198	.0104	.0054	.0028
34	.0757	.0416	.0224	.0118	.0062	.0032
35	.0844	.0467	.0253	.0134	.0070	.0036
36	.0938	.0523	.0284	.0152	.0080	.0042
37	.1039	.0583	.0319	.0171	.0090	.0047
38	.1147	.0649	.0357	.0192	.0102	.0053
39	.1262	.0719	.0398	.0216	.0115	.0060
40	.1384	.0795	.0443	.0241	.0129	.0068
41	.1514	.0877	.0492	.0269	.0145	.0077
42	.1651	.0964	.0544	.0300	.0162	.0086
43	.1796	.1057	.0601	.0333	.0180	.0096
44	.1947	.1156	.0662	.0368	.0201	.0107
45	.2106	.1261	.0727	.0407	.0223	.0120
46	.2271	.1372	.0797	.0449	.0247	.0133
47	.2444	.1489	.0871	.0494	.0273	.0148
48	.2622	.1613	.0950	.0542	.0301	.0164
49	.2807	.1742	.1034	.0594	.0331	.0181
50	.2997	.1877	.1123	.0649	.0364	.0200
51	.3193	.2019	.1218	.0708	.0399	.0220
52	.3394	.2166	.1317	.0770	.0437	.0242
53	.3599	.2319	.1421	.0837	.0478	.0266
54	.3808	.2477	.1530	.0907	.0521	.0291
55	.4020	.2641	.1645	.0982	.0567	.0319
56	.4235	.2809	.1764	.1061	.0616	.0348
57	.4452	.2983	.1889	.1144	.0668	.0379
58	.4670	.3161	.2019	.1231	.0723	.0413
59	.4890	.3343	.2153	.1323	.0782	.0448
60	.5110	.3529	.2293	.1419	.0844	.0487
61	.5330	.3718	.2437	.1519	.0909	.0527
62	.5548	.3910	.2585	.1624	.0978	.0570
63	.5765	.4104	.2738	.1733	.1051	.0615
64	.5980	.4301	.2895	.1846	.1127	.0664
65	.6192	.4500	.3056	.1964	.1206	.0715
66	.6401	.4699	.3221	.2086	.1290	.0768
67	.6606	.4900	.3389	.2211	.1377	.0825
68	.6807	.5100	.3559	.2341	.1467	.0884
69	.7003	.5301	.3733	.2475	.1562	.0947
70	.7193	.5500	.3910	.2613	.1660	.1012
71	.7378	.5699	.4088	.2754	.1762	.1081
72	.7556	.5896	.4268	.2899	.1868	.1153
73	.7729	.6090	.4450	.3047	.1977	.1227
74	.7894	.6282	.4633	.3198	.2090	.1305
75	.8053	.6471	.4816	.3353	.2207	.1387
76	.8204	.6657	.5000	.3509	.2327	.1471
77	.8349	.6839	.5184	.3669	.2450	.1559
78	.8486	.7017	.5367	.3830	.2576	.1650
79	.8616	.7191	.5550	.3994	.2706	.1744

TABLE H. Wilcoxon signed-rank distribution: $P(V \leqslant v)$ *(Continued)*

v \ N	15	16	17	18	19	20
80	.8738	.7359	.5732	.4159	.2839	.1841
81	.8853	.7523	.5912	.4325	.2974	.1942
82	.8961	.7681	.6090	.4493	.3113	.2045
83	.9062	.7834	.6267	.4661	.3254	.2152
84	.9156	.7981	.6441	.4831	.3397	.2262
85	.9243	.8123	.6611	.5000	.3543	.2375
86	.9323	.8258	.6779	.5169	.3690	.2490
87	.9397	.8387	.6944	.5339	.3840	.2608
88	.9465	.8511	.7105	.5507	.3991	.2729
89	.9527	.8628	.7262	.5675	.4144	.2853
90	.9584	.8739	.7415	.5841	.4298	.2979
91	.9635	.8844	.7563	.6006	.4453	.3108
92	.9681	.8943	.7707	.6170	.4609	.3238
93	.9723	.9036	.7847	.6331	.4765	.3371
94	.9760	.9123	.7981	.6491	.4922	.3506
95	.9794	.9205	.8111	.6647	.5078	.3643
96	.9823	.9281	.8236	.6802	.5235	.3781
97	.9849	.9351	.8355	.6953	.5391	.3921
98	.9872	.9417	.8470	.7101	.5547	.4062
99	.9892	.9477	.8579	.7246	.5702	.4204
100	.9910	.9533	.8683	.7387	.5856	.4347
101	.9925	.9584	.8782	.7525	.6009	.4492
102	.9938	.9630	.8877	.7659	.6160	.4636
103	.9949	.9673	.8966	.7789	.6310	.4782
104	.9958	.9712	.9050	.7914	.6457	.4927
105	.9966	.9747	.9129	.8036	.6603	.5073

TABLE I. Kruskal-Wallis upper-tail probabilities: $P = P(K \geqslant c)$; *(3 groups; sample sizes in parentheses)*

(1, 1, 4)

c	P
3.571	.200

(1, 1, 5)

c	P
3.857	.143

(1, 2, 2)

c	P
3.600	.200

(1, 2, 3)

c	P
3.857	.133
4.286	.100

(1, 2, 4)

c	P
3.750	.133
4.018	.114
4.500	.076
4.821	.057

(1, 2, 5)

c	P
3.783	.131
4.050	.119
4.200	.095
4.450	.071
5.000	.048
5.250	.036

(1, 3, 3)

c	P
4.000	.129
4.571	.100
5.143	.043

(1, 3, 4)

c	P
3.764	.136
3.889	.129
4.056	.093
4.097	.086
4.208	.079
4.764	.071
5.000	.057
5.208	.050
5.389	.036
5.833	.021

(1, 3, 5)

c	P
3.378	.143
3.484	.135
3.804	.131
3.840	.123
4.018	.095
4.284	.083
4.338	.079
4.551	.075
4.711	.056
4.871	.052
4.960	.048
5.404	.044
5.440	.036
5.760	.028
6.044	.020
6.400	.012

(1, 4, 4)

c	P
3.867	.121
3.900	.108
4.067	.102
4.167	.083
4.267	.070
4.800	.067
4.867	.054
4.967	.048
5.100	.041
5.667	.035
6.000	.029
6.167	.022
6.667	.010

(1, 4, 5)

c	P
3.524	.146
3.595	.138
3.682	.132
3.813	.110
3.960	.102
3.987	.098
4.206	.095
4.222	.087
4.287	.071
4.549	.067
4.636	.063
4.724	.060
4.833	.059
4.860	.056
4.986	.044
5.078	.041
5.160	.038
5.515	.037
5.558	,035
5.596	.033
5.733	.027
5.776	.025
5.858	.024
5.864	.022
5.967	.021
6.431	.019
6.578	.016
6.818	.013
6.840	.011
6.954	.008
7.364	.005

(1, 5, 5)

c	P
3.527	.141
3.600	.132
3.636	.116
3.927	.113
4.036	.105
4.109	.086
4.182	.082
4.400	.076
4.546	.074
4.800	.056
4.909	.053
5.127	.046
5.236	.039
5.636	.033
5.709	.030
5.782	.027
6.000	.022
6.146	.019
6.509	.018
6.546	.015
6.582	.014
6.727	.012
6.836	.011
7.309	.009
7.527	.008
7.746	.005
8.182	.002

(2, 2, 2)

c	P
4.571	.067

(2, 2, 3)

c	P
4.464	.105
4.500	.067
4.714	.048
5.357	.029

(2, 2, 4)

c	P
4.167	.105
4.458	.100
4.500	.090
5.125	.052
5.333	.033
5.500	.024
6.000	.014

(2, 2, 5)

c	P
4.093	.148
4.200	.138
4.293	.122
4.373	.090
4.573	.085
4.800	.063
4.893	.061
5.040	.056
5.160	.034
5.693	.029
6.000	.019
6.133	.013
6.533	.008

(2, 3, 3)

c	P
4.111	.129
4.250	.121
4.556	.100
4.694	.093
5.000	.075
5.139	.061
5.361	.032
5.556	.025
6.250	.011

(2, 3, 4)

c	P
4.000	.149
4.078	.140
4.200	.137
4.278	.124
4.311	.108
4.378	.105
4.444	.102
4.511	.098
4.544	.086
4.611	.083
4.711	.079
4.811	.076
4.878	.073
4.900	.071
4.978	.059
5.078	.057
5.144	.054
5.378	.052
5.400	.051
5.444	.046
5.500	.040
5.611	.032
5.800	.030
6.000	.024
6.111	.021
6.144	.014
6.300	.011
6.444	.008
7.000	.005

TABLE I. Kruskal-Wallis upper-tail probabilities: $P = P(K \geqslant c)$; *(3 groups; sample sizes in parentheses) (Continued)*

(2, 3, 5)

c	P
3.942	.146
3.996	.139
4.058	.137
4.069	.132
4.204	.129
4.214	.125
4.233	.122
4.258	.120
4.331	.117
4.378	.113
4.494	.101
4.651	.091
4.694	.089
4.724	.087
4.727	.085
4.814	.071
4.869	.067
4.913	.063
4.942	.062
5.076	.060
5.087	.053
5.106	.052
5.251	.049
5.349	.046
5.513	.044
5.524	.043
5.542	.041
5.727	.037
5.742	.034
5.786	.033
5.804	.033
5.949	.026
6.004	.025
6.033	.024
6.091	.021
6.124	.020
6.294	.017
6.386	.016
6.414	.015
6.818	.012
6.822	.010
6.909	.009
6.949	.006
7.182	.004
7.636	.002

(2, 4, 4)

c	P
4.009	.142
4.364	.125
4.418	.120
4.446	.103
4.554	.098
4.582	.094
4.691	.080
4.773	.075
4.854	.071
4.991	.065
5.127	.057
5.236	.052
5.454	.046
5.509	.044
5.536	.042
5.646	.039
5.727	.034
5.946	.028
6.082	.025
6.327	.024
6.409	.022
6.546	.020
6.600	.017
6.627	.016
6.873	.011
7.036	.006
7.282	.004
7.854	.002

(2, 4, 5)

c	P
3.818	.148
3.823	.145
3.864	.143
4.041	.139
4.064	.135
4.073	.133
4.091	.130
4.141	.128
4.154	.126
4.200	.123
4.223	.121
4.250	.119
4.323	.116
4.364	.114
4.368	.112
4.404	.110
4.500	.104
4.518	.101
4.541	.098
4.614	.090
4.664	.088
4.768	.079
4.791	.078
4.800	.076
4.818	.074
4.841	.072
4.868	.071
4.950	.063
5.073	.061
5.154	.059
5.164	.053
5.254	.052
5.268	.051
5.273	.049
5.300	.048
5.314	.046
5.414	.045
5.518	.043
5.523	.042
5.564	.038
5.641	.037
5.664	.036
5.754	.035
5.823	.034
5.891	.032
5.954	.030
5.973	.029
6.004	.026
6.041	.025
6.068	.025
6.118	.024
6.141	.023
6.223	.022
6.368	.021
6.391	.021
6.473	.020
6.504	.020
6.541	.017
6.550	.017
6.564	.016
6.654	.016
6.723	.015
6.904	.014
6.914	.013
7.000	.013
7.018	.012
7.064	.012
7.118	.010
7.204	.009
7.254	.009
7.291	.008
7.450	.007
7.500	.007
7.568	.006
7.573	.005
7.773	.004
7.814	.003
8.018	.002
8.114	.001
8.591	.001

(2, 5, 5)

c	P
3.862	.150
3.885	.146
4.015	.136
4.069	.132
4.131	.130
4.138	.127
4.231	.124
4.254	.114
4.438	.106
4.477	.103
4.508	.100
4.623	.097
4.685	.092
4.754	.084
4.808	.081
4.846	.073
4.877	.068
4.992	.066
5.054	.060
5.177	.057
5.238	.054
5.246	.051
5.338	.047
5.546	.045
5.585	.041
5.608	.040
5.615	.039
5.708	.037
5.731	.036
5.792	.032
5.915	.030
5.985	.028
6.077	.027
6.231	.026
6.346	.025
6.354	.021
6.446	.020
6.469	.019
6.654	.017
6.692	.016
6.815	.015
6.838	.014
6.969	.013
7.023	.013
7.185	.012
7.208	.011
7.269	.010
7.338	.010
7.392	.009
7.462	.008
7.577	.007
7.762	.007
7.923	.006
8.008	.006
8.077	.006
8.131	.005
8.169	.003
8.292	,003
8.377	.002
8.562	.002
8.685	.001
8.938	.001
9.423	.000

(3, 3, 3)

c	P
4.267	.139
4.356	.132
4.622	.100
5.067	.086
5.422	.071
5.600	.050
5.689	.029
5.956	.025
6.489	.011
7.200	.004

(3, 3, 4)

c	P
3.836	.150
3.973	.143
4.046	.132
4.091	.126
4.273	.123
4.336	.117
4.382	.111
4.564	.106
4.700	.101
4.709	.092
4.818	.085
4.846	.081
5.000	.074
5.064	.070
5.109	.068
5.254	.064
5.436	.062
5.500	.056
5.573	.053
5.727	.050
5.791	.046
5.936	.036
5.982	.034
6.018	.027
6.154	.025
6.300	.023
6.564	.017
6.664	.014
6.709	.013
6.746	.010
7.000	.006
7.318	.004
7.436	.002
8.018	.001

TABLE I. **Kruskal-Wallis upper-tail probabilities:** $P = P(K \geqslant c)$; (*3 groups; sample sizes in parentheses*) (*Continued*)

(3, 3, 5)	
c	P
3.927	.149
4.012	.144
4.048	.139
4.170	.135
4.194	.126
4.242	.122
4.303	.117
4.315	.113
4.412	.109
4.533	.097
4.679	.094
4.776	.090
4.800	.087
4.848	.085
4.861	.082
4.909	.079
5.042	.077
5.079	.069
5.103	.067
5.212	.065
5.261	.062
5.346	.058
5.442	.055
5.503	.053
5.515	.051
5.648	.049
5.770	.047
5.867	.042
6.012	.040
6.061	.033
6.109	.032
6.194	.027
6.303	.026
6.315	.021
6.376	.020
6.533	.019
6.594	.019
6.175	.014
6.776	.013
6.861	.012
6.982	.011
7.079	.009
7.333	.008
7.467	.008
7.503	.006
7.515	.005
7.636	.004
7.879	.003
8.048	.002
8.242	.001
8.727	.001

(3, 4, 4)	
c	P
3.848	.150
3.932	.145
3.962	.140
4.144	.135
4.167	.131
4.212	.129
4.296	.125
4.303	.121
4.326	.116
4.348	.113
4.409	.106
4.477	.102
4.546	.099
4.576	.097
4.598	.093
4.712	.090
4.750	.087
4.894	.084
5.053	.078
5.144	.073
5.182	.068
5.212	.066
5.296	.063
5.303	.061
5.326	.058
5.386	.054
5.500	.052
5.576	.051
5.598	.049
5.667	.047
5.803	.045
5.932	.043
5.962	.041
6.000	.040
6.046	.039
6.053	.035
6.144	.032
6.167	.031
6.182	.030
6.348	.027
6.386	.026
6.394	.025
6.409	.023
6.417	.022
6.546	.021
6.659	.020
6.712	.019
6.727	.018
6.962	.017
7.000	.016
7.053	.014
7.076	.011
7.136	.011
7.144	.010
7.212	.009
7.477	.006
7.598	.004
7.636	.004
7.682	.003
7.848	.003
8.227	.002
8.326	.001
8.909	.001

(3, 4, 5)	
c	P
3.831	.150
3.865	.148
3.876	.146
3.958	.144
4.015	.140
4.030	.137
4.060	.134
4.122	.132
4.154	.131
4.180	.125
4.195	.124
4.235	.121
4.241	.119
4.276	.117
4.318	.115
4.327	.112
4.368	.110
4.419	.109
4.426	.107
4.487	.106
4.522	.105
4.523	.103
4.549	.099
4.564	.097
4.645	.095
4.676	.093
4.754	.091
4.789	.089
4.810	.088
4.830	.083
4.856	.082
4.881	.081
4.891	.078
4.939	.075
4.953	.074
4.983	.073
5.041	.072
5.045	.071
5.106	.070
5.137	.068
5.158	.067
5.180	.065
5.291	.063
5.308	.062
5.342	.061
5.349	.061
5.353	.059
5.414	.058
5.426	.057
5.549	.054
5.568	.052
5.619	.051
5.631	.050
5.656	.049
5.660	.048
5.677	.047
5.718	.046
5.722	.045
5.753	.044
5.780	.043
5.804	.041
5.814	.040
5.862	.040
5.876	.039
5.964	.038
6.026	.038
6.030	.037
6.060	.037
6.087	.035
6.164	.035
6.173	.034
6.231	.033
6.265	.032
6.272	.030
6.337	.030
6.368	.029
6.369	.029
6.395	.026
6.410	.025
6.491	.025
6.522	.024
6.542	.023
6.580	.021
6.635	.020
6.676	.020
6.703	.019
6.780	.019
6.785	.018
6.799	.016
6.830	.016
6.891	.015
7.004	.015
7.010	.015
7.096	.014
7.106	.014
7.188	.013
7.195	.012
7.256	.012
7.260	.012
7.272	.012
7.291	.011
7.318	.011
7.395	.011
7.445	.010
7.465	.010
7.477	.009
7.523	.007
7.568	.007
7.641	.007
7.708	.006
7.753	.006
7.810	.006
7.876	.006
7.887	.006
7.906	.005
7.927	.005
8.030	.005
8.060	.004
8.077	.004
8.118	.004
8.122	.004
8.215	.003
8.256	.003
8.430	.002
8.446	.002
8.481	.002
8.503	.001
8.573	.001
8.626	.001
8.795	.001
9.035	.001
9.118	.001
9.199	.000
9.692	.000

TABLE I. Kruskal-Wallis upper-tail probabilities: $P = P(K \geqslant c)$; *(3 groups; sample sizes in parentheses) (Continued)*

(3, 5, 5)

c	P
3.807	.147
3.912	.144
3.965	.142
3.991	.139
4.114	.136
4.141	.135
4.150	.132
4.202	.127
4.220	.125
4.255	.117
4.308	.112
4.352	.110
4.378	.107
4.457	.105
4.466	.104
4.536	.102
4.545	.100
4.571	.098
4.694	.094
4.774	.092
4.826	.089
4.835	.088
4.888	.082
4.914	.079
4.941	.077
4.993	.075
5.020	.072
5.064	.070
5.152	.067
5.169	.065
5.222	.065
5.284	.063
5.363	.062
5.407	.059
5.486	.057
5.494	.056
5.521	.055
5.574	.053
5.600	.051
5.626	.051
5.706	.046
5.802	.045
5.837	.042
5.934	.040
5.943	.039
6.022	.038
6.048	.037
6.198	.035
6.207	.034
6.250	.034
6.259	.033
6.286	.031
6.312	.030
6.365	.030
6.391	.028
6.435	.027
6.488	.025
6.550	.024
6.593	.024
6.655	.022
6.734	.022
6.752	.021
6.866	.019
6.892	.018
6.945	.018
6.963	.017
6.998	.015
7.050	.015
7.121	.014
7.209	.014
7.226	.012
7.288	.012
7.306	.012
7.314	.011
7.437	.011
7.543	.010
7.578	.010
7.622	.009
7.736	.009
7.763	.008
7.780	.008
7.859	.007
7.894	.007
7.912	.007
8.026	.006
8.079	.006
8.106	.006
8.237	.005
8.264	.005
8.316	.005
8.334	.005
8.545	.004
8.571	.004
8.580	.004
8.650	.003
8.659	.003
8.791	.002
8.809	.002
8.950	.002
9.002	.002
9.055	.001
9.284	.001
9.336	.001
9.398	.001
9.521	.000
9.635	.000
9.916	.000
10.057	.000
10.550	.000

(4, 4, 4)

c	P
3.962	.145
4.154	.136
4.192	.131
4.269	.122
4.308	.114
4.500	.104
4.654	.097
4.769	.094
4.885	.086
4.962	.080
5.115	.074
5.346	.063
5.538	.057
5.654	.055
5.692	.049
5.808	.044
6.000	.040
6.038	.037
6.269	.033
6.500	.030
6.577	.026
6.615	.024
6.731	.021
6.962	.019
7.038	.018
7.269	.016
7.385	.015
7.423	.013
7.538	.011
7.654	.008
7.731	.007
8.000	.005
8.115	.003
8.346	.002
8.654	.001
8.769	.001
9.269	.001
9.846	.000

(4, 4, 5)

c	P
3.910	.146
3.986	.143
3.989	.141
4.025	.139
4.042	.134
4.068	.132
4.075	.130
4.118	.127
4.170	.125
4.200	.122
4.233	.121
4.253	.119
4.272	.117
4.289	.114
4.332	.112
4.381	.108
4.447	.106
4.497	.104
4.553	.102
4.619	.100
4.668	.098
4.685	.096
4.701	.094
4.711	.092
4.728	.091
4.747	.089
4.760	.088
4.813	.086
4.830	.084
4.833	.082
4.896	.081
4.975	.077
5.014	.076
5.024	.074
5.028	.073
5.090	.071
5.172	.069
5.196	.068
5.225	,066
5.344	.065
5.360	.063
5.370	.062
5.387	.061
5.410	.060
5.440	.059
5.476	.057
5.486	.056
5.489	.056
5.519	.054
5.568	.052
5.571	.051
5.618	.050
5.657	.049
5.687	.048
5.756	.047
5.782	.046
5.815	.045
5.819	.043
5.914	.042
6.003	.042
6.013	.041
6.030	.040
6.096	.039
6.119	.038
6.132	.037
6.201	.036
6.214	.034
6.228	.033
6.267	.032
6.310	.031
6.343	.030
6.382	.029
6.399	.028
6.462	.027
6.544	.027
6.547	.026
6.597	.026
6.672	.024
6.676	.024
6.804	.023
6.860	.022
6.870	.022
6.887	.021
6.890	.021
6.943	.020
6.953	.020
6.976	.019
7.058	.018
7.075	.017
7.101	.017
7.124	.016
7.190	.016
7.203	.015
7.233	.015
7.240	.014
7.256	.014
7.418	.014
7.467	.013
7.470	.013
7.497	.013
7.503	.012
7.586	.012
7.596	.012
7.714	.011
7.744	.011
7.760	.009
7.767	.009
7.797	.009
7.810	.009
7.833	.008
7.942	.007
7.981	.007
8.047	.006
8.113	.006
8.130	.006
8.140	.005
8.156	.005
8.189	.005
8.403	.004
8.440	.004
8.456	.004
8.525	.003
8.558	.003
8.571	.003
8.575	.003
8.604	.003
8.702	.003
8.723	.002
8.782	.002
8.868	.002
8.997	.001
9.053	.001
9.099	.001
9.129	.001
9.168	.001
9.396	.001
9.528	.001
9.590	.001
9.613	.000
9.758	.000
10.118	.000
10.187	.000
10.681	.000

TABLE I. **Kruskal-Wallis upper-tail probabilities:** $P = P(K \geqslant c)$; (*3 groups; sample sizes in parentheses*) (*Continued*)

(4, 5, 5)

c	P
3.883	.148
3.891	.144
3.906	.142
3.926	.140
3.951	.137
3.971	.135
4.043	.133
4.063	.131
4.166	.127
4.200	.124
4.203	.122
4.246	.120
4.271	.118
4.291	.115
4.303	.113
4.363	.111
4.383	.110
4.386	.108
4.486	.106
4.500	.105
4.520	.101
4.523	.099
4.531	.098
4.591	.096
4.611	.095
4.660	.093
4.706	.092
4.806	.089
4.843	.088
4.851	.086
4.866	.084
4.886	.083
4.911	.079
4.943	.078
4.980	.076
5.023	.075
5.071	.074
5.126	.073
5.163	.070
5.171	.069
5.186	.068
5.206	.067
5.231	.066
5.263	.064
5.323	.063
5.400	.061
5.446	.059
5.460	.058
5.483	.057
5.491	.056
5.526	.056
5.571	.055
5.583	.052
5.620	.051
5.643	.050
5.666	.049
5.711	.048
5.780	.048
5.803	.047
5.811	.046
5.871	.045
5.903	.043
5.963	.042
5.983	.042
5.986	.041
6.031	.040
6.086	.040
6.100	.038
6.123	.037
6.146	.037
6.166	.035
6.211	.035
6.223	.034
6.283	.034
6.303	.033
6.351	.032
6.406	.031
6.440	.030
6.451	.029
6.486	.029
6.531	.028
6.543	.028
6.603	.027
6.623	.026
6.626	.026
6.671	.025
6.760	.025
6.763	.024
6.771	.024
6.786	.023
6.806	.022
6.831	.022
6.900	.021
6.943	.020
7.000	.019
7.046	.019
7.080	.018
7.106	.018
7.171	.018
7.183	.017
7.220	.017
7.243	.017
7.266	.016
7.311	.015
7.320	.015
7.426	.015
7.446	.014
7.471	.014
7.491	.014
7.503	.013
7.563	.013
7.586	.012
7.631	.012
7.640	.011
7.686	.011
7.720	.011
7.766	.010
7.791	.010
7.823	.010
7.860	.010
7.903	.009
7.906	.009
8.006	.009
8.043	.009
8.051	.008
8.066	.008
8.086	.008
8.131	.008
8.143	.008
8.223	.007
8.226	.007
8.271	.007
8.280	.006
8.340	.006
8.363	.006
8.371	.005
8.386	.005
8.431	.005
8.463	.005
8.523	.005
8.543	.005
8.546	.004
8.683	.004
8.691	.004
8.726	.004
8.751	.004
8.771	.004
8.969	.003
8.980	.003
9.000	.003
9.011	.003
9.026	.003
9.071	.002
9.103	.002
9.163	.002
9.231	.002
9.286	.002
9.323	.001
9.411	.001
9.503	.001
9.506	.001
9.606	.001
9.643	.001
9.651	.001
9.686	.001
9.926	.001
9.986	.000
10.051	.000
10.063	.000
10.100	.000
10.260	.000
10.511	.000
10.520	.000
10.566	.000
10.646	.000
11.033	.000
11.083	.000
11.571	.000

(5, 5, 5)

c	P
3.860	.150
3.920	.145
3.980	.137
4.020	.132
4.160	.127
4.220	.123
4.340	.118
4.380	.110
4.460	.105
4.500	.102
4.560	.100
4.580	.096
4.740	.092
4.820	.089
4.860	.085
4.880	.084
4.940	.081
5.040	.075
5.120	.072
5.180	.070
5.360	.065
5.420	.063
5.460	.060
5.540	.055
5.580	.053
5.660	.051
5.780	.049
5.820	.048
5.840	.046
6.000	.044
6.020	.043
6.080	.040
6.140	.038
6.180	.036
6.260	.035
6.320	.033
6.480	.032
6.500	.031
6.540	.030
6.620	.028
6.660	.027
6.720	.026
6.740	.025
6.860	.024
6.980	.021
7.020	.020
7.220	.019
7.260	.018
7.280	.018
7.340	.016
7.440	.015
7.460	.015
7.580	.014
7.620	.013
7.740	.012
7.760	.012
7.940	.011
7.980	.011
8.000	.009
8.060	.009
8.180	.008
8.240	.008
8.340	.007
8.420	.007
8.540	.006
8.640	.006
8.660	.006
8.720	.005
8.780	.005
8.820	.005
8.880	.004
8.960	.004
9.060	.004
9.140	.003
9.260	.003
9.360	.003
9.380	.003
9.420	.002
9.500	.002
9.620	.002
9.680	.001
9.740	.001
9.780	.001
9.920	.001
9.980	.001
10.140	.001
10.220	.001
10.260	.000
10.500	.000
10.580	.000
10.640	.000
10.820	.000
11.060	.000
11.180	.000
11.520	.000
11.580	.000
12.020	.000
12.500	.000

TABLE J(a). χ^2 **upper-tail probabilities for** $\nu = 2, 3, 4, 5$ **degrees of freedom:** $\Psi_\nu(c) = P(\chi_\nu^{\,2} \geqslant c)$

c \ ν	2	3	4	5	c \ ν	2	3	4	5
0.06	.9704	.9962	.9996		6.40	.0408	.0937	.1712	.2692
0.07	.9656	.9952	.9994	.9999	6.60	.0369	.0858	.1586	.2521
0.08	.9608	.9941	.9992	.9999	6.80	.0334	.0786	.1468	.2359
0.09	.9560	.9930	.9990	.9999	7.00	.0302	.0719	.1359	.2206
0.10	.9512	.9918	.9988	.9998	7.20	.0273	.0658	.1257	.2062
0.20	.9048	.9776	.9953	.9991	7.40	.0247	.0602	.1162	.1926
0.30	.8607	.9600	.9898	.9976	7.60	.0224	.0550	.1074	.1797
0.40	.8187	.9402	.9825	.9953	7.80	.0202	.0503	.0992	.1676
0.50	.7788	.9189	.9735	.9921	8.00	.0183	.0460	.0916	.1562
0.60	.7408	.8964	.9631	.9880	8.20	.0166	.0421	.0845	.1456
0.70	.7047	.8732	.9513	.9830	8.40	.0150	.0384	.0780	.1355
0.80	.6703	.8495	.9385	.9770	8.60	.0136	.0351	.0719	.1261
0.90	.6376	.8254	.9246	.9702	8.80	.0123	.0321	.0663	.1173
1.00	.6065	.8013	.9098	.9626	9.00	.0111	.0293	.0611	.1091
1.10	.5769	.7771	.8943	.9541	9.20	.0101	.0267	.0563	.1013
1.20	.5488	.7530	.8781	.9449	9.40	.0091	.0244	.0518	.0941
1.30	.5220	.7291	.8614	.9349	9.60	.0082	.0223	.0477	.0874
1.40	.4966	.7055	.8442	.9243	9.80	.0074	.0203	.0439	.0811
1.50	.4724	.6823	.8266	.9131	10.00	.0067	.0186	.0404	.0752
1.60	.4493	.6594	.8088	.9012	10.50	.0052	.0148	.0328	.0622
1.70	.4274	.6369	.7907	.8889	11.00	.0041	.0117	.0266	.0514
1.80	.4066	.6149	.7725	.8761	11.50	.0032	.0093	.0215	.0423
1.90	.3867	.5934	.7541	.8628	12.00	.0025	.0074	.0174	.0348
2.00	.3679	.5724	.7358	.8492	12.50	.0019	.0059	.0140	.0285
2.20	.3329	.5319	.6990	.8208	13.00	.0015	.0046	.0113	.0234
2.40	.3012	.4936	.6626	.7915	13.50	.0012	.0037	.0091	.0191
2.60	.2725	.4575	.6268	.7614	14.00	.0009	.0029	.0073	.0156
2.80	.2466	.4235	.5918	.7308	14.50	.0007	.0023	.0059	.0127
3.00	.2231	.3916	.5578	.7000	15.00	.0006	.0018	.0047	.0104
3.20	.2019	.3618	.5249	.6692	15.50	.0004	.0014	.0038	.0084
3.40	.1827	.3340	.4932	.6386	16.00	.0003	.0011	.0030	.0068
3.60	.1653	.3080	.4628	.6083	16.50	.0003	.0009	.0024	.0056
3.80	.1496	.2839	.4338	.5786	17.00	.0002	.0007	.0019	.0045
4.00	.1353	.2615	.4060	.5494	17.50	.0002	.0006	.0015	.0036
4.20	.1225	.2407	.3796	.5210	18.00	.0001	.0004	.0012	.0029
4.40	.1108	.2214	.3546	.4934	18.50	.0001	.0003	.0010	.0024
4.60	.1003	.2035	.3309	.4666	19.00	.0001	.0003	.0008	.0019
4.80	.0907	.1870	.3084	.4408	19.50	.0001	.0002	.0006	.0016
5.00	.0821	.1718	.2873	.4159	20.00	.0000	.0002	.0005	.0012
5.20	.0743	.1577	.2674	.3920	21.00	.0000	.0001	.0003	.0008
5.40	.0672	.1447	.2487	.3690	22.00	.0000	.0001	.0002	.0005
5.60	.0608	.1328	.2311	.3471	23.00	.0000	.0000	.0001	.0003
5.80	.0550	.1218	.2146	.3262	24.00	.0000	.0000	.0001	.0002
6.00	.0498	.1116	.1991	.3062	25.00	.0000	.0000	.0001	.0001
6.20	.0450	.1023	.1847	.2872	26.00	.0000	.0000	.0000	.0001

TABLE J(*b*). Critical values c of χ^2 with ν degrees of freedom: $\alpha = P(\chi_\nu^2 \geqslant c)$

ν \ α	.001	.005	.01	.025	.05	.10	.20	.30
6	22.458	18.548	16.812	14.449	12.592	10.645	8.558	7.231
7	24.322	20.278	18.475	16.013	14.067	12.017	9.803	8.383
8	26.125	21.955	20.090	17.535	15.507	13.362	11.030	9.524
9	27.877	23.589	21.666	19.023	16.919	14.684	12.242	10.656
10	29.588	25.188	23.209	20.483	18.307	15.987	13.442	11.781
11	31.264	26.757	24.725	21.920	19.675	17.275	14.631	12.899
12	32.910	28.300	26.217	23.337	21.026	18.549	15.812	14.011
13	34.528	29.820	27.688	24.736	22.362	19.812	16.985	15.119
14	36.123	31.319	29.141	26.119	23.685	21.064	18.151	16.222
15	37.697	32.801	30.578	27.488	24.996	22.307	19.311	17.322
16	39.252	34.267	32.000	28.845	26.296	23.542	20.465	18.418
17	40.790	35.719	33.409	30.191	27.587	24.769	21.615	19.511
18	42.312	37.157	34.805	31.526	28.869	25.989	22.760	20.601
19	43.820	38.582	36.191	32.852	30.144	27.204	23.900	21.689
20	45.315	39.997	37.566	34.170	31.410	28.412	25.038	22.775
21	46.797	41.401	38.932	35.479	32.671	29.615	26.171	23.858
22	48.268	42.796	40.289	36.781	33.924	30.813	27.302	24.939
23	49.728	44.181	41.638	38.076	35.173	32.007	28.429	26.018
24	51.179	45.559	42.980	39.364	36.415	33.196	29.553	27.096
25	52.620	46.928	44.314	40.647	37.653	34.382	30.675	28.172
26	54.052	48.290	45.642	41.923	38.885	35.563	31.795	29.246
27	55.476	49.645	46.963	43.195	40.113	36.741	32.912	30.319
28	56.892	50.993	48.278	44.461	41.337	37.916	34.027	31.391
29	58.301	52.336	49.588	45.722	42.557	39.088	35.139	32.461
30	59.703	53.672	50.892	46.979	43.773	40.256	36.250	33.530
31	61.098	55.003	52.191	48.232	44.985	41.422	37.359	34.598
32	62.487	56.328	53.486	49.480	46.194	42.585	38.466	35.665
33	63.870	57.649	54.776	50.725	47.400	43.745	39.572	36.731
34	65.247	58.964	56.061	51.966	48.602	44.903	40.676	37.795
35	66.619	60.275	57.342	53.203	49.802	46.059	41.778	38.859
36	67.985	61.581	58.619	54.437	50.999	47.212	42.879	39.922
37	69.347	62.883	59.893	55.668	52.192	48.363	43.978	40.984
38	70.703	64.181	61.162	56.896	53.384	49.513	45.076	42.045
39	72.055	65.476	62.428	58.120	54.572	50.660	46.173	43.105
40	73.402	66.766	63.691	59.342	55.759	51.805	47.269	44.165
45	80.077	73.166	69.957	65.410	61.656	57.505	52.729	49.452
50	86.661	79.490	76.154	71.420	67.505	63.167	58.164	54.723
55	93.168	85.749	82.292	77.381	73.312	68.796	63.577	59.981
60	99.607	91.952	88.379	83.298	79.082	74.397	68.972	65.227
65	105.988	98.105	94.422	89.177	84.821	79.973	74.351	70.462
70	112.317	104.215	100.425	95.023	90.531	85.527	79.715	75.689
75	118.599	110.286	106.393	100.839	96.217	91.062	85.066	80.908
80	124.839	116.321	112.329	106.629	101.879	96.578	90.405	86.120
85	131.041	122.325	118.236	112.393	107.522	102.079	95.734	91.325
90	137.208	128.299	124.116	118.136	113.145	107.565	101.054	96.524
95	143.344	134.247	129.973	123.858	118.752	113.038	106.364	101.717
100	149.449	140.169	135.807	129.561	124.342	118.498	111.667	106.906

TABLE K. Upper-tail probabilities of Jonckheere's statistic: $P(J \geqslant c)$; *(for s groups of size n)*

	$s = 3$		$s = 4$			$s = 5$		$s = 6$
c	$n = 2$	$n = 4$	$n = 2$	$n = 3$	$n = 4$	$n = 2$	$n = 3$	$n = 2$
0	.5778	.5284	.5492	.5276	.5183	.5353	.5198	.5271
2	.4222	.4716	.4508	.4724	.4817	.4647	.4802	.4729
4	.2889	.4156	.3563	.4177	.4454	.3951	.4408	.4193
6	.1667	.3609	.2683	.3645	.4094	.3285	.4019	.3670
8	.0889	.3090	.1929	.3136	.3742	.2667	.3640	.3170
10	.0333	.2602	.1302	.2659	.3400	.2110	.3273	.2699
12	.0111	.2157	.0829	.2220	.3069	.1625	.2921	.2265
14		.1756	.0484	.1823	.2754	.1213	.2588	.1871
16		.1404	.0262	.1472	.2454	.0878	.2274	.1521
18		.1099	.0123	.1166	.2172	.0613	.1982	.1215
20		.0844	.0052	.0907	.1910	.0412	.1713	.0953
22		.0632	.0016	.0691	.1666	.0265	.1468	.0734
24		.0463	.0004	.0515	.1443	.0162	.1247	.0553
26		.0330		.0374	.1241	.0094	.1049	.0408
28		.0229		.0266	.1058	.0051	.0874	.0294
30		.0153		.0183	.0895	.0026	.0721	.0207
32		.0099		.0123	.0751	.0012	.0588	.0142
34		.0062		.0080	.0624	.0005	.0475	.0094
36		.0037		.0050	.0514	.0002	.0379	.0061
38		.0021		.0030	.0420	.0000	.0299	.0038
40		.0011		.0017	.0339	.0000	.0234	.0023
42		.0005		.0009	.0272		.0180	.0013
44		.0002		.0005	.0215		.0137	.0007
46		.0001		.0002	.0168		.0102	.0004
48		.0000		.0001	.0130		.0076	.0002
50				.0000	.0100		.0055	.0001
52				.0000	.0075		.0039	.0000
54				.0000	.0056		.0028	.0000
56					.0041		.0019	.0000
58					.0023		.0013	.0000
60					.0021		.0008	.0000
62					.0015		.0005	
64					.0010		.0003	
66					.0007		.0002	
68					.0005		.0001	
70					.0003		.0001	
72					.0002		.0000	
74					.0001		.0000	
76					.0001		.0000	
78					.0000		.0000	
80					.0000		.0000	
82					.0000		.0000	
84					.0000		.0000	
86					.0000		.0000	
88					.0000		.0000	
90					.0000		.0000	
92					.0000			
94					.0000			
96					.0000			

	$s = 3$	
c	$n = 3$	$n = 5$
1	.5000	.5000
3	.4155	.4589
5	.3339	.4182
7	.2595	.3783
9	.1940	.3396
11	.1387	.3025
13	.0946	.2672
15	.0613	.2340
17	.0369	.2032
19	.0208	.1748
21	.0107	.1489
23	.0048	.1256
25	.0018	.1049
27	.0006	.0867
29		.0708
31		.0572
33		.0456
35		.0359
37		.0279
39		.0214
41		.0161
43		.0120
45		.0087
47		.0063
49		.0044
51		.0030
53		.0020
55		.0013
57		.0009
59		.0005
61		.0003
63		.0002
65		.0001
67		.0001
69		.0000
71		.0000
73		.0000
75		.0000

TABLE L. Amalgamation probabilities for Chacko's test: $s!p_{i,s}$; (*s groups of equal size*)

i \ s	3	4	5	6	7	8	9	10
1	2	6	24	120	720	5,040	40,320	362,880
2	3	11	50	274	1,764	13,068	109,584	1,026,576
3	1	6	35	225	1,624	13,132	118,124	1,172,700
4		1	10	85	735	6,769	67,284	723,680
5			1	15	175	1,960	22,449	269,325
6				1	21	322	4,536	63,273
7					1	28	546	9,450
8						1	36	870
9							1	45
10								1
$s!$	6	24	120	720	5,040	40,320	362,880	3,628,800

TABLE M. Upper-tail probabilities of Friedman's statistic: $P(Q \geqslant c)$; (*s treatments and N blocks*)

$s = 3, N = 2$

c	P
3	.500
4	.167

$s = 3, N = 3$

c	P
4.667	.194
6.000	.028

$s = 3, N = 4$

c	P
4.5	.125
6.0	.069
6.5	.042
8.0	.005

$s = 3, N = 5$

c	P
4.8	.124
5.2	.093
6.4	.039
7.6	.024
8.4	.008
10.0	.001

$s = 3, N = 6$

c	P
4.33	.142
5.33	.072
6.33	.052
7.00	.029
8.33	.012
9.00	.008
9.33	.006
10.33	.002
12.00	.000

$s = 3, N = 7$

c	P
4.571	.112
5.429	.085
6.000	.051
7.143	.027
7.714	.021
8.000	.016
8.857	.008
10.286	.004
10.571	.003
11.143	.001
12.286	.000
14.000	.000

$s = 3, N = 8$

c	P
4.00	.149
4.75	.120
5.25	.079
6.25	.047
6.75	.038
7.00	.030
7.75	.018
9.00	.010
9.25	.008
9.75	.005
10.75	.002
12.00	.001
12.25	.001
13.00	.000
14.25	.000
16.00	.000

$s = 3, N = 9$

c	P
4.667	.107
5.556	.069
6.000	.057
6.222	.048
6.889	.031
8.000	.019
8.222	.016
8.667	.010
9.556	.006
10.667	.003
10.889	.003
11.556	.001
12.667	.001
13.556	.000
14.000	.000
14.222	.000
14.889	.000
16.222	.000
18.000	.000

$s = 3, N = 10$

c	P
4.2	.135
5.0	.092
5.4	.078
5.6	.066
6.2	.046
7.2	.030
7.4	.026
7.8	.018
8.6	.012
9.6	.007
9.8	.006
10.4	.003
11.4	.002
12.2	.001
12.6	.001
12.8	.001
13.4	.000
14.6	.000
15.0	.000
15.2	.000
15.8	.000
16.2	.000
16.8	.000
18.2	.000
20.0	.000

$s = 3, N = 11$

c	P
4.545	.116
4.909	.100
5.091	.087
5.636	.062
6.545	.043
6.727	.037
7.091	.027
7.818	.019
8.727	.013
8.909	.011
9.455	.007
10.364	.005
11.091	.002
11.455	.002
11.636	.002
12.182	.001
13.273	.001
13.636	.000
13.818	.000
14.364	.000
14.727	.000
15.273	.000
16.545	.000
16.909	.000
17.636	.000
18.182	.000
18.727	.000
20.182	.000
22.000	.000

$s = 3, N = 12$

c	P
4.167	.141
4.500	.123
4.667	.108
5.167	.080
6.000	.058
6.167	.050
6.500	.038
7.167	.028
8.000	.019
8.167	.017
8.667	.011
9.500	.008
10.167	.005
10.500	.004
10.667	.004
11.167	.002
12.167	.002
12.500	.001
12.667	.001
13.167	.001
13.500	.001
14.000	.000
15.167	.000
15.500	.000
16.167	.000
16.667	.000
17.167	.000
18.167	.000
18.500	.000
18.667	.000
19.500	.000
20.167	.000
20.667	.000
22.167	.000
24.000	.000

$s = 3, N = 13$

c	P
4.154	.145
4.308	.129
4.769	.098
5.538	.073
5.692	.064
6.000	.050
6.615	.038
7.385	.027
7.538	.025
8.000	.016
8.769	.012
9.385	.008
9.692	.007
9.846	.006
10.308	.004
11.231	.003
11.538	.002
11.692	.002
12.154	.001
12.462	.001
12.923	.001
14.000	.001
14.308	.000
14.923	.000
15.385	.000
15.846	.000
16.615	.000
16.769	.000
17.077	.000
17.231	.000
18.000	.000
18.615	.000
19.077	.000
19.538	.000
19.846	.000
20.462	.000
21.385	.000
22.154	.000
22.615	.000
24.154	.000
26.000	.000

$s = 3, N = 14$

c	P
4.000	.150
4.429	.117
5.143	.089
5.286	.079
5.571	.063
6.143	.049
6.857	.036
7.000	.033
7.429	.023
8.143	.018
8.714	.011
9.000	.010
9.143	.009
9.571	.007
10.429	.005
10.714	.003
10.857	.003
11.286	.003
11.571	.002
12.000	.002
13.000	.001
13.286	.001
13.857	.000
14.286	.000
14.714	.000
15.429	.000
15.571	.000
15.857	.000
16.000	.000
16.714	.000
17.286	.000
17.714	.000
18.143	.000
18.429	.000
19.000	.000
19.857	.000
20.571	.000
21.000	.000
21.143	.000
21.571	.000
22.286	.000
22.429	.000
23.286	.000
24.143	.000
24.571	.000
26.143	.000
28.000	.000

$s = 3, N = 15$

c	P
4.133	.136
4.800	.106
4.933	.096
5.200	.077
5.733	.059
6.400	.047
6.533	.043
6.933	.030
7.600	.022
8.133	.018
8.400	.015
8.533	.011
8.933	.010
9.733	.007

TABLE M. Upper-tail probabilities of Friedman's statistic: $P(Q \geqslant c)$; (*s treatments and N blocks*) (*Continued*)

c	P
10.000	.005
10.133	.005
10.533	.004
10.800	.003
11.200	.003
12.133	.002
12.400	.001
12.933	.001
13.333	.001
13.733	.001
14.400	.000
14.533	.000
14.800	.000
14.933	.000
15.600	.000
16.133	.000
16.533	.000
16.933	.000
17.200	.000
17.733	.000
18.533	.000
19.200	.000
19.600	.000
19.733	.000
20.133	.000
20.800	.000
20.933	.000
21.733	.000
22.533	.000
22.800	.000
22.933	.000
23.333	.000
24.133	.000
24.400	.000
25.200	.000
26.133	.000
26.533	.000
28.133	.000
30.000	.000

$s = 4, N = 2$

c	P
5.4	.167
6.0	.042

$s = 4, N = 3$

c	P
5.8	.148
6.6	.075
7.0	.054
7.4	.033
8.2	.017
9.0	.002

$s = 4, N = 4$

c	P
5.7	.141
6.0	.105
6.3	.094
6.6	.077
6.9	.068
7.2	.054
7.5	.052
7.8	.036
8.1	.033
8.4	.019
8.7	.014
9.3	.012
9.6	.007
9.9	.006
10.2	.003
10.8	.002
11.1	.001
12.0	.000

$s = 4, N = 5$

c	P
5.88	.119
6.12	.102
6.36	.089
6.84	.071
7.08	.067
7.32	.057
7.80	.049
8.04	.033
8.28	.032
8.76	.024
9.00	.021
9.24	.015
9.72	.011
9.96	.009
10.20	.008
10.68	.006
10.92	.003
11.16	.002
11.64	.002
11.88	.001
12.12	.001
12.60	.000
12.84	.000
13.08	.000
13.56	.000
14.04	.000
15.00	.000

$s = 4, N = 6$

c	P
5.6	.127
5.8	.113
6.2	.109
6.4	.088
6.6	.087
6.8	.073
7.0	.067
7.2	.063
7.4	.058
7.6	.043
7.8	.041
8.0	.036
8.2	.033
8.4	.031
8.6	.027
8.8	.021
9.0	.021
9.4	.017
9.6	.015
9.8	.015
10.0	.011
10.2	.010
10.4	.009
10.6	.008
10.8	.006
11.0	.006
11.4	.004
11.6	.003
11.8	.003
12.0	.002
12.2	.002
12.6	.001
12.8	.001
13.0	.001
13.2	.001
13.4	.001
13.6	.000
13.8	.000
14.0	.000
14.6	.000
14.8	.000
15.0	.000
15.2	.000
15.4	.000
15.8	.000
16.0	.000
16.2	.000
16.4	.000
17.0	.000
18.0	.000

$s = 4, N = 7$

c	P
5.571	.150
5.743	.122
5.914	.118
6.257	.101
6.429	.093
6.600	.081
6.943	.073
7.114	.062
7.286	.058
7.629	.051
7.800	.040
7.971	.037
8.314	.034
8.486	.032
8.657	.030
9.000	.024
9.171	.021
9.343	.018
9.686	.016
9.857	.014
10.029	.013
10.371	.009
10.543	.008
10.714	.008
11.057	.007
11.229	.006
11.400	.004
11.743	.004
11.914	.003
12.086	.003
12.429	.003
12.600	.002
12.771	.002
13.114	.001
13.286	.001
13.457	.001
13.800	.001
13.971	.001
14.143	.001
14.486	.000
14.657	.000
14.829	.000
15.171	.000
15.343	.000
15.514	.000
15.857	.000
16.029	.000
16.200	.000
16.543	.000
16.714	.000
16.886	.000
17.229	.000
17.400	.000
17.571	.000
17.914	.000
18.257	.000
18.771	.000
18.943	.000
19.286	.000
19.971	.000
21.000	.000

$s = 4, N = 8$

c	P
5.55	.144
5.70	.122
5.85	.120
6.00	.112
6.15	.106
6.30	.098
6.45	.091
6.75	.077
7.05	.067
7.20	.062
7.35	.061
7.50	.052
7.65	.049
7.80	.046
7.95	.043
8.10	.038
8.25	.037
8.55	.031
8.70	.028
8.85	.026
9.00	.023
9.15	.021
9.45	.019
9.60	.015
9.75	.015
9.90	.013
10.05	.013
10.20	.011
10.35	.010
10.50	.009
10.65	.008
10.80	.008
10.95	.008
11.10	.007
11.25	.007
11.40	.006
11.55	.005
11.85	.004
12.00	.004
12.15	.003
12.30	.003
12.45	.003
12.60	.003
12.75	.002
12.90	.002
13.05	.002
13.20	.002
13.35	.001
13.50	.001
13.65	.001
13.80	.001
13.95	.001
14.25	.001
14.55	.001
14.70	.000
†	

† The remaining 41 entries are all .000.

TABLE N. Distribution of Spearman's statistic: $D = \sum(T_i - i)^2$, $P[D \leqslant d]$ [*distribution is symmetric about* $(N^3 - N)/6$]

d	P	d	P	d	P	d	P	d	P	d	P	d	P
$N = 2$		12	.0240	60	.2504	80	.1927	64	.0334	.	.	116	.0729
0	.5000	14	.0331	62	.2682	82	.2050	66	.0367	.	.	118	.0773
		16	.0440	64	.2911	84	.2183	68	.0403	.	.	120	.0817
$N = 3$		18	.0548	66	.3095	86	.2315	70	.0441	12	.0000	122	.0865
0	.1667	20	.0694	68	.3323	88	.2467	72	.0481	14	.0000	124	.0913
2	.5000	22	.0833	70	.3517	90	.2603	74	.0524	16	.0001	126	.0964
		24	.1000	72	.3760	92	.2759	76	.0569	18	.0001	128	.1015
$N = 4$		26	.1179	74	.3965	94	.2905	78	.0616	20	.0001	130	.1070
0	.0417	28	.1333	76	.4201	96	.3067	80	.0667	22	.0002	132	.1125
2	.1667	30	.1512	78	.4410	98	.3218	82	.0720	24	.0003	134	.1183
4	.2083	32	.1768	80	.4674	100	.3389	84	.0774	26	.0003	136	.1242
6	.3750	34	.1978	82	.4884	102	.3540	86	.0831	28	.0005	138	.1304
8	.4583	36	.2222	84	.5116	104	.3718	88	.0893	30	.0006	140	.1365
10	.5417	38	.2488			106	.3878	90	.0956	32	.0007	142	.1431
		40	.2780	$N = 9$		108	.4050	92	.1022	34	.0009	144	.1496
$N = 5$		42	.2974	0	.0000	110	.4216	94	.1091	36	.0011	146	.1566
0	.0083	44	.3308	2	.0000	112	.4400	96	.1163	38	.0014	148	.1634
2	.0417	46	.3565	4	.0001	114	.4558	98	.1237	40	.0016	150	.1708
4	.0667	48	.3913	6	.0002	116	.4742	100	.1316	42	.0020	152	.1780
6	.1167	50	.4198	8	.0004	118	.4908	102	.1394	44	.0023	154	.1857
8	.1750	52	.4532	10	.0007	120	.5092	104	.1478	46	.0027	156	.1932
10	.2250	54	.4817	12	.0010			106	.1564	48	.0032	158	.2012
12	.2583	56	.5183	14	.0015	$N = 10$		108	.1652	50	.0037	160	.2091
14	.3417			16	.0023	0	.0000	110	.1744	52	.0043	162	.2174
16	.3917	$N = 8$		18	.0030	2	.0000	112	.1839	54	.0049	164	.2255
18	.4750	0	.0000	20	.0041	4	.0000	114	.1935	56	.0056	166	.2342
20	.5250	2	.0002	22	.0054	6	.0000	116	.2035	58	.0064	168	.2427
		4	.0006	24	.0069	8	.0001	118	.2135	60	.0072	170	.2517
$N = 6$		6	.0011	26	.0086	10	.0001	120	.2241	62	.0081	172	.2604
0	.0014	8	.0023	28	.0107	12	.0002	122	.2349	64	.0091	174	.2697
2	.0083	10	.0036	30	.0127	14	.0003	124	.2459	66	.0102	176	.2787
4	.0167	12	.0054	32	.0156	16	.0004	126	.2567	68	.0113	178	.2883
6	.0292	14	.0077	34	.0184	18	.0006	128	.2683	70	.0126	180	.2975
8	.0514	16	.0109	36	.0216	20	.0008	130	.2801	72	.0139	182	.3073
10	.0681	18	.0140	38	.0252	22	.0011	132	.2918	74	.0153	184	.3168
12	.0875	20	.0184	40	.0294	24	.0014	134	.3037	76	.0168	186	.3269
14	.1208	22	.0229	42	.0333	26	.0019	136	.3161	78	.0184	188	.3366
16	.1486	24	.0288	44	.0380	28	.0024	138	.3284	80	.0201	190	.3469
18	.1778	26	.0347	46	.0429	30	.0029	140	.3410	82	.0220	192	.3568
20	.2097	28	.0415	48	.0484	32	.0036	142	.3536	84	.0239	194	.3673
22	.2486	30	.0481	50	.0540	34	.0044	144	.3665	86	.0260	196	.3773
24	.2819	32	.0575	52	.0603	36	.0053	146	.3795	88	.0281	198	.3879
26	.3292	34	.0661	54	.0664	38	.0063	148	.3925	90	.0304	200	.3982
28	.3569	36	.0756	56	.0738	40	.0075	150	.4056	92	.0328	202	.4090
30	.4014	38	.0855	58	.0809	42	.0087	152	.4191	94	.0354	204	.4192
32	.4597	40	.0983	60	.0888	44	.0101	154	.4326	96	.0380	206	.4302
34	.5000	42	.1081	62	.0969	46	.0117	156	.4458	98	.0409	208	.4406
		44	.1215	64	.1063	48	.0134	158	.4592	100	.0438	210	.4516
$N = 7$		46	.1337	66	.1149	50	.0153	160	.4730	102	.0470	212	.4621
0	.0002	48	.1496	68	.1250	52	.0173	162	.4865	104	.0502	214	.4731
2	.0014	50	.1634	70	.1348	54	.0195	164	.5000	106	.0536	216	.4837
4	.0034	52	.1799	72	.1456	56	.0219			108	.0571	218	.4947
6	.0062	54	.1947	74	.1563	58	.0245	$N = 11$		110	.0609	220	.5053
8	.0119	56	.2139	76	.1681	60	.0272	0	.0000	112	.0647		
10	.0171	58	.2309	78	.1793	62	.0302	2	.0000	114	.0688		

ACKNOWLEDGMENTS

I gratefully acknowledge my indebtedness to:

Ellen Sherman for computing Table E and for modifying existing tables to produce Table F.

Howard D'Abrera for computing Table G and for modifying existing tables to produce Tables J(a) and J(b).

Howard D'Abrera and Greg Thompson for computing Table H.

The Trustees of the Biometrika Trust and the authors for permission to reproduce part of Table 3 of Jonckheere from *Biometrika*, Vol. 41, pp. 144/145 and a table of Miles from *Biometrika*, Vol. 46, p. 325 as Tables K and L.

The University of Montreal Press for permission to reproduce parts of Buckle, Kraft, and van Eeden: "Tables Prolongees de la Distribution de Wilcoxon-Mann-Whitney" as Table B.

The Macmillan Company for permission to reproduce part of Table F in Kraft and van Eeden: "A Nonparametric Introduction to Statistics" as Table I.

Addison-Wesley for permission to reproduce parts of Tables 13.2 and 14.1 from Owen: "Handbook of Statistical Tables" as Tables N and M.

Special thanks go to Howard D'Abrera for his supervision and painstaking proofreading of all the tables.

ANSWERS TO SELECTED PROBLEMS

CHAPTER 1

(A)

2. (i) 1225

4. (i) 210; (iii) 924

6. (i) 1/10; (iii) 6/10; (v) 3/10

8. (i) 1/352,716; (iii) 31,824/352,716

10. (i)

$W_s - 10$	0	1	2	3	4	5	6	7	8
P	1/15	1/15	2/15	2/15	3/15	2/15	2/15	1/15	1/15

12. (i) 0, 1/15, 2/15, 4/15, 6/15, 9/15, 11/15, 13/15, 14/15, 1

14. (i) 0; (iii) 2/21 = .095

16. (i) .0022 (iii) .0032

18. (i) .9978; (iii) .9913

20. (i) $W_r = 21 - W_s$; (iii) $W_r = 21 - W_s$

22. .6349

24. (i) Reject if $W_s \leqslant 52$ (level .0524); (iii) Reject if $W_s \leqslant 54$ (level .0464)

26. (i) $W_s \geqslant 53$ (level .0130); (iii) $W_s \geqslant 133$ (level .0177)

28. (i) .1711; (iii) .0184

30. (i) 1.2816; (iii) 1.2816

32.

	normal (with continuity correction)	exact
(i)	.0152	.0110
(iii)	.0073	.0040

36.

$m = 3;\ n =$	3	4	5
w	14	21	28
exact	.1000	.0571	.0714
normal (a)	.0952	.0558	.0680
normal (b)	.0633	.0386	.0505

38. (i) $W_s \leqslant 323$; (iii) $W_s \leqslant 2187$

40.

n	c	c'	Exact Probabilities $P(W_s \geqslant c')$	$P(W_s \geqslant c)$
6	52	53	.0130	.0206

42.

poor	4
indifferent	15.5
good	31
very good	38.5

44. (iii)

w	6.5	8.5	10.5	12.5	14.5
$P(W_s^* = w)$	2/20	4/20	8/20	4/20	2/20

(v)

w	6	9	12	15
$P(W_s^* = w)$	1/20	9/20	9/20	1/20

46. .0004 (approximately)

48. $\Phi(-3.756) \simeq .000086$

50.

	c	$P(W_s^* \geqslant c)$	normal approximation	normal approximation with a continuity correction
(i)	43.5	$\frac{40}{792} = .0505$	.0355	.0504

52.

	c	$P\{\lvert W_B - \frac{1}{2}n(N+1)\rvert \geqslant c\}$
(i)	27	.0280
(iii)	15	.0296

54.

	c (using continuity correction)	c'	Prob
(i)	20.584	20	.0554
(iii)	20.869	21	.0464

56. Suspend judgment if $20 < W_B < 35$

58.

	c_1	α_1	c_2	α_2
(i)	70	.0951	112	.0094
(iii)	76	.0966	114	.0093

60. (i) .1645

62.

T	6	7	8	9	10	11	12
Prob	1/35	4/35	7/35	8/35	9/35	4/35	2/35

64. (i) 7/20; (iii) 3/9

66. (i)

$D_{3,2}$	1/3	1/2	2/3	1
Prob	1/10	3/10	4/10	2/10

68.

	Table F	Table E
(i)	.9250	.9383
(iii)	.1868	.1886

(B)

72. (i) 1; (iii) N

74. $s = 1, 2, \ldots, N$ each with probability $1/N$

76. $\binom{N-k}{n-r} \Big/ \binom{N}{n}$

78. $k = \dfrac{N\left(c - \frac{1}{2}n(N+1)\right)}{nm}$

82. (i) $1 \Big/ \binom{N}{n}$; (iii) $2 \Big/ \binom{N}{n}$

90. 1/2

CHAPTER 2

(A)

2. (i) $$\begin{aligned} \Pi(\Delta) &= \tfrac{2}{20} : 0 \leqslant \Delta < 1 \\ &= \tfrac{3}{20} : 1 < \Delta < 2 \\ &= \tfrac{7}{20} : 2 < \Delta < 3 \\ &= \tfrac{10}{20} : 3 < \Delta < 5 \\ &= \tfrac{12}{20} : 5 < \Delta < 6 \\ &= \tfrac{16}{20} : 6 < \Delta < 7 \\ &= 1 : 7 < \Delta \end{aligned}$$

4.

$S_1^* + S_2^*$	3	4	5	6	7
Probability	p^2q^2	$2pq(p^2+q^2)$	$p^4+4p^2q^2+q^4$	$2pq(p^2+q^2)$	p^2q^2

6. (i) accept: $2P(W_c^* \leqslant 145) \sim .7724$

8. (i) .9444; (iii) .8767

10. (i) .3648; (iii) .9989

12. (i) $n \sim 13$; (iii) $n \sim 23$

14. (i)

n	6	11	16	21	31
$c_{n,n}(\alpha)$	2.764	2.528	2.457	2.423	2.390

16.

Δ/σ	.1	.2	.3	.4	.5	.6	.7	.8	.9
Wilcoxon	.096	.168	.268	.391	.526	.658	.773	.863	.924
t-test	.097	.172	.275	.401	.539	.672	.787	.874	.932

18. $\hat{\Delta} = -6$
$\bar{\Delta} = -6.4$

20. (i) $\hat{\Delta} = 25$ from C to B
$\bar{\Delta} = \bar{B} - \bar{C} = 25.129$
(ii) Reject hypothesis at $\alpha = .05$
since $2P(W_{XY} \leqslant 14) = .043$

22. (i) $D_{(11)} \leqslant \Delta < D_{(45)}$
(ii) $D_{(11)} = .5$
$D_{(45)} = 3.7$
t-test: $.559 \leqslant \Delta \leqslant 2.37$

24. $D_{(15)} = 1$

28. (i) $C(1;4) > 5$ (conditional level = 0)
$C(2;3,1) > 6$ (conditional level = 0)
$C(2;1,3) > 6$ (conditional level = 0)
$C(2;2,2) = 7$ (conditional level = 1/6)
(ii) For test in (i)
$l = p^2q^2$
Maximum is $l = \frac{1}{16}$ at $p = \frac{1}{2}$
Minimum is arbitrarily close to zero
(iii) $l' = 2pq(p^2+q^2) > p^2q^2$

45.

Δ	.5	1.0	1.5	2.0
Π	.287	.707	.943	.996

CHAPTER 3

(A)

2. (i) .2120; (iii) .9919

4. (i) .9102; (iii) .0214

6. .2905

8. .1939

10. $N = 6, a = 1$

exact	with continuity	without continuity
.1094	.1103	.0513

12. (i) $S_N \geqslant 17$; (iii) $S_N \geqslant 48$

14. (i) 1/16; (iii) 1/32

16. (i) 1/256; (iii) 70/256

18. (i)

Wilcoxon	0, 1/16, 2/16, 3/16, 5/16, 7/16, 9/16, 11/16, 13/16, 14/16, 15/16, 1
Sign	0, 1/16, 5/16, 11/16, 15/16, 1

20. (i) .8608; (iii) .2090

22. $N = 8$

Sign: 0, .0039, .0352

Wilcoxon: 0, .0039, .0078, .0117, .0195, .0273, .0391

24.

	Sign	Wilcoxon
(i)	.0547	.0186
(iii)	.0547	.0420

26. (i) .0072; (iii) .0112

28. $N = 5, v = 1$

exact	with continuity	without continuity
.0625	.0528	.0398

30. $N = 10$

c (normal)	c (Problem 25)	$P(V_S \geqslant c^1)$
48.15	48	.0186

34. Accept hypothesis;

	Significance Probability
Sign	.5892
Wilcoxon	.7655 (approx)

36. Approximate significance probability using two-sided Wilcoxon rank-sum test is .2211

38. (i) $V_s^* = 160.5$; (ii) .1016

40. $v = 1.5$, $P[V_s^* \leqslant v] = 3/32 = .09375$, Normal approx. = .0512

42. (i) .6449; (iii) .6254

44. (a) .00805; (b) .0092

48. (i) $\frac{237}{17,640} = .0134$; (ii) .0144

50. Significance probability is .0013

Reject hypothesis at usual levels

52. $\frac{20}{360} = .0556$

54. (i) .0361

56. (i) .00046; (ii) .00106

(B)

62. $c' = \frac{1}{2}N(N+1) - c$

64.

v	0	1	2	3	4	5	6	7	8	9	10
$\#(v, N)$	1	1	1	2	2	3	4	5	6	8	10

68. Minimum is 0

Maximum is $[N(N+1) - d_0(d_0+1)]/2$

72. $c' = -c + \frac{1}{2}\sum N_i$

CHAPTER 4

(A)

2. (i) .99986 approximately
(.99861 exactly)
(iii) (a) .9791
(c) .8139

4. $\Pi \sim 1-\Phi\left[\frac{1.1633-\sqrt{40}(p-.5)}{\sqrt{p\cdot q}}\right]$

6. (i) (a) 48 (b) 41

8. $\Pi \sim 1-\Phi\left[\frac{21.5-30p}{\sqrt{30p\cdot q}}\right]$

where $p=\Phi\left(\frac{\Delta}{\sigma}\right)$

10. (i) .3861
(ii) .6609

12. (i) .7354
(iii) .9832

14. 20

16. sign test $N \sim 101$
t-test $N \sim 63$

20. $\tilde{\theta}=41.7$
$\hat{\theta}=41.625$
$\bar{\theta}=41.568$

22. $\tilde{\theta}=.260$, $\hat{\theta}=.375$, $\bar{\theta}=.392$

24. (i)

a	.05	.10	.20	.25	.30	.40	.50	.75	1.00
Bin. Tables	.125	.247	.470	.568	.654	.790	.883	.981	.9997
Normal Approx.	.123	.243	.466	.565	.653	.795	.892	.989	.9997

(ii)

a	.05	.10	.20	.25	.30	.40	.50	.75	1.00
Bin. Tables	.307	.574	.900	.965	.992	.99986	1	1	1
Normal Approx.	.303	.571	.909	.975	.996	1	1	1	1

26. (i)

i	3	5	5	7	8	9	11	11	13
γ	.0037	.0592	.0592	.3036	.5000	.6964	.9408	.9408	.9963
Closest to	1%	5%	10%	25%	50%	75%	90%	95%	99%

(ii)

i	5	7	8	9	10	11	12	13	14	16
γ	.0059	.0577	.1316	.2517	.4119	.5881	.7483	.8684	.9423	.9941
Closest to	1%	5%	10%	25%	50%	50%	75%	90%	95%	99%

30.

N	10	14	19
Conf. interv.	$(Z_{(2)}, Z_{(9)})$	$(Z_{(4)}, Z_{(11)})$	$(Z_{(6)}, Z_{(14)})$
Conf. coeff.	.9786	.9426	.9364

32. (i)

	lower	upper
Wilcoxon	41.1	42.1
Binomial	41.0	42.2
t	40.5275	42.2198

34.

	N	$\underline{\mu}$	$\bar{\mu}$
(i)	25	$Z_{(8)}$	$Z_{(18)}$
(iii)	100	$Z_{(41)}$	$Z_{(60)}$

36. (i) $p = .2$, lower conf. pt. $= Z_{(13)}$, $\gamma = .9679$
(ii) $p = .2$ $(Z_{(13)}, Z_{(19)})$, $\gamma = .8987$

(B)

66. $\text{Var}(\hat{\theta}) = .3406$, $\text{Var}(\tilde{\theta}) = .4487$, $\text{Var}(\bar{\theta}) = 1/3$

CHAPTER 5

(A)

2. (i) (1, 4, 6) (1, 4, 8) (1, 4, 12) (1, 6, 8) (1, 6, 12)
(1, 8, 12) (4, 6, 8) (4, 6, 12) (4, 8, 12) (6, 8, 12)
(iii) (2, 5, 7) (2, 5, 8) (2, 5, 11) (2, 7, 8) (2, 7, 11)
(2, 8, 11) (5, 7, 8) (5, 7, 11) (5, 8, 11) (7, 8, 11)

4. (i) 15, 765, 750
(iii) 150, 150
(iv) 1, 261, 260

6. (i) 1/2520
(iii) 2/2520

8. $k = 2.3057$; not significant

10. $P(\chi_3^2 \geqslant 10.158) \simeq .017$

12. (i) $\Phi(-4.30) \simeq 9 \times 10^{-6}$
(ii) $2\Phi(-4.30) \simeq 1.8 \times 10^{-5}$
(iii) $P(\chi_2^2 \geqslant 18.58) \simeq .0001$

14. $P(\chi_3^2 \geqslant 5.462) \simeq .141$

16. $P(\chi_1^2 \geqslant 5.632) \simeq .0176$
Using formula (27) one-sided probability $\simeq .013$

18. Exact hypergeometric probability = .0466
normal approximation with cty correction .0487

20. $P(K^* \geqslant 79.7) \simeq 5 \times 10^{-18}$

22. Significance probabilities are:
Chlorpromazine 50 mg: $1-\Phi(3.65) = .0001$
Dimenhydrinate 100 mg: $1-\Phi(\ .5\ ) = .3085$
Pentobarbital 100 mg: $1-\Phi(1.26) = .1038$
Pentobarbital 150 mg: $1-\Phi(1.07) = .1423$
Only Chlorpromazine 50 mg is significantly better.

24. (i)

w	0	1	2	3	4	5	...
$P(W=w)$	$\frac{1}{56}$	$\frac{2}{56}$	$\frac{3}{56}$	$\frac{4}{56}$	$\frac{5}{56}$	$\frac{6}{56}$	...

25. $P(W^* \leqslant 97.5) \simeq .0228$

26. (i) .0175
(ii) .0544

28. Both significant
$P(W_{TS}^* \geqslant 349.5) \simeq .0055$
$P(W_{US}^* \geqslant 450) \simeq .0008$

29. (ii)

$14k'$	0	3	4	12	16	27	28	36	48	52	64
$P(K'=k')$	$\frac{40}{90}$	$\frac{12}{90}$	$\frac{2}{90}$	$\frac{14}{90}$	$\frac{2}{90}$	$\frac{6}{90}$	$\frac{2}{90}$	$\frac{1}{90}$	$\frac{8}{90}$	$\frac{2}{90}$	$\frac{1}{90}$

31. (i) No significant pairwise comparisons
$2P(W_{21} \geqslant 16) = .0286$
$2P(W_{31} \geqslant 14) = .1142$
$2P(W_{23} \geqslant 11) = .3429$

32. A is better than C:

$\underline{C\ B}$ A (with a line under $B\ A$)

$2P(W_{BA} \geqslant 31) = .1807$
$2P(W_{CA} \geqslant 34) = .0087$
$2P(W_{CB} \geqslant 34) = .0734$

47.

R_1^*	R_2^*	R_3^*	Probability
2	2	2	p^3+q^3
1.5	1.5	3	p^2q
⋮	⋮	⋮	⋮
2.5	2.5	1	pq^2

where $q = 1-p$

48. (ii)

$\max(R_1, R_2)$	3	4	5
Prob	$\frac{1}{3}$	$\frac{1}{3}$	$\frac{1}{3}$

CHAPTER 6

(A)

2. (i) $\dfrac{10}{3^4}$

4. (ii) $\left\{\dfrac{1}{s(s-1)}\right\}^N$

6. $Q^* = 3.6956$
$P(\chi_3^2 \geqslant 3.6956) \simeq .297$, no significant difference

8. .068

10. $Q^* = 1.345$, $P(\chi_3^2 \geqslant 1.345) \simeq .719$

12. (i) $Q^* = 3.534$, $P(\chi_2^2 \geqslant 3.534) \simeq .171$

CHAPTER 7

(A)

2. (ii) $\frac{1}{12}$

4. $\frac{1}{3}$

6. (i) $P\{\Sigma D_k \leqslant 206\} \simeq .812$
(ii) $P\{Q \geqslant 5.114\} = .402$

8. $P(D^* \leqslant 26.5) \simeq .0007$

10. (i) 30
(ii) $\frac{1}{10}$
(iii) $\frac{2}{5}$

12. (i)

d^*	2.5	7.5	12.5	17.5	22.5	27.5	32.5
$P(D^* = d^*)$	$\frac{1}{10}$	$\frac{1}{10}$	$\frac{2}{10}$	$\frac{2}{10}$	$\frac{2}{10}$	$\frac{1}{10}$	$\frac{1}{10}$

16. *w/o* cty correction $P(D \leqslant 455) \simeq .0302$
with cty correction $P(D \leqslant 456) \simeq .0307$

18. $P(D^* \leqslant 550) \simeq .2338$

20. Six of the eight possible tables have prob. 2/20, the other two have prob. 4/20 each.

22. (i)

k	$\frac{10}{27}$	$\frac{34}{27}$	$\frac{76}{27}$	$\frac{90}{27}$
$P(K^* = k)$	$\frac{1}{10}$	$\frac{4}{10}$	$\frac{4}{10}$	$\frac{1}{10}$

28. (i) .101
(ii) .022
(iii) .458

30. (i) .259
(ii) .193

32. (i) $L = 4.0381$, $P(\chi_4^2 \geqslant 4.0381) \simeq .401$
(ii) significance probability $\simeq \Phi(.6394) = .7387$

DATA GUIDE

The following table provides titles for the data presented in the text. (Full references are given in the text.)

Effect of discouragement in IQ tests, 4
Effect of hypnotic suggestion on respiratory function, 27, 108
Nutritive value of butter fat and vegetable fat, 29, 108, 207, 234, 255
Two drugs for relief of postoperative pain, 37, 47
Transfer of learning, 45
A guinea pig study of the effect of vitamin C, 45
The effect of vitamin B_1 on mushroom growth, 47
Psychological factors and human cancer, 61
A comparison of two routes, 83
Augmenters and reducers, 92
Televised vs live instruction, 107
Sutured vs taped wounds, 126, 148
Effect of thiamine on learning, 131
Speech patterns under hypnosis, 142
A new drug for schizophrenia, 148, 193
Group therapy for juvenile delinquents, 149
Effect of familiarity with the examiner on IQ tests, 149
Comparison of measurements in different laboratories, 150, 209, 240, 251
Growth of children during first year after birth, 176, 192, 319
Effect of brief exercise on muscle strength, 183, 193
Anxiety and dependence, 215
The effect of color in mail advertising, 218
IQ scores at four universities, 221
Birth conditions and IQ, 228
Effect of added information on the ranking of candidates, 236, 237, 254
A comparison of three kinds of rat food, 251
Genesis of the cat's response to the rat, 251, 252
Some drugs for nausea, 252, 306, 307, 308
Experimental transmission of the common cold, 252

Determination of the gravitational constant *G*, 252
A psychological hypothesis in appendectomy, 253
Frequency of left handedness in boys and girls, 253
Criminal justice in three California counties, 253
First and second calls in an election poll, 253, 320
Effect of experience and training on making a diagnosis, 254
Physiological effects of hypnotically requested emotions, 264, 270
Effect of emotional associations on memory, 265, 283
Different methods of soothing babies, 267, 282, 283
Comparing medications for cough relief, 282
Age regression under hypnosis, 282
Different media for growing diphtheria bacilli, 284
Pollution of Lake Michigan, 290
Trend of illegitimate births, 295, 318
Performance in language and arithmetic, 298
Relation of crying in infancy to later development, 302
Age trend in IQ, 318
Success of mediation and initial hostility, 319
Predicting potential from psychological tests, 319
Psychological tests in twins, 319
Family size and income, 320
Effect of age on certain opinions, 320
Study conditions and quality of work, 321

AUTHOR INDEX

Abrahamson, 105, 114
Adichie, 313, 322
Aiyar, 311, 312, 322, 391
Alling, 41, 52
Altham, 268, 285
Andrews, D., 189, 199
Andrews, F., 247, 248, 257
Ansari, 34, 52, 114
Anscombe, 199
Arbuthnot, vii, xi
Armitage, 322, 323
Arnold, 185, 187, 199

Bahadur, 99, 114, 186, 199
Barlow, 103, 236, 257
Barnard, 322, 326
Bartholomew, 237, 257
Barton, 42, 52, 115, 315, 323
Basu, 104, 114
Bauer, 102, 114
Behnen, 100, 114, 317, 322
Bell, 98, 115
Benard, 279, 285
Bennett, 215
Beran, 102, 115, 313, 322
Berk, 106, 115
Best, 316, 322
Bhapkar, 317, 322
Bhattacharya, 313, 316, 322
Bhuchongkul, 316, 317, 322
Bickel, 40, 52, 99, 115, 128, 153, 199, 313, 322, 362, 405
Billingsley, 314, 322, 387, 405
Birnbaum, 92, 115
Bishop, 317, 322
Blomquist, 316, 323
Blum, 316, 323
Blyth, 186, 199
Bradley, 34, 52, 106, 114, 115, 119
Bremmer, 257
Brown, G., 175
Brown, S., 286
Brunk, 257
Buckle, 8, 17, 41, 52
Bühler, 100, 115
Burr, 316, 323
Butler, 145, 153

Capon, 100, 103, 105, 115
Chackó, 236, 257
Chanda, 100, 115
Chapman, 317, 323, 325
Chernoff, viii, xi, 97, 115, 186, 189, 199
Claypool, 128, 153
Cochran, 99, 115, 186, 199, 267, 281, 285, 308, 317, 323
Conover, 145, 153
Cox, 66, 92, 104, 115
Cramér, 143, 153, 308, 323, 347, 348, 387, 406
Crouse, 249, 257

D'Abrera, 237, 257, 281, 285
Daniels, 323, 324
Darling, 106, 115
David, F. N., 42, 52, 96, 115, 314, 315, 323
David, H. A., 143, 153
Davies, 104, 115
Dixon, 105, 115, 173, 185, 186, 199
Doksum, 68, 101, 115, 187, 189, 199, 201, 280, 285, 342, 406
Drion, 36
Dunn, 257, 259
Dupac, 97, 116
Durbin, 38, 52, 279, 285

Edgington, 315, 323
Eisenhart, 314, 326
Eisenstat, 186, 200

Farlie, 316, 323
Feller, 36, 347, 348, 350, 406
Fellingham, 128, 154
Felsenstein, 213, 258
Fieller, 316, 323
Fienberg, 317, 323
Fine, 102, 116
Finney, 215
Fisher, 43, 52, 96, 116
Fix, 17, 40, 53, 247, 257
Folks, 281, 285
Fraser, 199
Friedman, vii, xi, 264, 265, 285

Gabriel, 207, 238, 245, 257
Ganeshalingham, 115
Gastwirth, 97, 116, 172, 187, 189, 190, 199
Geertsema, 188, 200, 246, 258
Gehan, 41, 53
Ghosh, 106, 116, 188, 201, 246, 258
Gibbons, 185, 200
Gilbert, J., 317, 323
Gilbert, R., 272, 277, 285
Glasser, 292
Gnedenko, 36, 53
Gokhale, 316, 317, 323
Goldberg, 3, 53
Goodman, 314, 317, 323
Govindarajulu, 98, 111, 116, 186, 187, 200, 201, 247, 258
Gridgeman, 144, 154
Grizzle, 317, 324, 326
Gross, 105, 116
Guillier, 311, 312, 324
Gupta, 246, 247, 258

Haberman, 317, 324
Haigh, 355
Hájek, ix, xi, 36, 53, 97, 104, 116, 200, 250, 258, 349, 352, 353, 362
Haldane, 295, 324
Hall, 106, 116
Halperin, 41, 53
Hampel, 189, 190, 199, 200
Harter, 53, 54, 96, 104, 115, 116, 200
Hartigan, 188, 200
Hartley, 108
Haynam, 98, 111, 116, 247, 258
Hemelrijk, 144, 154
Hettmansperger, 237, 25o
Hoadley, 99, 116
Hodges, viii, xi, 3, 17, 38, 40, 41, 42, 53, 78, 97, 116, 118, 143, 154, 186, 200, 201, 247, 257, 373, 377
Hoeffding, 100, 106, 116, 143, 154, 189, 200, 316, 324, 362
Holbert, 128, 153
Holland, 317, 323
Hollander, 96, 116, 279, 285, 313, 324
Hotelling, vii, xi, 324
Høyland, 84, 101, 117, 177, 190, 200
Hsu, 215
Huber, 189, 190, 199, 200
Hwang, 186, 200

Jacobson, 53
Jaeckel, 189, 190, 191, 200, 398
James, 104, 117
Jennrich, 36, 39, 53
Johns, 189, 199
Johnson, 41, 53, 316, 322
Jonckheere, 233, 258
Jureckova, 313, 324

Kaarsemaker, 316, 324
Katti, 8, 54, 127, 155
Kendall, vii, xi, 235, 258, 264, 265, 285, 300, 312, 316, 324
Kermack, 315, 324
Khinchine, 346
Kiefer, 250, 258, 316, 323
Kim, 36, 38, 39, 53, 115
Klotz, 40, 53, 96, 98, 102, 103, 105, 117, 173, 186, 187, 200
Knüsel, 101, 102, 117
Koch, 317, 322, 324
Kolmogorov, 38, 53, 54
Konijn, 316, 324
Korolyuk, 36, 53
Koul, 313, 324
Kraemer, 316, 324
Kraft, 8, 17, 41, 52, 53, 101, 117, 206, 258, 264, 285
Krauth, 100, 117, 144, 154
Kruskal, vii, xi, 16, 53, 76, 117, 206, 208, 258, 300, 316, 317, 324, 325

Lachenbruch, 207, 258
Lancaster, 98, 117, 308, 317, 325
Latscha, 215
Lehman, 20, 53
Leone, 247, 258
Lepage, 95, 117
Lev, 286

Levene, 315, 325
Lewontin, 213, 258
Light, 317, 325
Lin, 96, 116
Lindeberg, 350
Littell, 281, 285

Mallows, 42, 52
Mann, 12, 54, 76, 117, 312, 325
Mantel, 317, 325
Marascuilo, 249, 258
Mardia, 317, 325
Margolin, 317, 325
Massey, 105, 117
McCormack, 127, 154
McDonald, B., 244, 258, 279, 286
McDonald, G., 246, 247, 258
McKendrick, 315, 324
McLaughlin, 201
McNemar, 268, 286, 325
McSweeney, 249, 258
Mehra, 277, 279, 286, 373
Mehrotra, 41, 53
Meng, 317, 325
Merchant, 106, 115
Merrington, 115
Miles, 236, 258
Miller, 102, 229, 241, 244, 245, 258, 279, 286
Milton, 8, 54, 98, 117, 118
Mitra, 326
Mood, 315, 325
Moore, 315, 326
Moran, 324, 325
Moser, 98, 115
Moses, 3, 54, 93, 117, 118, 183, 200, 265, 286
Mosteller, 3, 54, 269, 286, 317, 325

Nam, 268, 286, 323, 325
Neave, 316, 322
Nemenyi, 33, 54, 102, 247, 257, 258, 279, 286
Neyman, 247, 259
Nievergelt, 97, 119
Noether, 93, 117, 118, 145, 154, 183, 188, 200, 277, 279, 285

Oakford, 3, 54
Okamoto, 316, 326
Olmstead, 315, 316, 325
Olshen, 313, 325, 373
O'Neill, 245, 259
Oosterhoff, 154, 281, 286
Owen, 53, 54, 78, 96, 118, 264, 286

Pabst, vii, xi, 324
Patnaik, 247, 259
Pearson, 108, 215, 316, 323, 326
Peritz, 238, 259
Pirie, 279, 286
Pitman, viii, xi, 43, 54, 118, 186, 201
Plackett, 316, 317, 326
Pledger, 96, 116
Polya, 348
Potthoff, 313, 326
Pratt, 95, 118, 145, 154
Proschan, 103
Puri, M. L., ix, xi, 145, 154, 246, 249, 259
Puri, P. S., 246, 259
Putter, 100, 118, 144, 154, 164, 201

Radhakrishna, 281, 286
Raghavachari, 103, 105, 118
Ramachandramurty, 101, 118
Ramsey, 98, 118
Rao, 41, 54, 308, 326
Rhodes, 106, 119
Richter, 244, 259
Rijkort, 208, 259, 265, 286
Rizvi, 246, 259
Robbins, 106, 115, 118
Rogers, 199
Rose, 244, 259
Rosenblatt, 316, 323
Rourke, 3, 54
Roy, 326
Rubin, 172, 190, 200
Ruist, 189, 201

Sarangi, 277, 279, 286, 373
Savage, vii, viii, xi, 41, 54, 97, 99, 104, 106, 115, 118
Saw, 41, 54
Scheffé, vii, xi
Scholz, 103, 118
Sen, ix, xi, 116, 118, 188, 190, 201, 238, 249, 258, 259, 277, 279, 286, 313, 326
Serfling, 96, 118
Sethuraman, 106, 118
Sherman, 259
Shorack, 33, 54, 102, 236, 259, 279, 286
Sidak, ix, xi, 116, 200, 250, 258
Siegel, 54
Sillito, 316, 326
Simon, 259
Smirnov, vii, xi, 35, 38, 54, 145, 154
Smith, B., 265, 285
Smith, C. A. B., 295, 324

Smith, M., 244, 259
Sobel, 41, 54, 99, 118
Spearman, 300, 326
Spjøtvoll, 248, 259
Starmer, 317, 324
Steel, 229, 241, 259
Stein, 190, 201
Stoker, 128, 154
Stone, 99, 118
Stuart, 311, 312, 326
Sugiura, 41, 54
Sukhatme, 105, 118
Sundrum, 99, 118
Swed, 314, 326

Takeuchi, 190, 201
Tate, 286
Teichroew, 103, 118
Terpstra, 233, 259
Terry, 100, 119
Theil, 313, 326
Thomas, D., 41, 53
Thomas, G., 54
Thompson, R., 98, 115, 187, 189, 199, 201
Thompson, W., 244, 258, 279, 286
Tobach, 244, 259
Tryon, 237, 259
Tukey, 92, 98, 119, 154, 175, 199, 201, 316, 317, 325

Van der Vaart, 95, 119
Van der Waerden, 97, 119
Van Eeden, 8, 17, 41, 52, 53, 101, 117, 206, 258, 264, 285
Van Elteren, 145, 154, 277, 279, 285
Van Wijngaarden, 316, 324
Van Zwet, 342
Verdooren, 40
Von Mises, 314, 325
Vorlickova, 100, 119, 145, 155

Wald, vii, xi, 105, 314, 326
Walker, 286
Wallace, 208, 259
Wallis, 16, 53, 206, 208, 258, 315, 326
Walsh, 164, 186, 187, 201
Wetherill, 95, 119, 245, 259
Whitney, 12, 54, 76, 117
Wijsman, 106, 116
Wilcox, 8, 54, 127, 155
Wilcoxon, vii, xi, 8, 54, 76, 106, 115, 119, 127, 144, 155
Wilkie, 279, 286
Williams, 317, 326
Winter, 292
Wise, 208, 259, 265, 286
Witting, 99, 119
Woinsky, 104, 119
Wolff, 97, 116
Wolfowitz, vii, xi, 314, 315, 325, 326
Woodworth, 99, 104, 114, 118, 119, 186, 187, 201, 246, 259, 317, 326

Yanagimoto, 316, 326
Yates, 96, 116
Yu, 105, 119

Zar, 292
Zaremba, 96, 119
Zelen, 281, 286

SUBJECT INDEX

Absolute error, 16
Absolute Normal Scores test, 186, 187
  optimum property of, 189, 199
Adaptive estimator, 190
Additivity of treatment effect, 66, 276
Aligning of observations, for two treatments, 138, 140
  for *s* treatments, 270, 275, 277, 279
Alternatives, 31, 65
  (*see also* Shift model)
  choice of, 31, 34, 39, 304, 309, 313
  different types, 31, 42
  nonparametric, 189
  omnibus, 34, 145, 211, 304
  one-sided, 26, 29, 211, 304
  ordered, 232, 241, 279, 305, 306
  two-sided, 23, 211, 304
Ansari-Bradley test, 34, 52, 114
Association, establishment of, 63
  (*see also* Independence)
  measures of, 300, 316
  positive, 299, 316
  test for, 221, 299, 315, 324
Asymptotic efficiency, 100, 190
Asymptotic normality, 348
Asymptotic relative efficiency (*see* Efficiency, Pitman)
Attributes, comparison of two, 60, 63, 64
  comparison of *s*, 220, 225, 310
  distinguished from treatments, 60, 220, 297
  testing independence, 298
Autoregressive alternative, 311
Average, as estimator of location, 175, 177, 181

Back dating, 41
Balance, in paired comparisons, 142
Behrens-Fisher problem, 95
Beta distribution, 344
Binomial coefficient, 3, 50, 151
  recursion formula for, 50
Binomial distribution, 122, 143, 159, 339
  comparison of several, 224, 225
  expectation and variance of, 331
  as limit of hypergeometric, 340
  normal approximation, 122, 159
  normal limit of, 348
  tables of, 122, 154, 416
Binomial sampling, 330
Bivariate distribution, independence in, 303, 315
  (*see also* Association, Independence, Multivariate normal distribution)
Bivariate normal distribution, 72, 167, 316
Blocking for homogeneity, 133, 145, 156, 162, 260, 279
  (*see also* Aligning of observations, Combining independent experiments)
Boundary of a set, 387

Categorical data, 21, 210
  (*see also* Attributes, Contingency table, Two by two table)
Cauchy distribution, 168, 175, 185, 342, 397, 403
Causal relationship, establishment of, 63, 298
  (*see also* Attributes)
Censored observations, 41
Center of symmetry, testing the value of, 162
Central limit theorem, 13, 347, 350
  multivariate, 387, 390, 393
Chacko-Shorack test, 236, 237
  table for, 430
Chebyshev inequality, 345
Chi distribution, 186, 344
Chi square distribution, 206, 209, 218, 343
  as a limit, 389
  table of, 207, 427, 428
Clustering of observations, 313

Cochran test, 267, 286
Combinations, 3
Combining independent experiments, 132, 145, 154, 281
(*see also* Blocking for homogeneity)
Comparison of treatments, suggested by the data, 242, 279
Concordant pairs of observations, 315
(*see also* Mann's trend statistic)
number of, 312
Conditional test, given configuration of ties, 59, 64, 220, 296, 301, 310
becoming unconditional in the limit, 60, 275
in contingency tables, 223, 310
Confidence bounds, 94, 185
Confidence coefficient, 93
Confidence intervals, for shift, 91, 93
(*see also* Simultaneous confidence intervals, Standard confidence points)
for center of symmetry, 181, 188
for median, 182, 188
for quantiles, 185
for regression coefficient, 312
sequential, 188
Configuration of ties, 59
Confounding, of treatment and type, 142
Consistency of a sequence of estimators, 89, 113, 181
Constant effect model, 42
Contingency table, viii, 210, 222, 238, 303, 317, 322, 326
(*see also* Attributes, Two by two table)
Continuity correction, 15, 20, 217
Continuity of distribution function, 57, 340, 397
Contrasts, 248, 280
Controls, 1, 120, 226, 279
comparison with several treatments, 226, 279
Convergence, in probability, 345, 386, 398
in law, 347, 348, 349, 387, 389, 398
Covariance, 332
structure, 385
Critical value, 6
Cumulative distribution function, 44, 66, 340
(*see also* Sample cumulative distribution function)
Cyclical trend, 313

Dependence, effect of, 172, 190
Dichotomous response, 218, 267, 294, 313
(*see also* Runs, Two by two table)
Dispersion, testing for equality of, 32, 102, 103, 104
(*see also* Ansari-Bradley test, Scale tests, Siegel-Tukey test)
Distribution (*see* Beta, Binomial, Bivariate, Cauchy, Chi, Chi square, Double exponential, Exponential, Logistic, Normal, Rectangular)
Distribution-free tests, 58, 77
confidence intervals, 91
Double exponential distribution, 80, 189, 342, 400, 401

Early termination in the Wilcoxon test, 41
Edgeworth series, for Wilcoxon rank-sum statistic, 40, 99
for sign test, 185
for Wilcoxon signed-rank statistic, 73
Efficacy, 373
Efficiency, viii, 78, 174
Bahadur, 99, 105, 186, 187, 317
Cochran, 99
of estimators, 180
Pitman, 80, 118, 174, 186, 187, 201, 312, 371, 375
in presence of ties, 100
Error, absolute, relative, 16
Error rate, 241
per comparison, 227, 240
per experiment, 228, 241
Estimation of treatment effect, 81, 101
(*see also* Confidence intervals)
of association, 300
of contrasts, 249, 280
of location parameter, 175
of median, 181
of regression, 311
of several treatment effects, 248, 279
Estimator, unbiased, 70, 86, 178
center of, 86, 178
consistent, 89, 113, 181
derived from rank test, 189
dispersion of, 87, 178
efficiency of, 180
maximum likelihood type, 189
robust, 189
Expectation, 327
consistent estimator of, 181
Expected number of false significance statements, 230
Exponential distribution, 80, 103, 111, 343, 400, 401, 404
Exponential scores test, 103

Factorial, 49, 289

F-distribution, test, 102, 248, 276
Finite population model, 58
Friedman randomized block test, 263
efficiency, 277
generalization of, 279
modification for prior ordering, 279
relation to sign test, 267
Friedman statistic, 263
aligned, 270, 392, 405
asymptotic distribution of, 264, 388
modification for ties, 265, 391
tables of, 264, 431

Goodness of fit, viii
Gross errors, 77, 84, 190, 199, 201, 341, 403

Haldane-Smith test, 295
Heavy tails, 81, 342
Histogram, 15
Homogeneity of subjects, 2, 120
(*see also* Blocking for homogeneity)
Hypergeometric distribution, 215, 333, 339
(*see also* Multiple hypergeometric distribution)
expectation and variance of, 333
limit of, 216, 340, 355, 397

Inconsistency in ranking several treatments, 245
Independence, alternatives to, 316
Independence, tests of, 298, 303, 311, 316
(*see also* Association)
power and efficiency of, 316
Integers, sum of, 50
sum of squares, 51

Joint ranking, 240, 244, 245
Jonckheere-Terpstra statistic, 233, 237, 281, 397, 402
modified for ties, 235
table of, 429

Kendall's tau, 316
Khinchine's theorem, 181, 346
Kruskal-Wallis statistic, 205, 258
applied to contingency tables, 212, 305
asymptotic distribution, 207, 208, 213, 396
combining several, 281
for dichotomous response, 218
modified for ties, 208, 220
in population model, 220
tables for, 206, 422
Kruskal-Wallis test, compared with Jonckheere-Terpstra test, 235, 238
Pitman, efficiency of, 248, 250
power of, 247

Large sample theory, ix
Law of large numbers, weak, 346
Level of significance, 6, 7
Life testing, 103
Lindeberg central limit theorem, 350, 390, 393
Linear combinations of order statistics, 189
Linear trend, test for, 313
Locally most powerful test, 100, 189
Location parameter, 175
Location-scale family of distributions, 343
Logistic distribution, 80, 100, 342, 401

Mann's trend statistic, 291, 312
(*see also* Concordant pairs of observations, Kendall's tau)
Mann-Whitney statistic, 12
(*see also* Wilcoxon rank-sum statistic)
asymptotic normality of, 69, 365
confidence procedures derived from, 93
estimator derived from, 82
expectation and variance of, 14, 70, 335
modification for ties, 22
Markov chain, 314
Maximin power, 101
Maximum likelihood type estimators, 189, 190
McNemar test, 268
Measurement model, 65, 289
Measures of association, 300, 316
Median, of a distribution, 162, 164, 198
of averages, 176, 181
confidence intervals for, 182, 188
as estimator of location, 176, 177, 181
test for, 162, 163
Median unbiasedness, 86
asymptotic, 113
Midranks, 18, 59, 329
(*see also* Signed midranks)
general expression for, 22
in trend problem, 293, 296
Minimax test, 101, 189
estimator, 190
Models, for comparing two treatments or populations, 64
(*see also* Constant effect model, Measurement model, Population model, Randomization model, Shift model)
comparing *s* treatments or populations, 223, 309
Multinomial coefficients, 203, 255, 256, 380
Multinomial distribution, 224, 226, 309, 380, 384, 405
comparison of two, 225, 310

Multiple hypergeometric distribution, 223, 381, 405
generalized, 382
Multivariate normal distribution, 385
(*see also* Bivariate normal distribution)
singular, 386
Multivariate techniques, viii

Nominal data, 214
Noncentral χ^2-distribution, 247, 257, 258
Nonparametric method test, vii, 58, 90
Normal approximation, 13, 16, 18
(*see also* Continuity correction)
Normal curve (standard), 13, 14
Normal distribution, 340
tables of, 411
Normal distribution, bivariate, multivariate
(*see* Bivariate normal distribution and Multivariate normal distribution)
Normal scale problem, 102
Normal Scores, viii, 96, 116, 145
in randomized blocks, 279
in scale test, 103
in *s*-sample problem, 249
for testing independence, 316, 317
Normal Scores test, 96, 98, 99, 100, 117
Null distribution of midranks, 19, 59, 64, 293, 296, 301
of aligned ranks, 139, 271
of signed midranks, 125, 169, 208
of signed ranks, 124, 125, 165
Null distribution of ranks, for two treatments, 4, 7, 58, 64, 134
for bivariate independence, 299, 302
for randomized blocks, 262
for *s* treatments, 204, 219, 222
for trend problem, 289
Null hypothesis, 7
Number of combinations, 3
Number of samples, 3

Omnibus tests (*see* Alternatives)
One-sample problem, 158
One-sided test, compared with two-sided, 26, 226
(*see also* Alternatives)
Order statistics, linear combinations of, 189
definition, 344
distribution of, 344, 398
joint distribution of, 404
Ordered alternatives (*see* Alternatives)
Ordered random sampling, 339
Ordinal data, 214
(*see also* Categorical data)

Paired comparisons, 121, 156
(*see also* Sign test, Signed ranks, Wilcoxon signed rank test)
balanced design for, 141
dichotomous response, 268
estimating treatment effect in, 175
omnibus alternatives, 145
in randomized blocks, 279
Pairwise ranking, 239, 245
Permutation tests, viii, 43, 106
Placebo, 1
Polya's theorem, 348
Population model, ix, 5, 362
for comparison of *s* treatments, 219, 223
for comparison of two treatments, 56, 64, 145
for contingency tables, 309
for hypothesis of independence, 302
for hypothesis of randomness, 289
for paired comparisons, 158, 161
for randomized blocks, 273
Positive association, 299, 316
Power, ix, 32, 42, 57, 65
(*see also* the various tests listed under Test)
limit of, 372
Power function, 66
Probability density, 340
Projection method, 362, 366, 401, 402

Quality control, 313
Quantiles, confidence intervals for, 185

Random assignment, 1, 3, 56, 121, 203
restricted, 134
Random digits, 3
Random sampling, 56, 332, 352
Randomization model, ix, 5, 64, 362
for comparing several treatments, 204, 219
for contingency tables, 223, 309
for paired comparisons, 121
Randomized blocks, 279
additive model, 276
complete, 260
incomplete, 279
Randomness, alternatives to, 311, 313
Randomness, hypothesis of, 289, 313
Rank correlation coefficient (*see* Spearman rank correlation coefficient)
Ranking several treatments, 238, 279
(*see also* Joint ranking, Pairwise ranking)
Ranks, vii, viii, 2, 134
(*see also* Midranks, Null distribution, Signed ranks)
aligned, 139

Rank-sum test, (*see* Wilcoxon rank-sum test)
Rectangular distribution, 80, 343, 400, 401
Recursion formula, for binomial coefficients, 50
  for Wilcoxon rank-sum distribution, 51
  for Wilcoxon signed-rank distribution, 152
Regression coefficients, 311
  comparison of several, 313
  multiple, 313
Related samples, 157
Relative error, 16
Reliability theory, 103
Restricted random assignment, 134
Robust estimation, 189, 313
Rounding, effect of, 86, 94, 114, 185, 199
Runs, 313
  above and below the median, 315
  up and down, 315

Sample cumulative distribution function, 34
Sample sum, asymptotic normality, 353
  expectation of, 332
  joint asymptotic distribution of several, 393
  variance of, 333
Samples, number of, 3
Savage test, 104
Scale tests, of Capon and Klotz, 103
  (*see also* Ansari-Bradley test, Dispersion, Siegel-Tukey test)
  of Savage, 104
  with unknown location, 104
Selection, of better of two treatments, 28
  of best of several treatments, 245
Sensitivity, of an estimator, 190
Sequential, methods, viii
  confidence intervals, 188
  rank tests, 105
  selection procedure, 246
Shift model, 66, 401
  for comparing several treatments with a control, 231
  for paired comparisons, 157, 161
Siegel-Tukey test, 32
  (*see also* Scale tests)
  efficiency of, 102
  in presence of ties, 39
Sign test, vii, 121
  (*see also* Wilcoxon signed rank test)
  different uses, 163
  efficacy, 379
  efficiency, 173, 174, 186, 379
  estimator based on, 176
  null distribution, 121, 159
  optimum property, 189
  power, 143, 159, 185, 194
  required sample size, 160, 196
  special case of Friedman's test, 267
  table, 416
  treatment of zeros, 123, 144
  unbiasedness, 195
Signed midranks, 129, 169
Signed ranks, 123, 164
Significance level, 6
  advantage of large number of, 97
Significance probability, 11
  for two-sided tests, 25
  relation of one- and two-sided tests, 52
Simple random sampling, 56
Simultaneous confidence intervals, for several treatment effects, 232
  for all contrasts, 249
  for all differences, 243, 244
Simultaneous point estimation, 248
Smirnov test, 35
  generalization to hypothesis of independence, 316
  generalization to several samples, 250
  power of, 98, 105
  in presence of ties, 39, 53
  survey of, 115
  table, 413, 415
Spearman rank correlation coefficient, 292, 300, 323
  (*see also* Sum of squared rank differences, Concordant pairs of observations)
  asymptotic normality of, 369
  efficiency of, 316, 317
  table for, 433
Squares, sum of, 51
Standard confidence points, 92, 182
Standard normal curve, 13
Standard normal distribution, 340
Standard units, 347
Standardized random variable, 347
Stochastically larger, 66, 68, 110
Stochastically positive, 195
Stratified sampling, 161, 274
Student's one-sample *t*-test, 171
  absolute Normal Scores in, 187
  efficacy of, 379
  efficiency of, 172, 379
  is asymptotically distribution-free, 171, 349
  lack of monotonicity, 197
  power of, 172, 196
Student's *t*-test for regression, 311

Student's two-sample t-test, 76
efficacy of, 376
efficiency of, 80, 99, 105, 376
is asymptotically distribution-free, 77
lack of monotonicity, 197
Normal Scores in, 116
permutation version, 106
power of, 78
Subpopulations, comparison of two, 61, 65
(*see also* Blocking for homogeneity)
Sum of integers, 50
Sum of squared midrank differences, 293, 301, 333, 360
Sum of squared rank differences, 291, 333, 359
(*see also* Mann's trend statistic)
asymptotic normality of, 292, 359
efficiency of, 311
in presence of ties, 293, 296
relation to Haldane-Smith test, 295
relation to Spearman's rank correlation coefficient, 300
table of, 433
Supplementary criterion to increase number of levels, 98
Symmetry, estimation of center of, 175, 189
(*see also* One-sample problem)
test for, 187

Test (*see* Absolute Normal Scores, Ansari-Bradley, Cochran, Conditional, Friedman, Haldane-Smith, Kruskal-Wallis, McNemar, Nonparametric method, Normal Scores, Permutation, Savage, Scale, Siegel-Tukey, Sign, Smirnov, Student's t, Sum of squared rank differences, Wilcoxon rank sum, Wilcoxon signed rank)
Ties, 18, 60, 129, 265, 293, 295, 301
(*see also* Contingency table, Midranks, Signed midranks, Zeros in sign test)
breaking at random, 39, 100, 144
a common error involving, 19
condition for absence of, 57, 340, 397
configuration of, 59
different methods of dealing with, 100
effect on confidence intervals, 94, 185, 199
effect on point estimation, 85
efficiency in presence of, 100
optimum test in the presence of, 100
reasons for avoiding, 60
Treatment effect, 66, 68
Treatment effect, estimation of, 81, 101, 113
(*see also* Contrasts)
changing with time, 287
in paired comparisons, 175
in randomized blocks, 276
suggested by the data, 242, 279
Treatments, comparing two, 1, 5, 60, 120
(*see also* Attributes, Controls, Estimation of treatment effect, Paired comparisons, Random assignment, Simultaneous confidence intervals)
comparing several, 202, 210, 223, 260
ranking several, 238
selecting the best, 245
selecting the better of two, 23
Trend, upward, 290, 324
(*see also* Mann's trend statistic, Sum of squared rank differences)
autoregressive, 311
case of several series, 295
cyclical, 313
efficiency of test for, 311
test for, 291, 293, 322
Trinomial distribution, 163
t-test (*see* Student's t-test)
Twin studies, 157
Two by two table, 214, 215, 217, 267, 268, 326
combining several, 281
Two-sample problem, 57, 60, 314
(*see also* Behrens-Fisher problem)
Two-sided test, 26
(*see also* Alternatives)

U-statistics, 362, 366, 402
asymptotic normality of, 364, 367, 369, 401
expectation of, 335, 336
joint limiting distribution of, 405
variance of, 335, 337
Unbiasedness, of test, 67, 76
of estimator, 70
Uncorrelated random variables, 332
Undetermined multipliers, 403
Unidentifiable parameters, 276
Uniform distribution (*see* Rectangular distribution)
Upper confidence bound, 94, 185

Variance, 328

Walsh confidence points, 188
Wilcoxon aligned-rank sum statistic, 140
Wilcoxon midrank-sum statistic, 19
asymptotic normality of, 20, 53, 60, 356, 401

Wilcoxon midrank-sum statistic, *contd*
expectation and variance of, 20, 333
Mann-Whitney form of, 22
optimum properties for, 100
use in Siegel-Tukey test, 39
Wilcoxon rank-sum statistic, 6
aligned, 139, 140, 399
asymptotic normality of, 14, 40, 355, 365
as basis of estimator, 82
in blocked experiments, 135, 145
Edgeworth expansion for, 99
expectation and variance of, 14, 70, 333
higher moments of, 99
relation to Mann-Whitney statistic, 12
tables of, 8, 53, 54, 408
Wilcoxon rank-sum test, vii, 6
application to categorical data, 21, 210
application to two by two tables, 214
for balanced paired comparisons, 142
in Behrens-Fisher problem, 95
for censored observations, 41
for comparing several treatments with a control, 227
for comparing two attributes, 21, 62
comparison with Smirnov test, 39
comparison with *t*-test, 78, 80, 81
early termination, 41
effect of dependence, 96
efficacy of, 374
efficiency, 78, 80, 99, 100, 376, 403
increasing the number of levels, 97
monotonicity, 67, 112
optimum properties, 42, 100
power, 42, 66, 71, 72, 75, 98, 111, 375
in ranking several treatments, 240
required sample size, 74, 112
in selecting a subgroup of treatments, 247
as trend test, 295
two-sided, 24
unbiasedness, 67, 69
Wilcoxon signed midrank statistic, 130
alternative interpretation of, 131
asymptotic normality of, 131, 352, 399
expectation and variance of, 130, 331
Wilcoxon signed rank statistic, 125
(*see also* Walsh confidence points)
alternative interpretation of, 129
asymptotic normality of, 128, 351, 368
confidence procedures derived from, 182
estimator derived from, 176
expectation and variance of, 128, 165, 166, 331, 336
recursion formula for, 152
table of, 127, 155, 418
Wilcoxon signed rank test, vii, 125
in blocked experiments, 133
comparison with sign and *t*-test, 172
efficacy, 379
efficiency of, 172, 187, 379, 403
monotonicity of, 197
power of, 143, 165, 166, 167, 187, 233
relation to Wilcoxon aligned-rank sum test, 141, 153
sample size for, 169, 196
treatment of zeros, 144
unbiasedness of, 195
Wilcoxon three-decision procedure, 27, 30, 68
asymmetric, 29

Zeros in sign test, 123, 144, 163
in signed-rank Wilcoxon test, 129, 144